Benchmark Papers in Electrical Engineering and Computer Science

Series Editor: John B. Thomas
Princeton University

Published Volumes and Volumes in Preparation

SYSTEM SENSITIVITY ANALYSIS / J. B. Cruz, Jr.
RANDOM PROCESSES: Multiplicity Theory and Canonical Decompositions / Anthony Ephremides and John B. Thomas
ALGEBRAIC CODING THEORY: History and Development / Ian F. Blake
DIGITAL FILTERS: Design, Theory, and Application / Bede Liu
COMPUTER AIDED DESIGN / S. W. Director
NATURAL LANGUAGE QUESTION-ANSWERING SYSTEMS / Robert Chien
CIRCUIT THEORY / M. E. Van Valkenburg
NONLINEAR SYSTEMS STABILITY / J. K. Aggarwal
DATA COMPRESSION / Lee Davisson
MODELING AND PHYSICAL FOUNDATIONS OF RANDOM PROCESSES / Robert Lugannani
NON-PARAMETRIC PATTERN CLASSIFICATION / T. J. Wagner
FUNDAMENTALS OF DATA COMMUNICATION / Lewis E. Franks
SEMICONDUCTOR DEVICE ANALYSIS / James F. Gibbons
PATTERN RECOGNITION: Introduction and Foundations / Jack Sklansky
INTERSYMBOL INTERFERENCE / Kung Yao
BANDLIMITED SIGNAL EXPANSION / Kung Yao
FORMAL LANGUAGES / Jack W. Carlyle and Sheila A. Greibach
N-PORT NETWORK THEORY / Nhan Levan
METHODS OF ENVIRONMENTAL MODELING / R. H. Pantell and D. Daetz
NONLINEAR SYSTEMS: Processing of Random Signals / A. H. Haddad
RANDOM PROCESSES—LINEAR ESTIMATION, Volumes I and II / Thomas Kailath
CLASSIC PAPERS IN CONTROL / George J. Thaler
SEMICONDUCTOR SURFACE PHENOMENA / Peter Mark

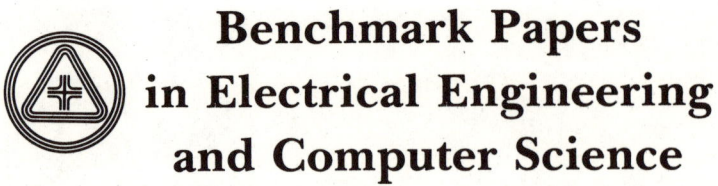

Benchmark Papers
in Electrical Engineering
and Computer Science

A *BENCHMARK*™ Books Series

ALGEBRAIC CODING THEORY:

History and Development

Edited by
IAN F. BLAKE
University of Waterloo

Dowden, Hutchinson & Ross, Inc.
Stroudsburg, Pennsylvania

Library of Congress Catalog Card Number: 73-9627
ISBN: 0-87933-038-4

Library of Congress Cataloging in Publication Data

Blake, Ian F comp.
Algebraic coding theory.

(Benchmark papers in electrical engineering and computer science)
Bibliography: p.
1. Coding theory. I. Title.
QA268.B55 519.4 73-9627
ISBN 0-87933-038-4

Manufactured in the United States of America.

Exclusive distributor outside the United States and Canada:
John Wiley & Sons, Inc.

Acknowledgments and Permissions

ACKNOWLEDGMENTS

Chiffres—"Codes correcteurs d'erreurs"
Doklady Akademiia Nauk SSSR—"Estimate of the Number of Signals in Error Correcting Codes"
Joint Publications Research Service, U.S. Department of Commerce—*Problems of Cybernetics*
"On Nongroup Close Packed Codes"

PERMISSIONS

The following papers have been reprinted with the permission of the authors and the copyright owners.

Academic Press, Inc.
Information and Control
"On a Class of Error Correcting Binary Group Codes"
"Further Results on Error Correcting Binary Group Codes"
"Power Moment Identities on Weight Distributions in Error Correcting Codes"
"An Optimum Nonlinear Code"
"On Linear and Nonlinear Single-Error-Correcting q-nary Perfect Codes"
"A Class of Optimum Nonlinear Double-Error-Correcting Codes"
Journal of Combinatorial Theory
"On Tactical Configurations and Error-Correcting Codes"

American Telephone and Telegraph Company—*Bell System Technical Journal*
"Error Detecting and Error Correcting Codes"
"A Comparison of Signalling Alphabets"
"A Class of Binary Signaling Alphabets"
"Some Further Theory of Group Codes"
"A Theorem on the Distribution of Weights in a Systematic Code"
"The MacWilliams Identities for Nonlinear Codes"

Institute of Electrical and Electronics Engineers, Inc.
IEEE Transactions on Information Theory
"Shift-Register Synthesis and BCH Decoding"
"A Class of Majority Logic Decodable Codes"
"New Generalizations of the Reed–Muller Codes; Part I: Primitive Codes"
"New Generalizations of the Reed–Muller Codes; Part II: Nonprimitive Codes"
"Polynomial Codes"
"A Class of Constructive Asymptotically Good Algebraic Codes"
IRE Transactions on Electronic Computers
"Application of Boolean Algebra to Switching Circuit Design and to Error Detection"
IRE Transactions on Information Theory
"A Class of Multiple-Error-Correcting Codes and the Decoding Scheme"
"Binary Coding"
"Notes on the Penny-Weighing Problem, Lossless Symbol Coding with Nonprimes, etc."
"Lossless Symbol Coding with Nonprimes"
"A Note on Two Binary Signaling Alphabets"

"Encoding and Error Correction Procedures for the Bose–Chaudhuri Codes"
"Error-Free Coding"
Proceedings of the IRE
"Notes on Digital Coding"

The Society for Industrial and Applied Mathematics—*Journal of the Society for Industrial and Applied Mathematics*
"Polynomial Codes over Certain Finite Fields"
"A Class of Error-Correcting Codes in p^m Symbols"
"A New Treatment of Bose–Chaudhuri Codes"

John Wiley & Sons, Inc.—*Error Correcting Codes*
"On a Class of Cyclic Codes"

Series Editor's Preface

The "Benchmark Papers in Electrical Engineering and Computer Science" series is aimed at sifting, organizing, and making readily accessible to the reader the vast literature that has accumulated on both subjects. Although the series is not intended as a complete substitute for a study of this literature, it will serve at least three major critical purposes. First, it provides a practical point of entry into a given area of research. Each volume offers an expert's selection of the critical papers on a given topic as well as his views on its structure, development, and present status. Second, the series provides a convenient and time-saving means for study in areas related to but not contiguous with one's principal interests. Last, but by no means least, the series allows the collection, in a particularly compact and convenient form, of the major works on which present research activities and interests are based.

Each volume in the series has been collected, organized, and edited by an authority in the area to which it pertains. In order to present a unified view of the area, the volume editor has prepared an introduction to the subject, has included his comments on each article, and has provided a subject index to facilitate access to the papers.

We believe that this series will provide a manageable working library of the most important technical articles in electrical engineering and computer science. We hope that it will be equally valuable to students, teachers, and researchers.

This volume, *Algebraic Coding Theory—History and Development*, has been edited by Professor Ian Blake of the University of Waterloo. It contains those thirty-five papers which have exerted, in his opinion, the most significant influence on the development of algebraic coding theory. The collection includes one paper in French and translations of two Russian papers which have not been readily available. This volume traces the origin of many of the concepts of modern coding theory and attempts to place them in their proper historical perspective. In addition, because of the youth of the field, a majority of the papers will be found still useful to current researchers.

John B. Thomas

Contents

Acknowledgments and Permissions v
Series Editor's Preface vii
Contents by Author xiii

Introduction 1

I. THE GOLAY, HAMMING, AND REED–MULLER CODES; BOUNDS ON CODE DICTIONARIES

Editor's Comments on Papers 1 Through 9 6

1. Golay, M. J. E.: "Notes on Digital Coding" **9**
   Proc. I.R.E., **37**, 657 (1949)
2. Hamming, R. W.: "Error Detecting and Error Correcting Codes" **10**
   Bell System Tech. J., **29**, 147–160 (1950)
3. Gilbert, E. N.: "A Comparison of Signalling Alphabets" **24**
   Bell System Tech. J., **31**, 504–522 (1952)
4. Muller, D. E.: "Application of Boolean Algebra to Switching Circuit Design and to Error Detection" **43**
   I.R.E. Trans. Electron. Computers, **EC-3**, 6–12 (1954)
5. Reed, I. S.: "A Class of Multiple-Error-Correcting Codes and the Decoding Scheme" **50**
   I.R.E. Trans. Inform. Theory, **PGIT-4**, 38–49 (1954)
6. Colay, M. J. E.: "Binary Coding" **62**
   I.R.E. Trans. Inform. Theory, **PGIT-4**, 23–28 (1954)
7. Varshamov, R. R.: "Estimate of the Number of Signals in Error Correcting Codes" **68**
   Translated from *Dokl. Akad. Nauk SSSR,* **117**, 739–741 (1957)
8. Golay, M. J. E.: "Notes on the Penny-Weighing Problem, Lossless Symbol Coding with Nonprimes, etc." **72**
   I.R.E. Trans. Inform. Theory, **IT-4**, 103–109 (1958)
9. Cocke, J.: "Lossless Symbol Coding with Nonprimes" **79**
   I.R.E. Trans. Inform. Theory, **IT-5**, 33–34 (1959)

II. LINEAR CODES

Editor's Comments on Papers 10, 11, and 12 82

10. Slepian, D.: "A Class of Binary Signaling Alphabets" **83**
    Bell System Tech. J., **35**, 203–234 (1956)
11. Slepian, D.: "A Note on Two Binary Signaling Alphabets" **115**
    I.R.E. Trans. Inform. Theory, **IT-2**, 84–86 (1956)
12. Slepian, D.: "Some Further Theory of Group Codes" **118**
    Bell System Tech. J., **39**, 1219–1252 (1960)

III. BCH AND REED–SOLOMON CODES

Editor's Comments on Papers 13 Through 18 **154**

13. Hocquenghem, A.: "Codes correcteurs d'erreurs" **155**
Chiffres, **2**, 147–156 (1959)

14. Bose, R. C., and D. K. Ray-Chaudhuri: "On a Class of Error Correcting Binary Group Codes" **165**
Inform. Contr., **3**, 68–79 (1960)

15. Bose, R. C., and D. K. Ray-Chaudhuri: "Further Results on Error Correcting Binary Group Codes" **177**
Inform. Contr., **3**, 279–290 (1960)

16. Reed, I. S., and G. Solomon: "Polynomial Codes over Certain Finite Fields" **189**
J. Soc. Ind. Appl. Math., **8**, 300–304 (1960)

17. Gorenstein, D. C., and N. Zierler: "A Class of Error-Correcting Codes in p^m Symbols" **194**
J. Soc. Ind. Appl. Math., **9**, 207–214 (1961)

18. Mattson, H. F., and G. Solomon: "A New Treatment of Bose–Chaudhuri Codes" **202**
J. Soc. Inc. Appl. Math., **9**, 654–669 (1961)

IV. DECODING

Editor's Comments on Papers 19 and 20 **220**

19. Peterson, W. W.: "Encoding and Error-Correction Procedures for the Bose–Chaudhuri Codes" **221**
I.R.E. Trans. Inform. Theory, **IT-6**, 459–470 (1960)

20. Massey, J. L.: "Shift-Register Synthesis and BCH Decoding" **233**
I.E.E.E. Trans. Inform. Theory, **IT-15**, 122–127 (1969)

V. WEIGHT ENUMERATION

Editor's Comments on Papers 21, 22, and 23 **240**

21. MacWilliams, F. J.: "A Theorem on the Distribution of Weights in a Systematic Code" **241**
Bell System Tech. J., **42**, 79–94 (1963)

22. Pless, V.: "Power Moment Identities on Weight Distributions in Error Correcting Codes" **257**
Inform. Contr., **6**, 147–152 (1963)

23. MacWilliams, F. J., Mrs., N. J. A. Sloane, and J.-M. Goethals: "The MacWilliams Identities for Nonlinear Codes" **263**
Bell System Tech. J., **51**, 803–819 (1972)

VI. CODING AND COMBINATORICS

Editor's Comments on Papers 24, 25, and 26 **282**

24. Rudolph, L. D.: "A Class of Majority Logic Decodable Codes" **284**
I.E.E.E. Trans. Inform. Theory, **IT-13**, 305–307 (1967)

25. Lin, S.: "On a Class of Cyclic Codes" **288**
Error Correcting Codes, edited by H. Mann, 131–148 (1968)

26. Assmus, E. F., Jr., and H. F. Mattson, Jr.: "On Tactical Configurations and Error-Correcting Codes" **306**
J. Combinatorial Theory, **2**, 243–257 (1967)

VII. GENERALIZED CODES

Editor's Comments on Papers 27, 28, and 29 322

27. Kasami, T., S. Lin, and W. W. Peterson: "New Generalizations of the Reed–Muller Codes; Part I: Primitive Codes" 323
I.E.E.E. Trans. Inform. Theory, **IT-14**, 189–199 (1968)

28. Weldon, E. J.: "New Generalizations of the Reed–Muller Codes: Part II: Nonprimitive Codes" 334
I.E.E.E. Trans. Inform. Theory, **IT-14**, 199–205 (1968)

29. Kasami, T., S. Lin, and W. W. Peterson: "Polynomial Codes" 341
I.E.E.E. Trans. Inform. Theory, **IT-14**, 808–814 (1968)

VIII. NONLINEAR CODES

Editor's Comments on Papers 30 Through 33 350

30. Vasil'yev, Y. L.: "On Nongroup Close-Packed Codes" 351
Prob. Cybernet., **8**, 337–339 (1962)

31. Nordstrom, A. W., and J. P. Robinson: "An Optimum Nonlinear Code" 358
Inform. Contr., **11**, 613–616 (1967)

32. Schönheim, J.: "On Linear and Nonlinear Single-Error-Correcting q-nary Perfect Codes" 362
Inform. Contr., **12**, 23–26 (1968)

33. Preparatá, F. P.: "A Class of Optimum Nonlinear Double-Error-Correcting Codes" 366
Inform. Contr., **13**, 378–400 (1968).

IX. SHANNON CODES

Editor's Comments on Papers 34 and 35 390

34. Elias, P.: "Error-Free Coding"
I.R.E. Trans. Inform. Theory, **IT-4**, 29–37 (1954) 391

35. Justesen, J.: "A Class of Constructive Asymptotically Good Algebraic Codes"
I.E.E.E. Trans. Inform. Theory, **IT-18**, 652–656 (1972) 400

References 405
Author Citation Index 407
Subject Index 409

Contents by Author

Introduction

It is difficult to overestimate the influence of the work of Shannon (1948)† on modern communication theory since so many of the fundamental theories arose from his work. Coding theory, in particular, has its origins in his famous theorem which guarantees the existence of codes that can transmit information at rates close to capacity with a vanishingly small probability of error. One purpose of algebraic coding theory is to devise methods for the construction and decoding of such codes. The purpose of this volume is to present those papers which have contributed substantially to coding theory as it has developed over the last twenty-five years. They represent, I hope, the history and development of algebraic coding theory.

A very brief history of the subject might be appropriate at this point. More detailed comments are included in the commentaries accompanying the papers. In the three years immediately following the appearance of Shannon's paper, the highly significant work of Golay [1] and Hamming [2] was published. The constructions of their codes have had a lasting influence on coding theory. In most of this early work the concept of parity-check equations was predominant. I am not sure who first cast the problem of coding in terms of a packing problem in a vector space over a finite field. Certainly the idea of viewing a code as a subgroup of the Abelian group of the 2^n binary n-tuples appears in the work of Reed [5] and Slepian [10]. Moreover, Golay [1] certainly recognized the significance of working over the integers modulo a prime. By 1961 the class of linear or group codes as subspaces of vector spaces over finite fields was well established. The work of Slepian [10,12] provided a solid theoretical base for the investigation of such codes. The introduction of cyclic

---

†As a matter of notation, any reference to a paper included in this collection will be by the author's name followed by square brackets containing the number that I have assigned it. Any other reference will be by the author's name followed by the year of publication in parentheses.

codes by Prange (1957) preceded the discovery of the Bose–Chaudhuri–Hocquenghem [13–15] codes, which for many reasons, are perhaps still the single most important class of codes known. That the BCH codes are cyclic was first shown by Peterson [19].

It had already been shown by Elias (1955) that there exist binary linear codes which can transmit information at rates arbitrarily close to capacity with an arbitrarily small probability of error. In this sense it was not a severe restriction to consider linear codes. The cyclic concept, although somewhat of a restriction, proved to be a great stimulus to the subject and has led to the discovery of a great many classes of codes over the last twelve years.

Some attention was also being given to the problem of decoding. The standard array of Slepian [10], while conceptually easy, proved impractical for large codes. The discovery of efficient decoding algorithms for BCH codes by Peterson [19], Gorenstein and Zierler [17], and Berlekamp (1966) has emphasized the importance of these codes. The algorithms associate each coordinate place in a codeword with an element of a Galois field, which reduces the decoding operation to solving a set of equations over a finite field. Such an operation requires considerably less computation than would a word-by-word search for the codeword closest to the received word.

The majority-logic decoding scheme, first employed by Reed to decode Muller's codes, was recognized as being particularly simple and efficient, and variants of the algorithm were considered further by Massey (1963) and Rudolph (1964). As there were very few classes of codes that could be decoded by such a method, efforts to find new classes that were majority-logic-decodable were made. These efforts resulted in the projective and Euclidean geometry codes.

As a class of codes, perhaps the most general are the polynomial codes of Kasami, Lin, and Peterson [29]. In some ways they are a natural conclusion of the efforts of many coding theorists during the sixties. Their construction is more complicated than some of the previous classes of codes, many of which they contain as subclasses.

During the last fifteen years many aspects of coding other than the construction of good codes, have been considered. Special-purpose codes, such as codes to correct a burst of errors rather than randomly occurring ones, were investigated. Codes and recurring sequences for synchronization purposes were also recognized as important for many applications. Arithmetic codes, used mainly for machine computation, were developed. Investigation of codes that were neither linear nor cyclic provided interesting results with the discovery of classes of nonlinear codes. Convolutional codes and their decoding algorithms received a great deal of attention, and their future appears promising. Techniques to analyze the performance of codes on various types of real and simulated channels were examined. The weight-enumeration problem for codes was prominent in this work.

While selecting the papers for this volume it became clear to me that there are two quite distinct periods to the development of coding theory, periods that may be described roughly as the pre-BCH and post-BCH eras. From the pre-BCH era, which I consider as covering the period from 1948 to 1961, it was a relatively simple matter to choose those papers which contributed substantially to coding theory. From the post-BCH era, however, the choice was not so clear, partly because the utility of much of the work in this era has yet to be demonstrated, and partly because of the large number of interesting papers. The selection from this latter time period, then, was largely subjective. Therefore, I adopted the practice of choosing those papers which led to the construction and analysis of good codes and the most efficient decoding algorithms. While such a constructionist point of view is narrow, I believe it to be the most reasonable and defensible under the given constraints of space. Of necessity I have ignored, almost completely, contributions in such areas as product, quasi-cyclic, shortened, concatenated, and other codes resulting from the modification or extension of more basic codes; codes for synchronization; burst-error-correcting codes; mechanizations for codes; coding for the Lee metric; the existence of perfect codes; and the performance of codes on various types of channels.

In reviewing the papers selected, I notice that all but two of the papers are from the Western Hemisphere, and that only three were not published originally in English. This is regrettable since, at least until recently, coding theorists of other parts of the world, in particular those of the U.S.S.R., have not received sufficient attention in the West. With the appearance, in the *IEEE Transactions on Information Theory,* of survey papers on coding and information theory in the U.S.S.R. [Kautz and Levitt (1969), and Dobrushin (1972)], the situation seems to be improving. These surveys indicate, however, that, with a few exceptions, work on coding theory in the U.S.S.R. has been, until recently, more concerned with the construction of special-purpose codes and the performance of codes on channels. It has thus been largely of an orientation different from that of this volume.

Unfortunately, many important concepts which have been introduced to coding theory did not appear in a form appropriate for inclusion in this collection. I refer, in particular, to the work of Prange, who introduced the idea of a cyclic code, which presently dominates all coding theory; the work of Massey on threshold decoding, which is really the only unified and comprehensive treatment of the subject yet available [although the books of Berlekamp (1968), and Weldon and Peterson (1972) present the application of threshold decoding to more recently discovered codes]; the work of Rudolph on majority-logic-decodable codes, which formed his Master's thesis; the work of Forney on concatenated codes; and finally, the work of Berlekamp on a decoding algorithm for BCH codes. The work of Massey and Forney has appeared in M.I.T. research monographs, that of Prange in Air Force Cambridge Research Laboratories Technical Reports, and that of Berlekamp in his book [Berlekamp (1958)] and in a University of North Carolina report [Berlekamp (1966)].

The papers in this volume have been arranged in chronological order, with only those exceptions required for grouping the papers by subject matter as well. The development of the subject permitted this, and it seems a more interesting method than that of strict adherence to chronology. The papers have been collected under nine subtitles, for each of which I have written a brief introduction. The purpose of these introductions, or commentaries, is to place the chosen papers in their proper setting and to justify my selection of them. In writing these commentaries it was difficult not to discourse at length on matters pertaining to the papers at hand and on those excellent papers that would have been included had space not been limited.

As a rather lengthy volume would have evolved, I have, as a rule, kept my comments to the minimum necessary. I have also kept the list of additional references small.

Finally, it is my pleasure to acknowledge the assistance of James Massey, Jack Stiffler, and Neil Sloane in the preparation of this volume. The careful reading of the original version of the commentary and the discussion on the selection of papers by James Massey and Jack Stiffler, in particular, was greatly appreciated. Their influence is present in this final version.

Golay, Hamming, and Reed–Muller Codes; Bounds on Code Dictionaries

I

Editor's Comments on Papers 1 Through 9

The first nontrivial example of an error-correcting code appears, appropriately enough, in the classical paper of Shannon in 1948. This code would today be called the (7,4) Hamming code, containing $16 = 2^4$ codewords of length 7, and its construction was credited to Hamming by Shannon. Golay [1] gives a construction that generalizes this code to codes over GF(p), p a prime number, of length $(p^n - 1)/(p - 1)$ for some positive integer n. Hamming [2] also obtained the same generalizations of his example of codes of length $(2^n - 1)$ over GF(2) and investigates their structure and decoding in some depth. The codes of both Golay and Hamming are now designated as Hamming codes. The interest of Golay was in perfect codes, which have also been called lossless, or close-packed, codes. Since he mentions the binary repetition codes and gives explicit constructions for his remarkable (23,11) binary and (11,6) ternary codes, it is not stretching a point to say that in the first paper written specifically on error-correcting codes, a paper that occupied, in its entirety, only half a journal page, Golay found essentially all the linear perfect codes which are known today. The only ones missing are the Hamming codes over GF(q), q a nonunity power of a prime. It is surely the most remarkable paper in coding theory. Its importance has recently been further enhanced by the surprising work of Tietäväinen and Perko (1971) and Tietäväinen (1973). They have been able to show that the above-mentioned codes, together with the nonlinear codes of Vasil'yév [30] and Schönheim [32] and the trivial nonlinear perfect codes consisting of cosets of perfect linear codes, are, in fact, the only perfect codes that exist over a finite field. The multiple-error-correcting perfect codes of Golay, now called the Golay codes, have inspired enough papers to fill a separate volume.

The paper by Hamming [2] forms a very readable introduction to the philosophy of coding. He introduces such concepts as systematic code, code equivalence, and redundancy. The metric he uses, now known as the Hamming metric, is used almost universally in coding work, although for some code alphabets and channels a different metric may be more appropriate. He also obtained an upper bound (called the Hamming, volume, or sphere-packing bound) on the number of codewords possible in a code dictionary with given block length and given minimum distance.

Gilbert [3] considered the performance of simple codes for both the binary symmetric and Gaussian channels. The lower bound on the minimum distance of a block code, which he obtains as a function of block length and code size, is still the strongest bound available. A similar result was found independently by Varshamov [7], and it is interesting to note that no one has yet been able to construct codes that approach this lower bound. The work of Justesen [35], however, has been an encouraging development. A great deal of effort has since been expended in tightening the upper bound of Hamming, which is very weak at low rates.

Muller [4], while investigating the application of Boolean algebras to switching circuit design, observed that if both the length of the code and the desired distance are a power of 2, say 2^l and 2^m, respectively, then his results could be applied to the construction of a code with 2^k codewords, where

$$k = \sum_{i=m}^{l} \binom{l}{i}$$

While the code and its construction is given in Muller [4], it was left to Reed [5] to provide a most remarkable decoding algorithm. This algorithm circumvented the need to go through the steps of locating the errors in a received codeword, correcting it, and then extracting the message. Reed's algorithm for Muller's codes extracts the message directly from the received word. It is also the precursor of majority-logic-decoding algorithms, of which it is a simple example. In turn, Muller's codes are the precursors of the finite geometry codes. It is interesting to note the mathematical framework of Reed [5]. He endows the space of the 2^n binary n-tuples with a scalar multiplication to form a vector space. That he referred to it as a module implies that he viewed GF(2) as a ring. He further introduced a component-wise vector multiplication to obtain a Boolean ring or algebra. It was the first organized framework in which a coding theorist had attempted to work, and it paved the way for later work. In view of this structure, perhaps it is not so surprising that Berman (1967) was able to interpret the Reed–Muller codes as Abelian codes.

The work of Golay [6] on perfect codes continued in his 1954 paper, in which he is more concerned with demonstrating the nonexistence of such codes for certain code parameters, and was carried on in his 1958 paper, in which he was able to give a systematic construction of Hamming codes over GF(p^2) for any prime p. The general construction of Hamming codes over GF(q) was left to Cocke [9]. The techniques of these last two papers are quite different from the more recent cyclic approach.

It should be mentioned that many results in coding were actually discovered earlier in the fields of statistics and group theory, although in a different context. I have not included these papers because of their orientation and the limited space available, although it would have been interesting to observe the coding results in a different setting. Thus Fisher (1942), in his study of experimental designs, gives a construction for a (15,11) Hamming code, along with its weight enumeration. On a not unrelated problem, Rao (1947) gives the Hamming sphere-packing bound. Zaremba (1949, 1952), in his study of Abelian groups, established the existence of Hamming codes over GF(p) (1951) and GF(q) (1952). In coding-theory terms, he established the existence of a subgroup (Hamming code) such that in the decomposition of the Abelian group [vector space of n-tuples over GF(q)] every vector of Hamming weight 1 may appear as a coset leader in a distinct coset.

Finally, it has recently been noticed that Reed–Muller codes were discovered by Mitani (1951) [see Peterson and Welden (1972, 2nd edition)]. This paper, however, was not available for inclusion.

Reprinted from *Proc. I.R.E.*, **37**, 657 (1949)

Correspondence 1

Notes on Digital Coding*

The consideration of message coding as a means for approaching the theoretical capacity of a communication channel, while reducing the probability of errors, has suggested the interesting number theoretical problem of devising lossless binary (or other) coding schemes serving to insure the reception of a correct, but reduced, message when an upper limit to the number of transmission errors is postulated.

An example of lossless binary coding is treated by Shannon[1] who considers the case of blocks of seven symbols, one or none of which can be in error. The solution of this case can be extended to blocks of 2^n-1-binary symbols, and, more generally, when coding schemes based on the prime number p are employed, to blocks of $p^n-1/p-1$ symbols which are transmitted, and received with complete equivocation of one or no symbol, each block comprising n redundant symbols designed to remove the equivocation. When encoding the message, the n redundant symbols x_m are determined in terms of the message symbols Y_k from the congruent relations

$$E_m \equiv X_m + \sum_{k=1}^{k=(p^n-1)/(p-1)-n} a_{mk} Y_k \equiv 0 \pmod{p}.$$

In the decoding process, the E's are recalculated with the received symbols, and their ensemble forms a number on the base p which determines univocally the mistransmitted symbol and its correction.

In passing from n to $n+1$, the matrix with n rows and $p^n-1/p-1$ columns formed with the coefficients of the X's and Y's in the expression above is repeated p times horizontally, while an $(n+1)$ st row added, consisting of $p^n-1/p-1$ zeroes, followed by as many one's etc. up to $p-1$; an added column of n zeroes with a one for the lowest term completes the new matrix for $n+1$.

If we except the trivial case of blocks of $2S+1$ binary symbols, of which any group comprising up to S symbols can be received in error which equal probability, it does not appear that a search for lossless coding schemes, in which the number of errors is limited but larger than one, can be systematized so as to yield a family of solutions. A necessary but not sufficient condition for the existence of such a lossless coding scheme in the binary system is the existence of three or more first numbers of a line of Pascal's triangle which add up to an exact power of 2. A limited search has revealed two such cases; namely, that of the first three numbers of the 90th line, which add up to 2^{12} and that of the first four numbers of the 23rd line, which add up to 2^{11}. The first case does not correspond to a lossless coding scheme, for, were such a scheme to exist, we could designate by r the number of E_m ensembles corresponding to one error and having an odd number of 1's and by $90-r$ the remaining (even) ensembles. The odd ensembles corresponding to two transmission errors could be formed by re-entering term by term all the conbinations of one even and one odd ensemble corresponding each to one error, and would number $r(90-r)$. We should have $r+r(90-r)=2^{11}$, which is impossible for integral values of r.

On the other side, the second case can be coded so as to yield 12 sure symbols, and the a_{mk} matrix of this case is given in Table I. A second matrix is also given, which is that of the only other lossless coding scheme encountered (in addition to the general class mentioned above) in which blocks of eleven ternary symbols are transmitted with no more than 2 errors, and out of which six sure symbols can be obtained.

It must be mentioned that the use of the ternary coding scheme just mentioned will always result in a power loss, whereas the coding scheme for 23 binary symbols and a maximum of three transmission errors yields a power saving of 1½ db for vanishing probabilities of errors. The saving realized with the coding scheme for blocks of 2^n-1 binary symbols approaches 3 db for increasing n's and decreasing probabilities of error, but a loss is always encountered when $n=3$.

MARCEL J. E. GOLAY
Signal Corps Engineering Laboratories
Fort Monmouth, N. J

TABLE I

	Y_1	Y_2	Y_3	Y_4	Y_5	Y_6	Y_7	Y_8	Y_9	Y_{10}	Y_{11}	Y_{12}
X_1	1	0	0	1	1	1	0	0	0	1	1	1
X_2	1	0	1	0	1	1	0	1	1	0	0	1
X_3	1	0	1	1	0	1	1	0	1	0	1	0
X_4	1	0	1	1	1	0	1	1	0	1	0	0
X_5	1	1	0	0	1	1	1	0	1	1	0	0
X_6	1	1	0	1	0	1	1	1	0	0	0	1
X_7	1	1	0	1	1	0	0	1	1	0	1	0
X_8	1	1	1	0	0	1	0	1	0	1	1	0
X_9	1	1	1	0	1	0	1	0	0	0	1	1
X_{10}	1	1	1	1	0	0	0	0	1	1	0	1
X_{11}	0	1	1	1	1	1	1	1	1	1	1	1

	Y_1	Y_2	Y_3	Y_4	Y_5	Y_6
X_1	1	1	1	2	2	0
X_2	1	1	2	1	0	2
X_3	1	2	1	0	1	2
X_4	1	2	0	1	2	1
X_5	1	0	2	2	1	1

* Received by the Institute, February 23, 1949.
[1] C. E. Shannon, "A mathematical theory of communication," *Bell Sys. Tech. Jour.*, vol. 27, p. 418; July, 1948.

Reprinted from *Bell System Tech. J.*, **29**, 147–160 (1950)

2

Error Detecting and Error Correcting Codes

By R. W. HAMMING

1. Introduction

THE author was led to the study given in this paper from a consideration of large scale computing machines in which a large number of operations must be performed without a single error in the end result. This problem of "doing things right" on a large scale is not essentially new; in a telephone central office, for example, a very large number of operations are performed while the errors leading to wrong numbers are kept well under control, though they have not been completely eliminated. This has been achieved, in part, through the use of self-checking circuits. The occasional failure that escapes routine checking is still detected by the customer and will, if it persists, result in customer complaint, while if it is transient it will produce only occasional wrong numbers. At the same time the rest of the central office functions satisfactorily. In a digital computer, on the other hand, a single failure usually means the complete failure, in the sense that if it is detected no more computing can be done until the failure is located and corrected, while if it escapes detection then it invalidates all subsequent operations of the machine. Put in other words, in a telephone central office there are a number of parallel paths which are more or less independent of each other; in a digital machine there is usually a single long path which passes through the same piece of equipment many, many times before the answer is obtained.

In transmitting information from one place to another digital machines use codes which are simply sets of symbols to which meanings or values are attached. Examples of codes which were designed to detect isolated errors are numerous; among them are the highly developed 2 out of 5 codes used extensively in common control switching systems and in the Bell Relay

Computers,[1] the 3 out of 7 code used for radio telegraphy,[2] and the word count sent at the end of telegrams.

In some situations self checking is not enough. For example, in the Model 5 Relay Computers built by Bell Telephone Laboratories for the Aberdeen Proving Grounds,[1] observations in the early period indicated about two or three relay failures per day in the 8900 relays of the two computers, representing about one failure per two to three million relay operations. The self-checking feature meant that these failures did not introduce undetected errors. Since the machines were run on an unattended basis over nights and week-ends, however, the errors meant that frequently the computations came to a halt although often the machines took up new problems. The present trend is toward electronic speeds in digital computers where the basic elements are somewhat more reliable per operation than relays. However, the incidence of isolated failures, even when detected, may seriously interfere with the normal use of such machines. Thus it appears desirable to examine the next step beyond error detection, namely error correction.

We shall assume that the transmitting equipment handles information in the binary form of a sequence of 0's and 1's. This assumption is made both for mathematical convenience and because the binary system is the natural form for representing the open and closed relays, flip-flop circuits, dots and dashes, and perforated tapes that are used in many forms of communication. Thus each code symbol will be represented by a sequence of 0's and 1's.

The codes used in this paper are called *systematic* codes. Systematic codes may be defined[3] as codes in which each code symbol has exactly n binary digits, where m digits are associated with the information while the other $k = n - m$ digits are used for error detection and correction. This produces a *redundancy* R defined as the ratio of the number of binary digits used to the minimum number necessary to convey the same information, that is,

$$R = n/m.$$

This serves to measure the efficiency of the code as far as the transmission of information is concerned, and is the only aspect of the problem discussed in any detail here. The redundancy may be said to lower the effective channel capacity for sending information.

The need for error correction having assumed importance only recently, very little is known about the economics of the matter. It is clear that in

[1] Franz Alt, "A Bell Telephone Laboratories' Computing Machine"—I, II. Mathematical Tables and Other Aids to Computation, Vol. 3, pp. 1–13 and 60–84, Jan. and Apr. 1948.

[2] S. Sparks, and R. G. Kreer, "Tape Relay System for Radio Telegraph Operation," *R.C.A. Review*, Vol. 8, pp. 393–426; (especially p. 417), 1947.

[3] In Section 7 this is shown to be equivalent to a much weaker appearing definition.

using such codes there will be extra equipment for encoding and correcting errors as well as the lowered effective channel capacity referred to above. Because of these considerations applications of these codes may be expected to occur first only under extreme conditions. Some typical situations seem to be:

a. unattended operation over long periods of time with the minimum of standby equipment.
b. extremely large and tightly interrelated systems where a single failure incapacitates the entire installation.
c. signaling in the presence of noise where it is either impossible or uneconomical to reduce the effect of the noise on the signal.

These situations are occurring more and more often. The first two are particularly true of large scale digital computing machines, while the third occurs, among other places, in "jamming" situations.

The principles for designing error detecting and correcting codes in the cases most likely to be applied first are given in this paper. Circuits for implementing these principles may be designed by the application of well-known techniques, but the problem is not discussed here. Part I of the paper shows how to construct special minimum redundancy codes in the following cases:

a. single error detecting codes
b. single error correcting codes
c. single error correcting plus double error detecting codes.

Part II discusses the general theory of such codes and proves that under the assumptions made the codes of Part I are the "best" possible.

PART I

SPECIAL CODES

2. Single Error Detecting Codes

We may construct a single error detecting code having n binary digits in the following manner: In the first $n - 1$ positions we put $n - 1$ digits of information. In the n-th position we place either 0 or 1, so that the entire n positions have an even number of 1's. This is clearly a single error detecting code since any single error in transmission would leave an odd number of 1's in a code symbol.

The redundancy of these codes is, since $m = n - 1$,

$$R = \frac{n}{n-1} = 1 + \frac{1}{n-1}.$$

It might appear that to gain a low redundancy we should let n become very large. However, by increasing n, the probability of at least one error in a

3

symbol increases; and the risk of a double error, which would pass undetected, also increases. For example, if $p \ll 1$ is the probability of any error, then for n so large as $1/p$, the probability of a correct symbol is approximately $1/e = 0.3679\ldots$, while a double error has probability $1/2e = 0.1839\ldots$.

The type of check used above to determine whether or not the symbol has any single error will be used throughout the paper and will be called a *parity check*. The above was an *even* parity check; had we used an odd number of 1's to determine the setting of the check position it would have been an odd parity check. Furthermore, a parity check need not always involve all the positions of the symbol but may be a check over selected positions only.

3. Single Error Correcting Codes

To construct a single error correcting code we first assign m of the n available positions as information positions. We shall regard the m as fixed, but the specific positions are left to a later determination. We next assign the k remaining positions as check positions. The values in these k positions are to be determined in the encoding process by even parity checks over selected information positions.

Let us imagine for the moment that we have received a code symbol, with or without an error. Let us apply the k parity checks, in order, and for each time the parity check assigns the value observed in its check position we write a 0, while for each time the assigned and observed values disagree we write a 1. When written from right to left in a line this sequence of k 0's and 1's (to be distinguished from the values assigned by the parity checks) may be regarded as a binary number and will be called the *checking number*. We shall require that this checking number give the position of any single error, with the zero value meaning no error in the symbol. Thus the check number must describe $m + k + 1$ different things, so that

$$2^k \geq m + k + 1$$

is a condition on k. Writing $n = m + k$ we find

$$2^m \leq \frac{2^n}{n + 1}$$

Using this inequality we may calculate Table I, which gives the maximum m for a given n, or, what is the same thing, the minimum n for a given m.

We now determine the positions over which each of the various parity checks is to be applied. The checking number is obtained digit by digit, from right to left, by applying the parity checks in order and writing down the corresponding 0 or 1 as the case may be. Since the checking number is

TABLE I

n	m	Corresponding k
1	0	1
2	0	2
3	1	2
4	1	3
5	2	3
6	3	3
7	4	3
8	4	4
9	5	4
10	6	4
11	7	4
12	8	4
13	9	4
14	10	4
15	11	4
16	11	5
	Etc.	

to give the position of any error in a code symbol, any position which has a 1 on the right of its binary representation must cause the first check to fail. Examining the binary form of the various integers we find

$$\begin{aligned} 1 &= 1 \\ 3 &= 11 \\ 5 &= 101 \\ 7 &= 111 \\ 9 &= 1001 \\ &\text{Etc.} \end{aligned}$$

have a 1 on the extreme right. Thus the first parity check must use positions

$$1, 3, 5, 7, 9, \cdots .$$

In an exactly similar fashion we find that the second parity check must use those positions which have 1's for the second digit from the right of their binary representation,

$$\begin{aligned} 2 &= 10 \\ 3 &= 11 \\ 6 &= 110 \\ 7 &= 111 \\ 10 &= 1010 \\ 11 &= 1011 \\ &\text{Etc.,} \end{aligned}$$

the third parity check

$$
\begin{aligned}
4 &= 100\\
5 &= 101\\
6 &= 110\\
7 &= 111\\
12 &= 1100\\
13 &= 1101\\
14 &= 1110\\
15 &= 1111\\
20 &= 10100\\
&\text{Etc.}
\end{aligned}
$$

It remains to decide for each parity check which positions are to contain information and which the check. The choice of the positions 1, 2, 4, 8, $\cdots$ for check positions, as given in the following table, has the advantage of making the setting of the check positions independent of each other. All other positions are information positions. Thus we obtain Table II.

TABLE II

Check Number	Check Positions	Positions Checked
1	1	1, 3, 5, 7, 9, 11, 13, 15, 17,$\cdots$
2	2	2, 3, 6, 7, 10, 11, 14, 15, 18,$\cdots$
3	4	4, 5, 6, 7, 12, 13, 14, 15, 20,$\cdots$
4	8	8, 9, 10, 11, 12, 13, 14, 15, 24,$\cdots$
.	.	.
.	.	.
.	.	.

As an illustration of the above theory we apply it to the case of a seven-position code. From Table I we find for $n = 7$, $m = 4$ and $k = 3$. From Table II we find that the first parity check involves positions 1, 3, 5, 7 and is used to determine the value in the first position; the second parity check, positions 2, 3, 6, 7, and determines the value in the second position; and the third parity check, positions 4, 5, 6, 7, and determines the value in position four. This leaves positions 3, 5, 6, 7 as information positions. The results of writing down all possible binary numbers using positions 3, 5, 6, 7, and then calculating the values in the check positions 1, 2, 4, are shown in Table III.

Thus a seven-position single error correcting code admits of 16 code symbols. There are, of course, $2^7 - 16 = 112$ meaningless symbols. In some applications it may be desirable to drop the first symbol from the code to avoid the all zero combination as either a code symbol or a code symbol plus a single error, since this might be confused with no message. This would still leave 15 useful code symbols.

Table III

Position							Decimal Value of Symbol
1	2	3	4	5	6	7	
0	0	0	0	0	0	0	0
1	1	0	1	0	0	1	1
0	1	0	1	0	1	0	2
1	0	0	0	0	1	1	3
1	0	0	1	1	0	0	4
0	1	0	0	1	0	1	5
1	1	0	0	1	1	0	6
0	0	0	1	1	1	1	7
1	1	1	0	0	0	0	8
0	0	1	1	0	0	1	9
1	0	1	1	0	1	0	10
0	1	1	0	0	1	1	11
0	1	1	1	1	0	0	12
1	0	1	0	1	0	1	13
0	0	1	0	1	1	0	14
1	1	1	1	1	1	1	15

As an illustration of how this code "works" let us take the symbol 0 1 1 1 1 0 0 corresponding to the decimal value 12 and change the 1 in the fifth position to a 0. We now examine the new symbol

0 1 1 1 0 0 0

by the methods of this section to see how the error is located. From Table II the first parity check is over positions 1, 3, 5, 7 and predicts a 1 for the first position while we find a 0 there; hence we write a

1 .

The second parity check is over positions 2, 3, 6, 7, and predicts the second position correctly; hence we write a 0 to the left of the 1, obtaining

0 1 .

The third parity check is over positions 4, 5, 6, 7 and predicts wrongly; hence we write a 1 to the left of the 0 1, obtaining

1 0 1 .

This sequence of 0's and 1's regarded as a binary number is the number 5; hence the error is in the fifth position. The correct symbol is therefore obtained by changing the 0 in the fifth position to a 1.

4. Single Error Correcting Plus Double Error Detecting Codes

To construct a single error correcting plus double error detecting code we begin with a single error correcting code. To this code we add one more position for checking all the previous positions, using an even parity check. To see the operation of this code we have to examine a number of cases:

1. No errors. All parity checks, including the last, are satisfied.
2. Single error. The last parity check fails in all such situations whether the error be in the information, the original check positions, or the last check position. The original checking number gives the position of the error, where now the zero value means the last check position.
3. Two errors. In all such situations the last parity check is satisfied, and the checking number indicates some kind of error.

As an illustration let us construct an eight-position code from the previous seven-position code. To do this we add an eighth position which is chosen so that there are an even number of 1's in the eight positions. Thus we add an eighth column to Table III which has:

TABLE IV

0
0
1
1
1
1
0
0
1
1
0
0
0
0
1
1

PART II

GENERAL THEORY

5. A Geometrical Model

When examining various problems connected with error detecting and correcting codes it is often convenient to introduce a geometric model. The model used here consists in identifying the various sequences of 0's and 1's which are the symbols of a code with vertices of a unit n-dimensional cube. The code points, labelled $x, y, z, \cdots$, form a subset of the set of all vertices of the cube.

Into this space of 2^n points we introduce a *distance*, or, as it is usually called, a *metric*, $D(x, y)$. The definition of the metric is based on the observation that a single error in a code point changes one coordinate, two errors, two coordinates, and in general d errors produce a difference in d coordinates.

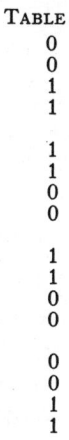

Thus we define the distance $D(x, y)$ between two points x and y as the number of coordinates for which x and y are different. This is the same as the least number of edges which must be traversed in going from x to y. This distance function satisfies the usual three conditions for a metric, namely,

$$D(x, y) = 0 \quad \text{if and only if } x = y$$

$$D(x, y) = D(y, x) > 0 \quad \text{if } x \neq y$$

$$D(x, y) + D(y, z) \geq D(x, z) \text{ (triangle inequality).}$$

As an example we note that each of the following code points in the three-dimensional cube is two units away from the others,

0 0 1
0 1 0
1 0 0
1 1 1 .

To continue the geometric language, a sphere of radius r about a point x is defined as all points which are at a distance r from the point x. Thus, in the above example, the first three code points are on a sphere of radius 2 about the point (1, 1, 1). In fact, in this example any one code point may be chosen as the center and the other three will lie on the surface of a sphere of radius 2.

If all the code points are at a distance of at least 2 from each other, then it follows that any single error will carry a code point over to a point that is *not* a code point, and hence is a meaningless symbol. This in turn means that any single error is detectable. If the minimum distance between code points is at least three units then any single error will leave the point nearer to the correct code point than to any other code point, and this means that any single error will be correctable. This type of information is summarized in the following table:

TABLE V

Minimum Distance	Meaning
1	uniqueness
2	single error detection
3	single error correction
4	single error correction plus double error detection
5	double error correction
	Etc.

Conversely, it is evident that, if we are to effect the detection and correction listed, then all the distances between code points must equal or exceed the minimum distance listed. Thus the problem of finding suitable codes is the same as that of finding subsets of points in the space which maintain at least the minimum distance condition. The special codes in sections 2, 3, and 4 were merely descriptions of how to choose a particular subset of points for minimum distances 2, 3, and 4 respectively.

It should perhaps be noted that, at a given minimum distance, some of the correctability may be exchanged for more detectability. For example, a subset with minimum distance 5 may be used for:

a. double error correction, (with, of course, double error detection).

b. single error correction plus triple error detection.

c. quadruple error detection.

Returning for the moment to the particular codes constructed in Part I we note that any interchanges of positions in a code do not change the code in any essential way. Neither does interchanging the 0's and 1's in any position, a process usually called complementing. This idea is made more precise in the following definition:

Definition. Two codes are said to be *equivalent* to each other if, by a finite number of the following operations, one can be transformed into the other:

1. The interchange of any two positions in the code symbols.
2. The complementing of the values in any position in the code symbols.

This is a formal equivalence relation ($\sim$) since $A \sim A$; $A \sim B$ implies $B \sim A$; and $A \sim B$, $B \sim C$ implies $A \sim C$. Thus we can reduce the study of a class of codes to the study of typical members of each equivalence class.

In terms of the geometric model, equivalence transformations amount to rotations and reflections of the unit cube.

6. Single Error Detecting Codes

The problem studied in this section is that of packing the maximum number of points in a unit n-dimensional cube such that no two points are closer than 2 units from each other. We shall show that, as in section 2, 2^{n-1} points can be so packed, and, further, that any such optimal packing is equivalent to that used in section 2.

To prove these statements we first observe that the vertices of the n-dimensional cube are composed of those of two $(n - 1)$-dimensional cubes. Let A be the maximum number of points packed in the original cube. Then one of the two $(n - 1)$-dimensional cubes has at least $A/2$ points. This cube being again decomposed into two lower dimensional cubes, we find that one of them has at least $A/2^2$ points. Continuing in this way we come to a two-dimensional cube having $A/2^{n-2}$ points. We now observe that a square can have at most two points separated by at least two units; hence the original n-dimensional cube had at most 2^{n-1} points not less than two units apart.

To prove the equivalence of any two optimal packings we note that, if the packing is optimal, then each of the two sub-cubes has half the points. Calling this the first coordinate we see that half the points have a 0 and half have a 1. The next subdivision will again divide these into two equal groups having 0's and 1's respectively. After $(n - 1)$ such stages we have, upon reordering the assigned values if there be any, exactly the first $n - 1$ positions of the code devised in section 2. To each sequence of the first $n - 1$ coordinates there exist $n - 1$ other sequences which differ from it by one coordinate. Once we fix the n-th coordinate of some one point, say the origin which has all 0's, then to maintain the known minimum distance of two units between code points the n-th coordinate is uniquely determined for all other code points. Thus the last coordinate is determined within a complementation so that any optimal code is equivalent to that given in section 2.

It is interesting to note that in these two proofs we have used only the assumption that the code symbols are all of length n.

7. Single Error Correcting Codes

It has probably been noted by the reader that, in the particular codes of Part I, a distinction was made between information and check positions, while, in the geometric model, there is no real distinction between the various coordinates. To bring the two treatments more in line with each other we redefine a *systematic* code as a code whose symbol lengths are all equal and

1. The positions checked are independent of the information contained in the symbol.
2. The checks are independent of each other.
3. We use parity checks.

This is equivalent to the earlier definition. To show this we form a matrix whose i-th row has 1's in the positions of the i-th parity check and 0's elsewhere. By assumption 1 the matrix is fixed and does not change from code symbol to code symbol. From 2 the rank of the matrix is k. This in turn means that the system can be solved for k of the positions expressed in terms of the other $n - k$ positions. Assumption 3 indicates that in this solving we use the arithmetic in which $1 + 1 = 0$.

There exist non-systematic codes, but so far none have been found which for a given n and minimum distance d have more code symbols than a systematic code. Section 9 gives an example of a non-systematic code.

Turning to the main problem of this section we find from Table V that a single error correcting code has code points at least three units from each other. Thus each point may be surrounded by a sphere of radius 1 with no two spheres having a point in common. Each sphere has a center point and

n points on its surface, a total of $n + 1$ points. Thus the space of 2^n points can have at most:

$$\frac{2^n}{n+1}$$

spheres. This is exactly the bound we found before in section 3.

While we have shown that the special single error correcting code constructed in section 3 is of minimum redundancy, we cannot show that all optimal codes are equivalent, since the following trivial example shows that this is not so. For $n = 4$ we find from Table I that $m = 1$ and $k = 3$. Thus there are at most two code symbols in a four-position code. The following two optimal codes are clearly not equivalent:

$$\begin{matrix} 0\,0\,0\,0 \\ 1\,1\,1\,1 \end{matrix} \quad \text{and} \quad \begin{matrix} 0\,0\,0\,0 \\ 0\,1\,1\,1 \end{matrix}\,.$$

8. Single Error Correcting Plus Double Error Detecting Codes

In this section we shall prove that the codes constructed in section 4 are of minimum redundancy. We have already shown in section 4 how, for a minimum redundancy code of $n - 1$ dimensions with a minimum distance of 3, we can construct an n dimensional code having the same number of code symbols but with a minimum distance of 4. If this were not of minimum redundancy there would exist a code having more code symbols but with the same n and the same minimum distance 4 between them. Taking this code we remove the last coordinate. This reduces the dimension from n to $n - 1$ and the minimum distance between code symbols by, at most, one unit, while leaving the number of code symbols the same. This contradicts the assumption that the code we began our construction with was of minimum reduncancy. Thus the codes of section 4 are of minimum redundancy.

This is a special case of the following general theorem: To any minimum redundancy code of N points in $n - 1$ dimensions and having a minimum distance of $2k - 1$ there corresponds a minimum redundancy code of N points in n dimensions having a minimum distance of $2k$, and conversely. To construct the n dimensional code from the $n - 1$ dimensional code we simply add a single n-th coordinate which is fixed by an even parity check over the n positions. This also increases the minimum distance by 1 for the following reason: Any two points which, in the $n - 1$ dimensional code, were at a distance $2k - 1$ from each other had an odd number of differences between their coordinates. Thus the parity check was set oppositely for the two points, increasing the distance between them to $2k$. The additional coordinate could not decrease any distances, so that all points in the code are now at a minimum distance of $2k$. To go in the reverse direction we simply

12

drop one coordinate from the n dimensional code. This reduces the minimum distance of $2k$ to $2k - 1$ while leaving N the same. It is clear that if one code is of minimum redundancy then the other is, too.

9. Miscellaneous Observations

For the next case, minimum distance of five units, one can surround each code point by a sphere of radius 2. Each sphere will contain

$$1 + C(n, 1) + C(n, 2)$$

points, where $C(n, k)$ is the binomial coefficient, so that an upper bound on the number of code points in a systematic code is

$$\frac{2^n}{1 + C(n, 1) + C(n, 2)} = \frac{2^{n+1}}{n^2 + n + 2} \geq 2^m.$$

This bound is too high. For example, in the case of $n = 7$, we find that $m = 2$ so that there should be a code with four code points. The maximum possible, as can be easily found by trial and error, is two.

In a similar fashion a bound on the number of code points may be found whenever the minimum distance between code points is an odd number. A bound on the even cases can then be found by use of the general theorem of the preceding section. These bounds are, in general, too high, as the above example shows.

If we write the bound on the number of code points in a unit cube of dimension n and with minimum distance d between them as $B(n, d)$, then the information of this type in the present paper may be summarized as follows:

$$B(n, 1) = 2^n$$

$$B(n, 2) = 2^{n-1}$$

$$B(n, 3) = 2^m \leq \frac{2^n}{n + 1}$$

$$B(n, 4) = 2^m \leq \frac{2^{n-1}}{n}$$

$$B(n - 1, 2k - 1) = B(n, 2k)$$

$$B(n, 2k - 1) = 2^m \leq \frac{2^n}{1 + C(n, 1) + \cdots + C(n, k - 1)}$$

While these bounds have been attained for certain cases, no general methods have yet been found for contructing optimal codes when the minimum distance between code points exceeds four units, nor is it known whether the bound is or is not attainable by systematic codes.

We have dealt mainly with systematic codes. The existence of non-systematic codes is proved by the following example of a single error correcting code with $n = 6$.

0 0 0 0 0 0
0 1 0 1 0 1
1 0 0 1 1 0
1 1 1 0 0 0
0 0 1 0 1 1
1 1 1 1 1 1

The all 0 symbol indicates that any parity check must be an even one. The all 1 symbol indicates that each parity check must involve an even number of positions. A direct comparison indicates that since no two columns are the same the even parity checks must involve four or six positions. An examination of the second symbol, which has three 1's in it, indicates that no six-position parity check can exist. Trying now the four-position parity checks we find that

1 2 5 6
2 3 4 5

are two independent parity checks and that no third one is independent of these two. Two parity checks can at most locate four positions, and, since there are six positions in the code, these two parity checks are not enough to locate any single error. The code is, however, single error correcting since it satisfies the minimum distance condition of three units.

The only previous work in the field of error correction that has appeared in print, so far as the author is aware, is that of M. J. E. Golay.[4]

[4] M. J. E. Golay, Correspondence, Notes on Digital Coding, *Proceedings of the I.R.E.*, Vol. 37, p. 657, June 1949.

Reprinted from *Bell System Tech. J.*, **31**, 504–522 (1952)

3

A Comparison of Signalling Alphabets

By E. N. GILBERT

(Manuscript received March 24, 1952)

Two channels are considered; a discrete channel which can transmit sequences of binary digits, and a continuous channel which can transmit band limited signals. The performance of a large number of simple signalling alphabets is computed and it is concluded that one cannot signal at rates near the channel capacity without using very complicated alphabets.

INTRODUCTION

C. E. Shannon's encoding theorems[1] associate with the channel of a communications system a capacity C. These theorems show that the output of a message source can be encoded for transmission over the channel in such a way that the rate at which errors are made at the receiving end of the system is arbitrarily small provided only that the message source produces information at a rate less than C bits per second. C is the largest rate with this property.

Although these theorems cover a wide class of channels there are two channels which can serve as models for most of the channels one meets in practice. These are:

1. The binary channel

This channel can transmit only sequences of binary digits 0 and 1 (which might represent hole and no hole in a punched tape; open-line and closed line; pulse and no pulse; etc.) at some definite rate, say one digit per second. There is a probability p (because of noise, or occasional equipment failure) that a transmitted 0 is received as 1 or that a transmitted 1 is received as 0. The noise is supposed to affect different digits independently. The cpacity of this channel is

$$C = 1 + p \log p + (1 - p) \log (1 - p) \qquad (1)$$

bits per digit. The log appearing in Equation (1) is log to the base 2; this convention will be used throughout the rest of this paper.

[1] C. E. Shannon, "A Mathematical Theory of Communication," *Bell System Tech. J.*, **27**, p. 379–423 and pp. 623–656, 1948, theorems 9, 11, and 16 in particular.

2. The low-pass filter

The second channel is an ideal low-pass filter which attenuates completely all frequencies above a cutoff frequency W cycles per second and which passes frequencies below W without attenuation. The channel is supposed capable of handling only signals with average power P or less. Before the signal emerges from the channel, the channel adds to it a noise signal with average power N. The noise is supposed to be white Gaussian noise limited to the frequency band $|\nu| < W$. The capacity of this channel is

$$C = W \log\left(1 + \frac{P}{N}\right) \tag{2}$$

bits per second.

Shannon's theorems prove that encoding schemes exist for signalling at rates near C with arbitrarily small rates of errors without actually giving a constructive method for performing the encoding. It is of some interest to compare encoding systems which can easily be devised with these ideal systems. In Part I of this paper some schemes for signalling over the binary channel will be compared with ideal systems. In Part II the same will be done for the low-pass filter channel.

Part I

THE BINARY CHANNEL

1. Error-Correcting Alphabets

Imagine the message source to produce messages which are sequences of letters drawn from an alphabet containing K letters. We suppose that the letters are equally likely and that the letters which the source produces at different times are independent of one another. (If the source given is a finite state source which does not fit this simple description, it can be converted into one which approximately does by a preliminary encoding of the type described in Shannon's Theorem 9.) To transmit the message over the binary channel we construct a new alphabet of K letters in which the letters are different sequences of binary digits of some fixed length, say D digits. Then the new alphabet is used as an encoding of the old one suitable for transmission over the channel. For example, if the source produced sequences of letters from an alphabet of 3 letters, a typical encoding with $D = 5$ might convert the message into a binary sequence composed of repetitions of the three letters

00000
11100
and 00111.

If $K = 2^D$, the alphabet consists of all binary sequences of length D and hence if any of the digits of a letter is altered by noise the letter will be misinterpreted at the receiving end of the channel. If K is somewhat smaller than 2^D it is possible to choose the letters so that certain kinds of errors introduced by the noise do not cause a misinterpretation at the receiver. For example, in the three letter alphabet given above, if only one of the five digits is incorrect there will be just one letter (the correct one) which agrees with the received sequence in all but one place. More generally if the letters of the alphabet are selected so that each letter differs from every other in at least $2k + 1$ out of the D places, then when k or fewer errors are made the correct interpretation of the received sequence will be the (unique) letter of the alphabet which differs from the received sequence in no more than k places. An alphabet with this property will be called a *k error correcting alphabet*[2].

Error correcting alphabets have the advantage over the random alphabets which Shannon used to prove his encoding theorems that they are uniformly reliable whereas Shannon's alphabets are reliable only in an average sense. That is, Shannon proved that the probability that a letter *chosen at random* shall be received incorrectly can be made arbitrarily small. However, a certain small fraction of the letters of Shannon's alphabets are allowed a much higher probability of error than the average. This kind of alphabet would be undesirable in applications such as the signalling of telephone numbers; one would not want to give a few subscribers telephone numbers which are received incorrectly more often than most of the others. It is only conjectured that the rate C can be approached using error correcting alphabets. The alphabets which are to be considered here are all error correcting alphabets.

A geometric picture of an alphabet is obtained by regarding the D digits of a sequence as coordinates of a point in Euclidean D dimensional space. The possible received sequences are represented by vertices of the unit cube. A k error correcting alphabet is represented by a set of vertices, such that each pair of vertices is separated by a distance at least $\sqrt{2k + 1}$

Let $K_0(D, k)$ be the largest number of letters which a D dimensional

[2] R. W. Hamming, "Error Detecting and Error Correcting Codes," *Bell System Tech. J.*, **29**, pp. 147–160, 1950.

k error correcting alphabet can contain. Except when $k = 1$, there is no general method for constructing an alphabet with $K_0(D, k)$ letters, nor is $K_0(D, k)$ known as a function of D and k. Crude upper and lower bounds for $K_0(D, k)$ are given by the following theorem.

Theorem 1. The largest number of letters $K_0(D, k)$ satisfies

$$\frac{2^D}{N(D, 2k)} \leq K_0(D, k) \leq \frac{2^D}{N(D, k)} \tag{3}$$

where

$$N(D, k) = \sum_{r=0}^{k} C_{D,r}$$

is the number of sequences of D digits which differ from a given sequence in $0, 1, \cdots$, or k places.

Proof

The upper bound is due to R. W. Hamming and is proved by noting that for each letter S of a k error correcting alphabet there are $N(D, k)$ possible received sequences which will be interpreted as meaning S. Hence $N(D, k)\, K_0(D, k) \leq 2^D$, the total number of sequences.

The lower bound is proved by a random construction method. Pick any sequence S_1 for the first letter. There remain $2^D - N(D, 2k)$ sequences which differ from S_1 in $2k + 1$ or more places. Pick any one of these S_2 for the second letter. There remain at least $2^D - 2N(D, 2k)$ sequences which differ from both S_1 and S_2 in $2k + 1$ or more places. As the process is continued, there remain at least $2^D - rN(D, 2k)$ sequences, which differ in $2k + 1$ or more places from $S_1, \cdots, S_r$, from which S_{r+1} is chosen. If there are no choices available after choosing S_K, then $2^D - KN(D, 2k) \leq 0$ so the alphabet $(S_1, \cdots, S_K)$ has at least as many letters as the lower bound (3).

For all the simple cases (D and k not very large) investigated so far the upper bound is a better estimate of $K_0(D, k)$ than the lower bound. The upper and lower bounds differ greatly, as may be seen from a quick inspection of Table I. For example, in the case of a ten dimensional two error correcting alphabet, the bounds are 2.7 and 18.3.

2. Efficiency Graph

The first step in constructing an efficiency graph for comparing alphabets is to decide on what constitutes reliable transmission. The criterion used here is that on the average no more than one letter in 10^4 shall be misinterpreted.

TABLE I

TABLE OF $2^D/N(D, k)$

k =	1	2	3	4	5	6	7
D = 3	2						
4	3.2						
5	5.3	2					
6	9.1	2.9					
7	16	4.4	2.9				
8	28.4	6.9	2.8				
9	51.2	11.1	3.9	2			
10	93.1	18.3	5.8	2.7			
11	170.7	30.6	8.8	3.6	2		
12	315.8	51.8	13.7	5.2	2.6		
13	585.2	89.0	21.6	7.5	3.4	2	
14	1092.3	154.4	34.9	11.1	4.7	2.5	
15	2048	270.8	56.8	16.8	6.6	3.3	2

Missing entries are numbers between 1 and 2.

This sort of criterion might be appropriate for a channel transmitting English text. For other messages it is not always appropriate. For example, if the messages are telephone numbers, one would naturally require that the probability of mistaking a telephone number be small, say less than 10^{-4}. If the telephone numbers are L decimal digits long, and if the alphabet has K different letters in it (so that it takes about $L \log 10/\log K$ letters to make up a telephone number) the probability of making a mistake in a single letter should be required to be less than about

$$\frac{10^{-4} \log K}{L \log 10}$$

which gives alphabets with large K an advantage over alphabets with small K.

Since the probability that exactly r binary digits out of D shall be received incorrectly is $C_{D,r}p^r (1 - p)^{D-r}$, we achieve the required reliability with a D-dimensional k-error correcting alphabet provided p satisfies

$$\sum_{r=k+1}^{D} C_{D,r}p^r(1 - p)^{D-r} \leq 10^{-4}. \tag{4}$$

The value of p which makes the inequality hold with the equals sign determines the noisiest channel over which the alphabet can be used safely.

Let K be the number of different letters in the alphabet. Then the rate in bits per digit at which information is being recieved is

$$R = \frac{\log K}{D}. \tag{5}$$

In Equation (5) we have neglected a term which takes account of the information lost due to channel noise. This is legitimate because all but 10^{-4} of the letters are received correctly.

The worst tolerable probability p of (4) and the rate R of Equation (5) determine the noise combating ability of an alphabet. To compare different alphabets one may represent them as points on an efficiency graph of R versus p. Fig. 1 is an efficiency graph on which the values (p, R) for a number of simple error correcting alphabets have been plotted. Each point on the graph is labelled with the two numbers k, D in that order. The alphabets represented were not found by any systematic process and are not all proved to be best possible (i.e., to have the largest K) for the stated values of k and D. Fortunately, R depends on K only logarithmically so that it is not likely the points representing the best possible alphabets lie far away from the plotted points.

The solid line represents the curve

$$R = C = 1 + p \log p + (1 - p) \log (1 - p).$$

According to Shannon's theorems, all alphabets are represented by points lying below this line.

The efficiency graph only partially orders the alphabets according to

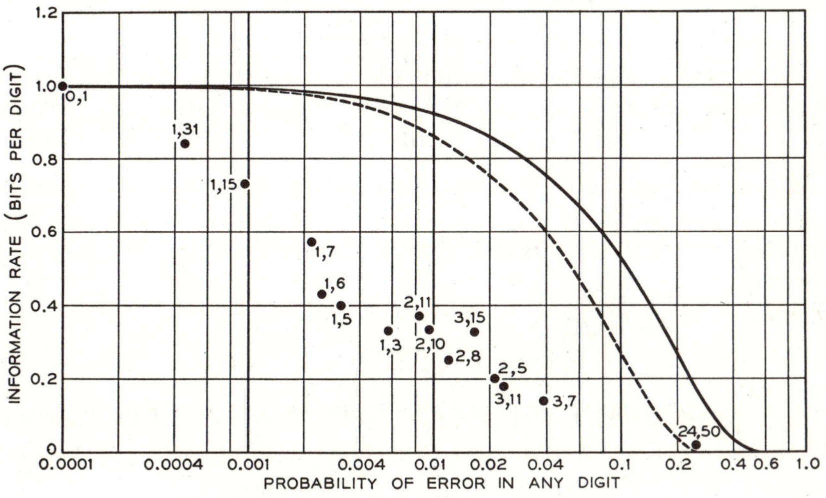

Fig. 1—Probability of error in a letter is 10^{-4}.

their invulnerability to noise. For example, it is clear that the alphabet 3, 15 is better than 2, 8. However, without further information about the channel, such as knowledge of p, there is no reasonable way of choosing between 3, 15 and 3, 7.

3. Large Alphabets

We have been unable to prove that there are error correcting alphabets which signal at rates arbitrarily close to C while maintaining an arbitrarily small probability of error for any letter. A result in this direction is the following theorem.

Theorem 2. Let any positive ϵ and δ be given. Given a channel with $p < \frac{1}{4}$ there exists an error correcting alphabet which can signal over the channel at a rate exceeding $R_0 - \epsilon$ where

$$R_0 = 1 + 2p \log 2p + (1 - 2p) \log (1 - 2p)$$

bits per digit and for which the probability of error in any letter is less than δ.

Proof

The probability of error in any letter is the sum on the left of (4). This is a sum of terms from a binomial distribution which, as is well known, tends to a Gaussian distribution with mean Dp and variance $Dp(1 - p)$ for large D. Hence there is a constant $A(\delta)$ such that all k error correcting alphabets with sufficiently large D have a letter error probability less than δ provided

$$k \geq Dp + A(\delta)\,(Dp(1 - p))^{1/2}. \tag{6}$$

Let $k(D)$ be the smallest integer which satisfies (6) and consider an alphabet which corrects $k(D)$ errors and contains $K_0(D, k(D))$ letters. By Equation (5) and the lower bound of Theorem 1, this alphabet signals at a rate $R(D)$ satisfying

$$1 - \frac{1}{D} \log N(D, 2k(D)) \leq R(D).$$

Since $p < \frac{1}{4}$, $2k(D) < D/2$ for large D and hence

$$N(D, 2k(D)) < (2k(D) + 1)C_{D,2k(D)}.$$

Then an application of Stirling's approximation for factorials shows that as $D \rightarrow \infty$

$$1 - \frac{1}{D} \log N(D, 2k(D)) \rightarrow R_0\,.$$

Hence by taking D large enough one obtains an alphabet with rate exceeding $R_0 - \epsilon$ and letter error probability less than δ.

The rate R_0 appears on the efficiency graph as a dotted line.

It has not been shown that no error-correcting alphabet has a rate exceeding R_0. In fact, one alphabet which exceeds R_0 in rate is easy to construct. If the noise probability p is greater than $\frac{1}{4}$, then $R_0 = 0$. The alphabet with just two letters

$$0\,0\,0\,0 \ldots 0$$

and

$$1\,1\,1\,1 \ldots 1$$

will certainly transmit information at a (small) positive rate, and with a 10^{-4} probability of errors if D is large enough, as long as $p < \frac{1}{2}$.

Using a more refined lower bound for $K_0(D, k)$ it might be shown that there are error-correcting alphabets which signal with rates near C. If one repeats the calculation that led to R_0 using the upper bound (3) (which seems to be a better estimate of the true $K_0(D, k)$) instead of the lower bound (3), one is led to the rate C instead of R_0.

The condition (4) is more conservative than necessary. The structure of the alphabet may be such that a particular sequence of more than k errors may occur without causing any error in the final letter. This is illustrated by the following simple example due to Shannon: the alphabet with just two letters

$$\begin{matrix} 0\,0\,0\,0\,0\,0 \\ 1\,1\,1\,0\,0\,0 \end{matrix}$$

corrects any single error but also corrects certain more serious errors such as receiving 0 0 1 1 1 1 for 0 0 0 0 0 0. An alphabet designed for practical use would make efficient enough use of the available sequences so that any sequence of much more than k errors causes an error in the final letter; the random alphabets constructed above probably do not. If this kind of error were properly accounted for, the rate R_0 could be improved, perhaps to C.

4. Other Discrete Channels

If instead of transmitting just 0's and 1's the channel can carry more digits

$$0, 1, 2, \cdots, n$$

8

a similar theory can be worked out. The simplest kind of noise in this channel changes a digit into any one of the n other possible numbers with probability p/n. Then the capacity of the channel is

$$C = \log (n + 1) + p \log \frac{p}{n} + (1 - p) \log (1 - p).$$

Error-correcting alphabets for this channel can also be constructed and the criterion (4) for good transmission remains unchanged. The proof of theorem 1 can be repeated with little change using

$$N(D, k) = \sum_{r=0}^{k} C_{D, r} n^r$$

as the number of sequences which can be reached after k or fewer errors [the terms 2^D in (1) and (3) are replaced by $(n + 1)^D$]. Once more, using the lower bound, one finds an expression for R_0 which is the same as the one for C but with p replaced by $2p$.

Part II

THE LOW PASS FILTER

1. Encoding and Detection

If $f(t)$ is a signal emerging from a low pass filter (so that its spectrum is confined to the frequency band $|\nu| < W$ cycles per second) then $f(t)$ has a special analytic form given by the sampling theorem[3]

$$f(t) = \sum_{m=-\infty}^{\infty} f\left(\frac{m}{2W}\right) \frac{\sin \pi (2Wt - m)}{\pi(2Wt - m)} \tag{7}$$

Thus the signal is completely determined by the sequence of sample values $f(m/2W)$. The average power of the signal $f(t)$ is measured by

$$P = \lim_{T\to\infty} \frac{1}{2T} \int_{-T}^{T} f^2(t)\, dt$$

which can be expressed in terms of the sample values as follows

$$P = \lim_{M\to\infty} \frac{1}{2M} \sum_{m=-M}^{M} f^2\left(\frac{m}{2W}\right). \tag{8}$$

As in Part I, consider a message source producing a sequence of letters from an alphabet of K equally likely letters. To transmit this information over the low pass filter we must encode the sequence into a function

[3] C. E. Shannon, "*Communication in the Presence of Noise*," *Proc. I. R. E.*, **37**, pp. 10–21, Jan. 1949.

$f(t)$ of the form (7), or in other words into a sequence of sample values $f(m/2W)$. To do this, we construct a new alphabet containing K letters which are different sequences of real numbers of some fixed length, say D places. When we let the letters of the new alphabet correspond to letters of the old one the message is translated into a sequence of real numbers which we use for the sequence $f(m/2W)$.

If the K letters of the sequence alphabet are

$$\begin{array}{l} S_1\colon\ a_{11},\ \cdots,\ a_{1D} \\ S_2\colon\ a_{21},\ \cdots,\ a_{2D} \\ \quad\vdots \qquad \vdots \qquad\quad \vdots \\ S_K\colon\ a_{K1},\ \cdots,\ a_{KD}, \end{array}$$

the expression (8) for the average power of the function $f(t)$ becomes

$$P = \frac{1}{DK}\,(d_1^2 + d_2^2 + \cdots + d_K^2) \tag{9}$$

where

$$d_i^2 = \sum_{j=1}^{D} a_{ij}^2 .$$

If the D numbers in the sequence S_i are regarded as coordinates of a point in Euclidean D dimensional space, d_i^2 represents the square of the distance from the point representing S_i to the origin.

When $f(t)$ is transmitted, the received signal will be $f(t) + n(t)$ where $n(t)$ is some (unknown) white Gaussian noise signal. The noise signals $n(t)$ are characterized by the fact that their sample values $n(m/2W)$ are independently distributed according to Gaussian laws. That is,

$$\text{Prob}\left(n\left(\frac{m}{2W}\right) \leq X\right) = \frac{1}{\sqrt{2\pi}\sigma}\int_{-\infty}^{X} e^{-y^2/2\sigma^2}\,dy. \tag{10}$$

The variance σ^2 of the distribution of noise samples is, by an application of (8), the power of this ensemble of noise signals.

At the receiving end of the channel, there is a detector which observes each block of D sample values $f(m/2W) + n(m/2W)$ and tries to decide which one of the K letters $S_1, \cdots, S_K$ was sent. In terms of the geometric picture, the detector divides all of D dimensional space into K non-overlapping regions $U_1, \cdots, U_K$ with the property that, if the D received sample values are represented by a point in U_i, the detector decides that S_i was sent. By Equation (10), the probability that the detector picks the wrong letter when S_i is sent is

$$p_i = \frac{1}{(2\pi)^{D/2}\sigma^D} \int \int_{\bar{U}_i} \cdots \int e^{-r_i^2/2\sigma^2} \, dy_1 \cdots dy_D \tag{11}$$

where $\overline{U}_i$ is the set of all points not in U_i and r_i is the distance from $(y_1, \cdots, y_D)$ to the point representing S_i.

For any given alphabet the best possible detector (in the sense that it minimizes the average probability of making an error in guesssing a letter) is called a *maximum likelihood detector*. The region U_i for a maximum likelihood detector consists of all points $(y_1, \cdots, y_D)$ which are closer to the point S_i than to any other letter point $S_j(r_i < r_j$ for all $j \neq i)$. To prove that this choice of U_i is best possible consider any other detector such that U_i contains a set V of points in which $r_i > r_j$. A direct calculation shows that the detector obtained by removing V from U_i and making V part of U_j has a smaller probability of error per letter. The set of points equidistant from two given points is a hyperplane. The region U_i of a maximum likelihood detector is a convex region bounded by segments of the hyperplanes

$$r_i = r_1, \qquad r_i = r_2, \cdots.$$

To compare signalling alphabets under the most favorable possible circumstances, we always compute letter error probabilities assuming that the detector is a maximum likelihood detector.

2. Computation of error probabilities

Exact evaluation of the letter error probability integral (11) is impossible except in a few special cases. Fortunately we are only interested in (11) when σ is small enough in comparison to the size of U_i to make the integral small. Then fairly accurate approximate formulas can be derived.

Theorem 3. Let R_{ij} be the distance between letter points S_i and S_j. Then

$$1 - \prod_{j \neq i} (1 - Q_{ij}) \leq p_i \leq \sum_{j \neq i} Q_{ij} \tag{12}$$

where

$$Q_{ij} = \frac{1}{\sqrt{2\pi}} \int_{R_{ij}/2\sigma}^{\infty} e^{-x^2/2} \, dx.$$

The proof of Theorem 3 follows from the fact that Q_{ij} is the probability that, when S_i is transmitted, the received sequence will be closer to S_j than to S_i.

In the cases to be computed Q_{ij} is a rapidly decreasing function of R_{ij} and the only terms worth keeping in (12) are the ones for which R_{ij} is the smallest of the numbers $R_{i1}, \cdots, R_{iK}$. Moreover since the Q_{ij} are all small enough so that the upper and lower bounds differ only by a few per cent, the upper bound is a good approximation to p_i. Then a simple approximate formula for the average letter error probability $p = (p_1 + \cdots + p_K)/K$ is

$$p = \frac{N}{\sqrt{2\pi}} \int_{r_0/\sigma}^{\infty} e^{-x^2/2} \, dx \qquad (13)$$

where $2r_0$ is the smallest of the $K(K - 1)/2$ distances R_{ij} and N is the average over all letters in the alphabet of the number of letter points which are a distance $2r_0$ away.

3. Efficiency graph

The efficiency graph to be described was constructed originally to compare alphabets for signalling telephone numbers of length equal to ten decimal digits. It was desired that on the average only one telephone number in 10^4 should be received incorrectly. As described in Part I section 2, if the telephone numbers are encoded into sequences of letters from an alphabet of K letters, we must require that the average probability of error in any letter be

$$p = 10^{-5} \log_{10} K \qquad (14)$$

or smaller.

Given an alphabet, one can compute with the help of (13) and (14) and a table of the error integral the largest value of the noise power σ^2 which can be tolerated. The average power of the transmitted signal is P given by Equation (9). Hence we can compute the smallest signal to noise ratio

$$Y = P/\sigma^2 \qquad (15)$$

which will be satisfactory.

A letter containing $\log K$ bits of information is transmitted during an interval of $D/2W$ seconds. Hence the rate at which information is received is

$$R = \frac{2W \log K}{D} \qquad (16)$$

bits per second. Again Equation (16) ignores a term representing in-

12

formation lost due to channel noise which is negligible because the error probability is low.

The efficiency graph, Fig. 2, is a chart on which the signal to noise ratio Y in db [computed from Equation (15)] is plotted against the signalling rate per unit bandwidth $R/W = (2 \log K)/D$ for different alpha-

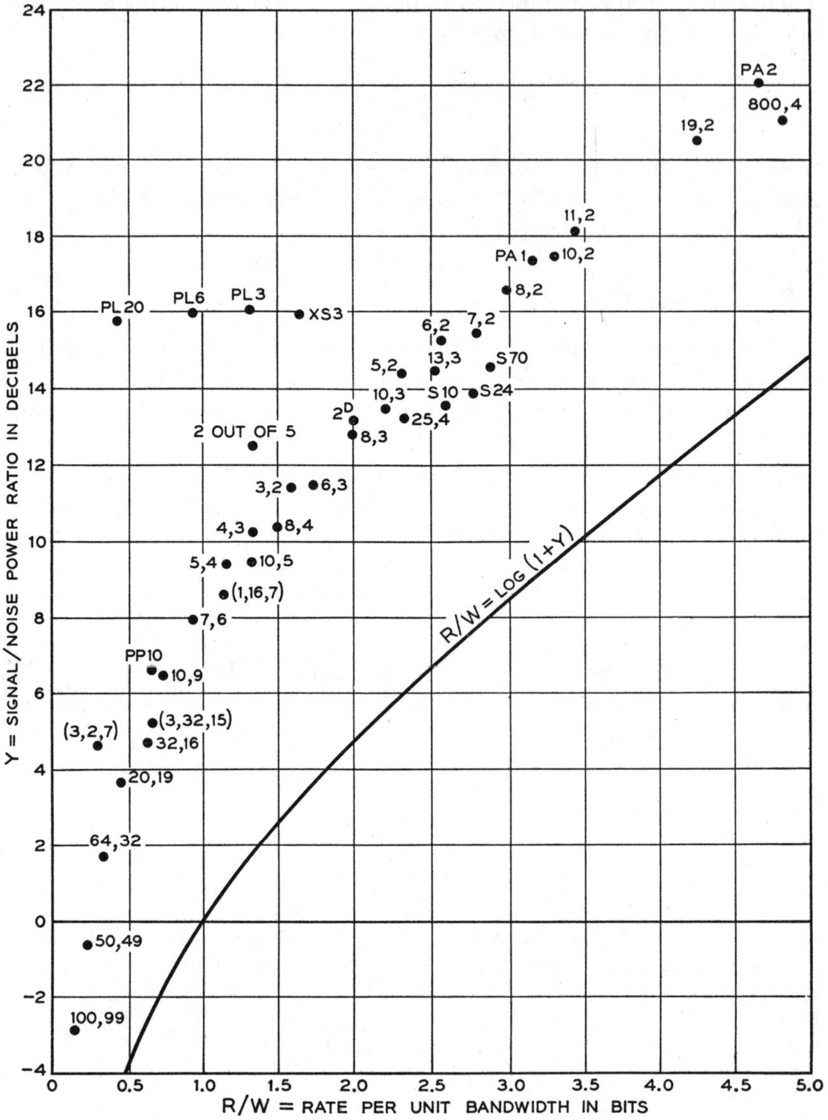

Fig. 2—Probability is 10^{-4} that an error is made in a 10 digit decimal number.

13

bets. An alphabet is considered poor if its point on the efficiency graph lies far above the ideal curve $R/W = C/W = \log(1 + Y)$.

4. The alphabets

The alphabets which appear on the efficiency graph are the following:

excess three (*XS3*): the ten sequences of 4 binary digits which represent 3, 4, $\cdots$, and 12 in binary notation;

two out of five: the ten sequences of five binary digits which contain exactly two ones;

pulse position (*PP10*): the ten sequences of ten binary digits which contain exactly one one;

2^D *binary*: all of 2^D sequences of D binary digits.

pulse amplitude (*PAn*): the $2n + 1$ sequences of length 1 consisting of $-n, -n + 1, \cdots, n$. This alphabet gives rise to a sort of quantized amplitude modulation.

pulse length (*PLn*): the $n + 1$ sequences of n binary digits of the form $11 \cdots 10 \cdots 0$, i.e., a run of ones followed by a run of zeros.

Minimizing alphabets (K, D): The above alphabets are taken from actual practice. They are convenient because, aside from PAn, they require a signal generator with only two amplitude levels. If we ignore ease of generating the signals as a factor, a great many geometric arrangements of points suggest themselves as possible good alphabets. The principle by which one arrives at good alphabets may be described as follows. When a D and K have been determined which give the desired information rate R [by Equation (16)] try to arrange the K letter points in D dimensional space in such a way that the distances between pairs of points are all greater than some fixed distance and that the average of the K squared distances to the origin is minimized. By Equations (9) and (13) it is seen that, apart from the small influence of the factor N, this process must minimize the signal to noise ratio Y required.

Ordinarily it is difficult to prove that a configuration is a minimizing one. Even to recognize a configuration which leads to a relative minimum (*i.e.* a minimum over all nearby configurations) is not always easy. The eight vertices of a cube, for example, do not give a relative minimum. Consequently, most of the alphabets to be described are only conjectured to be "best possible." Each of them satisfies one necessary requirement of minimizing alphabets that the centroid of the point configuration (assuming a unit mass at each letter point) lies at the origin. That this condition is necessary follows from the easily derived identity

$$r_2^2 = r_1^2 - R_0^2$$

14

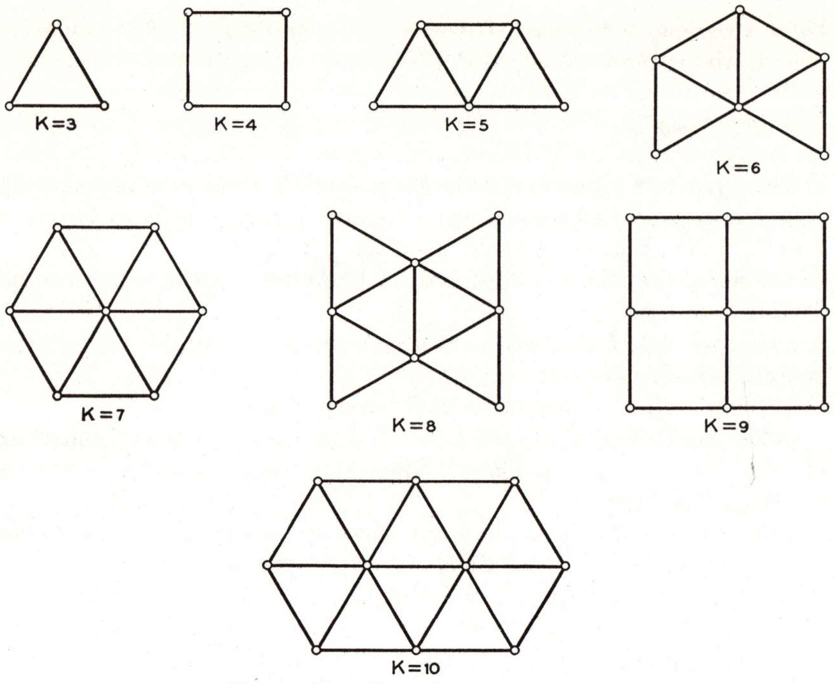

Fig. 3—Two dimensional alphabets.

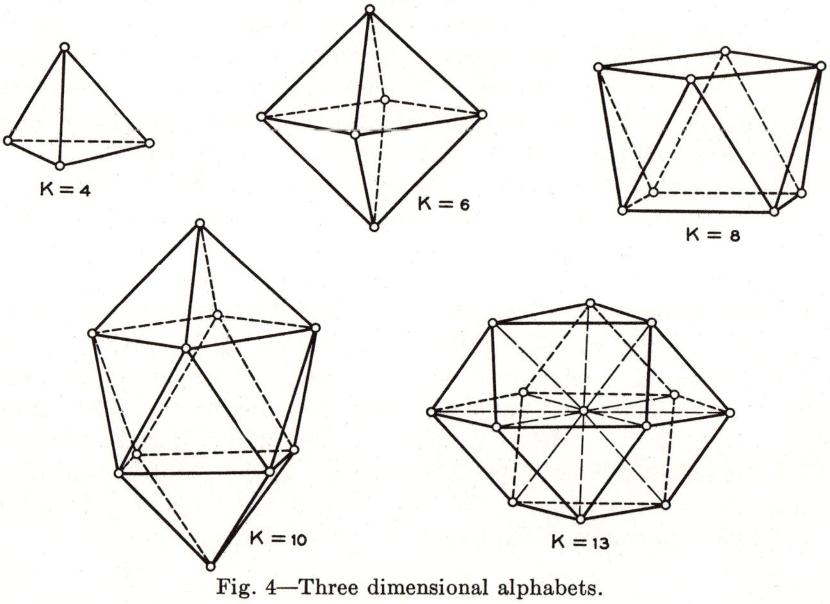

Fig. 4—Three dimensional alphabets.

where r_1 is the rms distance from the origin to the points of a configuration A, R_0 is the distance from the origin to the centroid of A, and r_2 is the rms distance from the points of A to the centroid of A.

In plotting points on the efficiency graph the notation K, D is used for the best K-letter D-dimensional alphabet which has been found. The arrangement of points for various $K, 2$ and $K, 3$ alphabets is given in Figs. 3 and 4. In these figures two points are joined by a straight line if the distance between them is 1 (which is the value we have adopted for the minimum allowed separation $2r_0$). Although not shown, the origin is always at the centroid of the figure. To aid interpretation of these diagrams we have included Fig. 5 which demonstrates how all the signals of a typical alphabet can be generated. The functions of time shown in

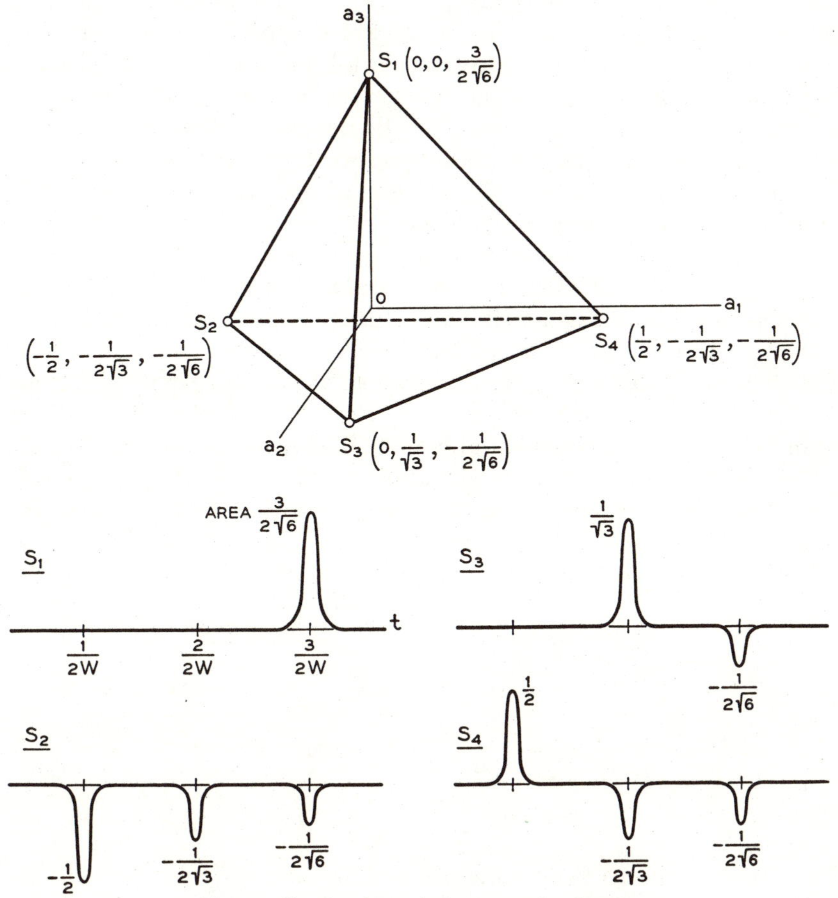

Fig. 5—Generation of the 4,3 code signals.

Fig. 5 are not the code signals themselves but impulse functions which are to be passed through a low pass filter with cutoff at W c.p.s. to form the code signals.

The best possible higher dimensional alphabets can be described more easily verbally than pictorially. In four dimensions we have found four alphabets.

The *25,4* alphabet consists of the origin and all 24 points in 4 dimensional space having two coordinates equal to zero and the remaining two equal to $1/\sqrt{2}$ or $-1/\sqrt{2}$. Each of the 24 points lies a unit distance away from the origin and its 10 other nearest neighbors; they are, in fact, the vertices of a regular solid. This alphabet has an advantage beyond its high efficiency. The code signals are composed entirely of positive and negative pulses of fixed energy and so should be easier to generate than most of the other codes which appear in this paper.

The *800, 4* alphabet is constructed in the following way: Consider a lattice of points throughout the entire 4-dimensional space formed by taking all the linear combinations with integer coefficients of a basic set of four vectors. That is, the lattice points are of the form $C_1v_1 + C_2v_2 + C_3v_3 + C_4v_4$ where $C_1, \cdots, C_4$ are integers and the v_i are the four given vectors. In connection with our problem it is of interest to know what lattice, (i.e. what choice of v_1, v_2, v_3, v_4) has all lattice points separated at least unit distance from one another and at the same time packs as many points as possible into the space per unit volume. When a solution to this "packing problem" is known, it is clear that a good alphabet can be obtained just by using all the lattice points which are contained inside a hypersphere about the origin as the letter points. Many of the two dimensional alphabets illustrated in the sketches are related in this way to the corresponding two dimensional packing problem (which is solved by letting v_1 and v_2 be a pair of unit vectors 60° apart). A solution to the four dimensional packing problem is afforded by

$$v_1 = \frac{1}{\sqrt{2}},\ \frac{1}{\sqrt{2}},\ 0,\ 0$$

$$v_2 = \frac{1}{\sqrt{2}},\ 0,\ \frac{1}{\sqrt{2}},\ 0$$

$$v_3 = \frac{1}{\sqrt{2}},\ 0,\ 0,\ \frac{1}{\sqrt{2}}$$

$$v_4 = 0,\ \frac{1}{\sqrt{2}},\ \frac{1}{\sqrt{2}},\ 0.$$

This lattice contains two points per unit volume (twice as dense as the cubic lattice in which $v_1, \cdots, v_4$ are orthogonal to one another) and each point has 18 nearest neighbors. A hypersphere of radius 3 about the origin has a volume $(\pi^2/2)3^4$, about 400. Thus it contains about 800 lattice points. Take these as the code points of the 800, 4 code. Their average squared distances from the origin can be estimated as

$$\frac{\int_0^3 r^5\,dr}{\int_0^3 r^3\,dr} = \frac{2}{3}(3)^2 = 6.$$

N in Equation (13) may be estimated at 18; this is conservative because some lattice points outside the sphere are being counted.

The two remaining four dimensional alphabets belong to two families of D-dimensional alphabets.

The 4, 3; 5, 4; $\cdots$; $D + 1$, D $\cdots$ alphabets are the vertices of the simplest regular solid in D-dimensional space. For example, 4, 3 is a tetrahedron. Such a solid can be constructed from $D + 1$ vertices whose coordinates are the first $D + 1$ rows of the scheme

$$\begin{array}{cccccc}
0 & 0 & 0 & 0 & 0 & \cdots \\
1 & 0 & 0 & 0 & 0 & \cdots \\
\frac{1}{2} & \frac{3}{2\sqrt{3}} & 0 & 0 & 0 & \cdots \\
\frac{1}{2} & \frac{1}{2\sqrt{3}} & \frac{4}{2\sqrt{6}} & 0 & 0 & \cdots \\
\frac{1}{2} & \frac{1}{2\sqrt{3}} & \frac{1}{2\sqrt{6}} & \frac{5}{2\sqrt{10}} & 0 & \cdots \\
\frac{1}{2} & \frac{1}{2\sqrt{3}} & \frac{1}{2\sqrt{6}} & \frac{1}{2\sqrt{10}} & \frac{6}{2\sqrt{15}} & \cdots \\
\cdot & \cdot & \cdot & \cdot & \cdot & \cdots \\
\cdot & \cdot & \cdot & \cdot & \cdot & \cdots \\
\cdot & \cdot & \cdot & \cdot & \cdot & \cdots
\end{array}$$

The vertices all lie a distance $\sqrt{D/2(D + 1)}$ from the centroid of the figure.

6, 3; 8, 4; $\cdots$; $2D$, D, $\cdots$ are obtained by placing a point wherever any positive or negative coordinate axis intersects the sphere of radius

$1/\sqrt{2}$ about the origin. Thus it follows that 6, 3 consists of the vertices of an octohedron.

Error correcting alphabets ((*k*, *K*, *D*)): The error correcting alphabets discussed in Part I can be converted into good alphabets for this channel by replacing all digits which equalled 0 by -1. Three error correcting alphabets appear on the chart; each is labelled by three numbers signifying (*k*, *K*, *D*).

Slepian alphabets (*SD*): Using group theoretic methods, D. Slepian has attempted to construct families of alphabets which signal at rates approaching *C*. Although this goal has not yet been reached, families of alphabets depending on the parameter *D* have been found which approach the ideal curve to within 6.2 db and then get worse as $D \rightarrow \infty$. In the simplest of these families of alphabets, $D = 2m$ is even and the letters consist of all the $2^m C_{2m,\, m}$ sequences containing *m* zeros, the remaining places being filled by ± 1. The best alphabet in this family is the one with $D = 24$. It lies 6.23 db away from the ideal curve and contains 1.1×10^{10} letters. The alphabets of this family for $D = 10, 24$, and 70 appear on the efficiency graph labelled *S*10, *S*24, and *S*70.

The conclusion to which one is forced as a result of this investigation is that one cannot signal over a channel with signal to noise level much less than 7 db above the ideal level of Equation (2) without using an unbelievably complicated alphabet. No ten digit alphabet tolerates less than 7.7 db more than the ideal signal to noise ratio.

It would be interesting to know more about good higher dimensional alphabets. They are very much more difficult to obtain. The regular solids, which provided some good alphabets in 3 and 4 dimensions, provide nothing new in 5 or more dimensions; there are only three of them and they correspond to our $D + 1, D$; $2D, D$, and 2^D binary alphabets. Worse still, the packing problem also becomes unmanageable after dimension 5.

ACKNOWLEDGMENT

The author wishes to thank R. W. Hamming, L. A. MacColl, B. McMillan, C. E. Shannon, and D. Slepian for many helpful suggestions during the investigation summarized by this paper.

Reprinted from *I.R.E. Trans. Electron. Computers,* **EC-3,** 6–12 (1954)

4

Application of Boolean Algebra to Switching Circuit Design and to Error Detection

D. E. MULLER*

***Summary*—A solution is sought to the general problem of simplifying switching circuits that have more than one output. The mathematical treatment of the problem applies only to circuits that may be represented by "polynomials" in Boolean algebra. It is shown that certain parts of the multiple output problem for such circuits may be reduced to a single output problem whose inputs are equal in number to the sum of the numbers of inputs and outputs in the original problem. A particularly simple reduction may be effected in the case of two outputs.**

Various techniques are described for simplifying Boolean expressions, called "+ polynomials," in which the operation "exclusive or" appears between terms. The methods described are particularly suitable for use with an automatic computer, and have been tested on the Illiac.

An unexpected metric relationship is shown to exist between the members of certain classes of "+ polynomials" called "nets." This relationship may be used for constructing error-detecting codes, provided the number of bits in the code is a power of two.

FOLLOWING the work of Shannon,[1] design of switching circuits has leaned heavily upon logical algebra, and systematic methods have been developed by Burkhart, Kalin, Aiken, Quine,[2,3] and others for reducing polynomial expressions in logical algebra. Much of the effectiveness of the application of these techniques has depended on the skill of the designer and upon the amount of time he is willing to spend in the manipulation of algebraic expressions which are obtained after having applied systematic reduction procedures. This has been especially true in the frequently encountered case in which more than one output is required from a particular circuit. Here, systematic methods for treating the single output circuit will be extended to the multiple output case.

Multiple Output Circuits

A switching circuit will be defined as a circuit in which voltage (or current) at any point in the circuit may take either of two possible values. These values may be arbitrarily described by the symbols 0 and 1. Such a circuit will be assumed to have p points $X^1, X^2, X^3, \cdots, X^p$ at which input voltages will be applied and q other points $Z^1, Z^2, Z^3, \cdots, Z^q$ from which outputs may be taken. It will be further assumed that all voltages in the circuit will be uniquely determined by the combined effect of the p inputs. If each of the q outputs is specified for each admissible combination of values at the p inputs, then the logical specifications for the circuit have been completely given and each output may be expressed as a logical function of the inputs

$$\begin{aligned} Z^1 &= Z^1(X^1, X^2, \cdots, X^p) \\ Z^2 &= Z^2(X^1, X^2, \cdots, X^p) \\ &\cdots\cdots\cdots\cdots \\ Z^q &= Z^q(X^1, X^2, \cdots, X^p). \end{aligned} \tag{1}$$

In general, certain combinations of values at the inputs will never occur, and for this reason the inputs will not be entirely independent. Such a relation will be expressed by the subsidiary condition

$$g(X^1, X^2, \cdots, X^p) = 0. \tag{2}$$

Those combinations of input values which never occur are just those for which $g=1$. Hence condition (2) completely specifies those combinations.

Algebraic manipulations may now be carried out to simplify the functional expressions (1) while making use of the subsidiary condition (2). These manipulations should tend to simplify the switching circuit corresponding to (1) according to prescribed criteria of simplicity, while maintaining the logical specifications for the circuits. Such manipulations, if carried out empirically, may be quite difficult and tedious. Often it is necessary to expand the functions Z^i so as to make them more complex before they can be simplified later. Systematic methods have therefore been developed to relieve the designer of some of the tedious work involved in reducing the functions Z^i.

A function Z^i of the inputs $X^1, X^2, \cdots, X^p$ may be expressed in canonical form

$$\begin{aligned} Z^i = {} & Z_0{}^iX^pX^{p-1}\cdots X^2X^1 \vee Z_1{}^iX^pX^{p-1}\cdots X^2\overline{X}^1 \\ & \vee \cdots \vee Z^i_{2^p-1}\overline{X}^p\,\overline{X}^{p-1}\cdots\overline{X}^2\overline{X}^1 \end{aligned} \tag{3}$$

where $\overline{X}^i$ represents the complement (or negation) of X^i and the symbol "$\vee$" represents the logical operation "or." In a particular term the inputs and their complements are connected by the logical operation "and." The coefficients $Z_j{}^i$ of the $j+1$ term is a constant having either the value 0 or 1, and serves to define the value of Z^i when the input values are such that the other factors in the $j+1$ term are all 1.

Expansion (3) is a special case of what may be called a polynomial in Boolean algebra. In a general polynomial, however, it will not be necessary for a term to depend on all inputs but it may be represented by a product of less than p of the inputs and complements of inputs. Thus $X^1 \vee \overline{X}^2 X^3$ and $\overline{X}^1 \vee X^4$ would also be re-

* Digital Computer Lab., University of Illinois, Urbana, Illinois.

[1] C. E. Shannon, "A symbolic analysis of relay and switching circuits" *Trans. A.I.E.E.*, vol. 57, pp. 713–723; 1938.

[2] "The Synthesis of Electronic Computing and Control Circuits," vol. XXVII. Annals of the Computation Laboratory of Harvard University, Harvard University Press, Cambridge, Mass.; 1951.

[3] W. V. Quine, "The problem of symplifying truth functions," *Amer. Math. Monthly*, vol. 59, p. 521; October, 1952.

garded as polynomials. A different type of polynomial may be formed if the operation "exclusive or" (designated by "+") is used between terms. Such a polynomial will be referred to as a + polynomial while the previous type will be referred to as a $\vee$ polynomial. Expansion (3) may also be written as a + polynomial since all terms in (3) are disjoint (i.e., never more than one term may equal 1) and "+" may be used to replace "$\vee$" wherever it appears, giving

$$Z^i = Z_0^i X^p X^{p-1} \cdots X^2 X^1 + Z_1^i X^p X^{p-1} \cdots X^2 \overline{X}^1 + \cdots + Z^i_{2^p-1} \overline{X}^p \overline{X}^{p-1} \cdots \overline{X}^2 \overline{X}^1. \quad (4)$$

In discussing multiple output functions, results will be valid for both + polynomials and $\vee$ polynomials and the symbol "+" will be used to refer to both operations. Furthermore, the term polynomial will mean either type of polynomial. Functions in Boolean algebra such as Z^i and g will also be expressed in polynomial form.

If a suitable reduction of the functions Z^i has been achieved in polynomial form, a set of polynomials must be specified, each one of which will be used when manufacturing certain z functions. These polynomials will be written $Mj_1, j_2, \cdots, j_q$ where j_i will have the value 0 if $Mj_1, j_2, \cdots, j_q$ is used in the construction of Z^i and the value 1 if it is not. There are 2^q-1 of these polynomials, in general, since it is not possible for j_i to be 1 for all i. In reduced form the functions Z^i will be written.

$$Z^i = \sum_{j_i=0} Mj_1, j_2, \cdots, j_q \quad (5)$$

where the sum is taken over all $Mj_1, j_2, \cdots, j_q$ for which $j_i=0$. In this sum the operation between the polynomials is either "$\vee$" or "+" depending upon whether $\vee$ polynomials or + polynomials are being used. In forming the switching circuit having outputs Z^i each of the polynomials $Mj_1, j_2, \cdots, j_q$ will be manufactured first and then combined according to (5) giving the Z^i.

The problem of reducing the switching circuit producing the outputs Z^i may now be considered in two parts:

(a) The problem of simultaneously minimizing the set of polynomials $Mj_1, j_2, \cdots, j_q$.

(b) The problem of minimizing the number of connectives between such polynomials in (5). When none of the polynomials $Mj_1, j_2, \cdots, j_q$ are zero, part (b) may be ignored, since the structure of the connectives is unalterable. On the other hand, if the number of outputs is large and the number of inputs small, part (b) tends to assume importance comparable to part (a).

Theorem 1: The problem of simultaneously minimizing the polynomials used in constructing a circuit having p inputs and q outputs may be replaced by the problem of finding a minimal polynomial to represent a certain single output circuit having $p+q$ inputs.

Proof: It is not necessary to define precisely the meaning of minimization for the purposes of this theorem since the two processes are merely to be shown to be equivalent.

The imaginary single output circuit described in the above theorem is assumed to use q inputs $y^1, y^2, \cdots, y^q$ in addition to inputs $X^1, X^2, \cdots, X^p$ which are used in the multiple output circuit. The single output F of the imaginary circuit is defined as

$$F = \sum_{i=1}^{q} \bar{y}^i Z^i(X^1, X^2, \cdots, X^p). \quad (6)$$

Just as the inputs $X^1, X^2, \cdots, X^p$ are assumed to be restricted by the condition $g(X^1, X^2, \cdots, X^p)=0$ the artificial inputs y^i will be restricted by the conditions:

$$\bar{y}^i \bar{y}^j = 0 \quad \text{when} \quad i \neq j$$
$$y^1 y^2 \cdots y^q = 0. \quad (7)$$

In order to express all of these conditions as a single condition they may be added, giving

$$g(X^1, X^2, \cdots, X^p) \vee y^1 y^2 \cdots y^q \vee \sum_{i \neq j} \bar{y}^i \bar{y}^j = 0. \quad (8)$$

The sum used in this last expression is understood to use the "$\vee$" operation while all other sums in this proof represent either "$\vee$" or "+" depending upon which type of polynomial is being considered.

If F as defined by (6) is minimized subject to condition (8) it will be represented as a polynomial P. From (7) it may be shown that

$$\bar{y}^i = y^1 y^2 \cdots y^{i-1} y^{i+1} \cdots y^q \quad (9)$$

and each $\bar{y}^i$ appearing in P may be replaced accordingly. After this has been done the resulting polynomial may be written.

$$F = P \equiv \sum (y^1)^{j_1} (y^2)^{j_2} \cdots (y^q)^{j_q} Mj_1, j_2, \cdots, j_q \quad (10)$$

where $Mj_1, j_2, \cdots, j_q$ is a polynomial involving $X^1, X^2, \cdots, X^p$ and the notation $(y^i)^{j_i}$ is defined by

$$(y^i)^{j_i} = 1 \quad \text{if} \quad j_i = 0$$
$$(y^i)^{j_i} = y^i \quad \text{if} \quad j_i = 1.$$

The sum in expression (10) is taken over the 2^q-1 combinations of values of the j_i in which not all of them are 1.

It now remains to identify the $Mj_1, j_2, \cdots, j_q$ in (10) with the $Mj_1, j_2, \cdots, j_q$ in (5). From (6) and (7) it may be seen that if $\bar{y}^i$ assumes the value 1 relation (6) will become $F=Z^i$. Eq. (10) then turns into (5) since terms in (10) containing y^i vanish. If (5) have not been minimized by this process then a more reduced set of $Mj_1, j_2, \cdots, j_q$ exists satisfying (5). These equations may be substituted in (6) and using (7) may be manipulated into form (10) thus giving a more reduced version of (10). Since (10) was assumed to be minimal this contradicts the hypothesis and the theorem is proved.

Eqs. (5) now represent the multiple output circuit made up of minimal polynomials $Mj_1, j_2, \cdots, j_q$.

Theorem 1 specializes in a convenient fashion when $q=2$, and Theorem 2 expresses this special case. Eqs. (7) then become

$$\bar{y}^1 \bar{y}^2 = 0, \qquad y^1 y^2 = 0.$$

Both of these equations will be automatically satisfied if the single condition $\bar{y}^2 = y^1$ is used. y^2 therefore may be eliminated by this equation and no subsidiary conditions are required.

Theorem 2: The problem of minimizing a two output circuit having p inputs, which is to be expressed in polynomial form, may be replaced by the problem of minimizing a single output circuit having $p+1$ inputs.

Equations may be specialized as follows:

$$Z^1 = M_{0,1} + M_{0,0}$$

$$Z^2 = M_{1,0} + M_{0,0}, \quad (5')$$

$$F = \bar{y}^1 Z^1 + y^1 Z^2, \quad (6')$$

$$g(X^1, X^2, \cdots, X^p) = 0, \quad (8')$$

$$F = P \equiv \bar{y}^1 M_{0,1} + y^1 M_{1,0} + M_{0,0}. \quad (10')$$

In (6′) y^1 was substituted for $\bar{y}^2$ of (6), and in (10′) $\bar{y}^1$ was substituted for y^2. In this way (7) becomes unnecessary.

By way of example the set of equations

$$Z^1 = X^3X^2X^1 + X^3\bar{X}^2\bar{X}^1 + \bar{X}^3X^2\bar{X}^1 + \bar{X}^3\bar{X}^2X^1$$

$$Z^2 = X^3X^2X^1 + X^3X^2\bar{X}^1 + X^3\bar{X}^2X^1 + \bar{X}^3X^2X^1$$

may be used with no subsidiary conditions. These equations represent a single stage binary adder where Z^1 is the output and Z^2 is the carry. Eq. (6′) becomes

$$F = \bar{y}^1X^3X^2X^1 + \bar{y}^1X^3\bar{X}^2\bar{X}^1 + \bar{y}^1\bar{X}^3X^2\bar{X}^1 + \bar{y}^1\bar{X}^3\bar{X}^2X^1 + y^1X^3X^2X^1 + y^1X^3X^2\bar{X}^1 + y^1X^3\bar{X}^2X^1 + y^1\bar{X}^3X^2X^1.$$

Again no subsidiary conditions are to be used. Using + polynomial reduction techniques, to be described in the next section, this polynomial may be reduced to

$$F = \bar{y}^1\bar{X}^3\bar{X}^2 + \bar{y}^1\bar{X}^1 + y^1X^3X^1 + y^1X^2X^1 + X^3X^2.$$

This gives

$$M_{0,1} \equiv \bar{X}^3\bar{X}^2 + \bar{X}^1$$

$$M_{1,0} \equiv X^3X^1 + X^2X^1$$

$$M_{0,0} \equiv X^3X^2$$

to complete the construction.

The multiple output circuit is represented by

$$Z^1 = \bar{X}^3\bar{X}^2 + \bar{X}^1 + X^3X^2$$

$$Z^2 = X^3X^1 + X^2X^1 + X^3X^2.$$

These expressions are not necessarily the simplest forms for Z^1 and Z^2. Further reduction, by replacing "+" with "∨" and by factoring, are outside the realm of the present discussion since the resulting expressions would then no longer be polynomials.

Reduction of + Polynomials

Reduction of ∨ polynomials has been completely analyzed by Quine[3] and by Burkhart, Kalin, and Aiken.[2] Extension of these methods to include the possibility of subsidiary conditions has been carried out by I. S. Reed.[4] Applying these methods to theorem 1 permits multiple output circuits to be treated also.

Circuit reduction by use of methods involving + polynomials presents an alternative process which usually yields considerably different results from those involving ∨ polynomials. By way of review, the important properties of the operation "+" are:

$$\begin{aligned} &\text{i)} & a + b &= a\bar{b} \vee \bar{a}b \\ &\text{ii)} & a + b &= b + a \\ &\text{iii)} & a + (b + c) &= (a + b) + c \\ &\text{iv)} & a(b + c) &= ab + ac \\ &\text{v)} & a + a &= 0 \\ &\text{vi)} & a + \bar{a} &= 1. \end{aligned} \quad (11)$$

If property (i) is taken as a definition the other properties may be directly deduced. Because of rule (v), it is evident that one need never retain duplicate terms in a polynomial. For this reason it will be assumed that duplicate terms are always to be combined in any polynomial representation. Operations may be performed upon polynomials which leave them equal to the same Boolean function but change their form. Two polynomials, P_1 and P_2 will be regarded as equivalent only if they are termwise equivalent. Such a relation will be written $P_1 \equiv P_2$ while $P_1 = P_2$ will be taken to mean that the two polynomials equal the same Boolean function, but are not necessarily equivalent. The symbol $P_1 + P_2$ will represent a polynomial containing the terms of both P_1 and P_2 with the exception that duplicate terms are combined according to (v). P_1P_2 will represent the expanded product of the two polynomials and $P_1 \cdot P_2$ will represent a polynomial having only those terms which are common to P_1 and P_2.

A general operator R_j which may be used to alter the form of a + polynomial without changing it functionally is defined by the relation

$$R_jP \equiv P + X^jM + \bar{X}^jM + M$$

where M is a + polynomial which depends on R_j and may or may not depend on P. If M is independent of P the operator R_j is its own inverse since $R_jR_jP \equiv P$. Special operators of this type may be formed in various ways. Four operators of type R_j may be defined by writing:

$$P \equiv X^jM_0 + \bar{X}^jM_1 + M_2$$

where M_0, M_1 and M_2 are polynomials which are independent of X^j. They are:

$$A_jP \equiv P + X^jM_2 + \bar{X}^jM_2 + M_2$$

$$B_jP \equiv P + X^jM_1 + \bar{X}^jM_1 + M_1$$

$$C_jP \equiv P + X^jM_0 + \bar{X}^jM_0 + M_0$$

$$D_jP \equiv P + X^j(M_0 + M_1 + M_2) + \bar{X}^j(M_0 + M_1 + M_2) + (M_0 + M_1 + M_2).$$

The symbol Q_j will be used to denote any one of these four operators, and can be shown to possess the algebraic properties

[4] Technical Memo No. 23, Lincoln Lab., M.I.T.

1) $Q_j(P_1 + P_2) \equiv Q_jP_1 + Q_jP_2$

2) $Q_jQ_kP \equiv Q_kQ_jP$

3) $Q_jR_jP \equiv Q_jP \quad (\text{any } R_j).$

Theorem 3: The operator $A_pA_{p-1} \cdots A_1$ reduces a polynomial P to its canonical form (4).

Proof: A_jP is a polynomial in which each term contains either X^j or $\overline{X}^j$. To see this one may write

$$\begin{aligned} A_jP &\equiv P + X^jM_2 + \overline{X}^jM_2 + M_2 \\ &\equiv X^jM_0 + \overline{X}^jM_1 + M_2 + X^jM_2 + \overline{X}^jM_2 + M_2 \\ &\equiv X^j(M_0 + M_2) + \overline{X}^j(M_1 + M_2). \end{aligned}$$

If all terms in the polynomial P contain either X^k or $\overline{X}^k$, then A_jP also possesses this property since no terms containing neither X^k nor $\overline{X}^k$ are introduced. Hence every term of $A_pA_{p-1} \cdots A_1P$ contains either each input or its complement. Therefore $A_pA_{p-1} \cdots A_1P$ has the form of (4) which is a unique canonical form for each function. Henceforth the expression $A_pA_{p-1} \cdots A_1$ will be abbreviated A.

Theorem 4: If $P_1 = P_2$ then it is possible to transform P_1 into P_2 by p operations of the general type R_j.

Proof: If P_3 represents the canonical form of P_1 and P_2, the relations $AP_1 \equiv P_3$ and $AP_2 \equiv P_3$ are satisfied. The polynomials M_2 in the expression $A_jP \equiv P + X^jM_2 + \overline{X}^jM_2 + M_2$ may now be regarded as constants which do not depend on P since they are defined by the relations $AP_1 \equiv P_3$ and $AP_2 \equiv P_3$. The resulting operators may no longer be regarded as of the type A_j since the M's involved are constants. These operators will be written $S_pS_{p-1} \cdots S_1P_1 \equiv P_3$ and $T_pT_{p-1} \cdots T_1P_2 \equiv P_3$.

The operator S_j has the same effect as A_j when used in this equation but when applied to a different polynomial it would not have the same effect since the polynomial M_2 would be altered in the case of A_j and not in the case of S_j.

Thus if

$$P \equiv X^jM_0 + \overline{X}^jM_1 + M_2$$

and

$$P' \equiv X^jM_0' + \overline{X}^jM_1' + M_2'$$

then

$$A_jP \equiv P + X^jM_2 + \overline{X}^jM_2 + M_2$$

and

$$S_jP \equiv P + X^jM_2 + \overline{X}^jM_2 + M_2$$

but

$$A_jP' \equiv P' + X^jM_2' + \overline{X}^jM_2' + M_2'$$

while

$$S_jP' \equiv P' + X^jM_2 + \overline{X}^jM_2 + M_2.$$

Since operators using constant M commute and are their own inverses it is evident that

$$P_3 \equiv S_1S_2 \cdots S_pP_1$$

and

$$\begin{aligned} P_2 &\equiv T_pT_{p-1} \cdots T_1S_1S_2 \cdots S_pP_1 \\ &\equiv T_pS_pT_{p-1}S_{p-1} \cdots T_1S_1P_1. \end{aligned}$$

The operator T_jS_j may, however, be regarded as a single operator since the M's involved may be added, and the theorem is proved.

From theorem 4 it may be seen that operations of the type R_j are sufficient to reduce any arbitrary polynomial P_1 to its minimal form P_2. Such a reduction is not in general possible simply because the required operations cannot usually be found without a knowledge of P_2.

If operators of type Q_j are combined, a variety of characteristic forms are obtained. Theorem 5 proves the existence of these forms.

Theorem 5: If $P_1 = P_2$ then $Q_pQ_{p-1} \cdots Q_1P_1 \equiv Q_pQ_{p-1} \cdots Q_1P_2$ where Q_j may represent different ones of the four operators A_j, B_j, C_j or D_j for each j, but must have the same meaning on the two sides of the equation.

Proof: Let $AP_1 \equiv AP_2 \equiv P_3$.
Then $Q_pQ_{p-1} \cdots Q_1P_3 \equiv Q_pQ_{p-1} \cdots Q_1A_pA_{p-1} \cdots A_1P_1 \equiv Q_pQ_{p-1} \cdots Q_1P_1$ by properties 1 and 3. Similarly, $Q_pQ_{p-1} \cdots Q_1P_3 \equiv Q_pQ_{p-1} \cdots Q_1P_2$ and hence $Q_pQ_{p-1} \cdots Q_1P_1 \equiv Q_pQ_{p-1} \cdots Q_1P_2$.

From this property the operator $Q_pQ_{p-1} \cdots Q_1$ may be said to yield a "characteristic" polynomial. Since four possible choices are available for each operator Q_j (it may be either A_j, B_j, C_j, or D_j) the number of such expansions is 4^p. A particularly symmetrical expansion of this type is the one produced by the operator $D_pD_{p-1} \cdots D_1$. Other expansions such as that produced by $B_pB_{p-1} \cdots B_1$ have singular metric properties which will be described later.

Simplification of polynomials is carried out with the help of operators of the type Q_j but principally one must rely on the mathematically less interesting operator H_j defined by

$$\begin{aligned} H_jP \equiv P &+ X^j(M_0 \cdot M_1 + M_1 \cdot M_2 + M_2 \cdot M_0) \\ &+ \overline{X}^j(M_0 \cdot M_1 + M_1 \cdot M_2 + M_2 \cdot M_0) \\ &+ (M_0 \cdot M_1 + M_1 \cdot M_2 + M_2 \cdot M_0) \end{aligned}$$

where M_0, M_1 and M_2 do not involve X_j or $\overline{X}_j$ and are defined by the relation $P \equiv X^jM_0 + \overline{X}^jM_1 + M_2$. Polynomials represented by $M_0 \cdot M_1$ etc., are defined, as before, to be those containing terms common to M_0 and M_1 etc. H_j has none of the convenient algebraic properties of the Q_j's but it tends to reduce the number of terms in the polynomial to which it is applied. If "a" represents one term in a polynomial, H_j effects the following types of simplifications.

$$\begin{aligned} H_j(X^ja + \overline{X}^ja) &\equiv a \\ H_j(X^ja + a) &\equiv \overline{X}^ja \\ H_j(\overline{X}^ja + a) &\equiv X^ja \\ H_j(X^ja + \overline{X}^ja + a) &\equiv 0. \end{aligned} \tag{12}$$

One of the most elementary types of simplifications which can be applied to a polynomial is therefore $H_pH_{p-1} \cdots H_1P$ which will be denoted by HP. Although the operator H tends to simplify the polynomial to which it is applied, it will usually yield a result which is far more complex than that attained by more refined methods. In order to attain greater simplification than is possible merely by use of the H operator, one may expand each term by reversing one of the first three operations (12) whenever subsequent application of the H operator effects a still greater simplification. Such a process which will be called Method I may be explained, stepwise, as follows:

1) One starts with a polynomial P_1 to be simplified. It is first reduced to canonical form.

$$P_2 \equiv AP_1.$$

2) This result is simplified initially by use of H.

$$P_3 \equiv HP_2.$$

3) A simplification operator C_j' is constructed according to the definition

$$C_j'P \equiv H[\overline{X}^j\{H(M_0 + M_1)\} + \{M_0 + M_2\}].$$

Successive application of C_j' yields

$$P_4 \equiv C_p'C_{p-1}' \cdots C_1'P_3.$$

4) An operator B_j' is constructed according to the definition

$$B_j'P \equiv H[X^j(M_0 + M_1) + H(M_2 + M_1)]$$

and the final result P_5 is given by

$$P_5 \equiv B_p'B_{p-1}' \cdots B_1'P_4.$$

In this process the operators B_j' and C_j' have the effect of expanding the polynomial whenever it may be simplified later by application of the H operator.

An operator H' which is more effective than the H operator may be formed by use of a gate polynomial KP. If $P_1 \nabla P_2$ represents a polynomial having terms which are in either or both of the polynomials P_1 and P_2, then the polynomial K_jP may be defined as

$$K_jP \equiv (X^j \nabla \overline{X}^j \nabla 1)(M_0 \nabla M_1 \nabla M_2)$$

and

$$KP \equiv K_pK_{p-1} \cdots K_1P.$$

For purposes of notation let

$$KP \equiv X^jN_0 + \overline{X}^jN_1 + N_2$$

and let

$$J_j^0P \equiv P + X^j(M_0 \cdot N_1 \cdot N_2) + \overline{X}^j(M_0 \cdot N_1 \cdot N_2) + (M_0 \cdot N_1 \cdot N_2)$$

$$J_j^1P \equiv P + X^j(N_0 \cdot M_1 \cdot N_2) + \overline{X}^j(N_0 \cdot M_1 \cdot N_2) + (N_0 \cdot M_1 \cdot N_2)$$

$$J_j^2P \equiv P + X^j(N_0 \cdot N_1 \cdot M_2) + \overline{X}^j(N_0 \cdot N_1 \cdot M_2) + (N_0 \cdot N_1 \cdot M_2)$$

and let

$$G_j^0P \equiv H_{j-1}H_{j-2} \cdots H_1H_p \cdots H_{j+1}J_j^0P$$

$$G_j^1P \equiv H_{j-1}H_{j-2} \cdots H_1H_p \cdots H_{j+1}J_j^1P$$

$$G_j^2P \equiv H_{j-1}H_{j-2} \cdots H_1H_p \cdots H_{j+1}J_j^2P.$$

Then

$$G^0P \equiv G_p^0G_{p-1}^0 \cdots G_1^0P$$

$$G^1P \equiv G_p^1G_{p-1}^1 \cdots G_1^1P$$

$$G^2P \equiv G_p^2G_{p-1}^2 \cdots G_1^2P$$

and finally

$$H'P \equiv G^2G^1G^0HP.$$

Method II may now be described as Method I with H' substituted for H wherever it appears. Method II has the advantage of forming as many as two expansions provided later contractions makes this advantageous, and thus yields a more effective, but more time-consuming process.

Justification for the choice of these processes rather than others which involve expansion and later simplification is based mainly upon their efficiency in reducing randomly chosen polynomials. A set of twenty randomly chosen functions of 5 inputs was used for comparison of different processes. Taking the number of terms in the final polynomial as a convenient measure of the effectiveness of the processes, the results obtained from Method I, Method II and the simple HA operator are compared in Table I.

TABLE I

TWENTY RANDOM FUNCTIONS OF FIVE INPUTS WERE USED TO TEST THREE SIMPLIFICATION PROCESSES. THE NUMBER OF TERMS IN THE SIMPLIFIED EXPANSION IS LISTED IN EACH CASE

Using Only HA Operator	Method I	Method II
11	7	7
9	7	6
6	5	5
9	6	6
8	7	6
10	7	6
8	7	7
9	7	7
8	6	6
7	7	6
7	7	6
9	7	7
9	7	7
10	8	7
10	8	7
8	7	7
7	5	5
8	7	7
9	6	6
8	5	5

Following polynomial-type simplifications such as Methods I and II, nonsystematic manipulation may be used to further simplify the result. Two types are especially useful:

a) Between pairs of terms of the form X^ia and $\overline{X}^ib$ the operation "+" may be replaced by "$\vee$" and between triples of the form $ab+bc+ca$ one may make the same substitution. By use of skill one should attempt to make the combination of substitutions which leave the fewest "+" operations to be performed.

b) Using skill, factor the result so as to reduce the resulting expression as much as possible.

Systematic polynomial reduction processes may conveniently be carried out by the use of high speed computers. Programs for the ILLIAC have been prepared to reduce $\vee$ polynomials using the Harvard method and to reduce + polynomials using Methods I and II described here. As yet none of these processes have been modified to permit the inclusion of subsidiary conditions. These programs make use of an interpretive subroutine which makes it possible to manipulate polynomials conveniently in the machine. In the memory of the machine a polynomial is represented as a set of 3^p binary digits. The position of each binary digit specifies the term. If the digit is 1 it is regarded as being present in the polynomial, and if the digit is 0 it is regarded as absent. No distinction need be made between $\vee$ polynomials and + polynomials. To each input is allotted a digit of a number written in the ternary system. This digit is 0, 1, or 2 according to whether the input is present, complemented, or absent. The ternary number so obtained represents the relative position of the binary digit corresponding to a term in a polynomial. Thus to the term $\overline{X}^3X^2$ corresponds the number 102 written in the ternary system. Since 0 represents $X^3X^2X^1$ it would be placed in the first relative position and $\overline{X}^3X^2$ would be represented by a 1 in the twelfth relative position.

Using the interpretive routine it is possible to extract just those digits of the polynomial $P \equiv X^jM_0 + \overline{X}^jM_1 + M_2$ corresponding to one of the M's, say X^jM_0. By shifting these digits to a new relative position it is possible to form from these extracted digits either X^jM_0, $\overline{X}^jM_0$ or M_0. Assume that $\overline{X}^jM_0$ is formed. This result may then be combined algebraically with some other polynomial P' to form $P' + \overline{X}^jM_0$, $P' \cdot \overline{X}^jM_0$ or $P' \nabla X^jM_0$. The polynomial containing all terms not in $\overline{X}^jM_0$ may also be used when performing these combinations. Such a sequence of operations as the one described will be produced by one order in the interpretive routine. By a series of such orders the operators of the simplification processes may be formed. Special control transfer orders allow repeating a process using logical input indices 1, 2, $\cdots$, $p-1$, p and other orders permit algebraic operations to be performed without extractions. Various "red tape" orders are also provided.

Error Detection

In the theory of error detecting codes one deals with sequences of n binary digits. Such a set "a" may be written as a vector $a = (a_0, a_1, \cdots, a_{n-1})$ where a_i may take on values 0 or 1. A metric $L(a, b)$ has been defined with respect to two such vectors "a" and "b" as the number of components in which "a" and "b" differ.[5] This metric may be shown to possess all the usual metric properties. The problem of finding an error detecting code consists of finding a set of r vectors $r^0, r^1, \cdots, r^j, \cdots, r^{t-1}$ such that $L(r^j, r^k) \geqq d$ for $j \neq k$ when one is given a number d called the order of the code.

The theory of + polynomials in Boolean algebra may be applied directly, when n and d are powers of two. n and d were not assumed to be restricted in this fashion in the definition of the general problem. If $n = 2^p$ and $d = 2^m$ the solutions so obtained give $t = 2^{C_p{}^p + C_{p-1}{}^p + \cdots + C_m{}^p}$ where $C_q{}^p$ is a binomial coefficient. Components "a_i" of the vector "a" may be identified with the coefficients of the terms in expansion (4), the canonical expansion of a corresponding function "a" of p inputs in Boolean algebra. Such a canonical expansion may be regarded as a polynomial P_1.

$$a = P_1 \equiv a_0X^pX^{p-1} \cdots X^1 + a_1X^pX^{p-1} \cdots \overline{X}^1 + \cdots + a_{2^p-1}\overline{X}^p\overline{X}^{p-1} \cdots \overline{X}^1. \quad (13)$$

A characteristic polynomial P_2 may be formed by successive application of the operators Bj described previously.

$$P_2 \equiv B_pB_{p-1} \cdots B_1P_1 \equiv BP_1. \quad (14)$$

By an argument similar to that given in the proof of theorem 3, it may be seen that none of the inputs appearing in the terms of P_2 are complemented. Each input therefore is either present or absent and P_2 may be written

$$P_2 \equiv g_0X^pX^{p-1} \cdots X^1 + g_1X^pX^{p-1} \cdots X^2 + \cdots + g_{2^p-1} \quad (15)$$

The coefficients $g_0, g_1, \cdots, g_{2^p-1}$ are each either 0 or 1, and depend uniquely on $a_0, a_1, \cdots, a_{2^p-1}$. It is interesting to note that the transformation of the coefficients $a_0, a_1, \cdots, a_{2^p-1}$ to $g_0, g_1, \cdots, g_{2^p-1}$ is its own inverse. This may be seen by interchanging the role of X^j and the absence of X^j or $\overline{X}^j$ in the terms.

In expression (15) it is possible to group those terms containing a given number of inputs. There may exist $C_k{}^p$ terms having k inputs but certain terms in (15) may vanish because their coefficients are zero.

Definition: A net of logical functions of order d is defined as all those functions whose expansions (as given in (15)) contain no terms having more than $p-m$ inputs, where m is defined by the relation $2^m = d$.

Theorem 6: If $r^1, r^2, \cdots, r^t$ are members of a net of order d then $L(r^i, r^j) \geqq d$ for all pairs r^i, r^j with $i \neq j$.

Proof: The theorem is proved by induction. It is true in case $p = \log_2 d$, since then the functions 0 and 1 are the only net members. Assume it is true when $p = k$ for all allowable d. It will be shown to be true when $p = k+1$.

From expansion (15) it may be noted that the members of a net are closed under the operation "+" since no terms containing more $p-m$ inputs can be generated by adding expansions having no such terms. Thus there is an r^l in the net such that $r^i + r^j = r^l$ for every pair r^i, r^j in the net. From the definition of the operation "+"

$$L(r^j, r^i) = L(r^i + r^j, 0) = L(r^l, 0).$$

Thus it is only necessary to prove that $L(r^l, 0) \geqq d$ for each non-zero member of the net r^l. If r^l, a member of the net of order d, is a function of $k+1$ inputs and is expressed in the form of (15), the $(k+1)$st input may be factored out giving

$$r^l = f^1 + X^{k+1}f^2$$

f^1 and f^2 are functions of k inputs, which are not both zero. f^1 is a member of the net of order $d/2$ and f^2 is a

[5] R. W. Hamming, "Error detecting and error correcting codes," *Bell Sys. Tech. Jour.*, vol. 29, pp. 147–160; April, 1950.

member of the net of order d. Four cases must be considered:

a) If the function f^2 is zero, then $r^l = f^1$. Regarding f^1 as a function of $k+1$ inputs the function r^l may be written as a sum of disjoint parts: $r^l = \overline{X}^{k+1}f^1 + X^{k+1}f^1$. If f^1 is written in the form of expansion 13 then it may seen that the separate parts $\overline{X}^{k+1}f^1$ and $X^{k+1}f^1$ of r^l contribute independently to $L(r^l, 0)$. Since $L(\overline{X}^{k+1}f^1, 0) \geqq d/2$ and $L(X^{k+1}f^1, 0) \geqq d/2$ the result

$$L(r^l, 0) = L(\overline{X}^{k+1}f^1, 0) + L(X^{k+1}f^1, 0) \geqq d$$

follows.

b) If the function f^1 is zero, then $r^l = X^{k+1}f^2$, and

$$L(X^{k+1}f^2, 0) \geqq d.$$

c) If f^1 and f^2 are not zero, but $f^1 = f^2$ then

$$r^l = f^2 + X^{k+1}f^2 = \overline{X}^{k+1}f^2$$

and

$$L(\overline{X}^{k+1}f^2, 0) \geqq d.$$

d) If f^1 and f^2 are not zero and not equal then

$$r^l = f^1 + X^{k+1}f^2 = \overline{X}^{k+1}f^1 + X^{k+1}f^3$$

where $f^3 = f^1 + f^2$ is not zero since $f^1 \neq f^2$ and is a member of the net of order $d/2$ by closure. Hence $L(\overline{X}^{k+1}f^1, 0) \geqq d/2$ and $L(X^{k+1}f^3, 0) \geqq d/2$ giving $L(r^l, 0) \geqq d$ since as before the expansions $\overline{X}^{k+1}f^1$ and $X^{k+1}f^3$ are disjoint.

Error detecting codes of order d may be formed therefore by use of vectors whose components are the coefficients of terms in the expansion (13) of net members. The information carried in such a code depends upon the number of coefficients in expansion (15) which are not forced to be zero by the net requirement. This number is $C_p{}^p + C_{p-1}{}^p + \cdots + C_m{}^p$ so that the number of such vectors available is $2^{C_p{}^p + C_{p-1}{}^p + \cdots + C_m{}^p}$.

For convenience expansion (15) may be used for interpreting information, and expansion (13) for transmission.

It has been shown that the members[6] of the net of order d do not always give the most numerous set of functions satisfying the relation $L(r^i, r^j) \geqq d$ and including the net members. When $d = 2^p, 2^{p-1}, 4, 2, 1$ the net does always give the most numerous set. When d takes on any other allowable value it can be shown that a more numerous set always exists which includes the net members, if sufficiently large p is used.

The case $d = 8$, $p = 5$ was investigated by use of the ILLIAC. In this case it was shown that no larger set of functions than the net members exists which satisfies $L(r^i, r^j) \geqq d$, and contains all net members. Hence the first case of a more numerous set must have $p \geqq 6$.

[6] D. E. Muller, "Metric Properties of Boolean Algebra and Their Application to Switching Circuits." Internal Report No. 46, Univ. of Illinois Graduate College, Digital Computer Laboratory.

5

Reprinted from *I.R.E. Trans. Inform. Theory,* **PGIT-4,** 38–49 (1954)

A CLASS OF MULTIPLE-ERROR-CORRECTING CODES AND THE DECODING SCHEME

Irving S. Reed

Lincoln Laboratory - Massachusetts Institute of Technology
Cambridge, Massachusetts

I. Introduction

A procedure for constructing one-error-correcting and two-error-detecting systematic codes was introduced in a recent study by R. W. Hamming.[1] It is the purpose of this paper to exhibit some examples of n-error-correcting and (n + 1) error-detecting systematic codes for the cases where both the code length and (n + 1) are powers of two. The class of codes to be considered was developed by D. E. Muller in his recent work.[2]

The decoding scheme presented in this paper differs from Hamming's scheme in that the encoded message will be extracted directly from the possibly corrupted received code by a majority testing of the redundant relations within the code. Hamming's scheme for n = 1 was dependent first on the location of a possible digit error in the code; secondly, on the correction of that digit; and lastly, on the extraction of the message from the corrected code. By circumventing Hamming's step of error location and correction, which is quite a severe problem when n is not equal to one, we have arrived at a decoding scheme that makes a natural use of the redundancy within the code as well as being conceptually simple.

In this paper, some of the mathematical proofs of the methods discussed will be avoided for the sake of brevity of exposition. A more detailed mathematical analysis will appear elsewhere.

II. Some Mathematical Preliminaries

A code having n binary digits may be considered the element of a space, consisting of 2^n elements of the form

$$f = (f_0, \ldots f_{n-1})$$

where

$$(f_j = 0, 1) \quad \text{for} \quad (j = 0, 1, 2, \ldots n-1) \ .$$

This space is technically an Abelian group if the sum of any two elements f and g in the space is defined as follows:

$$f \oplus g = (f_0, f_1, \ldots f_{n-1}) \oplus (g_0, g_1, \ldots g_{n-1}) = (f_0 \oplus g_0, f_1 \oplus g_1, \ldots f_{n-1} \oplus g_{n-1}) \ ,$$

where $f_j \oplus g_j$ is the sum modulo two of the binary digits f_j and g_j for $(j = 0,1,2, \ldots n-1)$. If multiplication by the binary scalar α is allowed as

$$\alpha f = \alpha(f_0, f_1, \ldots f_{n-1}) = (\alpha f_0, \alpha f_1, \ldots \alpha f_{n-1}) \ ,$$

the Abelian group may be termed a generalized vector space of n-dimensions or a module. Finally, if the product operation

$$f \cdot g = (f_0, f_1, \ldots f_{n-1}) \cdot (g_0, g_1, \ldots g_{n-1}) = (f_0 g_0, f_1 g_1, \ldots f_{n-1} g_{n-1})$$

for f and g in the module is introduced, the space is a Boolean ring. The prime operation is defined to be

$$f' = f \oplus I$$

for f in the ring, and where I is the identity vector (1, 1, 1, ... 1).

Into this space one may further introduce a norm or length of a vector as follows:

$$\|f\| = \sum_{i=0}^{n-1} f_i$$

where Σ refers to ordinary addition. It is not difficult to see that the norm of the sum of two elements f and g in the ring or $\|f \oplus g\|$ is precisely the Hamming distance $D(f,g)$ as defined in Ref. 1.

Now let n the dimension of the vector space be a power of two or $n = 2^m$. Let a vector of this space be of the form

$$f = (f_0, f_1, \ldots f_{2^m-1}) ,$$

where f_j is a binary digit for $(j = 0, 1, \ldots 2^m-1)$. Now the vector f may be clearly expressed as

$$f = f_0 I_0 \oplus f_1 I_1 \oplus \cdots f_{2^m-1} I_{2^m-1} , \qquad (1)$$

where I_j is a unit vector with the digit one in j-th coordinate of the vector and zeros elsewhere for $(j = 0,1, \ldots 2^m-1)$. Further, each unit vector I_j can be determined as a product of m vectors from the set of 2m vectors $x_1, x_2, x_3, \ldots x_m, x_1', x_2', x_3', \ldots x_m'$, where x_1 is a vector consisting of alternating zeros and ones, beginning with zero; x_2 is a vector consisting of alternating zero pairs and one pairs, beginning with a zero pair, and so forth, as follows:

$$\begin{aligned} x_1 &= (0\,1\,0\,1\,0\,1\,0\,1 \ldots 0\,1) , \\ x_2 &= (0\,0\,1\,1\,0\,0\,1\,1 \ldots 1\,1) , \\ x_3 &= (0\,0\,0\,0\,1\,1\,1\,1 \ldots 1\,1) , \\ &\vdots \\ x_m &= (0\,0\,0\,0\,0\,0\,0\,0 \ldots 1\,1) . \end{aligned} \qquad (2)$$

If $x_k^{i_k}$ is defined to be x_k' for $i_k = 0$ and x_k for $i_k = 1$, then by the rules of Boolean algebra,

$$I_j = x_1^{i_1} x_2^{i_2} \ldots x_m^{i_m} , \qquad (3)$$

where

$$j = \sum_{k=1}^{m} i_k 2^{k-1} \quad \text{with } (i_k = 0,1) \text{ for } (j = 0,1, \ldots m - 1) .$$

Combining Eqs. (1) and (3), we have

$$f = \mathop{\mathbf{\Sigma}}_{j=0}^{2^m-1} f_j x_1^{i_1} x_2^{i_2} \ldots x_m^{i_m} , \qquad (4)$$

where $i_1, i_2, \ldots i_m$ are the digits of the binary representation of j, and where the summation sign $\mathbf{\Sigma}$ is with respect to the sum operation $\oplus$. Equation (4) is the canonical expansion of any vector f in the Boolean algebra of 2^m dimensional vectors, consisting of binary digits.

If the identity $x_j' = I \oplus x_j$ and the distributive law of algebra is used, Eq.(4) may be expanded to obtain the following polynomial in the x_j's:

$$f = g_0 \oplus g_1 x_1 \oplus \cdots \oplus g_m x_m \oplus g_{12} x_1 x_2 \oplus \cdots g_{m-1,m} x_{m-1} x_m \oplus \cdots$$

$$\cdots \oplus g_{12 \ldots m} x_1 x_2 \ldots x_m . \qquad (5)$$

Equation (5) can be written more explicitly as

$$f = f(0,\dots 0) \oplus \Delta_{1} f(0,\dots 0)x_1 \oplus \dots \oplus \Delta_{m} f(0,\dots 0)x_m \oplus \Delta^{2}_{12} f(0,\dots 0)\, x_1x_2$$

$$\oplus \dots \oplus \Delta^{m}_{12\dots m} f(0,\dots 0)x_1x_2\dots x_m \quad , \tag{6}$$

where

$$f(i_1,\dots i_m) = f_j \quad \text{when} \quad j = \sum_{k=1}^{m} i_k\, 2^{k-1} \quad \text{for } i_k = 0,1 \ ,$$

and the Δ's are multiple partial differences, for example,

$$\Delta_{1} f(0) = f(1,0,0,\dots) \oplus f(0,0,0,\dots 0) \ ,$$

$$\Delta^{2}_{12} f(0) = [f(1,1,0,\dots) \oplus f(0,1,0,\dots)] \oplus [f(1,0,0,\dots) \oplus f(0,0,0,\dots)] \ ,$$

and so forth. The polynomial representation in Eq.(6) of the vector f supplies the relations between the coefficients of Eq.(5) and the scalars f_j of Eq.(4) for ($j = 0,1,2,\dots 2^m-1$). This definition of the Δ's will be expanded in another section of this paper.

III. The Generation of the Multiple Error Allowing Codes

Suppose that the dimension of the space considered in the previous section is 2^m. Consider the set Φ^m_r of all polynomials of the form (5) of degree less than or equal to r where $r \leq m$. Each such polynomial must have the form

$$g_0 \oplus g_1x_1 \oplus \dots \oplus g_mx_m \oplus \dots \oplus g_{12\dots r}\, x_1\dots x_r \oplus \dots \oplus g_{m-r+1,\dots m}$$

$$x_{m-r+1}\dots x_m \ , \tag{7}$$

and the sum of any two such polynomials is a member of the same set. This implies that Φ^m_r the set of all polynomials of type (7) or of degree less than or equal to r forms an Abelian group or submodule of the Boolean ring of 2^m dimensional vectors. Since Φ^m_r is a module, the Hamming distance between any two elements of Φ^m_r is the norm of a third element of Φ^m_r. This fact was exploited by D. E. Muller[2] in proving his Theorem 25. Muller's Theorem 25, in our terminology, may be expressed as follows:

<u>Theorem A</u>:- The norms of all non-zero vectors f of Φ^m_r satisfy

$$\| f \| \geq 2^{m-r} \quad \text{for } (m = 0,1,2,\dots) \quad \text{and } r \leq m \ .$$

We shall not prove this theorem here. It suffices to say that Muller proved the theorem by an induction on m and r and the properties of the Hamming distance.

By the above theorem there is at least a distance 2^{m-r} between two elements of Φ^m_r and, as a consequence, there is an open Hamming sphere of radius 2^{m-r-1} about each element of Φ^m_r in Φ^m_m (the whole vector space) which does not intersect any other such sphere. This means that it is possible to associate each element of such a sphere with the element defining the sphere or what is the same to associate an element of Φ^m_r which is less than a distance 2^{m-r-1} from an element f of Φ^m_r with f.

In order to illustrate how a message may be coded into an error-detecting code of the type described above, consider the following example: Let m = 4 and r = 1, by (7) the vectors of Φ^4_1 are of the form

$$g_0 \oplus g_1x_1 \oplus g_2x_2 \oplus g_3x_3 \oplus g_4x_4 \ . \tag{8}$$

Let the message consist of the five binary digits $(g_0, g_1, g_2, g_3, g_4)$. The code space Φ_1^4 may be regarded as generated by the four vectors x_1, x_2, x_3, x_4 and the identity vector I which may be written explicitly as follows:

$$\begin{aligned} x_1 &= (0\,1\,0\,1\,0\,1\,0\,1\,0\,1\,0\,1\,0\,1\,0\,1) \quad , \\ x_2 &= (0\,0\,1\,1\,0\,0\,1\,1\,0\,0\,1\,1\,0\,0\,1\,1) \quad , \\ x_3 &= (0\,0\,0\,0\,1\,1\,1\,1\,0\,0\,0\,0\,1\,1\,1\,1) \quad , \\ x_4 &= (0\,0\,0\,0\,0\,0\,0\,0\,1\,1\,1\,1\,1\,1\,1\,1) \quad , \\ I &= (1\,1\,1\,1\,1\,1\,1\,1\,1\,1\,1\,1\,1\,1\,1\,1) \quad . \end{aligned} \tag{9}$$

The 32 vector codes of Φ_1^4 can be obtained by scalar multiplication of the vectors of (9) by the message digits g_0, g_1, g_2, g_3, g_4 in accordance with (8). For example, the message (0 1 1 0 0) has the code vector $g_1 x_1 \oplus g_2 x_2$ or

$$(0\,1\,1\,0\,0\,1\,1\,0\,0\,1\,1\,0\,0\,1\,1\,0) \quad .$$

Each of the 32 codes will be a distance of at least eight from each other.

In order to practically generate the above code, one should note that the vector x_1 is the sequence of digits generated by the least significant binary stage B_1 of a binary counter of scale sixteen; x_2 is obtained from the second stage B_2; x_3 from the third stage B_3; and x_4 from the final stage B_4, as the counter goes through one period of its operation. If the message $(g_0, g_1, g_2, g_3, g_4)$ is stored in a binary register with stages A_0, A_1, A_2, A_3, A_4, then the switching function

$$C = A_0 \oplus A_1 B_1 \oplus A_2 B_2 \oplus A_3 B_3 \oplus A_4 B_4$$

will generate the code sequentially during one period of operation of the binary counter.

If one of the above codes of Φ_1^4 is corrupted during transmission so that no more than three errors are made, it is evidently possible by the previous discussion of this section to somehow extract the original message from the corrupted received code. The method by which this extraction may be accomplished will be shown by example in the next section and in general in the last section. It should be clear from the above example how the vectors of Φ_r^m may be generated for arbitray r and m where $r \leq m$.

IV. Decoding Corrupted Codes of Φ_r^m by a Majority Testing of Redundancy Relations

Let us first consider the coding space Φ_1^3. By (7), the vector of this space has the form

$$g_0 I \oplus g_1 x_1 \oplus g_2 x_2 \oplus g_3 x_3 \quad . \tag{10}$$

The message will consist of the four binary digits (g_0, g_1, g_2, g_3), and the generating vectors of the space are

$$\begin{aligned} x_1 &= (0\,1\,0\,1\,0\,1\,0\,1) \quad , \\ x_2 &= (0\,0\,1\,1\,0\,0\,1\,1) \quad , \\ x_3 &= (0\,0\,0\,0\,1\,1\,1\,1) \quad , \\ I &= (1\,1\,1\,1\,1\,1\,1\,1) \quad . \end{aligned} \tag{11}$$

By (6) we have the following set of relations for the message digits g_j in terms of f_k, the code digits.

$$g_0 = f(0, \ldots 0) = f_0 \quad , \qquad \underset{12}{\Delta} f(0\ldots) = f_0 \oplus f_1 \oplus f_2 \oplus f_3 = 0 \; ,$$

$$g_1 = \underset{1}{\Delta} f(0\ldots) = f_0 \oplus f_1 \quad , \qquad \underset{13}{\Delta} f(0\ldots) = f_0 \oplus f_1 \oplus f_4 \oplus f_5 = 0 \; ,$$

$$g_2 = \underset{2}{\Delta} f(0\ldots) = f_0 \oplus f_2 \quad , \qquad \underset{23}{\Delta} f(0\ldots) = f_0 \oplus f_2 \oplus f_4 \oplus f_6 = 0 \; ,$$

$$g_3 = \underset{3}{\Delta} f(0\ldots) = f_0 \oplus f_4 \quad , \qquad \underset{123}{\Delta} f(0\ldots) = \sum_{i=0}^{7} f_i = 0 \; . \tag{12}$$

By (12) there are four relations which g_1 satisfies,

$$g_1 = f_0 \oplus f_1 = f_2 \oplus f_3 = f_4 \oplus f_5 = f_2 \oplus f_3 \oplus f_4 \oplus f_5 \oplus f_6 \oplus f_7 \; .$$

By substituting the second and third relations into the fourth relation, we have

$$g_1 = g_1 \oplus g_1 \oplus f_6 \oplus f_7 = 0 \oplus f_6 \oplus f_7 = f_6 \oplus f_7 \; .$$

Thus we obtain the four independent and disjoint relations for g_1 ,

$$g_1 = f_0 \oplus f_1 = f_2 \oplus f_3 = f_4 \oplus f_5 = f_6 \oplus f_7 \; .$$

These four relations are disjoint in the sense that no two of the relations have variables in common. In a similar manner, we may obtain four independent and disjoint relations for both g_2 and g_3 so that g_1, g_2, g_3 may be expressed as

$$g_1 = f_0 \oplus f_1 = f_2 \oplus f_3 = f_4 \oplus f_5 = f_6 \oplus f_7 \; ,$$

$$g_2 = f_0 \oplus f_2 = f_1 \oplus f_3 = f_4 \oplus f_6 = f_5 \oplus f_7 \; ,$$

$$g_3 = f_0 \oplus f_4 = f_1 \oplus f_5 = f_2 \oplus f_6 = f_3 \oplus f_7 \; .$$

Let us now suppose that the received code is the vector $(f_0, f_1, \ldots f_7)$. If there were no error in transmission of the code, all of the above relations would hold. If there were one error, three out of four of the relations would hold. If there were two errors, at least two of the g_j's would have two out of four incorrect relations. Then g_1, g_2, g_3 may be determined uniquely if one or no error occurred during transmission, and two errors may always be detected by making a majority test on the arithmetic sum of the values of the four relations for each $g_j (j = 1,2,3)$. In order to state this criterion more explicitly, let the values of the four relations for g_j be denoted by $r_{j1}, r_{j2}, r_{j3}, r_{j4}$ for $(j = 1,2,3)$, and let S_j be the arithmetic sum of $r_{j1}, r_{j2}, r_{j3}, r_{j4}$ or

$$S_j = \sum_{i=1}^{4} r_{ji} \; .$$

Then the majority decision test for g_j is

$$\begin{aligned} g_j &= 0 && \text{if } 0 \le S_j < 2 \; , \\ g_j &\text{ is indeterminate} && \text{if } S_j = 2 \; , \\ g_j &= 1 && \text{if } 2 < S_j \le 4 \text{ for } (j = 1,2,3) \; . \end{aligned} \tag{13}$$

With the assumption that the received code is no more than two digits in error, the majority test (13) will determine g_1, g_2, g_3 uniquely for only one or no errors, and reject the code as meaningless in the case of two errors. In the case of one error or less, g_1, g_2, g_3 may be assumed now to be determined; it remains to determine g_0. In order to find g_0, note that if, as g_1, g_2, g_3 are found, the vectors $g_1 x_1, g_2 x_2, g_3 x_3$ are added successively to the received vector, by (10) we will end with either the vector $g_0 I$ in the case of no error or with a vector of distance one from $g_0 I$. Thus to detect g_0 the following majority decision test will suffice:

$$g_0 = 0 \text{ if } \sum_{i=0}^{7} m_i < 4 ,$$

$$= 1 \text{ if } \sum_{i=0}^{7} m_i > 4 , \qquad (14)$$

where m_i are the digits of the code after extraction of digits g_1, g_2, g_3 in accordance with the above procedure.

The above method of decoding may be illustrated by the following example: Suppose that the message sent was (1 0 1 1), and that during transmission an error was made in the fifth digit of the original code (1 1 0 0 0 0 1 1) so that the received code had the form (1 1 0 0 1 0 1 1). We first test for g_1, g_2, g_3 by (12) and find $g_1 = 0$, $g_2 = 1$ and $g_3 = 1$. Using (11), we add $g_1 x_1 \oplus g_2 x_2 \oplus g_3 x_3$ to the code, obtaining

$$0(0\ 1\ 0\ 1\ 0\ 1\ 0\ 1) \oplus (0\ 0\ 1\ 1\ 0\ 0\ 1\ 1) \oplus (0\ 0\ 0\ 0\ 1\ 1\ 1\ 1) \oplus (1\ 1\ 0\ 0\ 1\ 0\ 1\ 1)$$

$$= (1\ 1\ 1\ 1\ 0\ 1\ 1\ 1) = (m_0, m_1, m_2, \ldots m_3) .$$

Finally, by (14)

$$g_0 = 1 , \text{ since } \sum_{i=0}^{7} m_i = 7 > 4 .$$

Although Φ_1^3 is none other than an example of a set of one-error-correcting and two-error-detecting codes of the type described by Hamming in Ref. 1, the method of decoding considered above is different. Our procedure of decoding is advantageous in that it may be generalized in a natural way to include any of the coding spaces Φ_r^m of the second section of this paper. Before we consider the generalization by further examples, let us note a tabular way of representing the redundancy relations.

If the digits or variables of each relation are connected by lines for each of the vectors x_1, x_2, x_3 as

$$x_1 = (0\ 1\ 0\ 1\ 0\ 1\ 0\ 1) ,$$
$$x_2 = (0\ 0\ 1\ 1\ 0\ 0\ 1\ 1) ,$$
$$x_3 = (0\ 0\ 0\ 0\ 1\ 1\ 1\ 1) .$$

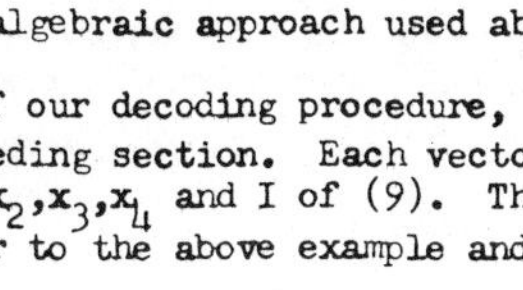

$$(15)$$

the relations of (12) become almost self-evident by their simplicity with respect to order and symmetry. This simplicity makes it possible to discover the redundancy relations for more general spaces Φ_r^m without resorting to the algebraic approach used above.

As a second example of our decoding procedure, consider the coding space Φ_1^4 introduced in the latter part of the preceding section. Each vector of this space has the form of (8), where the generating vectors are x_1, x_2, x_3, x_4 and I of (9). The first-degree redundancy relations may be determined in a manner similar to the above example and represented in a tabular manner similar to (15) as follows:

$$x_1 = (0\ 1\ 0\ 1\ 0\ 1\ 0\ 1\ 0\ 1\ 0\ 1\ 0\ 1\ 0\ 1) \quad ,$$

$$x_2 = (0\ 0\ 1\ 1\ 0\ 0\ 1\ 1\ 0\ 0\ 1\ 1\ 0\ 0\ 1\ 1) \quad ,$$

$$x_3 = (0\ 0\ 0\ 0\ 1\ 1\ 1\ 1\ 0\ 0\ 0\ 0\ 1\ 1\ 1\ 1) \quad ,$$

$$x_4 = (0\ 0\ 0\ 0\ 0\ 0\ 0\ 0\ 1\ 1\ 1\ 1\ 1\ 1\ 1\ 1) \quad . \tag{16}$$

For instance, the eight independent and disjoint relations for g_1 are

$$g_1 = f_{2i} \oplus f_{2i+1} \quad \text{for } (i = 0,1, \ldots 7) \quad .$$

If the eight values of the redundancy relations for g_j are labeled $r_{j1}, r_{j2}, \ldots r_{j8}$ for $(j = 1,2,3,4)$, and S_j is defined by

$$S_j = \sum_{i=1}^{8} r_{ji} \quad ,$$

then, by an argument similar to that used in the previous example, the majority decision test for g_j is as follows:

$$\begin{aligned} g_j &= 0 && \text{if } 0 \le S_j < 4 \quad , \\ g_j &\text{ is indeterminate} && \text{if } S_j = 4 \quad , \\ g_j &= 1 && \text{if } 4 < S_j \le 8 \quad \text{for } (j = 1,2,3,4). \end{aligned} \tag{17}$$

In order to determine g_0, we first add the determined vectors $g_j \mathbf{x}_j$ to the received message, assuming, of course, that no g_j is indeterminate, and we are left with the zero-degree polynomial Φ_0^4 , possibly corrupted by errors. If there had been no errors, there would be sixteen zero-degree relations which g_0 satisfies, or

$$g_0 = m_j \quad \text{for } (j = 0,1,2, \ldots 15) \quad ,$$

where, as in (14), m_j are the digits of the code after extraction of g_1, g_2, g_3 and g_4. Thus g_0 is determined by the majority decision test

$$\begin{aligned} g_0 &= 0 \ \text{ if } \sum_{i=0}^{15} m_i < 8 \quad , \\ &= 1 \ \text{ if } \sum_{i=0}^{15} m_i > 8 \quad . \end{aligned} \tag{18}$$

For the above example three errors may be made in the code and the correct message obtains. If four errors are made, some of the message digits are indeterminate. It is of some interest to note that, for some cases of five errors in the code, the message may be extracted correctly. For example, suppose that the message was (0 0 0 0 0) and that the received code was (1 1 0 0 1 0 1 0 1 0 0 0 0 0 0 0). Clearly, the correct message will be extracted from this code by the above procedure.

As a final example of coding and decoding scheme, consider Φ_2^4. This space is generated by x_1, x_2, x_3, x_4 of (16) and I, as well as the quadratic variables $x_1x_2, x_1x_3, x_1x_4, x_2x_3, x_2x_4, x_3x_4$. The latter six vectors may be presented in the following tabular manner.

$$x_1x_2 = (0\ 0\ 0\ 1\ \ 0\ 0\ 0\ 1\ \ 0\ 0\ 0\ 1\ \ 0\ 0\ 0\ 1) \quad ,$$

$$x_1x_3 = (0\ 0\ 0\ 0\ \ 0\ 1\ 0\ 1\ \ 0\ 0\ 0\ 0\ \ 0\ 1\ 0\ 1) \quad ,$$

$$x_1x_4 = (0\ 0\ 0\ 0\ \ 0\ 0\ 0\ 0\ \ 0\ 1\ 0\ 1\ \ 0\ 1\ 0\ 1) \quad ,$$

$$x_2x_3 = (0\ 0\ 0\ 0\ \ 0\ 0\ 1\ 1\ \ 0\ 0\ 0\ 0\ \ 0\ 0\ 1\ 1) \quad ,$$

$$x_2x_4 = (0\ 0\ 0\ 0\ \ 0\ 0\ 0\ 0\ \ 0\ 0\ 1\ 1\ \ 0\ 0\ 1\ 1) \quad ,$$

$$x_3x_4 = (0\ 0\ 0\ 0\ \ 0\ 0\ 0\ 0\ \ 0\ 0\ 0\ 0\ \ 1\ 1\ 1\ 1) \quad . \qquad (19)$$

The messages for this example will be 11 binary digit numbers of the form $(g_0, g_1, g_2, g_3, g_4, g_{12}, g_{13}, g_{14}, g_{23}, g_{24}, g_{34})$. Each code will be sent as a vector of the form

$$g_0 \oplus g_1x_1 \oplus g_2x_2 \oplus g_3x_3 \oplus g_4x_4 \oplus g_{12}x_1x_2 \oplus g_{13}x_1x_3 \oplus g_{14}x_1x_4$$

$$\oplus\ g_{23}x_2x_3 \oplus g_{24}x_2x_4 \oplus g_{34}x_3x_4 \quad .$$

The second-degree coefficients g_{ij} of the received message are extracted first with a majority decision based on the redundancy relations illustrated in (19). Next, assuming that no indeterminacy occurred in the second-degree coefficients, the vectors $g_{ij}x_ix_j$ are added to the received code, after which we are left with a residual code from which the first-degree coefficient g_0 may be determined by test (18) after adding the vectors $g_1x_1, g_2x_2, g_3x_3, g_4x_4$ to the residual code.

This example illustrates the general principle of decoding the particular class of codes under consideration. The highest degree coefficients of a received code are extracted first; then these terms of the polynomial are subtracted out of the code, thereby leaving a residual code of the next lower degree than the original code in the special case of no errors. The operation is repeated over and over on the successive residual codes until either an indeterminacy occurs or until g_0 is extracted.

The relations of (19) illustrate the fact that there are four redundancy relations each of four variables for the second-degree coefficients g_{ij}. For example, the redundancy relations for g_{12} are

$$g_{12} = f_{4i} \oplus f_{4i+1} \oplus f_{4i+2} \oplus f_{4i+3} \quad \text{for } (i = 0, 1, 2, 3) \quad . \qquad (20)$$

In general, these relations will allow only one error; two errors will lead to indeterminacy. This is another example of Hamming's one-error-correction and two-error-detection codes.

It should be noted that the majority decision tests used in the above examples were, in general, overdeterminate. For instance, in the first example, if one error had been made, no more than one error would remain in the residual code after determining g_1, g_2, g_3. On the other hand, if two errors had occurred, the process of extraction would have ended before g_0 could be determined. Thus a test of only the following type would be necessary:

$$g_0 = 0 \quad \text{if } m_{i1} + m_{i2} + m_{i3} \le 1 \quad ,$$

$$g_1 = 1 \quad \text{if } m_{i1} + m_{i2} + m_{i3} \ge 2 \quad ,$$

where i_1, i_2, i_3 are any three distinct numbers between zero and seven, inclusive. Refinements such as this, however, do not destroy the validity of the previous tests.

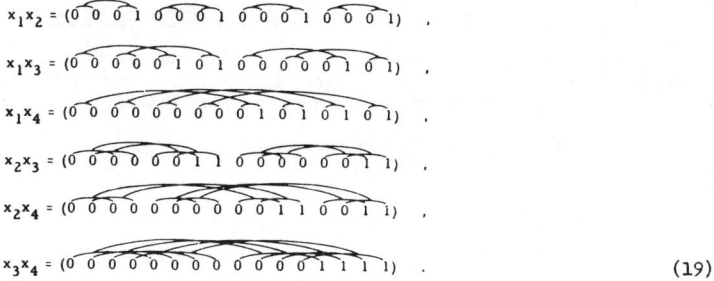

V. THE GENERAL DECODING PRINCIPLE

To study the general decoding scheme, illustrated by example in Section IV, it will be necessary to consider the general multinomial expansion formula (6) more carefully. Let us first define the multiple differences, used in (6) in more detail.

As in (6), $f(i_1, \dots i_m)$ is defined as

$$f(i_1,\dots i_m) = f_j \quad \text{when } j = \sum_{k=1}^{m} i_k\, 2^{k-1} \quad \text{for } (i_k = 0, 1) \quad . \tag{21}$$

The general multiple partial difference

$$\overset{p}{\underset{k_1,k_2,\dots k_p}{\Delta}} f(i_1,i_2,\dots i_m)$$

is defined inductively as

$$\underset{k}{\Delta} f(i_1,\dots i_m) = f(i_1,\dots i_{k-1}, i_k \oplus 1, i_{k+1},\dots i_m) \oplus f(i_1,\dots i_k,\dots i_m)$$

$$\overset{p}{\underset{k_1 k_2,\dots k_p}{\Delta}} f(i_1,\dots i_m) = \overset{p-1}{\underset{k_1,\dots k_{p-1}}{\Delta}} f(i_1,\dots i_{k_{p-1}}, i_{k_p} \oplus 1, i_{k_{p+1}},\dots i_m)$$

$$\oplus \overset{p-1}{\underset{k_1,\dots k_{p-1}}{\Delta}} f(i_1,\dots i_m) \tag{22}$$

With these definitions it is possible to prove by induction the validity and uniqueness of expansion (6) for any Boolean algebra of m variables, and in particular, for the Boolean algebra of 2^m dimensional vectors as described in Section II.

One evident consequence of (21) is the identity

$$f(i_1,\dots i_{k-1} i_k \oplus 1, i_{k+1},\dots i_m) = f_{i+(-1)^{i_k} 2^{k-1}} \quad . \tag{23}$$

By the use of (23) it is possible to write (22) explicitly in terms of the f_i as

$$\underset{k}{\Delta} f(i_1,\dots i_m) = f_i \oplus f_{i+(-1)^{i_k} 2^{k-1}}$$

and

$$\overset{p}{\underset{k_1,k_2,\dots k_p}{\Delta}} f(i_1,\dots i_m) = \sum_{i=1}^{2^{p-1}} f_{j_i} \oplus \sum_{i=1}^{2^{p-1}} f_{j_i+(-1)^{i_{k_p}} 2^{k_p-1}}$$

where

$$\overset{p-1}{\underset{k_1,k_2,\dots k_{p-1}}{\Delta}} f(i_1,\dots i_m) = \sum_{i=1}^{2^{p-1}} f_{j_i} \quad \text{and} \quad j_i \neq j_s+(-1)^{i_{k_p}} 2^{k_p-1}$$

$$\text{for } (i,s = 1,\dots 2^{p-1}) \quad . \tag{24}$$

We are now in a position to prove the following fundamental theorem on which the general decoding principle of the class of codes under consideration rests.

Theorem B:- Each highest or r-th degree coefficient of any vector or polynomial f of Φ_r^m satisfies exactly 2^{m-r} disjoint relations where each relation has precisely the form

$$\sum_{k=1}^{2^r} f_{i_k} \quad ,$$

where i_k are distinct numbers from the set $(0,1,2,\ldots 2^m - 1)$ for $(k = 1,2,\ldots 2^r)$. Disjointness of relations means that no two relations have variables f_i in common.

Proof:- Choose m and r. By (6),(7) and (24), the highest degree coefficients for an f of Φ_r^m are

$$g_{k_1\ldots k_r} = \overset{r}{\underset{k_1 k_2\ldots k_r}{\Delta}} f(0,\ldots 0) = \sum_{i=1}^{2^r} f_{j_i} \quad , \tag{25}$$

where k_j are distinct integers from the set $(1,2,\ldots m)$ for $(j = 1,\ldots r)$, and j_i are distinct integers from the set $(0,1,\ldots 2^m - 1)$ for $(i = 1,2,\ldots 2^r)$. Moreover,

$$\underset{k_1\ldots k_r n_1 n_2\ldots n_t}{\Delta} f(0,\ldots 0) = 0 \tag{26}$$

for $t \geq 1$, and k_j and n_1 are distinct integers from the set $(1,2,\ldots m)$ for $(j = 1,\ldots t)$.

Let $k_1,k_2\ldots k_r$ be a distinct set of integers from the set $(1,2,\ldots m)$. Then by (26) and (22),

$$\overset{r+1}{\underset{k_1\ldots k_r n_1}{\Delta}} f(0,\ldots 0) = \overset{r}{\underset{k_1\ldots k_r}{\Delta}} f(0,\ldots 0) \oplus \overset{r}{\underset{k_1\ldots k_r}{\Delta}} f(0,\ldots 1,\ldots 0) = 0 \tag{27}$$

where n_1 is any one of the m-r integers from the set $(1,2,\ldots m)$ which is distinct from the integers $(k_1,k_2,\ldots k_r)$. Thus, by (24) and (25), we have exhibited m-r new relations of the form required by the theorem. Each of these new relations is distinguished by the fact that the digit one appears only in the n_1-th position of the function $f(i_1,\ldots i_m)$ operated on by

$$\overset{r}{\underset{k_1\ldots k_r}{\Delta}}$$

Now define $f\left[n_1,n_2,\ldots n_t\right]$ to be $f(i_1,i_2,\ldots i_m)$ with $i_k = 1$ for $k = n_1,n_2,\ldots n_t$ and $i_k = 0$ otherwise. The theorem will be proved by induction on the subscript of n. Assume therefore that

$$\overset{r+s-1}{\underset{k_1 k_2\ldots k_r n_1 n_2\ldots n_{s-1}}{\Delta}} f(0,0,\ldots 0) = \overset{r}{\underset{k_1\ldots k_r}{\Delta}} f(0,0,\ldots 0)$$

$$\oplus \overset{r}{\underset{k_1\ldots k_r}{\Delta}} f\left[n_1,n_2\ldots n_{s-1}\right] \quad . \tag{28}$$

47

Now, by (22) and (26) and the induction hypothesis (28),

$$\Delta^{r+s}_{k_1 \ldots k_r n_1 \ldots n_s} f(0,0,\ldots 0) = \Delta_{n_s}\left(\Delta^{r+s-1}_{k_1 \ldots k_r n_1 \ldots n_{s-1}} f(0,0,\ldots 0)\right)$$

$$= \Delta_{n_s}\left(\Delta^{r}_{k_1 \ldots k_r} f(0,\ldots 0) \oplus \Delta^{r}_{k_1 \ldots k_r} f\left[n_1,\ldots n_{s-1}\right]\right)$$

$$= \Delta^{r}_{k_1, \ldots k_r} f(0,0,\ldots 0) \oplus \Delta^{r}_{k_1 \ldots k_r} f\left[n_1,\ldots n_{s-1}\right]$$

$$\oplus \Delta^{r}_{k_1 \ldots k_r} f\left[n_s\right] \oplus \Delta^{r}_{k_1 \ldots k_r} f\left[n_1,\ldots n_s\right] = 0 \quad .$$

Now, by (27) and (28), the two middle terms are equal to

$$\Delta^{r}_{k_1 \ldots k_r} f(0,0,\ldots 0) \quad ,$$

and therefore their sum modulo 2 is zero. Hence

$$\Delta^{r+s}_{k_1 \ldots n_s} f(0,\ldots 0) = \Delta^{r}_{k_1 \ldots k_r} f(0,\ldots 0) \oplus \Delta^{r}_{k_1 \ldots k_r} f\left[n_1,\ldots n_s\right] = 0 \quad ,$$

and the induction is complete. The theorem is proved when we observe that the relation

$$\Delta^{r+s}_{k_1 \ldots n_s} f(0,\ldots 0) = 0 \text{ contributes } \binom{m-r}{s} \text{ distinct relations,}$$

$$\Delta_{k_1 \ldots k_r} f(0,\ldots 0) = \Delta_{k_1 \ldots k_r} f\left[n_1, n_2 \ldots n_s\right] \quad ,$$

since there are $\binom{m-r}{s}$ ways of choosing s integers from m-r integers. Using all the relations (26) for the particular set $k_1 \ldots k_r$ and $t = 1$ to $t = m-r$ and the relation (25), we get

$$1 + \sum_{t=1}^{m-r} \binom{m-r}{t} = 2^{m-r}$$

distinct relations for $g_{k_1, k_2, \ldots k_r}$. Since these relations exhaust all variables f_{i_k}, the theorem is proved.

The above theorem shows that the generalization of the decoding principle, discussed in the last section obtains. The majority decision test for the general case can clearly be used to extract the r-th degree coefficients of Φ_r^m, where the relations used for the test are the 2^{m-r} relations of Theorem B. The (r-1)-th degree coefficients are then extracted the same way after the determined r-th order terms have been subtracted or added into the received code. This process is continued for the r-2,r-3,... degree coefficients until the message is extracted or an indeterminacy is reached.

VI. Concluding Remarks

Since there are $\binom{m}{j}$ j-th degree coefficients $g_{i_1 i_2 \ldots i_j}$ in expansion (5), there must be

$$N = \sum_{i=0}^{r} \binom{m}{i}$$

coefficients in each polynomial (7) of the coding space Φ_r^m. The coefficients of (7) constitute the message sent, thus each code of Φ_r^m contains N bits of message information. Since each element of Φ_r^m is a vector of dimension 2^m, there are 2^m-N bits of the code used to supply redundancy.

In order to illustrate the relationship of the number of message bits to number of errors corrected, consider the coding space Φ_4^7. By (29) each code of (29) has 99 bits of message information for a code of 128 bits. By Section III at least

$$2^{m-r-1} - 1 = 2^{7-4-1} - 1 = 3$$

bits of error in the code can be corrected. By Section IV and Section V four bits of error will lead undoubtedly to an indeterminacy in the message and it is likely that in some cases of five errors the correct message will be extracted by the majority decision process. Further examples of the numerical relationship of message bits to number of errors corrected may be constructed in a similar manner.

Attempts have been made with little success to investigate the structure of the complete convex set S of points, containing an element σ of Φ_r^m, whose points correspond to the element σ under the majority decision test procedure of Section V. As the second example of Section IV shows, there are in general more points in S than in a Hamming sphere of radius 2^{m-r-1} containing σ. These attempts were motivated by a desire to show that the coding system discussed here would satisfy Shannon's fundamental theorem for a discrete channel with noise (Theorem 11 in Ref. 3). So far, this fact has not been shown.

There are two generalizations of the codes discussed in this paper. In Ref. 2 Muller discusses generalizations of the binary codes, discussed here, for lengths other than 2^m. Another generalization is possible where the polynomials considered here are considered over a field of characteristic other than two; i.e., ternary codes, etc. It will not be the purpose of this paper to investigate these generalizations.

ACKNOWLEDGMENTS

The author expresses his appreciation to E. B. Rawson for his assistance in the construction of the second example of Section 4; to G. P. Dinneen for his help in the simplification of Theorem B; and to T. A. Kalin, W. B. Davenport, D. E. Muller, and O. G. Selfridge for several useful discussions.

REFERENCES

1. R. W. Hamming, Bell System Tech. J. 29, No. 2, 147 (April 1950).

2. D. E. Muller, "Metric Properties of Boolean Algebra and Their Application to Switching Circuits," Report No. 46, Digital Computer Laboratory, Univ. of Illinois (April 1953).

3. D. E. Shannon, "A Mathematical Theory of Communication," Bell System Tech. J. 27, (July,October 1948).

Reprinted from *I.R.E. Trans. Inform. Theory,* **PGIT-4,** 23–28 (1954)

BINARY CODING

Marcel J. E. Golay
Signal Corps Engineering Laboratories
Fort Monmouth, New Jersey

INTRODUCTION

The upper bound given by Shannon[1] to the transmission capacity of a noisy, discrete channel has challenged the mathematicians, who have accepted this challenge, to devise digital error correcting codes or coding systems approximating as close as possible this upper bound.

This mathematical effort has been concentrated in the binary system and has had the aim to devise codes which are as efficient as possible, in the sense that, given an upper limit to the number m of errors during the transmission and reception of a block or message of n binits, the following obtains: (a) All messages are received in all cases without equivocation; (b) The number of transmittable messages approaches as close as possible the value $2^n \Big/ \sum_{m=0}^{m=e} \frac{n!}{(n-m)!\,m!}$, the sum in the denominator being the sum of the (e + 1)st numbers of the nth line of Pascal's triangle, and representing the number of ways in which any one transmitted message can be received when transmission errors in any number from zero to e can occur.

Codes in which the upper limit is exactly reached will be termed lossless codes in what follows, and it may be worth noting here the paradoxical circumstance that while the existence proof for codes approaching indefinitely Shannon's upper bound was based on the assumption of codes consisting of random messages, the search for efficient or lossless codes has been successful to the extent that codes were discovered which were characterized by deeply seated, entwined symetries.

It is the purpose of this discussion to explore certain aspects of this circumstance, and to describe some group-theoretical approaches to coding problems.

The first example of a symbol correcting code was given by Shannon[2] who quotes Hamming's lossless coding of a seven binit message, none or one of which can be received in error. This case was extended by the writer to blocks of 2^n-1 binary symbols, and, more generally, to blocks of $\frac{p^n-1}{p-1}$ p-nary symbols (p prime), none or one of which can be received in error.[3] With the exception of the trivial cases of (2n + 1) binit messages, up to n of which can be received in error, and of two special cases treated in the last paper cited, these are the only cases of lossless symbol coding known, and the possibility must be considered that others do not exist. Their impossibility will be demonstrated below for the case of lossless 2-error correcting symbol codes, and it will be shown also that the search for e-error correcting symbol codes need be a finite one only, because lossless symbol correcting codes become impossible beyond a determinable message length, for any one selected value of e.

These results will leave open the question of whether cases of two or more error-correcting lossless message codes exist (outside of the one mentioned above) because message codes form a more general class of codes than symbol codes, which form a sub-class of it only, and various examples of message codes which are more efficient than symbol codes will be cited, and their mode of formation illustrated. It is this mode of formation which is suggestive of the kind of group theoretical approach which the writer believes to be the most promising for the class of coding problems considered.

Symbol Correcting Codes

When a symbol correcting digital code exists for the transmission of n-binit messages, up to e of which can be received in error, and i of which (the X_m's) carry the message while the remaining j = n - i (the Y_k's) binits are redundant and are provided to remove the equivocation, the transmitted binits are related by the matrix:

$$E_m \equiv \sum_{k=1}^{k=j} a_{mk} Y_k + X_m \equiv 0 \ (\text{mod. } 2), \quad m = 1, 2, \cdots i$$

and the essential property of this matrix is that the E's recalculated from the partially erroneously received X_k's and Y_m's form a j-binit number E, which will be termed the corrector, and which determines univocally which symbols were received in error.

1 Bell System Technical Journal, July, 1948.

2 Loc. cit., p. 418.

3 Marcel J. E. Golay, "Notes on Digital Coding," Proc. I.R.E., vol. 37, p. 637; 1949.

When the code is lossless, a first condition must obtain, which stipulates that all possible cases of up to e errors are represented by all the possible values of the corrector:

$$\sum_{k=0}^{k=e} \frac{n!}{(n-k)!\,k!} = 2^{j} \tag{1}$$

Another condition can be obtained as follows:

Whenever all but one Y_k and all X_m's received are zero, the binits of the corrector E(k) consist of the series of a_{mk} values for the particular k considered, and will be termed the characteristic of Y_k.

Whenever all but one X_m and all Y_k's are zero, the binits of the corrector E(m) consist of zeroes with a single one corresponding to the particular m considered, and will be called the characteristic of X_m. In general, the corrector E consists of the j-binit number formed by adding modulo 2 the corresponding binits of the characteristics of the symbols received in error. A general condition for a lossless code is that all possible (boolean) additions thus made of up to e characteristics reproduce exactly, and only once, each of the 2^j possible values of the corrector.

If the parity of the characteristics or of the corrector is defined as zero when the number of ones in these numbers is even, and as one otherwise, it will be readily seen that the parity of the corrector will be the parity of the number of odd characteristics (parity one) required to form it. In a lossless code, all even correctors, 2^{j-1} in number, shall be formed from all possible additions of up to e characteristics in which the number of odd characteristics employed, 2s, is always even. Let r be the total number of odd characteristics. We shall have the other condition sought:

$$\sum_{k=0}^{k=e} \sum^{s} \frac{(n-r)!}{(n-r-k+2s)!\,(k-2s)!} \cdot \frac{r!}{(r-2s)!\,(2s)!} = 2^{j-1} \tag{2}$$

A corollary from (1) and (2) can be obtained by subtracting the second relation from the first, member by member. This operation yields the relation:

$$\sum_{k=0}^{k=e} \sum^{s} \frac{(n-r)!}{(n-k-r+2s-1)!\,(k-2s+1)!} \cdot \frac{r!}{(r-2s+1)!\,(2s-1)!} = 2^{j-1} \tag{3}$$

When e = 2, (1) and (2) can be written:

$$n^2 + n - 2 = 2^{j+1} \tag{1a}$$

$$(n - r + 1)\; r = 2^{j-1} \tag{2a}$$

These relations are satisfied for n = 5 and r = 2 or 4 (r = 2 does not correspond to any code, and r = 4 corresponds to the trivial case of a five binit message, up to two of which can be in error), but for larger values of n and r the approximation obtained by eliminating all but the highest degree terms in the left members of (1a) and (2a).

$$n^2 = 2^{j+1} \tag{1b}$$

$$(n - r)\; r \simeq 2^{j-1} \tag{2b}$$

indicates that $n \simeq 2r$.
(2a) requires that:

$$r = 2^{\frac{j-1}{2}} \quad \text{and} \quad n + 1 = 2^{\frac{j+1}{2}}$$

and substitution of the value for n derived from the last relation in (1a) yields n = 1, which contradicts the postulation of a large n.

In the general case where $e > 2$, it can be shown that the search for a lossless code need be a finite one only as follows:

The highest power of n in (1) is in the term $\frac{n^e}{e!}$. It is therefore, possible to rewrite (1) as follows:

$$n^e\,(1 + \varepsilon) = e!\; 2^j \tag{1c}$$

in which, for any given e, the quantity $\mathcal{E}$ can be made arbitrarily small for large values of n.

The difference between the number of even and odd correctors should be zero in a lossless code, and this condition can be expressed by the relation:

$$\sum_{k=0}^{k=e} \sum^{t} (-1)^t \frac{(n-r)!}{(n-r-k+t)!(k-t)!} \cdot \frac{r!}{(r-t)!\,t!} = 0 \qquad (4)$$

which is obtained by subtracting (3) from (2), member by member.

The highest power terms in n and r in the expression above are:

$$\sum^{s} (-1)^t \frac{(n-r)^{e-t}\, r^t}{(e-t)!\, t!} = \frac{(n-2r)^e}{e!} \qquad (5)$$

all other terms being of the form $n^a r^b$ where $a + b < e$. It is seen thus that (4) will be satisfied when n and r are related by an expression of the form:

$$n = 2r\,(1 + \gamma) \qquad (6)$$

in which, for any given e, γ can be made arbitrarily small by making n and r sufficiently large.

It will be noted now that r can be factored algebraically out of (3). The terms multiplied by r which are of the form $\frac{(r-1)!}{(r-2s+1)!\,(2s-1)!}$ could be fractional, but each term will be an integer if multiplied by the highest common denominator of r and 2s-1, h.c.d. (r,2s-1). Therefore, if the lowest common multiplier of all h.c.d. (r,2s-1)'s is factored out of r, l.c.m. (all h.c.d. (r,2s-1)'s) = r', the multiplication by r' of all terms multiplied by r in the left member of (4) will be integers in all cases, and in order to satisfy (4), it should be possible to write r in the form:

$$r = 2^a r'' \qquad r'' \leqslant r' \qquad (7)$$

It will be further noted that r', and hence r" also, have the upper bound:

$$r',\ r'' \leqslant \text{l.c.m. (all } (2s-1)\text{'s)},\ 2s-1 \leqslant e \qquad (8)$$

Elimination of n and r between (1c), (6) and (7) gives:

$$2^{e(a+1)} r''^{e} (1+\gamma)^e (1+\mathcal{E}) = e!\ 2^j \qquad (9)$$

For any given e, γ and $\mathcal{E}$ approach zero for increasing j, while r" has an upper bound. Therefore an upper bound for j exists, beyond which (9) will not be satisfiable, because either e! will contain odd prime factors not contained in the left member of (9), or the left member of (9) will contain a number of odd prime factors which is a multiple of e, and which exceeds the number of the same odd prime factors in e!.

While this demonstration indicates that the search for lossless two or more symbol correcting binary codes need be a limited one only for any chosen number of errors, a search for such codes has only revealed, outside of the trivial cases of n errors in a 2n+1 binit message, the case mentioned earlier of a 3-error out of a 23-binit message symbol correcting code.. Whether, with the exception of the trivial cases mentioned, this particular 3-error symbol correcting code is the only lossless binary code correcting more than one symbol, is a matter of speculation. The degree of rarity of the happenstance required for the satisfaction of both relations (1) and (2) suggests that it could be so indeed, and offers the challenge of finding a mathematical demonstration of the impossibility to satisfy (1) and (2) for any other case.

<u>Message Correcting Codes</u>

The demonstration above leaves open the question of whether there are lossless e-error message correcting binary codes for any length of message, for condition (1) only applies to these, while condition (2) does not, since it is predicated upon the existence of a lossless symbol correcting code. For instance, the question is left open, whether a 2-error correcting 90 binit message code exists, since condition (1) is satisfied for this case.

The possibility that lossless message correcting codes exist where lossless symbol correcting codes do not is thus predicated upon the circumstance that message correcting codes form a more general class of codes. While no lossless binary message correcting codes are known, for which there are no corresponding lossless symbol correcting codes, examples will be given below of lossy message correcting codes which are more efficient than the available symbol correcting codes for the same number of of message symbols and maximum allowable number of errors.

Some of these examples will be derived from the a_{mk} matrix already published in the referenced Letter to the Editor, and the formation of the top ten symbols in the Y_2 to Y_{12} columns will be explained briefly first.

If we consider five straight lines in a plane, A, B, C, D and E and order their respective intersections as follows: AB, AC, AD, AE, BC, BD, BE, CD, CE, DE, then Y_2 is formed by associating a 0 with the four intersections represented by the products in the expression:

$$A(B + C + D + E)$$

and a 1 with all other positions. Y_3, Y_4, Y_5, and Y_6 are formed likewise by associating a 0 with the intersections of B, C, D and E respectively with all four remaining lines.

Y_7 is formed by associating a 0 with the 5 intersections AB, BE, ED, DC, and CA of neighboring lines (including the first and last) in the operator:

(ABEDC)

which will be designated to represent the ensemble of 5 intersections listed above.

There are 4! cyclical permutation of the five lines, which can be separated into two groups of 12, the members of any one group being derivable from the other 11 members by an even number of interchanges of elements so that they can be said to be of the same parity. Within each group of 12 there are 6 pairs of permutations which differ only by their order, so that both members of each pair determine the same ensemble of 5 intersections. Thus, there will be only 6 district ensembles of 5 intersections having the same parity, and those belonging to the parity of the operator written above for Y_6 will determine the 0's of the upper 10 places of the Y_6 to Y_{12}.

It can be verified by inspection that the upper 10 symbols of Y_2 to Y_{12}, as well as the Boolean additions of any two of these ensembles, differ from all others in at least three places. Thus we can form the ensemble of 66 10-symbol messages written in Table I, which, together with the all 0's and all 1's messages form 68 10-symbol messages which differ in at least three places, and are therefore 1-error correcting messages.

TABLE I

```
001110001010001110:101110010110101010:101110010101100110100111010111
011101100001010010:010101001110011011:100010100110111101111101011001
010101010010101001:110010110010011101:011101100110001110101010111101
010010100110010101:101101100101011100:110101011010100101011011001111
100011101000100011:011011011001001110:110100001101011011101101101110
101011010100011000:100100011001110111:010011110001010111111011110010
100100011001010101:110011100101100011:101011011010011010011100110111
110000000101101110:001111101010100101:001111101001101001110110101011
111111111111111111:000000000000000000:111111111111111111000000000000
111111111111111111:111111111111111111:000000000000000000000000000000
```

Likewise, the two smaller blocks of 36 9-symbol messages and 18 8-symbol messages, shown within the dashed enclosures, indicate that, together with the 2-all 0's and all 1's messages, there are 38-1 error correcting 9-symbol messages, and 20 1-error correcting 8-symbol messages.

On the other hand, it can be easily verified that there are only 16, 32 and 64 messages possible on the basis of 1 error correcting symbol codes for 8, 9 and 10-symbol messages respectively, because the number of cases of zero or one error are 9, 10 and 11 respectively in these three cases, which requires the assignment of 4 redundant symbols to the removal of the equivacation, thus leaving only 4, 5 and 6 symbols respectively for the message transmission.

It will also be noted that the upper 10 places of all Y_2 to Y_{12}, plus the all 0's message, form an ensemble of 12 10-symbol messages each of which differ from all others by at least five symbols, and are, therefore, 2-error correcting messages. The number of ways in which 0, 1 or 2 errors can occur in 10 places is: $1 + 10 + 45 = 55$, which indicates that a minimum of 6 redundant symbols should be assigned to the removal of the equivacation, thus leaving at most 4 symbols for the message transmission. However, it can be verified by inspection that it is impossible to form 4 6-symbol characteristics, which together with the 6 correctors for redundant X's constitute an ensemble in which any member of which, and any sum of two of which differ from all other single members or sums of two.

On the other hand, it is possible to assign 7 symbols to the removal of the equivocation, and to have 3 7-symbol characteristics satisfying the conditions required so that only 3 symbols become thus available for the transmission of only 8 possible messages. Thus, here again, a larger number of messages can be transmitted by message coding than by symbol coding.

When the formation of 2-error correcting message codes is extended to 15-binit messages, in which the 15 intersections of 5 straight lines are associated in various ways with the message symbols, more care is required for the selection of favorable symetries.

Thus, we may associate the five intersections:

$$A(B + C + D + E + F)$$

with five 0's and the 6 groups of intersections of any one of the six lines with all others gives us 6 messages sufficiently distant from each other for 2-error correction.

We may consider next the 15 groups of intersections given by the various products of the form

$$(A + B) \quad (C + D + E + F)$$

and we can verify that these vary in at least 5 symbol positions with each other and with those of the preceding form.

The 10 groups of intersections determined by expressions of the form

$$(A + B + C) \quad (D + E + F)$$

can be added likewise to the other groups while satisfying the required criterion of a minimum of 5-symbol separations for 2-error correcting messages.

The 6 groups of 5 0's represented by the operator

$$(A\ B\ C\ D\ E)$$

and the five other operators of the same parity derivable from it can be verified to represent messages sufficiently distant from all others to permit 2-error correction. The letter F can be substituted for any and all other letters provided any other two letters are interchanged whenever a substitution is made, to provide more messages satisfying the 5 symbol distance criterion. Thus, 36 messages of this last type can be formed.

The total of messages satisfying the 5-symbol distance criterion which can be formed as indicated above is therefore:

$$6 + 15 + 10 + 36 = 67$$

It can be verified further that the 67 new messages formed by the boolean addition of the all 1's message to these satisfy the criterion with the 67 old messages. Adding the all 0's and all 1's messages gives us the total of 136 messages for the case of 2-error correcting 15-symbol messages.

Up to 2 errors can occur in a 15 symbol message in $1 + 5 + 105 = 121$ ways, and the upper bound to the number of theoretically possible is therefore $\frac{2^{15}}{121}$. The number of possible messages found above, 136, is seen to be slightly over half the number given by that upper bound. With a symbol-correcting code, 128 messages, i.e. slightly less than half the upper bound stated, could be transmitted by means of the a_{mk} matrix given in Table II.

TABLE II

Matrix for 2-error Symbol Correction of 15-symbol Messages

1	1	1	0	1	0	0
1	1	1	0	0	0	1
1	1	0	1	0	0	0
1	1	0	0	1	1	1
1	0	1	1	0	1	0
1	0	1	0	0	1	1
1	0	0	1	1	1	0
1	0	0	1	1	0	1

The examples of message coding given above suggest the question of whether the procedures described could be made methodical and be extended to longer messages. This question cannot be answered at this stage; instead circumstances will be pointed out which make such an answer difficult.

In the case just examined of a 15-symbol code in which the symbol positions were associated with the 15 intersections of 6 straight lines in a plane, a restricted number only of line groupings were studied. For instance, messages in which the 0's or 1's are given by the intersection of elements not above each other in the two lines of the matrix $\begin{Bmatrix} ABC \\ DEF \end{Bmatrix}$ constitute another symetrical grouping, which examination indicated not to be useful in building 2-error correcting 15-symbol message codes, but which could be useful in other codes. Thus, a yet unsystematized selection of favorable groups must be made.

Codes may be based on the restricted class of n (2n-1) symbol messages ($n \nless 3$) which can be formed by assigning the symbols 0 or 1 to the positions determined by the intersections of 2n lines, A_1, ^{A},....A_{2n}, given by all expressions of the form:

$$(A_1 + \ldots\ldots A_{2m}) \quad (A_{2m} + 1 - \ldots + A_{2n})$$

and by assigning 1 versus 0 to all other points.

Together with the 2 all 0's and all 1's messages, these number 2^{n-1} and are n-3 error correcting. Thus, 15, 28 and 45 symbol messages will be in number 2^5, 2^7, and 2^9 and will be 3, 5 and 7 error correcting respectively. This code equals the Reed code in the case of 15-symbol messages, and is inferior to it for longer messages. A short examination of codes formed by considering the intersections of the planes A_1, A_2, - - - - A_n, in a 3 dimensional space which are of the form:

$$(A_1 + - - + A_k) \quad (A_{k+1} + - - + A_m) \quad (A_{m+1} + - - + A_n)$$

has not indicated that an extension of this attack to multidimensional spaces is promising. There again, a selection of proper symetries is required.

Another circumstance to be pointed out here is the completely symetrical part played by all straight lines in the formation of the 1 or 2-error correcting 10-symbol messages and 2-error correcting 15-symbol messages described above. By contrast, an examination of the lossless 1-error symbol correcting 15 symbol message code can be seen to be expressible in terms of the intersections of 6 lines in which 4 lines play symetrical roles, but the other 2 do not. This may permit the speculation that an approach to the problems of building a 2-error correcting 90 symbol message codes of 2^{78} messages may be to consider the 90 intersections of 14 lines in which 2 lines, the intersection of which is not counted, play a part not symetrical with that played by the 12 others.

CONCLUSION

It has been shown that lossless e) symbol correcting message codes can exist only for message lengths which have an upper bound, and it can be speculated whether any exist, outside of the cases of 1-error correcting 2^n-1 symbol messages, n-error correcting 2n + 1 symbol messages, and 3-error correcting 23-symbol messages.

It has also been shown by examples that the more general class of message correcting codes permits a higher coding efficiency than symbol correcting codes, and the existence of lossless message correcting codes not included within the lossless symbol correcting codes mentioned above appears less improbable.

While the only systematic message correcting codes described in the text is less efficient than the Reed Code, it is suggested that an attack along these lines may prove more fruitful than if restricted to the sub-class of symbol correcting codes, when attempts are made to design systematic codes approaching Shannon's upper bound.

Translated from *Dokl. Akad. Nauk SSSR,* **117,** 739–741 (1957)
Acknowledgment is made of the assistance of the National Translations Center, John Crerar Library, Chicago, Illinois, in obtaining a translation of this paper.

7

Estimate of the Number of Signals in Error Correcting Codes

R. R. VARSHAMOV†

Coding systems where the signals are sequences consisting of elementary messages of two kinds have received widespread use. Such a sequence can be denoted by a sequence of ones and zeros, such as 1101001101, say. The set D^n of $N = 2^n$ sequences of the form $a = (a_1, a_2, \ldots, a_n)$, where each symbol a_i can take on only two values: 0 or 1, can naturally be considered as an n-dimensional vector space over the field D of residues modulo 2 (this field consists of two elements, 0 and 1). It is natural to introduce the norm $|a|$ in the space D^n, which equals the number of ones in the sequence a, and $\rho(a', a'') = |a' - a''|$, which is considered the distance between the elements a' and a''.

If the signals are transmitted with errors and it is desired to correct these errors, then a certain subset of M signals rather than all N possible signals a is used to transmit the information. It is known [1] that in order to be able to correct r erroneous symbols, it is necessary and sufficient that the pairwise distances between the signals being used should not be less than $d = 2r + 1$. In this connection, the question arises of for which n, r, and M it is possible to find M signals, from among the N signals, with pairwise distances not less than d.‡ In addition to more special results, the necessary condition

$$M \leqslant \frac{N}{S_n^r} \tag{1}$$

and the sufficient condition

$$M \leqslant \frac{N}{S_n^{d-1}}, \tag{2}$$

†Translated by Morris D. Friedman, Lincoln Laboratory.

‡This problem can be set up for even d, but the case of even d is reduced trivially to the odd $- d$ case. Here, $d = 2r + 1$ is always odd.

where

$$S_n^q = 1 + C_n^1 + \cdots + C_n^q; \quad C_n^p = \frac{n!}{p!(n-p)!}$$

are known for this [2].

If the signals a are used to transmit the message $b = (b_1, b_2, \ldots, b_m)$ from D^m, then $M = 2^m$.

Assuming $n = m + k$, conditions (1) and (2) can be written for this special case as

$$S_n^r \leqslant 2^k, \tag{1a}$$

$$S_n^{d-1} \leqslant 2^k. \tag{2a}$$

The fundamental result, which will be proved later, is that the sufficient condition (2a) can be weakened as follows:

$$S_{n-1}^{d-2} = S_k^d C_{m-1}^{d-2} + S_k^1 C_{m-1}^{d-3} + \cdots + S_k^{d-3} C_{m-1}^1 + S_k^{d-2} C_{m-1}^0 < 2^k. \tag{3}$$

The necessary condition (1a) and the sufficient condition (3) coincide in the $r = 1$ case (correction of one error) and reduce to the inequality

$$n+1 \leqslant 2^k, \tag{4}$$

which can also be written as

$$m \leqslant 2^k - k - 1. \tag{4a}$$

(The relation between the number of symbols m in the message being transmitted and the number k of "additional" symbols to be appended for error correction possibilities is seen directly.)

Hamming [2] mentioned the coding method which would correct one error under condition (4). We obtain the Hamming result as a particular case of a general coding method with the possibility of correcting r errors under the condition (3).

Both the Hamming method and our method of coding are linear; i.e., if the messages $b' \in D^m$, $b'' \in D^m$ are transmitted by the signals a' and a'', then the message $b' + b''$ is transmitted by the signal $a' + a''$. In order to give a linear coding method, it is sufficient to indicate the signals $a^1, a^2, \ldots, a^m$ by which the "fundamental" messages $e^1, e^2, \ldots, e^m$ are transmitted, where

$$e^i = (e_1^i, e_2^i, \ldots, e_m^i), e_j^i = 1 \text{ for } i=j; e_j^i = 0 \text{ for } i \neq j.$$

Let us first solve an auxiliary problem: To select in D^k a set of elements C among m so that the following inequalities will be satisfied for any pairwise different $c^1, c^2, \ldots, c^{d-1}$ from C:

$$\begin{aligned} |c^1| &\geqslant d-1, \\ |c^1 + c^2| &\geqslant d-2, \\ &\cdots\cdots \\ |c^1 + c^2 + \cdots + c^{d-1}| &\geqslant 1. \end{aligned} \tag{5}$$

Let us solve this problem by a successive choice of the elements c^q one after the other. It is easy to calculate that when $q - 1$ elements c^p have already been chosen, there will be not more than

$$C_{q-1}^{d-2} + S_k^1 C_{q-1}^{d-3} + \cdots + S_k^{d-3} C_{q-1}^1 + S_k^{d-2}$$

elements $c \in D^k$ which it is not possible to select as c^q because of the limitations imposed. In order to be able to select all elements of c^q including the mth, it is sufficient that the number of forbidden elements for the last choice of the mth element be less than the total number of elements in D^k, i.e., less than 2^k. This is our condition (3).

The elements a^q we need from D^n are constructed as follows:

$$\begin{aligned} a^1 &= (1, 0, \ldots, 0, c_1^1, c_2^1, \ldots, c_k^1) \\ a^2 &= (0, 1, \ldots, 0, c_1^2, c_2^2, \ldots, c_k^2) \\ &\cdots\cdots\cdots\cdots\cdots\cdots\cdots\cdots \\ a^m &= (0, 0, \ldots, 1, c_1^m, c_2^m, \ldots, c_k^m) \end{aligned}$$

The set $A \subseteq D^n$ of elements a used to transmit the messages consists of elements such as

$$a = \sum_{q=1}^{m} b_q a^q. \tag{6}$$

The distance between the two elements a' and a'' from A equals the norm $|a|$ of the difference $a = a'' - a'$, which also belongs to A. It is easy to see that for an a such as (6), $|a| = |b| + |c|$, where

$$c = \sum_{q=1}^{m} b_q c^q.$$

If $|b| \geqslant d$, then $|a| \geqslant d$. If $|b| < d$, then c is the sum of $|b|$ of vectors c^q which differ from zero and $|c| = d - |b|$, i.e., $|a| \geqslant d$, because of condition (5). This completes the proof that the distance between two elements of A cannot be less than d.

Conditions (1a), (2a), and (3) enable estimates of the lower and upper bounds to be obtained for the minimum number $k_d(m)$ of additional symbols which would permit messages of m symbols to be transmitted by signals with pairwise distances $\geqslant d$, i.e., with the possibility of correcting r errors. It is easy to establish that the lower bound, corresponding to (1a), will be

$$k_d(m) \geqslant k_d^a(m) \sim r \log_2 m. \tag{7}$$

The upper bounds, obtained from (2a) and (3), are, respectively,

$$k_d^m \leqslant k_d^{-a}(m) \sim (d-1)\log_2 m, \tag{8}$$

$$k_d^m \leqslant \bar{k}_d(m) \sim (d-2)\log_2 m, \tag{9}$$

where $f \sim g$ denotes $f{:}g \to 1$.

June 10, 1957

References

1. A. A. Kharkevich: *Outline of General Communication Theory,* 1955.
2. R. W. Hamming, *Bell System Tech. J.,* **29** (2), 147 (1950).

8

Reprinted from *I.R.E. Trans. Inform. Theory*, IT-4, 103–109 (1958)

Notes on the Penny-Weighing Problem, Lossless Symbol Coding with Nonprimes, Etc.*

MARCEL J. E. GOLAY†

***Summary*—The method of construction of lossless symbol coding matrices for one-error correction is illustrated for the case when the prime symbol order is three, and the application of this matrix to the penny-weighing problem is described. This method is then extended to those cases in which the symbol order is 2^2, 2^3, 2^4, 2^6, 3^2, 3^3, 3^4, 3^5, 5^2, 5^3, 5^4, 7^2, 7^3, and p^2, where p is any higher prime. This extension is based on the concept of the master iterating matrix. These matrices are given for the first thirteen cases cited, and their existence is demonstrated for p^2.**

This paper concludes with a short description of Zaremba's condition, and its application to various problems, and more particularly to the hypothetical one-error correcting close-packed code with the symbol order 6.

Introduction

In a former note,[1] the writer described the construction of lossless symbol coding[2] matrices for one-error correction when the symbol order is prime, and two additional singular cases of close-packed coding for two and three-error correction, respectively.

An elaboration of the one-error correcting binary code was published in an industrial journal,[3] but the only additional error correcting lossless codes (with no restriction placed on the individual symbol errors) published since 1949 are due to Zaremba,[4] who showed on group-theoretical grounds that a close-packed code book for message coding[2] always exists for one-error correction, when the symbol order is a power of a prime. The essential portion of these notes will be devoted to those cases of lossless nonprime symbol coding for which coding matrices can be constructed systematically.

In the first part of the discussion, a recapitulation is given of the matrix construction formerly described,[1] for lossless one-error correction coding when the symbol order is three, and the application of this particular code to the penny-weighing problem is described.

In the second part, it is shown how the construction of lossless symbol coding matrices may be extended from code symbols of prime orders to code symbols of orders 2^2, 2^3, 2^4, 2^6, 3^2, 3^3, 3^4, 3^5, 5^2, 5^3, 5^4, 7^2, 7^3, and any square of an odd prime number.

The third part of the discussion contains observations on various solved or unsolved coding problems.

The Penny-Weighing Problem

Given a balance which can be used to determine whether two weights are equal, or which is heavier, and given a certain quantity of pennies, of which at most one may be heavier or lighter than the standard weight, it is asked to determine what paired assemblies of pennies should have their weight compared with each other, in order to find, with a minimum number of operations, which penny, if any, is too heavy or too light. It is also required that the weighing program be completely predetermined, and thus be not affected by the results of the successive weighings.

Each penny may be in one of three possible states, too light, of correct weight, and too heavy, and its state is thus expressible by a ternary symbol. The information gathered from each weighing is also expressible by a ternary symbol. These circumstances suggest an analogy with the transmission and reception of messages composed of ternary symbols, of which one at most may be received in error. It may be surmised that, for instance, three weighings, yielding $\log_2 27$ information bits, should determine which of 13 pennies, if any, is too light or too heavy. A form of lossless coding appears required to solve this problem, since each penny may be off-standard in two ways, thus yielding 26 possibilities, to which must be added the twenty-seventh possibility of all being of standard weight.

Consider now the coding matrices described in the former publication,[1] for the case $p = 3$. The simplest matrix, corresponding to $n = 1$, covers the trivial case of no information message and a single transmitted ternary check symbol, X_1. The coding matrix for this case consists of a single "1":

$$\begin{array}{c} X_1 \\ \hline 1 \end{array} \tag{1}$$

and since there are no information symbols, the transmitted check symbol will be zero.

The passage from $n = 1$ to $n = 2$ is made by writing the single term matrix (1) three times, writing below the numbers 2, 1 and 0, and adding a $\begin{smallmatrix}0\\1\end{smallmatrix}$ column at the end:

* Manuscript received by the PGIT, February 25, 1938; revised manuscript received June 15, 1958.

† Philco Corporation, Philadelphia, Pa.

[1] M. J. E. Golay, "Notes on digital coding," Proc. IRE, vol. 37, p. 657; June, 1949.

[2] The expressions "symbol coding," "message coding," "corrector," "characteristic," etc., are given the same meaning as in a former publication (M. J. E. Golay, "Binary coding," IRE Trans. on Information Theory, no. PGIT-4, pp. 23–28; September, 1954). A "lossless symbol coding matrix" yields a "close-packed code," but cases are conceivable—although unknown—in which a close-packed code exists but symbol coding is impossible.

[3] R. W. Hamming, "Error detecting and correcting codes," *Bell Sys. Tech. J.*, vol. 29, pp. 147–161; April, 1950. Mention should be made of an earlier article by Shannon, in which the "somewhat artificial" case of coding seven-bit words against one error by means of three parity checks is described and attributed to Hamming (C. E. Shannon, "A mathematical theory of communication," *Bell Sys. Tech. J.*, vol. 27, p. 418; July, 1948.)

[4] S. K. Zaremba, "Covering problems concerning abelian groups," *J. Lond. Math. Soc.*, vol. 27, pp. 242–246; April, 1952.

$$\begin{array}{cc|cc} Y_1 & Y_2 & X_1 & X_2 \\ \hline 1 & 1 & 1 & 0 \\ 2 & 1 & 0 & 1 \end{array} \quad (2)$$

The passage from $n = 2$ to $n = 3$ is accomplished similarly:

$$\begin{array}{cccccccccc|ccc} Y_1 & Y_2 & Y_3 & Y_4 & Y_5 & Y_6 & Y_7 & Y_8 & Y_9 & Y_{10} & X_1 & X_2 & X_3 \\ \hline 1 & 1 & 1 & 0 & 1 & 1 & 1 & 0 & 1 & 1 & 1 & 0 & 0 \\ 2 & 1 & 0 & 1 & 2 & 1 & 0 & 1 & 2 & 1 & 0 & 1 & 0 \\ 2 & 2 & 2 & 2 & 1 & 1 & 1 & 1 & 0 & 0 & 0 & 0 & 1 \end{array} \quad (3)$$

and so forth.

The coding equation:[1]

$$E_m \equiv X_m + \sum^{k} a_{mk} Y_k \equiv 0 \quad (\text{mod. } p) \quad (4)$$
$$k = 1, \cdots \frac{p^n - 1}{p - 1} - n, \quad m = 1, \cdots n$$

is used to calculate the X check symbols at the transmitting end, and to calculate the corrector E_m at the receiving end; the terms of matrix (3) are the coefficients of the Y's and X's in (4), for the case $n = 3$, $p = 3$.

The 13 ternary numbers written vertically in the 13 columns of (3), which are termed the "base characteristics" of the X's or Y's, and the other 13 numbers obtained by multiplying the individual digits of the base characteristics by 2 modulo 3, which are termed the "derived characteristics" of the X's or Y's, are all different and represent the two ways in which one of the 13 X's or Y's could have been received in error. Inspection of the corrector, when it is not $\begin{matrix}0\\0\\0\end{matrix}$, indicates which symbol was received in error, and whether the error was +1 (base characteristic) or +2 (derived characteristic).

The one-to-one correspondence between any nonzero corrector, and the base or derived characteristic of one of the 13 message symbols constitutes the lossless property of this one-error correction code, and it will be readily seen that the iteration process described above conserves this property when passing from n to $n + 1$.

Consider now that the X's and Y's represent 13 pennies, all equal when shipped, but one of which may have had its weight altered upward or downward before being received. This corresponds to the case of a known transmitted message which is not delivered as such to the addressee. Instead, the addressee is given the E_m's obtained from the three weighing operations, and asked which penny, if any, had its weight increased or decreased.

Inspection of the coding matrix indicates a difficulty in determining the E_m's by weighing various assemblies of pennies: none of the matrix lines contains 1's and 2's in equal number. On the other hand, if an extra penny of correct weight is available—a "catalyzer" conveying no information as such—it can be placed on one side of the scale together with Y_1, Y_2, Y_3 and Y_4, while Y_5, Y_6, Y_7, Y_8 and X_3 are placed on the other. The result of this weighing operation clearly yields the addition modulo 3 required to determine E_3.

The weighing operations required to determine E_2 and E_1 would require three extra and nine extra (good) pennies, respectively, but we may obtain instead the sums E_2E_3 and E_1E_3. When the elements of E_2 and E_3 are added modulo 3, the following 13 numbers are obtained under $Y_1 \cdots X_3$: 1020021221011. Likewise, the E_1E_3 numbers are: 0002222111101. This indicates that the determination of E_2E_3 and E_1E_3 can be effected by weighing operations in which, as for E_3, five received pennies are weighed against four other received pennies and the extra (good) penny. These three weighing operations determine completely which penny, if any, is too light or too heavy.

Lossless Symbol Coding with Powers of Primes

When the order of the message symbols is a power of a prime, p^q, n check symbols should cover $p^{qn} - 1$ possibilities of transmission errors.

Any received symbol may be in error in $p^q - 1$ ways, and if at most one symbol per message may be received in error, a first condition for lossless coding is that the total number of information and check symbols in a message be

$$\frac{p^{qn} - 1}{p^q - 1}.$$

The coding equation for such a case has the general form:

$$E_{mi} \equiv X_{mi} + \sum^{k,j} a_{mi}^{kj} Y_{kj} \equiv 0 \quad (\text{mod. } p) \quad (5)$$
$$k = 1, \cdots \frac{p^{qn} - 1}{p^q - 1} - n, \; m = 1, \cdots n, \; i, j = 1, \cdots q$$

where $X_{m1} \cdots X_{mq}$ and $Y_{n1} \cdots Y_{nq}$ are the q elements of the X_m symbol and Y_n symbol, respectively:

$$X_m = (X_{m1}, X_{m2}, \cdots X_{mq})$$
$$Y_k = (Y_{k1}, Y_{k2}, \cdots Y_{kq}) \quad (5a)$$
$$E_m = (E_{m1}, E_{m2}, \cdots E_{mq}).$$

The matrix formed by the coefficients of the Y's and X's in (5) will be termed the (n, p, q) matrix. When $p = q = 2$, the trivial case of a message containing no information and a single check symbol is represented by the (1, 2, 2) matrix:

$$\begin{array}{|cc} X_{11} & X_{12} \\ \hline 1 & 0 \\ 0 & 1 \end{array} \quad (6)$$

The passage from the (1, 2, 2) matrix to the (2, 2, 2) matrix is analogous to that described earlier and in the former publication[1] for the cases $q = 1$. The matrix (6) is first written four times horizontally and the four "iteration matrices"

$$\begin{matrix}1 & 1\\ 1 & 0,\end{matrix}\quad \begin{matrix}1 & 0\\ 0 & 1,\end{matrix}\quad \begin{matrix}0 & 1\\ 1 & 1\end{matrix}\quad \text{and}\quad \begin{matrix}0 & 0\\ 0 & 0\end{matrix} \tag{7}$$

are written below the four identical matrices first written. The (2, 2, 2) matrix is completed by a double column consisting of an all-zero matrix above a (1, 2, 2) matrix.

$$\begin{matrix} Y_{11} & Y_{12} & Y_{21} & Y_{22} & Y_{31} & Y_{32} & X_{11} & X_{12} & X_{21} & X_{22}\\ 1 & 0 & 1 & 0 & 1 & 0 & 1 & 0 & 0 & 0\\ 0 & 1 & 0 & 1 & 0 & 1 & 0 & 1 & 0 & 0\\ 1 & 1 & 1 & 0 & 0 & 1 & 0 & 0 & 1 & 0\\ 1 & 0 & 0 & 1 & 1 & 1 & 0 & 0 & 0 & 1. \end{matrix} \tag{8}$$

It can be verified by inspection of matrix (8) that all 15 possible nonzero correctors are reproduced by the 10 columns of (8), which are the 10 base characteristics of the 5 pairs of Y's and X's, and by the 5 derived characteristics formed by the sums of the two base characteristics of a $Y_{k1}Y_{k2}$ pair or of an $X_{m1}X_{m2}$ pair. Thus, all 15 possible cases of an error in one of 5 transmitted biquadratic symbols are covered exactly once, and the code calculable with (8) is close packed.

When passing from biquadratic to quaternary symbols, the following convention will be used:

$$00 \rightarrow 0 \quad 01 \rightarrow 1 \quad 10 \rightarrow 2 \quad 11 \rightarrow 3 \tag{9}$$

With this convention, the 64 messages belonging to the (2, 2, 2) code, and determined by giving the Y symbols all combinations of numerical values and calculating the X's with matrix (8), can be written compactly:

00000	02022	10012	12030	20023	22001	30031	32013	
00113	02131	10101	12123	20130	22112	30122	32100	
00221	02203	10233	12211	20202	22220	30210	32232	
00332	02310	10320	12302	20311	22333	30303	32321	(10)
01011	03033	11003	13021	21032	23010	31020	33002	
01102	03120	11110	13132	21121	23103	31133	33111	
01230	03212	11222	13200	21213	23231	31201	33223	
01323	03301	11331	13313	21300	23322	31312	33330.	

When passing from the (2, 2, 2) matrix to the (3, 2, 2) matrix, the (2, 2, 2) matrix is written four times horizontally and a third double line is added, consisting of the four iteration matrices, repeated each five times. A twenty-first double column, consisting of two zero matrices above a (1, 2, 2) matrix completes the (3, 2, 2) matrix.

It has been observed that the 15 nonzero correctors associated with the (2, 2, 2) matrix are reproduced by the single columns of the (2, 2, 2) matrix or by the sums of column pairs. A set of three further observations will now be made, namely: that the first column of the four iteration matrices are all different from each other $\left(\begin{matrix}1 & 1 & 0 & \text{and } 0\\ 1, & 0, & 1 & 0\end{matrix}\right)$; that the second column of all iteration matrices are all different from each other, $\left(\begin{matrix}1 & 0 & 1 & \text{and } 0\\ 0, & 1, & 1 & 0\end{matrix}\right)$, and that the sums of the column pairs of the four iteration matrices are all different from each other $\left(\begin{matrix}0 & 1 & 1 & \text{and } 0\\ 1, & 1, & 0 & 0\end{matrix}\right)$. Thus any base or derived characteristic of the (2, 2, 2) matrix is continued, in the (3, 2, 2) matrix, by one of the four biquadratic elements $\begin{matrix}1 & 1 & 0 & \text{or } 0\\ 1, & 0, & 1 & 0\end{matrix}$. Therefore, it is concluded from these four observations that the 20 first pairs of X and Y columns of the (3, 2, 2) matrix, and the 20 sums of the two columns of a pair, constitute the 60 different base and derived characteristics of the (3, 2, 2) matrix in which the first four binary symbols do not all vanish. The three additional characteristics of the (3, 2, 2) matrix in which the first four binary symbols vanish are the two columns of the added twenty-first pair, and their sum. Thus, we obtain in all the $p^{qn} - 1 = 2^{2 \cdot 3} - 1 = 63$ characteristics of the (3, 2, 2) matrix.

It is seen readily that the iteration process just described is valid for the passage from any $(n, 2, 2)$ matrix thus formed to another lossless $(n + 1, 2, 2)$ symbol coding matrix for one-error correction. It is also seen that the iteration matrices, which play a part analogous to that played by the single symbols of the last line when passing from an $(n, p, 1)$ matrix to an $(n + 1, p, 1)$ matrix, constitute the key to the iteration process. These iteration matrices, and some associated concepts, are now examined in some detail.

Iteration Matrices

Designate by b_1, b_2 and b_1b_2 the two base elements, $\begin{matrix}1\\0\end{matrix}$ and $\begin{matrix}0\\1\end{matrix}$, of the iteration matrices, and their sum, $\begin{matrix}1\\1\end{matrix}$. With this convention, the three nonvanishing iteration matrices (7) may be written:

$$b_1b_2 \; b_1, \qquad b_1 \; b_2 \quad \text{and} \quad b_2 \; b_1b_2.$$

Consider now the matrix:

$$\begin{matrix} b_1 & b_2\\ b_2 & b_1b_2 \end{matrix}\;. \tag{11}$$

It will be observed that the second and third iteration matrices utilized above are the two lines of matrix (11), and that the first iteration matrix is the sum modulo 2 of these two lines.

TABLE I

Master Iteration Matrices

	$q=2$	$q=3$		$q=4$		$q=5$		$q=6$	
MAT. IND.	1 2 12	1 2 3 12 23	1 2 3 13 123	1 2 3 4 12 23 34	1 2 3 4 14 124 1234	1 2 3 4 5 12 23 34 45	1 2 3 4 5 15 125 1235 12345	1 2 3 4 5 6 12 23 34 45 56	1 2 3 4 5 6 16 126 1236 12346 123456
$p=2$	MIM	MIM	MIM	MIM	MIM	α's: 1 0 1 1 0 0 1 0 1 1 β's: 1 1 1 0 0 1 0 0 1 0 1 0 1 0 1	α's: 1 1 0 1 0 0 1 1 0 1 β's: 1 1 1 0 0 1 0 0 1 0 1 0 1 0 1	MIM	MIM
$p=3$	MIM	MIM	α's: 1 2 2 β's: 1 1 0 1 0 2	MIM	MIM	MIM	α's: 1 2 1 2 2 β's: 1 1 0 0 0 1 0 2 0 0 1 0 0 1 0 1 0 0 0 2 also α's: 1 0 2 2 1 1 1 0 2 0 β's: 1 0 0 1 0 0 1 0 0 1 1 0 1 0 1		
$p=5$	α's: 1 3 β's: 1 3	α's: 1 4 2 β's: 1 2 0 1 0 1	MIM	MIM	MIM				
$p=7$	MIM	α's: 1 4 5 β's: 1 0 5 1 3 3	MIM						

Designate by p_{ij} the element at the intersection of the ith line and jth column of matrix (11), and let α_i and β_j designate factors which for $p = 2$ may be 1 or 0. It will be verified by inspection that the set of three observations made earlier is equivalent to the observation that no two sums of the form

$$\sum^{i,j} \alpha_i \beta_j p_{ij} \tag{12}$$

are identical unless all the α's and β's of the two sums are also identical. This last observation is also equivalent to the observation that no sum of the form (12) vanish, unless all α's and β's vanish. It will be convenient to state that matrix (11) fulfils for $p = q = 2$ the general condition:

$$\sum^{i,j} \alpha_i \beta_j p_{ij} \not\equiv 0 \quad (\text{mod. } p)$$

when

$$\sum^{i} \alpha_i \not\equiv 0, \quad \sum^{j} \beta_j \not\equiv 0 \tag{13}$$

$$i,j = 1, 2, \cdots q \quad 0 \leq \alpha_i < p, 0 \leq \beta_j < p$$

and any matrix satisfying this condition will be termed a "master iteration matrix."[5] A set of p^q iteration matrices is formed from a master iteration matrix by giving the α_i's in $\sum \alpha_i p_{ij}$ all p^q combinations of values. When passing from the (n, p, q) coding matrix to the $(n + 1, p, q)$ coding matrix, the (n, p, q) matrix is written horizontally p^q times, and each iteration matrix is written $(p^{qn} - 1)/(p^q - 1)$ times below each (n, p, q) matrix. The $(n + 1, p, q)$ matrix is completed with a q-tuple column consisting of n all-zero matrices above a $(1, p, q)$ matrix (*i.e.*, a $q \times q$ matrix of zeros except for a main diagonal of 1's).

Table I has been prepared to list, in compact form, several matrices which, depending upon the value of p, may or may not constitute master iteration matrices (MIM).

All matrices listed in Table I are symmetric with respect to their main diagonal, and each sucessive line of these matrices is shifted to the left one element with re-

[5] A master iteration matrix is three-dimensional, since it is a two-dimensional array of one-dimensional sequences. Two-dimensional matrices may be derived from it with the operation $\sum^i \alpha_i p_{ij}$ and one-dimensional sequences may be derived from it with the operation $\sum^{i,j} \alpha_i \beta_j p_{ij}$.

spect to the line above it. They are therefore fully characterized by their first line and last column. Furthermore, the indices only of the q-dimensioned base elements p_r forming the p_{ij}'s of the first line and last column have been indicated in the matrix indices (MAT. IND.) lines. For instance, the first of the two matrices at the $p = 3$, $q = 3$ intersection is the ternary master iteration matrix:

$$\begin{matrix} t_1 & t_2 & t_3 \\ t_2 & t_3 & t_1t_2 \\ t_3 & t_1t_2 & t_2t_3 \end{matrix} \qquad (14)$$

where

$$t_1 = \begin{matrix}1\\0,\\0\end{matrix} \quad t_2 = \begin{matrix}0\\1,\\0\end{matrix} \quad t_3 = \begin{matrix}0\\0,\\1\end{matrix} \quad t_1t_2 = \begin{matrix}1\\1,\\0\end{matrix} \quad t_2t_3 = \begin{matrix}0\\1.\\1\end{matrix}$$

Giving the α_i's in $\sum \alpha_i p_{ij}$ the successive 27 combinations of values:

$$\begin{matrix}2&2&2\\2,&2,&2,\\2&1&0\end{matrix} \quad \cdots \quad \begin{matrix}0&0&0\\0,&0,&0\\2&1&0\end{matrix}$$

yields the 27 successive iteration matrices:

$$\begin{matrix}222&212&202\\211,&200,&222,\\221&120&022\end{matrix} \quad \cdots \quad \begin{matrix}020&010&000\\022,&011,&000.\\202&101&000\end{matrix}$$

When a matrix is not a master iteration matrix, the single set or the several sets of values of the α_i's and β_j's for which (13) is not fulfilled are listed in ascending index order. For instance, the second matrix at the $p = 3$, $q = 3$ intersection is not a master iteration matrix because (13) is not satisfied when

$$\alpha_1 = 1, \quad \alpha_2 = 2, \quad \alpha_3 = 2$$

and when

$$\beta_1 = 1, \quad \beta_2 = 1, \quad \beta_3 = 0.$$

Alternately, the same α values and

$$\beta_1 = 1, \quad \beta_2 = 0, \quad \beta_3 = 2$$

do not satisfy (13), nor again the same α values (or the α values multiplied by 2 modulo 3) and any linear combination modulo 3 of the β values given above, such as

$$\beta_1 = 2, \quad \beta_2 = 1, \quad \beta_3 = 2$$

or

$$\beta_1 = 0, \quad \beta_2 = 1, \quad \beta_3 = 1, \quad \textit{etc.}$$

The remarks made above apply to all sets of values of the α's and β's given in Table I which do not satisfy (13), and since the matrices of Table I are symmetric with respect to their main diagonal, the α's and β's may be interchanged in all sets of values given.

The case $p = 2$, $q = 5$ is particularly noteworthy, since it is the only case examined for which a master iteration matrix was not found. Six other symmetric matrices have been examined, and none has yielded a master iteration matrix for this case. The indices of the first and last line of the matrices thus examined are:

$$\begin{matrix}1&2&3&4&5,&1&2&3&4&5,&1&2&3&4&5,\\5&13&24&35&14&5&12&34&15&23&5&123&234&345&145\end{matrix}$$

$$\begin{matrix}1&2&3&4&5,&1&2&3&4&5,&1&2&3&4&5\\5&12&123&1234&12345&5&12345&2345&345&45&5&125&123&234&345.\end{matrix}$$

It has been noted[2] that no case of lossless coding is known, for which symbol coding is impossible, but the failure to find a master iterating matrix for $p = 2$, $q = 5$, among the eight matrices examined, suggests the interesting possibility that symbol coding may be proven impossible for this case, which is known to have a close-packed code (see *Note*).

Consider now the extension of the foregoing coding matrix construction to the cases in which the symbol order is the square ($q = 2$) of any odd prime. We may, without loss of generality, postulate the following matrix:

$$\begin{matrix} p_1 & ap_2 \\ p_2 & -bp_1 \end{matrix} \qquad (15)$$

where $p_1 = \begin{matrix}1\\0\end{matrix}$, $p_2 = \begin{matrix}0\\1\end{matrix}$. In order to determine under what conditions matrix (15) is a master iteration matrix, condition (13) is applied to (15), and we obtain the requirements

$$\begin{matrix} \alpha_1\beta_1 - \alpha_2\beta_2 b \not\equiv 0 \\ \text{or} \quad \alpha_1\beta_2 a + \alpha_2\beta_1 \not\equiv 0 \end{matrix} \quad (\text{mod. } p) \qquad (16)$$

or both.

If we consider the β's as the variables with respect to which the set of homogeneous congruences of (16) must be unsolvable, condition (16) can be replaced by the equivalent necessary and sufficient condition that the determinant of the coefficients of the β's do not vanish:

$$D = \begin{vmatrix} \alpha_1 & -b\alpha_2 \\ \alpha_2 & a\alpha_1 \end{vmatrix} = a\alpha_1^2 + b\alpha_2^2 \not\equiv 0 \quad (\text{mod. } p). \qquad (17)$$

It is obvious that condition (17) is satisfied whenever

$$a \not\equiv 0, \quad b \not\equiv 0,$$

and

$$\alpha_1 \not\equiv 0, \quad \alpha_2 \equiv 0, \quad (\text{mod. } p)$$

or

$$\alpha_1 \equiv 0, \quad \alpha_2 \not\equiv 0.$$

and we examine now the cases in which

$$\begin{matrix} a \not\equiv 0, & b \not\equiv 0, \\ \alpha_1 \not\equiv 0, & \alpha_2 \not\equiv 0. \end{matrix} \quad (\text{mod. } p)$$

It is known[6] that when $p \equiv 3 \pmod{4}$ p cannot divide numbers of the form $x^2 + 1$. Therefore, if we make $a = b = 1$ in (17) and set $\alpha_1 \equiv \alpha_2 x$, we obtain

$$D = \alpha_1^2 + \alpha_2^2 \equiv \alpha_2^2(x^2 + 1) \not\equiv 0. \quad (\text{mod. } p = 4n + 3)$$

When we make $a = 1$, $b = 2$ in (17), we obtain a quadratic expression of the form $\alpha_1^2 + 2\,\alpha_2^2$, and since numbers of the form $x^2 + 2$ are not divisible by primes[7] $\equiv 5 \pmod{8}$, we obtain, setting $\alpha_1 \equiv \alpha_2 x$:

$$D = \alpha_1^2 + 2\alpha_2^2 \equiv \alpha_2^2(x^2 + 2) \not\equiv 0. \quad (\text{mod. } p = 8n + 5)$$

Finally, when we make $a = 2$, $b = 3$ in (17), we obtain an expression of the form $2\alpha_1^2 + 3\alpha_2^2$, and since numbers of the form $2x^2 + 3$ cannot be divided by primes[8] of the form $24n + 1$ or $24n + 17$, we obtain, setting $\alpha_1 \equiv \alpha_2 x$:

$$D = 2\alpha_1^2 + 3\alpha_2^2 \equiv \alpha_2^2(2x^2 + 3) \not\equiv 0. \quad (\text{mod. } p = 24n + 1, 24n + 17).$$

It is seen readily that the last two cases cover all primes of the form $4n + 1$, and since the case $a = b = 1$, covers all primes of the form $4n + 3$, all odd primes are covered. Therefore, master iteration matrices can be constructed for all squares of odd primes.

As an example, consider the case $p = 5$. Condition (17) is satisfied when $a = 1$, $b = 2$, and the matrix

$$\begin{matrix} v_1 & v_2 \\ v_2 & -2v_1 \end{matrix} \tag{18}$$

is a master iteration matrix modulo 5. On the other hand, matrix (11) is not a master iteration matrix modulo 5, because (12) vanishes modulo 5 when $\alpha_1 = \beta_2 = 1$, $\alpha_2 = 3$, $\beta_1 = 2$, and when the p_{ij}'s are the b's of (11).

The question of whether the search for master iteration matrices for values of q higher than 2 can be systematized is proposed as an interesting unsolved problem of coding or group theory.

Miscellaneous Notes

An interesting condition has been utilized by Zaremba[9] for proving the nonexistence of certain codes. This condition will be described in connection with its application to the once surmised close-packed code for correcting up to two errors in 90-binary symbol messages. These numbers satisfy the condition previously published[1] that the first three numbers of the corresponding line of Pascal's triangle add up to a power of 2:

$$1 + 90 + \frac{90.89}{2} = 2^{12}.$$

If such a code were to exist, the code book would contain 2^{7b} code messages with the minimum distance 5, and with the allowable assumption that the all-zero message belongs to the code book, the remaining code messages with the greatest number of 0's would contain 85 0's and 5 1's. For the code to be close packed, every message with 87 0's and 3 1's should be obtainable in only one way by replacing 2 0's by 1's in the 85–0 and 5–1 code messages, and this condition determines the number of the 85–0 and 5–1 code messages. A single error in the 0's of these would produce 86–0 and 4–1 messages, and the number of remaining 86–0 and 4–1 messages determines the number of 84–0 and 6–1 code messages. This procedure can be extended to determine all the numbers of code messages with an increasing number of 1's. The numbers thus found should be integers, and this necessary but not sufficient condition for the existence of a close packed code constitutes Zaremba's condition. In the particular instance examined here, it is found that the number of 83–0 and 7–1 code messages is fractional, and this demonstrates anew the impossibility of the close-packed code surmised above.

When Zaremba's condition is applied to the singular 3-error correcting code with 12 binary information symbols and 11 check symbols built with the matrix previously published, it is found that there are: 1, 253, 506, 1288, 1288, 506, 253 and 1 messages with 0, 7, 8, 11, 12, 15, 16 and 23 1's, respectively.

Of course, the addition of a check digit serves to cluster the code messages even more into 1, 759, 2576, 759 and 1 message with 0, 8, 12, 16 and 24 1's, respectively.

The application of Zaremba's condition to the other singular 2-error correcting close packed code with 6 information and 5 check ternary symbols shows that there are 1, 132, 132, 330, 110 and 24 code messages with 11, 6, 5, 3, 2 and no 0's, respectively.

The application of Zaremba's condition to the hypothetical 1 error correcting code with 5 information and and 2 check symbols of order 6 does not disprove the existence of this code, for we find that the code book should contain 1, 175, 525, 1890, 3010 and 2175 code messages having, respectively, 7, 4, 3, 2, 1 and no 0's.

It is not known at present whether this code exists or not, but if it exists, it can be shown not to have the following property exhibited by the (2, 2, 2) code tabulated earlier.

Consider any code message, and leaving unchanged any 0's it may have, add respectively 1 and 2 modulo 3 to all other symbols, in the number system of order 3 with the symbols 1, 2 and 3. The two new messages thus obtained form part of the code (for instance, from the code message 01323 we derive the other code message 02131 and 03212).

Consider now the hypothetical (2, 6, 1) code, and select the greatest number of code messages in which three given symbols are 0's. The most favorable assumption, for ease of code message "packing," is that this number be the same for all similar groups, and be equally broken down into code messages in which the other four positions are filled by three or four nonzero symbols. We obtain thus $525/35 = 15$ messages with no 0's for

[6] T. Nagell, "Introduction to Number Theory," John Wiley and Sons, Inc., New York, N. Y., p. 135; 1951.
[7] *Ibid.*, p. 136.
[8] *Ibid.*, p. 190.
[9] Private communication.

instance, in the first four positions, and 175/35 = 5 messages each with a single 0 in any one of these first four positions, with 0's in the last three positions for all.

Postulate now that to each message corresponds 4 additional messages obtained by adding 1, 2, 3 and 4 modulo 5 to the nonzero symbols, in the number system of order 5 in which the symbols are 1, 2, 3, 4 and 5. We may assume, without loss of generality, that the code messages selected by this process are:

1	1	1	1	0	0	0
1	2	3	4	0	0	0
1	3	x	x	0	0	0
1	4	x	0	0	0	0
1	5	0	3	0	0	0
1	0	x	x	0	0	0
0	1	x	x	0	0	0

and the 28 other messages obtained by adding 1, 2, 3 and 4 modulo 5 to the nonzero symbols. If the code exists, it should be possible to replace the x's by numbers which, for the whole rectangle, satisfy the condition that no four nonzero numbers exist at the apex of any rectangle, which have the property that the difference (mod. 5) of any two on a side is equal to the difference of the other two. This condition is imposed by the requirement that the minimum distance 3 exists between any two for the 35 code messages involved. It has been verified by inspection that this condition cannot be realized. Hence, if the (2, 6, 1) code exists, it cannot have the property postulated above.

List of Symbols

p = prime number base.
q = power of p in coding system of order p^q.
n = number of redundant digits in system of order p^q.
(n, p, q) = characterization of lossless code or coding matrix in system base p^q with n redundant digits.
X_m = redundant check digits in system of order p.
Y_k = information digits in system of order p.
E_m = check digits calculated after reception in system of order p.
X_{mi} = redundant check digits in system of order p^q.
Y_{ki} = information digits in system of order p^q.
E_{mi} = check digits calculated after reception in system of order p^q.
p_{ij} = q-dimensional term in matrices (whether master iteration matrices or not).
$b_1, b_2, \cdots, b_q$ = base elements $\begin{matrix} 1 & 0 & & 0 \\ 0 & 1 & & 0 \\ \vdots, & \vdots, & \cdots & \vdots \\ 0 & 0 & & 1 \end{matrix}$ in system of order 2^q.
$t_1, t_2, \cdots, t_q$ = same base elements in system of order 3^q.
$v_1, v_2, \cdots, v_q$ = same base elements in system of order 5^q.
$p_1, p_2, \cdots, p_q$ = same base elements in system of order p^q.
α_i and β_i = one-dimensional factors in system of order p.

Definition of Terms

Close-packed code: An e-error correcting code in which the number of words is given exactly by the expression:

$$\frac{q^{qn}}{\sum_{j=0}^{j=e} (p^q - 1)^j \binom{\frac{p^{qn} - 1}{p^q - 1}}{j}}.$$

Sum (p_1p_2 of ensembles p_1 and p_2): The ensemble formed by the addition modulo p of the homologous symbols of the individual ensembles. The symbol $\oplus$, sometimes used to designate such sums, is not used here in order to conserve space, as no confusion results from this convention.

Coding matrix: The matrix formed by the coefficients of (4) or (5).

Iteration matrix: One of p^q q-dimensional square matrices which are required to pass from the (n, p, q) coding matrix to the $(n + 1, p, q)$ coding matrix.

Master iterating matrix: A q-dimensional square matrix composed of q-dimensional elements, from which the p^q iteration matrices can be derived systematically, by the operation $\sum_{i=1}^{i=q} \alpha_i p_{ij}$.

Corrector: The ensemble of nq digits of order p obtained as a result of the checking operations defined by (4) and (5).

Base characteristic: The ensemble of nq digits of order p forming any column of a coding matrix below an X_{mi} or a Y_{ki}, (also simply the "characteristic" when $p = 2$, $q = 1$).

Derived characteristic: Any sum of the base characteristics below X_{mi}'s or Y_{ki}'s (i or j = 1, 2, $\cdots$ q), with one of the weights 0, 1, $\cdots$, $p - 1$ given to each base characteristic.

Note: After the submission of this article, Prof. F. L. Dennis discovered and communicated to the writer an MIM for the case $p = 2$, $q = 5$. Like the matrices listed in Table I, Dennis' matrix is symmetric with respect to the main diagonal, and its successive columns are shifted upward one element with respect to the preceding columns. Thus, Dennis' matrix is fully characterized by its first and fifth lines, the indices of which are:

1	2	3	4	5
5	13	24	35	134

9

Lossless Symbol Coding with Nonprimes*

Golay[1] has asked the question as to "whether the search for master iteration matrices, of the form p^n for values of n higher than 2 can be systematized." There exists a straight-forward method for constructing the coding matrix for realizing all the one-error correction codes whose existence was shown by Zaremba.[2] The ability to do this depends upon the fact that there exist finite fields of order p^n, where p is a prime, namely the Galois fields. The method is completely constructive since in order to construct the Galois field $GF(p^n)$ it is only necessary to find a polynomial of degree n which is irreducible over the field of integers mod p. This is constructive since there is only a finite number of these polynomials; if one can't find a better way to obtain an irreducible polynomial, one can test each of these polynomials for irreducibility. Assuming that a Galois field of order p^n has been constructed, the following method is one way of obtaining the coding matrix. Incidently, the method proves constructively the theorem of Zaremba as to the existence of such codes.

Let $0, 1, a_3, \cdots, a_{p^n}$ be the elements of a Galois field $GF(p^n)$ and H be the prime ideal defining said field. Suppose one wishes to construct a single-symbol correcting code with k check symbols. First write down all possible distinct column vectors v_i of length k deleting the vector of the form 0, and, although this is not necessary, make the first k vectors of the list of the form $v_i = \delta_{ij}$ (kronecker delta); this simplifies encoding and decoding. One has $p^{nk} - 1$ distinct vectors. Now systematically delete from the list all vectors which are of the form $v_j = av_i$ for $i < j$ and $a \in GF(p^n)$. Since a field has no zero divisors, one clearly has left $p^{nk} - 1/p^n - 1$ vectors. Now construct a code with $s = p^{nk} - 1/p^n - 1$ total symbols, k check symbols, and $s - k$ information symbols. Let A be the matrix composed of the vectors left after the deletion process. The column order is the same as left by the processes.

Given $s - k$ information symbols, form a column vector w of length s, the first k elements from the top being 0 and the remaining $s - k$ elements composing the given information symbols. Forming $Aw \equiv z \bmod (H)$, one obtains a column vector of length k. Since the elements of the vector are in a field, the additive inverse $-z$ of z exists. Form a vector x which has $-z$ as its first k elements and the information symbols as the remaining $s - k$ elements. Since the first k columns of the matrix are of the form $v_i = \delta_{ij}$, we see that $Ax \equiv 0 \bmod H$. This is now the encoded message. Suppose there is an error of amount b in the eth character. Then the received message is $x + \delta_{ej}b$. Form $A(x + \delta_{ej}b) \equiv A\delta_{ej}b \equiv v_eb \bmod (H)$.

Looking at v_eb we can determine that the eth character must be in error, for by construction of A, v_eb is not of the form v_ja for any $j \neq e$ and $a \in GF(p^n)$. If this were so we would have $v_eb = v_ja$ or $v_e = v_jab^{-1}$, and v_e would be the multiple of a vector v_j; and this is not the case by construction. Thus we know that the eth character is in error and, further solving $v_ey = v_eb$, we can obtain $y = b$ and thus correct any single symbol error. It might be noted that this is as far as this particular method can be extended since we are relying on the property that the elements form a field and every finite field is isomorphic to some Galois field.

In order to determine the Master Iteration Matrices (MIM) so that one can check the codes using the multiplication table of the integers mod p rather than the multiplication table of the Galois field, the following procedure is easily seen to work. Each element of a field $GF(p^n)$ can be represented as a polynomial of the form $\sum_{i=0}^{n-1} c_ix^i$ where c_i are the integers mod p. Make the correspondence $c = \sum_{i=0}^{n-1} c_ix^i \leftrightarrow (c_0, c_1 \cdots c_{n-1}) = c'$ making this a column vector. If $a \in GF(p^n)$ is a typical element of the coding matrix A, form the product $a \cdot c$ where $a = \sum_{i=0}^{n-1} a_ix^i$ and the a_i are the particular coefficients in a, and $c = \sum_{i=0}^{n-1} c_ix^i$ where the c_i are indeterminates. Then using the polynomial H to reduce all the powers we determine a matrix a^* which if

$$c = \sum_{i=0}^{n} c_ix^i \leftrightarrow (c_0c_1 \cdots c_{n-1}) = c'$$

and

$$ac = \sum a_ix^i \sum c_ix^i$$

$$\leftrightarrow (d_0 \cdots d_{n-1}) = d'$$

then $a^*c' = d'$. Thus matrix a^* is the MIM searched for. Replacing the elements of A by the corresponding MIM we obtain a new matrix A' which operates on vectors where each element of the message is represented by a column vector of the form $(c_0 \cdots c_{n-1})$ where the c_i's are integers mod p. This matrix is used to encode the message as before but the congruence is mod p instead of mod H.

Example: Let $0, 1, x, x + 1$ be the elements of $GF(2^2)$ where $H = x^2 + x + 1$. Then let $k = 2$; we calculate $s = 5$; a possible matrix turns out to be:

$$\begin{pmatrix} 1 & 0 & 1 & 1 & x+1 \\ 0 & 1 & 1 & x+1 & 1 \end{pmatrix}.$$

To form the MIM we see that $1 \cdot (c_1x + c_0) = c_1x + c_0$, so 1 gives $\binom{10}{01}$; and $(x + 1)(c_1x + c_0) = c_1(x^2 + x) + c_1 + c_0 = (xc_0) + c_0 + c_1$, so $(x + 1)$ gives $\binom{11}{10}$. Thus the matrix becomes

$$\begin{matrix} 1 & 0 & 0 & 0 & 1 & 0 & 1 & 0 & 1 & 1 \\ 0 & 1 & 0 & 0 & 0 & 1 & 0 & 1 & 1 & 0 \\ 0 & 0 & 1 & 0 & 1 & 0 & 1 & 1 & 1 & 0 \\ 0 & 0 & 0 & 1 & 0 & 1 & 1 & 0 & 0 & 1 \end{matrix}.$$

One thing that is interesting is that the Hamming Code has lossless points at $s = 2^i - 1 = 1, 3, 7, 15, 31, 63 \ldots$. For the lossless codes of radix 2^2 we obtain $s = 4^k 1\ 3 = 1, 5, 13 \ldots$ or, since each character is 2 bits, it is bitwise $1, 26, \ldots$. and the number of check bits would be 2, 4, 6. Comparing this with the Hamming code, notice for instance that for $s = 10$ one must use 4 bits so that what this code does to becone lossless is to correct 5 double errors. Thus for radix 2^2 and assuming that the errors are not correlated and that all have equal probability we are able to construct some best possible codes. This, of course, can be done easily by the proper interpretation of the Hamming check bits. However, it interested me that this code did it automatically.

John Cocke
IBM
Poughkeepsie, N. Y.

* Received by the PGIT, January 26, 1959.

[1] M. J. E. Golay, "Notes on the penny-weighting problem, lossless symbol coding with nonprimes, etc.," IRE Trans. on Information Theory, vol. IT-4, pp. 103–109; September, 1958.

[2] S. K. Zaremba, "Covering problems concerning Abelian groups," *J. London Math. Soc.*, vol. 27, pp. 242–246; April, 1952.

Linear Codes

II

Editor's Comments on Papers 10, 11, and 12

10 Slepian: *A Class of Binary Signaling Alphabets*

11 Slepian: *A Note on Two Binary Signaling Alphabets*

12 Slepian: *Some Further Theory of Group Codes*

The work of Slepian [10] was the first investigation of a general nature into the problem of block coding, all previous papers having been concerned with construction and decoding techniques of a very specific nature. Although the notions of parity-check codes, systematic codes, and the equivalence of codes were well established by 1956, the introduction and investigation by Slepian [10] of the equivalent concept of linear codes provided a theoretical basis from which a deeper understanding and important advances could be made.

One of the significant properties of linear codes which was demonstrated by Slepian is that the set of distances from a fixed codeword to all other codewords is independent of the fixed codeword chosen. The channel thus treats all codewords in a similar manner. The standard array-decoding algorithm, which he showed to be maximum likelihood for the binary symmetric channel, yielded greater comprehension of the decoding operation. The important practical problem of determining optimal codes which maximize the probability of correct decoding was also considered.

In his 1960 paper, Slepian [12] investigates thoroughly the equivalence and decomposition of codes with particular attention to optimality of codes with respect to both the probability of correct decoding and maximum minimum distance. It is a fascinating study which is remarkable for the amount of insight it gives into the structure of codes. These two papers by Slepian [10,12] are, without question, among the most significant in coding theory. They provided a theoretical framework which profoundly affected the development of coding.

In a third important paper on coding theory, Slepian [11] generalizes Hamming codes to the binary nonprimitive case and provides a decoding algorithm for them. He further shows that the Reed–Muller codes are, in fact, linear and gives a parity-check formulation for them, along with an alternative decoding algorithm.

Reprinted from *Bell System Tech. J.*, **35**, 203–234 (1956)

10

A Class of Binary Signaling Alphabets

By DAVID SLEPIAN

(Manuscript received September 27, 1955)

A class of binary signaling alphabets called "group alphabets" is described. The alphabets are generalizations of Hamming's error correcting codes and possess the following special features: (1) all letters are treated alike in transmission; (2) the encoding is simple to instrument; (3) maximum likelihood detection is relatively simple to instrument; and (4) in certain practical cases there exist no better alphabets. A compilation is given of group alphabets of length equal to or less than 10 binary digits.

INTRODUCTION

This paper is concerned with a class of signaling alphabets, called "group alphabets," for use on the symmetric binary channel. The class in question is sufficiently broad to include the error correcting codes of Hamming,[1] the Reed-Muller codes,[2] and all "systematic codes".[3] On the other hand, because they constitute a rather small subclass of the class of all binary alphabets, group alphabets possess many important special features of practical interest.

In particular, (1) all letters of the alphabets are treated alike under transmission; (2) the encoding scheme is particularly simple to instrument; (3) the decoder — a maximum likelihood detector — is the best possible theoretically and is relatively easy to instrument; and (4) in certain cases of practical interest the alphabets are the best possible theoretically.

It has very recently been proved by Peter Elias[4] that there exist group alphabets which signal at a rate arbitarily close to the capacity, C, of the symmetric binary channel with an arbitrarily small probability of error. Elias' demonstration is an existence proof in that it does not show *explicitly* how to construct a group alphabet signaling at a rate greater than $C - \varepsilon$ with a probability of error less than δ for arbitrary positive δ and ε. Unfortunately, in this respect and in many others, our understanding of group alphabets is still fragmentary.

In Part I, group alphabets are defined along with some related concepts necessary for their understanding. The main results obtained up to the present time are stated without proof. Examples of these concepts are given and a compilation of the best group alphabets of small size is presented and explained. This section is intended for the casual reader.

In Part II, proofs of the statements of Part I are given along with such theory as is needed for these proofs.

The reader is assumed to be familiar with the paper of Hamming,[1] the basic papers of Shannon[5] and the most elementary notions of the theory of finite groups.[6]

Part I — Group Alphabets and Their Properties

1.1 INTRODUCTION

We shall be concerned in all that follows with communication over the symmetric binary channel shown on Fig. 1. The channel can accept either of the two symbols 0 or 1. A transmitted 0 is received as a 0 with probability q and is received as a 1 with probability $p = 1 - q$: a transmitted 1 is received as a 1 with probability q and is received as a 0 with probability p. We assume $0 \leqq p \leqq \frac{1}{2}$. The "noise" on the channel operates independently on each symbol presented for transmission. The capacity of this channel is

$$C = 1 + p \log_2 p + q \log_2 q \text{ bits/symbol} \tag{1}$$

By a *K-letter, n-place binary signaling alphabet* we shall mean a collection of K distinct sequences of n binary digits. An individual sequence of the collection will be referred to as a *letter* of the alphabet. The integer K is called the size of the alphabet. A letter is transmitted over the channel by presenting in order to the channel input the sequence of n zeros and ones that comprise the letter. A *detection scheme* or *detector* for

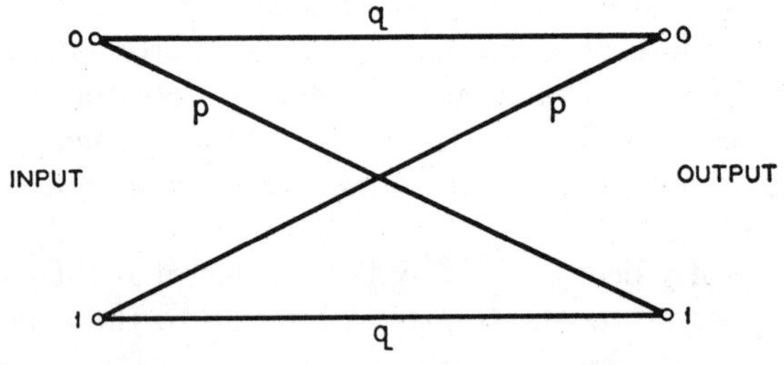

Fig. 1 — The symmetric binary channel.

a given K-letter, n-place alphabet is a procedure for producing a sequence of letters of the alphabet from the channel output.

Throughout this paper we shall assume that signaling is accomplished with a given K-letter, *n-place* alphabet by choosing the letters of the alphabet for transmission independently with equal probability $1/K$. Shannon[5] has shown that for sufficiently large n, there exist K-letter, n-place alphabets and detection schemes that signal over the symmetric binary channel at a rate $R > C - \varepsilon$ for arbitrary $\varepsilon > 0$ and such that the probability of error in the letters of the detector output is less than any $\delta > 0$. Here C is given by (1) and is shown as a function of p in Fig. 2. No algorithm is known (other than exhaustvie procedures) for the construction of K-letter, n-place alphabets satisfying the above inequalities for arbitrary positive δ and ε except in the trivial cases $C = 0$ and $C = 1$.

1.2 THE GROUP B_n

There are a totality of 2^n different n-place binary sequences. It is frequently convenient to consider these sequences as the vertices of a cube of unit edge in a Euclidean space of n-dimensions. For example the 5-place sequence 0, 1, 0, 0, 1 is associated with the point in 5-space whose

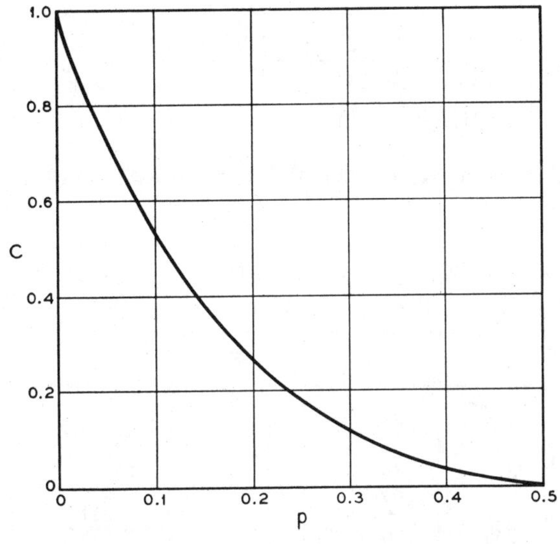

Fig. 2 — The capacity of the symmetric binary channel.
$C = 1 + p \log_2 p + (1 - p) \log_2 (1 - p)$

3

coordinates are (0, 1, 0, 0, 1). For convenience of notation we shall generally omit commas in writing a sequence. The above 5-place sequence will be written, for example, 01001.

We define the *product of two n-place binary sequences*, $a_1a_2 \cdots a_n$ and $b_1b_2 \cdots b_n$ as the n-place binary sequence

$$a_1 \dotplus b_1 , \quad a_2 \dotplus b_2 , \cdots , a_n \dotplus b_n$$

Here the a's and b's are zero or one and the $\dotplus$ sign means addition modulo 2. (That is $0 \dotplus 0 = 1 \dotplus 1 = 0, \quad 0 \dotplus 1 = 1 \dotplus 0 = 1$) For example, (01101) (00111) = 01010. With this rule of multiplication the 2^n n-place binary sequences form an Abelian group of order 2^n. The elements of the group, denoted by $T_1, T_2, \cdots, T_{2^n}$, say, are the n-place binary sequences; the identity element I is the sequence $000 \cdots 0$ and

$$IT_i = T_iI = T_i ; \qquad T_iT_j = T_jT_i ; \qquad T_i(T_jT_k) = (T_iT_j)T_k ;$$

the product of any number of elements is again an element; every element is its own reciprocal, $T_i = T_i^{-1}$, $T_i^2 = I$. We denote this group by B_n.

All subgroups of B_n are of order 2^k where k is an integer from the set $0, 1, 2, \cdots, n$. There are exactly

$$\begin{aligned} N(n, k) &= \frac{(2^n - 2^0)(2^n - 2^1)(2^n - 2^2) \cdots (2^n - 2^{k-1})}{(2^k - 2^0)(2^k - 2^1)(2^k - 2^2) \cdots (2^k - 2^{k-1})} \qquad (2) \\ &= N(n, n - k) \end{aligned}$$

distinct subgroups of B_n of order 2^k. Some values of $N(n, k)$ are given in Table I.

TABLE I — SOME VALUES OF $N(n, k)$, THE NUMBER OF SUBGROUPS OF B_n OF ORDER 2^k. $N(n, k) = N(n, n - k)$

n\k	0	1	2	3	4	5
2	1	3	1			
3	1	7	7	1		
4	1	15	35	15	1	
5	1	31	155	155	31	1
6	1	63	651	1395	651	63
7	1	127	2667	11811	11811	2667
8	1	255	10795	97155	200787	97155
9	1	511	43435	788035	3309747	3309747
10	1	1023	174251	6347715	53743987	109221651

1.3 GROUP ALPHABETS

An n-place *group alphabet* is a K-letter, n-place binary signaling alphabet whose letters form a subgroup of B_n. Of necessity the size of an n-place group alphabet is $K = 2^k$ where k is an integer satisfying $0 \leqq k \leqq n$. By an (n, k)-*alphabet* we shall mean an n-place group alphabet of size 2^k. Example: the $N(3, 2) = 7$ distinct (3, 2)-alphabets are given by the seven columns

(i)	(ii)	(iii)	(iv)	(v)	(vi)	(vii)
000	000	000	000	000	000	000
100	100	100	010	010	001	110
010	001	011	001	101	110	011
110	101	111	011	111	111	101

(3)

1.4 STANDARD ARRAYS

Let the letters of a specific (n, k)-alphabet be $A_1 = I = 00 \cdots 0$, $A_2, A_3, \cdots, A_\mu$, where $\mu = 2^k$. The group B_n can be developed according to this subgroup and its cosets:

$$B_n = \begin{array}{llll} I, & A_2, & A_3, & \cdots, A_\mu \\ S_2, & S_2A_2, & S_2A_3, & \cdots, S_2A_\mu \\ S_3, & S_3A_2, & S_3A_3, & \cdots, S_3A_\mu \\ \cdot & & & \\ \cdot & & & \\ \cdot & & & \\ S_\nu, & S_\nu A_2, & S_\nu A_3, & \cdots, S_\nu A_\mu \end{array} \qquad (4)$$

$$\mu = 2^k, \qquad \nu = 2^{n-k}.$$

In this array every element of B_n appears once and only once. The collection of elements in any row of this array is called a *coset* of the (n, k)-alphabet. Here S_2 is any element of B_n not in the first row of the array, S_3 is any element of B_n not in the first two rows of the array, etc. The elements $S_2, S_3, \cdots, S_\nu$ appearing under I in such an array will be called the *coset leaders*.

If a coset leader is replaced by any element in the coset, the same coset will result. That is to say the two collections of elements

$$S_i, \quad S_iA_2, \quad S_iS_3, \cdots, S_iA_\mu$$

and

$$S_iA_k, \quad (S_iA_k)A_2, \quad (S_iA_k)A_3, \cdots (S_iA_k)A_\mu$$

are the same.

We define the *weight* $w_i = w(T_i)$ of an element, T_i, of B_n to be the number of ones in the n-place binary sequence T_i.

Henceforth, unless otherwise stated, we agree in dealing with an array such as (4) to adopt the following convention:

the leader of each coset shall be taken to be an element of minimal weight in that coset. (5)

Such a table will be called a *standard array*.

Example: B_4 can be developed according to the (4, 2)-alphabet 0000, 1100, 0011, 1111 as follows

0000	1100	0011	1111
1010	0110	1001	0101
1110	0010	1101	0001
1000	0100	1011	0111

(6)

According to (5), however, we should write, for example

0000	1100	0011	1111
1010	0110	1001	0101
0010	1110	0001	1101
1000	0100	1011	0111

(7)

The coset leader of the second coset of (6) can be taken as any element of that row since all are of weight 2. The leader of the third coset, however, should be either 0010 or 0001 since these are of weight one. The leader of the fourth coset should be either 1000 or 0100.

1.5 THE DETECTION SCHEME

Consider now communicating with an (n, k)-alphabet over the symmetric binary channel. When any letter, say A_j, of the alphabet is transmitted, the received sequence can be of any element of B_n. We agree to use the following detector:

if the received element of B_n lies in column i of the array (4), *the detector prints the letter A_i, $i = 1, 2, \cdots, \mu$. The array* (4) *is to be constructed according to the convention* (5). (8)

The following propositions and theorems can be proved concerning signaling with an (n, k)-alphabet and the detection scheme given by (8).

1.6 BEST DETECTOR AND SYMMETRIC SIGNALING

Define the *probability* $\ell_i = \ell(T_i)$ of an element T_i of B_n to be $\ell_i = p^{w_i}q^{n-w_i}$ where p and q are as in (1) and w_i is the weight of T_i. Let

Q_i, $i = 1, 2, \cdots, \mu$ be the sum of the probabilities of the elements in the ith column of the standard array (4).

Proposition 1. The probability that any transmitted letter of the (n, k)-alphabet be produced correctly by the detector is Q_1.

Proposition 2. The equivocation[5] per symbol is

$$H_y(x) = -\frac{1}{n} \sum_{i=1}^{\mu} Q_i \log_2 Q_i$$

Theorem 1. The detector (8) is a maximum likelihood detector. That is, for the given alphabet no other detection scheme has a greater average probability that a transmitted letter be produced correctly by the detector.

Let us return to the geometrical picture of n-place binary sequences as vertices of a unit cube in n-space. The choice of a K-letter, n-place alphabet corresponds to designating K particular vertices as letters. Since the binary sequence corresponding to any vertex can be produced by the channel output, any detector must consist of a set of rules that associates various vertices of the cube with the vertices designated as letters of the alphabet. We assume that every vertex is associated with some letter. The vertices of the cube are divided then into disjoint sets, $W_1, W_2, \cdots, W_K$ where W_i is the set of vertices associated with ith letter of the signaling alphabet. A maximum likelihood detector is characterized by the fact that every vertex in W_i is as close to or closer to the ith letter than to any other letter, $i = 1, 2, \cdots, K$. For group alphabets and the detector (8), this means that no element in the ith column of array (4) is closer to any other A than it is to A_i, $i = 1, 2, \cdots, \mu$.

Theorem 2. Associated with each (n, k)-alphabet considered as a point configuration in Euclidean n-space, there is a group of $n \times n$ orthogonal matrices which is transitive on the letters of the alphabet and which leaves the unit cube invariant. The maximum likelihood sets W_1, $W_2, \cdots W_\mu$ are all geometrically similar.

Stated in loose terms, this theorem asserts that in an (n, k)-alphabet every letter is treated the same. Every two letters have the same number of nearest neighbors associated with them, the same number of next nearest neighbors, etc. The disposition of points in any two W regions is the same.

1.7 GROUP ALPHABETS AND PARITY CHECKS

Theorem 3. Every group alphabet is a systematic[3] code: every systematic code is a group alphabet.[7]

We prefer to use the word "alphabet" in place of "code" since the latter has many meanings. In a *systematic alphabet,* the places in any letter can be divided into two classes: the information places — k in number for an (n, k)-alphabet — and the check positions. All letters have the same information places and the same check places. If there are k information places, these may be occupied by any of the 2^k k-place binary sequences. The entries in the $n - k$ check positions are fixed linear (mod 2) combinations of the entries in the information positions. The rules by which the entries in the check places are determined are called *parity checks.* Examples: for the (4, 2)-alphabet of (6), namely 0000, 1100, 0011, 1111, positions 2 and 3 can be regarded as the information positions. If a letter of the alphabet is the sequence $a_1a_2a_3a_4$, then $a_1 = a_2$, $a_4 = a_3$ are the parity checks determining the check places 1 and 4. For the (5, 3)-alphabet 00000, 10001, 01011, 00111, 11010, 10110, 01100, 11101 places 1, 2, and 3 (numbered from the left) can be taken as the information places. If a general letter of the alphabet is $a_1a_2a_3a_4a_5$, then $a_4 = a_2 \dotplus a_3$, $a_5 = a_1 \dotplus a_2 \dotplus a_3$.

Two group alphabets are called *equivalent* if one can be obtained from the other by a permutation of places. Example: the 7 distinct (3, 2)-alphabets given in (3) separate into three equivalence classes. Alphabets (i), (ii), and (iv) are equivalent; alphabets (iii), (v), (vi), are equivalent; (vii) is in a class by itself.

Proposition 3. Equivalent (n, k)-alphabets have the same probability Q_1 of correct transmission for each letter.

Proposition 4. Every (n, k)-alphabet is equivalent to an (n, k)-alphabet whose first k places are information places and whose last $n - k$ places are determined by parity checks over the first k places.

Henceforth we shall be concerned only with (n, k)-alphabets whose first k places are information places. The parity check rules can then be written

$$a_i = \sum_{j=1}^{k} \gamma_{ij} a_j, \qquad i = k + 1, \cdots, n \tag{9}$$

where the sums are of course mod 2. Here, as before, a typical letter of the alphabet is the sequence $a_1a_2 \cdots a_n$. The γ_{ij} are $k(n - k)$ quantities, zero or one, that serve to define the particular (n, k)-alphabet in question.

1.8 MAXIMUM LIKELIHOOD DETECTION BY PARITY CHECKS

For any element, T, of B_n we can form the sum given on the right of (9). This sum may or may not agree with the symbol in the ith place of

T. If it does, we say T satisfies the ith-place parity check; otherwise T fails the ith-place parity check. When a set of parity check rules (9) is given, we can associate an $(n - k)$-place binary sequence, $R(T)$, with each element T of B_n. We examine each check place of T in order starting with the $(k + 1)$-*st* place of T. We write a zero if a place of T satisfies the parity check; we write a one if a place fails the parity check. The resultant sequence of zeros and ones, written from left to right is $R(T)$. We call $R(T)$ the *parity check sequence* of T. Example: with the parity rules $a_4 = a_2 \dotplus a_3$, $a_5 = a_1 \dotplus a_2 \dotplus a_3$ used to define the (5, 3)-alphabet in the examples of Theorem 3, we find $R(11000) = 10$ since the sum of the entries in the second and third places of 11001 is not the entry of the fourth place and since the sum of $a_1 = 1$, $a_2 = 1$, and $a_3 = 0$ is $0 = a_5$.

Theorem 4. Let $I, A_2, \cdots A_\mu$ be an (n, k)-alphabet. Let $R(T)$ be the parity check sequence of an element T of B_n formed in accordance with the parity check rules of the (n, k)-alphabet. Then $R(T_1) = R(T_2)$ if and only if T_1 and T_2 lie in the same row of array (4). The coset leaders can be ordered so that $R(S_i)$ is the binary symbol for the integer $i - 1$.

As an example of Theorem 4 consider the (4, 2)-alphabet shown with its cosets below

0000	1011	0101	1110
0100	1111	0001	1010
0010	1001	0111	1100
1000	0011	1101	0110

The parity check rules for this alphabet are $a_3 = a_1$, $a_4 = a_1 \dotplus a_2$. Every element of the second row of this array satisfies the parity check in the third place and fails the parity check in the 4th place. The parity check sequence for the second row is 01. The parity check for the third row is 10, and for the fourth row 11. Since every letter of the alphabet satisfies the parity checks, the parity check sequence for the first row is 00. We therefore make the following association between parity check sequences and coset leaders

$$
\begin{aligned}
00 &\rightarrow 0000 = S_1\\
01 &\rightarrow 0100 = S_2\\
10 &\rightarrow 0010 = S_3\\
11 &\rightarrow 1000 = S_4
\end{aligned}
$$

1.9 INSTRUMENTING A GROUP ALPHABET

Proposition 4 attests to the ease of the encoding operation involved with the use of an (n, k)-alphabet. If the original message is presented as a long sequence of zeros and ones, the sequence is broken into blocks of length k places. Each block is used as the first k places of a letter of the signaling alphabet. The last n-k places of the letter are determined by fixed parity checks over the first k places.

Theorem 4 demonstrates the relative ease of instrumenting the maximum likelihood detector (8) for use with an (n, k)-alphabet. When an element T of B_n is received at the channel output, it is subjected to the n-k parity checks of the alphabet being used. This results in a parity check sequence $R(T)$. $R(T)$ serves to identify a unique coset leader, say S_i. The product S_iT is then formed and produced as the detector output. The probability that this be the correct letter of the alphabet is Q_1.

1.10 BEST GROUP ALPHABETS

Two important questions regarding (n, k)-alphabets naturally arise. What is the maximum value of Q_1 possible for a given n and k and which of the $N(n, k)$ different subgroups give rise to this maximum Q_1? The answers to these questions for general n and k are not known. For many special values of n and k the answers are known. They are presented in Tables II, III and IV, which are explained below.

The probability Q_1 that a transmitted letter be produced correctly by the detector is the sum, $Q_1 = \sum_1^{\nu} \ell(S_i)$ of the probabilities of the coset leaders. This sum can be rewritten as $Q_1 = \sum_{i=0}^{n} \alpha_i p^i q^{n-i}$ where α_i is the number of coset leaders of weight i. One has, of course, $\sum \alpha_i = \nu = 2^{n-k}$ for an (n, k)-alphabet. Also $\alpha_i \leqq \binom{n}{i} = \frac{n!}{i!(n-i)}$! since this is the number of elements of B_n of weight i.

The α_i have a special physical significance. Due to the noise on the channel, a transmitted letter, A_i, of an (n, k)-alphabet will in general be received at the channel output as some element T of B_n different from A_i. If T differs from A_i in s places, i.e., if $w(A_iT) = s$, we say that an s-tuple error has occurred. For a given (n, k)-alphabet, α_i is the number of i-tuple errors which can be corrected by the alphabet in question, $i = 0, 1, 2, \cdots, n$.

Table II gives the α_i corresponding to the largest possible value of Q_1 for a given k and n for $k = 2, 3, \cdots n - 1, n = 4 \cdots, 10$ along with a few other scattered values of n and k. For reference the binomial coefficients $\binom{n}{i}$ are also listed. For example, we find from Table II that the best group alphabet with $2^4 = 16$ letters that uses $n = 10$ places has a probability of correct transmission $Q_1 = q^{10} + 10q^9p + 39q^8p^2 + 14q^7p^3$. The alphabet corrects all 10 possible single errors. It corrects 39 of the possible $\binom{10}{2} = 45$ double errors (second column of Table II) and in addition corrects 14 of the 120 possible triple errors. By adding an additional place to the alphabet one obtains with the best (11, 4)-alphabet an alphabet with 16 letters that corrects all 11 possible single errors and all 55 possible double errors as well as 61 triple errors. Such an alphabet might be useful in a computer representing decimal numbers in binary form.

For each set of α's listed in Table II, there is in Table III a set of parity check rules which determines an (n, k)-alphabet having the given α's. The notation used in Table III is best explained by an example. A (10, 4)-alphabet which realizes the α's discussed in the preceding paragraph can be obtained as follows. Places 1, 2, 3, 4 carry the information. Place 5 is determined to make the mod 2 sum of the entries in places 3, 4, and 5 equal to zero. Place 6 is determined by a similar parity check on places 1, 2, 3, and 6; place 7 by a check on places 1, 2, 4, and 7, etc.

It is a surprising fact that for all cases investigated thus far an (n, k)-alphabet best for a given value of p is uniformly best for all values of p, $0 \leqq p \leqq \frac{1}{2}$. It is of course conjectured that this is true for all n and k.

It is a further (perhaps) surprising fact that the best (n, k)-alphabets are not necessarily those with greatest nearest neighbor distance between letters when the alphabets are regarded as point configurations on the n-cube. For example, in the best (7, 3)-alphabet as listed in Table III, each letter has two nearest neighbors distant 3 edges away. On the other hand, in the (7, 3)-alphabet given by the parity check rules 413, 512, 623, 7123 each letter has its nearest neighbors 4 edges away. This latter alphabet does not have as large a value of Q_1, however, as does the (7, 3)-alphabet listed on Table III.

The cases $k = 0, 1, n - 1, n$ have not been listed in Tables II and III. The cases $k = 0$ and $k = n$ are completely trivial. For $k = 1$, all $n > 1$ the best alphabet is obtained using the parity rule $a_2 = a_3 = \cdots = a_n = a_1$. If $n = 2j$,

$$Q_1 = \sum_0^{j-1} \binom{n}{i} p^i q^{n-i} + \frac{1}{2}\binom{n}{j} p^j q^j. \text{ If } n = 2j + 1,\ Q_1 = \sum_0^{j} \binom{n}{i} p^i q^{n-i}.$$

For $k = n - 1$, $n > 1$, the maximum Q_1 is $Q_1 = q^{n-1}$ and a parity rule for an alphabet realizing this Q_1 is $a_n = a_1$.

If the α's of an (n, k)-alphabet are of the form $\alpha_i = \binom{n}{i}$, $i = 0, 1,$

TABLE II — PROBABILITY OF NO ERROR WITH BEST ALPHABETS, $Q_1 = \sum \alpha_i p^i q^{n-i}$

	i	$\binom{n}{i}$	$k = 2$ a_i	$k = 3$ a_i	$k = 4$ a_i	$k = 5$ a_i	$k = 6$ a_i	$k = 7$ a_i	$k = 8$ a_i	$k = 9$ a_i	$k = 10$ a_i
$n = 4$	0	1	1								
	1	4	3								
	0	1	1	1							
$n = 5$	1	5	5	3							
	2	10	2								
	0	1	1	1	1						
$n = 6$	1	6	6	6	3						
	2	15	9	1							
	0	1	1	1	1	1					
$n = 7$	1	7	7	7	7	3					
	2	21	18	8							
	3	25	6								
	0	1	1	1	1	1	1				
$n = 8$	1	8	8	8	8	7	3				
	2	28	28	20	7						
	3	56	27	3							
	0	1	1	1	1	1	1	1			
	1	9	9	9	9	9	7	3			
$n = 9$	2	36	36	33	22	6					
	3	84	64	21							
	4	126	18								
	0	1	1	1	1	1	1	1	1		
	1	10	10	10	10	10	10	7	3		
$n = 10$	2	45	45	45	39	21	5				
	3	120	110	64	14						
	4	210	90	8							
	0	1	1	1	1		1	1	1	1	
	1	11	11	11	11		11	11	7	3	
$n = 11$	2	55	55	55	55		20	4			
	3	165	165	126	61						
	4	330	226	63							
	5	462	54								
	0	1	1	1				1	1	1	1
	1	12	12	12				12	12	7	3
$n = 12$	2	66	66	66				19	3		
	3	220	220	200							
	4	495	425	233							
	5	792	300								

2, $\cdots$, j, $\alpha_{j+1} = r$ some integer, $\alpha_{j+2} = \alpha_{j+3} = \cdots = \alpha_n = 0$, then there does not exist a 2^k-letter, n-place alphabet of any sort better than the given (n, k)-alphabet. It will be observed that many of the α's of Table II are of this form. It can be shown that

Proposition 5 if $n + \binom{n-k}{2} + \binom{n-k}{3} \geqq 2^{n-k} - 1$ there exists no 2^k-letter, n-place alphabet better than the best (n, k)-alphabet.

When the inequality of proposition 5 holds the α's are either $\alpha_0 = 1$, $\alpha_1 = 2^{n-k} - 1$, all other $\alpha = 0$; or $\alpha_0 = 1$, $\alpha_1 = \binom{n}{1}$, $\alpha_2 = 2^{n-k} - 1 - \binom{n}{1}$ all other $\alpha = 0$; or the trivial $\alpha_0 = 1$ all other $\alpha = 0$ which holds when $k = n$. The region of the $n - k$ plane for which it is known that (n, k)-alphabets cannot be excelled by any other is shown in Table IV.

1.11 A DETAILED EXAMPLE

As an example of the use of (n, k)-alphabets consider the not unrealistic case of a channel with $p = 0.001$, i.e., on the average one binary digit per thousand is received incorrectly. Suppose we wish to transmit messages using 32 different letters. If we encode the letters into the 32 5-place binary sequences and transmit these sequences without further encoding, the probability that a received letter be in error is $1 - (1 - p)^5 = 0.00449$. If the best (10, 5)-alphabet as shown in Tables II and III is used, the probability that a letter be wrong is $1 - Q_1 = 1 - q^{10} - 10q^9p - 21q^8p^2 = 24p^2 - 72p^3 + \cdots = 0.000024$. Thus by reducing the signaling rate by ½, a more than *one hundredfold* reduction in probability of error is accomplished.

A (10, 5)-alphabet to achieve these results is given in Table III. Let a typical letter of the alphabet be the 10-place sequence of binary digits $a_1a_2 \cdots a_9a_{10}$. The symbols $a_1a_2a_3a_4a_5$ carry the information and can be any of 32 different arrangements of zeros and ones. The remaining places are determined by

$$\begin{aligned}
a_6 &= a_1 \dot{+} a_3 \dot{+} a_4 \dot{+} a_5 \\
a_7 &= a_1 \dot{+} a_2 \dot{+} a_4 \dot{+} a_5 \\
a_8 &= a_1 \dot{+} a_2 \dot{+} a_3 \dot{+} a_5 \\
a_9 &= a_1 \dot{+} a_2 \dot{+} a_3 \dot{+} a_4 \\
a_{10} &= a_1 \dot{+} a_2 \dot{+} a_3 \dot{+} a_4 \dot{+} a_5
\end{aligned}$$

To design the detector for this alphabet, it is first necessary to determine the coset leaders for a standard array (4) formed for this alphabet.

TABLE III — PARITY CHECK RULES FOR BEST ALPHABETS

	$k = 2$	$k = 3$	$k = 4$	$k = 5$	$k = 6$	$k = 7$	$k = 8$	$k = 9$	$k = 10$
$n = 4$	3 2 4 1 2								
$n = 5$	3 1 2 4 2 5 1	4 1 2 5 1 3							
$n = 6$	3 2 4 1 2 5 1 6 1	4 1 2 5 1 3 6 2 3	5 1 2 3 6 1 2 4						
$n = 7$	3 1 4 1 5 1 6 1 2 7 2	4 1 3 5 1 2 6 1 2 3 7 1 2 3	5 1 3 4 6 1 2 4 7 1 2 3	6 1 7 1					
$n = 8$	3 1 4 1 5 2 6 2 7 1 2 8 1 2	4 1 5 1 2 6 1 3 7 2 3 8 1 2 3	5 1 3 4 6 1 2 4 7 1 2 3 8 1 2 3 4	6 1 3 4 7 1 2 4 8 1 2 3	7 1 8 1				
$n = 9$	3 1 4 1 5 1 6 2 7 2 8 1 2 9 1 2	4 1 5 2 6 1 2 7 1 3 8 2 3 9 1 2 3	5 1 3 4 6 1 2 4 7 1 2 3 8 1 2 3 9 1 2 3	6 1 3 4 5 7 1 2 4 5 8 1 2 3 5 9 1 2 3 4	7 1 3 4 8 1 2 4 9 1 2 3	8 1 9 1			

$n = 10$	3 1 4 1 5 1 6 2 7 2 8 1 2 9 1 2 10 1 2	4 1 5 2 6 3 7 1 2 8 1 3 9 2 3 10 1 2 3	5 3 4 6 1 2 3 7 1 2 4 8 1 3 4 9 2 3 4 10 1 2 3 4	6 1 3 4 5 7 1 2 4 5 8 1 2 3 5 9 1 2 3 4 10 1 2 3 4 5	7 1 3 4 5 8 1 2 4 5 9 1 2 3 5 6 10 1 2 3 4 6	8 1 3 4 9 1 2 4 10 1 2 3	9 1 10 1		
$n = 11$	3 1 4 1 5 1 6 2 7 2 8 2 9 1 2 10 1 2 11 1 2	4 3 5 3 6 2 7 1 3 8 1 3 9 1 2 10 1 2 3 11 1 2 3	5 1 3 6 2 4 7 1 4 8 2 3 9 1 3 4 10 2 3 4 11 1 2 3 4		7 1 3 4 5 6 8 1 2 4 5 6 9 1 2 3 5 6 10 1 2 3 4 6 11 1 2 3 4 5	8 1 3 4 5 9 1 2 4 5 7 10 1 2 3 5 6 11 1 2 3 4 6 7	9 1 3 4 10 1 2 4 11 1 2 3	10 1 11 1	
$n = 12$	3 1 4 1 5 1 6 1 7 2 8 2 9 2 10 2 11 1 2 12 1 2	4 1 5 2 6 3 7 1 2 8 1 2 9 1 3 10 2 3 11 1 2 3 12 1 2 3				8 1 3 4 5 6 9 1 2 4 5 6 10 1 2 3 5 6 7 11 1 2 3 4 6 7 12 1 2 3 4 5 7	9 1 2 3 5 6 7 8 10 1 2 3 4 6 11 1 2 4 5 7 12 1 3 4 5 8	10 1 2 3 11 1 2 4 12 1 3 4	11 1 12 1

Table IV — Region of the n-k Plane for Which it is Known that (n, k)-Alphabets Cannot Be Excelled

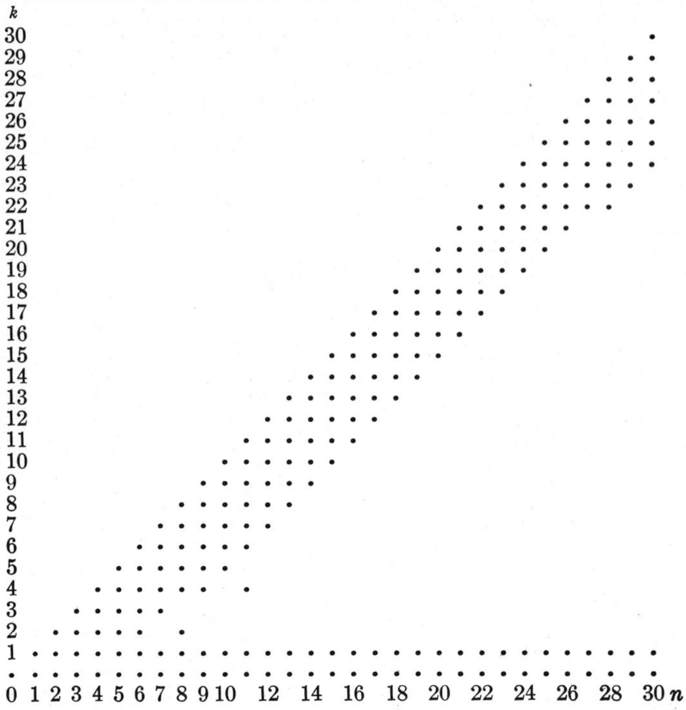

This can be done by a variety of special methods which considerably reduce the obvious labor of making such an array. A set of best S's along with their parity check symbols is given in Table V.

A maximum likelihood detector for the (10, 5)-alphabet in question forms from each received sequence $b_1 b_2 \cdots b_{10}$ the parity check symbol $c_1 c_2 c_3 c_4 c_5$ where

$$\begin{aligned} c_1 &= b_6 \dotplus b_1 \dotplus b_3 \dotplus b_4 \dotplus b_5 \\ c_2 &= b_7 \dotplus b_1 \dotplus b_2 \dotplus b_4 \dotplus b_5 \\ c_3 &= b_8 \dotplus b_1 \dotplus b_2 \dotplus b_3 \dotplus b_5 \\ c_4 &= b_9 \dotplus b_1 \dotplus b_2 \dotplus b_3 \dotplus b_4 \\ c_5 &= b_{10} \dotplus b_1 \dotplus b_2 \dotplus b_3 \dotplus b_4 \dotplus b_5 \end{aligned}$$

According to Table V, if $c_1 c_2 c_3 c_4 c_5$ contains less than three ones, the detector should brint $b_1 b_2 b_3 b_4 b_5$. The detector should print $(b_1 \dotplus 1) b_2 b_3 b_4 b_5$ if the parity check sequence $c_1 c_2 c_3 c_4 c_5$ is either 11111 or 11110; the de-

TABLE V — COSET LEADERS AND PARITY CHECK SEQUENCES FOR (10,5)-ALPHABET

$c_1c_2c_3c_4c_5$ ↔	S	$c_1c_2c_3c_4c_5$ ↔	S
00000	0000000000	11100	0000100001
10000	0000010000	11010	0001000001
01000	0000001000	11001	0001000010
00100	0000000100	10110	0010000001
00010	0000000010	10101	0010000010
00001	0000000001	10011	0010000100
11000	0000011000	01110	0100000001
10100	0000010100	01101	0100000010
10010	0000010010	01011	0100000100
10001	0000010001	00111	0100001000
01100	0000001100	11110	1000000001
01010	0000001010	11101	0000100000
01001	0000001001	11011	0001000000
00110	0000000110	10111	0010000000
00101	0000000101	01111	0100000000
00011	0000000011	11111	1000000000

tector should print $b_1(b_2 \dotplus 1)b_3b_4b_5$ if the parity check sequence is 01111, 00111, 01011, 01101, or 01110; the detector should print $b_1b_2(b_3 \dotplus 1)b_4b_5$ if the parity check sequence is 10111, 10011, 10101, or 10110; the detector should print $b_1b_2b_3(b_4 \dotplus 1)b_5$ if the parity check sequence is 11011, 11001, 11010; and finally the detector should print $b_1b_2b_3b_4(b_5 \dotplus 1)$ if the parity check sequence is 11101 or 11100.

Simpler rules of operation for the detector may possibly be obtained by choice of a different set of S's in Table V. These quantities in general are not unique. Also there may exist non-equivalent alphabets with simpler detector rules that achieve the same probability of error as the alphabet in question.

PART II — ADDITIONAL THEORY AND PROOFS OF THEOREMS OF PART I

2.1 THE ABSTRACT GROUP C_n

It will be helpful here to say a few more words about B_n, the group of n-place binary sequences under the operation of addition mod 2. This group is simply isomorphic with the abstract group C_n generated by n commuting elements of order two, say $a_1, a_2, \cdots, a_n$. Here $a_ia_j = a_ja_i$ and $a_i^2 = I$, $i, j = 1, 2, \cdots, n$, where I is the identity for the group. The eight distinct elements of C_3 are, for example, I, a_1, a_2, a_3, a_1a_2, a_1a_3, a_2a_3, $a_1a_2a_3$. The group C_n is easily seen to be isomorphic with the n-fold direct product of the group C_1 with itself.

It is a considerable saving in notation in dealing with C_n to omit the symbol "a" and write only the subscripts. In this notation for example, the elements of C_4 are I, 1, 2, 3, 4, 12, 13, 14, 23, 24, 34, 123, 124, 134, 234, 1234. The product of two or more elements of C_n can readily be written down. Its symbol consists of those numerals that occur an odd number of times in the collection of numerals that comprise the symbols of the factors. Thus, $(12)(234)(123) = 24$.

The isomorphism between C_n and B_n can be established in many ways. The most convenient way, perhaps, is to associate with the element $i_1 i_2 i_3 \cdots i_k$ of C_n the element of B_n that has ones in places $i_1, i_2, \cdots, i_k$ and zeros in the remaining $n - k$ places. For example, one can associate 124 of C_4 with 1101 of B_4; 14 with 1001, etc. In fact, the numeral notation afforded by this isomorphism is a much neater notation for B_n than is afforded by the awkward strings of zeros and ones. There are, of course, other ways in which elements of C_n can be paired with elements of B_n so that group multiplication is preserved. The collection of all such "pairings" makes up the group of automorphisms of C_n. This group of automorphisms of C_n is isomorphic with the group of non-singular linear homogenous transformations in a field of characteristic 2.

An element T of C_n is said to be *dependent* upon the set of elements $T_1, T_2, \cdots, T_j$ of C_n if T can be expressed as a product of some elements of the set $T_1, T_2, \cdots, T_j$; otherwise, T is said to be *independent* of the set. A set of elements is said to be independent if no member can be expressed solely in terms of the other members of the set. For example, in C_8, 1, 2, 3, 4 form a set of independent elements as do likewise 2357, 12357, 14. However, 135 depends upon 145, 3457, 57 since $135 = (145)(3457)(57)$. Clearly any set of n independent elements of C_n can be taken as generators for the group. For example, all possible products formed of 12, 123, and 23 yield the elements of C_3.

Any k independent elements of C_n serve as generators for a subgroup of order 2^k. The subgroup so generated is clearly isomorphic with C_k. All subgroups of C_n of order 2^k can be obtained in this way.

The number of ways in which k independent elements can be chosen from the 2^n elements of C_n is

$$F(n, k) = (2^n - 2^0)(2^n - 2^1)(2^n - 2^2) \cdots (2^n - 2^{k-1})$$

For, the first element can be chosen in $2^n - 1$ ways (the identity cannot be included in a non-trivial set of independent elements) and the second element can be chosen in $2^n - 2$ ways. These two elements determine a subgroup of order 2^2. The third element can be chosen as any element of the remaining $2^n - 2^2$ elements. The 3 elements chosen determine a subgroup of order 2^3. A fourth independent element can be chosen as any of the remaining $2^n - 2^3$ elements, etc.

Each set of k independent elements serves to generate a subgroup of order 2^k. The quantity $F(n, k)$ is not, however, the number of distinct subgroups of C_n of this order, for, a given subgroup can be obtained from many different sets of generators. Indeed, the number of different sets of generators that can generate a given subgroup of order 2^k of C_n is just $F(k, k)$ since any such subgroup is isomorphic with C_k. Therefore the number of subgroups of C_n of order 2^k is $N(n, k) = F(n, k)/F(k, k)$ which is (2). A simple calculation gives $N(n, k) = N(n, n - k)$.

2.2 PROOF OF PROPOSITIONS 1 AND 2

After an element A of B_n has been presented for transmission over a noisy binary channel, an element T of B_n is produced at the channel output. The element $U = AT$ of B_n serves as a record of the noise during the transmission. U is an n-place binary sequence with a one at each place altered in A by the noise. The channel output, T, is obtained from the input A by multiplication by U: $T = UA$. For channels of the sort under consideration here, the probability that U be any particular element of B_n of weight w is $p^w q^{n-w}$.

Consider now signaling with a particular (n, k)-alphabet and consider the standard array (4) of the alphabet. If the detection scheme (8) is used, a transmitted letter A_i will be produced without error if and only if the received symbol is of the form $S_j A_i$. That is, there will be no error only if the noise in the channel during the transmission of A_i is represented by one of the coset leaders. (This applies for $i = 1, 2, \cdots, \mu = 2^k$). The probability of this event is Q_1 (Proposition 1, Section 1.6). The convention (5) makes Q_1 as large as is possible for the given alphabet.

Let X refer to transmitted letters and let Y refer to letters produced by the detector. We use a vertical bar to denote conditions when writing probabilities. The quantity to the right of the bar is the condition. We suppose the letters of the alphabet to be chosen independently with equal probability 2^{-k}.

The equivocation $h(X \mid Y)$ obtained when using an (n, k)-alphabet with the detector (8) can most easily be computed from the formula

$$h(X \mid Y) = h(X) - h(Y) + h(Y \mid X) \qquad (10)$$

The entropy of the source is $h(X) = k/n$ bits per symbol. The probability that the detector produce A_j when A_i was sent is the probability that the noise be represented by $A_i A_j S_\ell$, $\ell = 1, 2, \cdots, \nu$. In symbols,

$$Pr(Y \to A_j \mid X \to A_i) = \sum_l Pr(N \to A_iA_jS_l) = Q(A_iA_j)$$

where $Q(A_i)$ is the sum of the probabilities of the elements that are in the same column as A_i in the standard array. Therefore

$$Pr(Y \to A_j) = \sum_i Pr(Y \to A_j \mid X \to A_i)Pr(X \to A_i) = \frac{1}{2^k} \sum_i Q(A_iA_j)$$

$$= \frac{1}{2^k}, \qquad \text{since} \sum_i Q(A_iA_j) = \sum_i Q(A_i) = 1.$$

This last follows from the group property of the alphabet. Therefore

$$h(Y) = -\frac{1}{n} \sum Pr(Y \to A_j) \log Pr(Y \to A_j) = \frac{k}{n} \text{ bits/symbol.}$$

It follows then from (10) that

$$h(X \mid Y) = h(Y \mid X)$$

The computation of $h(Y \mid X)$ follows readily from its definition

$$h(Y \mid X) = \sum_i Pr(X \to A_i)h(Y \mid X \to A_i)$$

$$= -\sum_{ij} Pr(X \to A_i)Pr(Y \to A_j \mid X \to A_i)$$

$$\log Pr(Y \to A_j \mid X \to A_i)$$

$$= -\frac{1}{2^k} \sum_{ij} \sum_l Pr(N \to A_iS_lA_j) \log \sum_m Pr(N \to A_iS_mA_j)$$

$$= -\frac{1}{2^k} \sum_{ij} Q(A_iA_j) \log Q(A_iA_j)$$

$$= -\sum_i Q(A_i) \log Q(A_i)$$

Each letter is n binary places. Proposition 2, then follows.

2.3 DISTANCE AND THE PROOF OF THEOREM 1

Let A and B be two elements of B_n. We define the *distance*, $d(A, B)$, between A and B to be the weight of their product,

$$d(A, B) = w(AB) \tag{11}$$

The distance between A and B is the number of places in which A and B differ and is just the "Hamming distance."[1] In terms of the n-cube, $d(A, B)$ is the minimum number of edges that must be traversed to go from vertex A to vertex B. The distance so defined is a monotone function of the Euclidean distance between vertices.

It follows from (11) that if C is any element of B_n then

$$d(A, B) = d(AC, BC) \tag{12}$$

This fact shows the detection scheme (8) to be a maximum likelihood detector. By definition of a standard array, one has

$$d(S_i, I) \leqq d(S_iA_j, I) \qquad \text{for all } i \text{ and } j$$

The coset leaders were chosen to make this true. From (12),

$$d(S_i, I) = d(S_iA_mS_i, I\,A_mS_i) = d(S_iA_m, A_m)$$

$$d(S_iA_j, I) = d(S_iA_jS_iA_m, I\,S_iA_m) = d(A_jA_m, S_iA_m)$$

$$= d(S_iA_m, A_\ell)$$

where $A_\ell = A_jA_m$. Substituting these expressions in the inequality above yields

$$d(S_iA_m, A_m) \leqq d(S_iA_m, A_\ell) \qquad \text{for all } i, m, \ell$$

This equation says that an arbitrary element in the array (4) is at least as close to the element at the top of its column as it is to any other letter of the alphabet. This is the maximum likelihood property.

2.4 PROOF OF THEOREM 2

Again consider an (n, k)-alphabet as a set of vertices of the unit n-cube. Consider also n mutually perpendicular hyperplanes through the centroid of the cube parallel to the coordinate planes. We call these planes "symmetry planes of the cube" and suppose the planes numbered in accordance with the corresponding parallel coordinate planes.

The reflection of the vertex with coordinates $(a_1, a_2, \cdots, a_i, \cdots, a_n)$ in symmetry plane i yields the vertex of the cube whose coordinates are $(a_1, a_2, \cdots, a_i \dotplus 1, \cdots, a_n)$. More generally, reflecting a given vertex successively in symmetry planes $i, j, k, \cdots$ yields a new vertex whose coordinates differ from the original vertex precisely in places $i, j, k \cdots$. Successive reflections in hyperplanes constitute a transformation that leaves distances between points unaltered and is therefore a "rotation." The rotation obtained by reflecting successively in symmetry planes i, j, k, etc. can be represented by an n-place symbol having a one in places i, j, k, etc. and a zero elsewhere.

We now regard a given (n, k)-alphabet as generated by operating on the vertex $(0, 0, \cdots, 0)$ of the cube with a certain collection of 2^k rotation operators. The symbols for these operators are identical with the sequences of zeros and ones that form the coordinates of the 2^k points. It is readily seen that these rotation operators form a group which is transitive on the letters of the alphabet and which leave the unit cube invariant. Theorem 2 then follows.

Theorem 2 also follows readily from consideration of the array (4). For example, the maximum likelihood region associated with I is the set of points $I, S_2, S_3, \cdots, S_\nu$. The maximum likelihood region associated with A_i is the set of points $A_i, A_iS_2, A_iS_3, \cdots, A_iS_\nu$. The rotation (successive reflections in symmetry planes of the cube) whose symbol is the same as the coordinate sequence of A_i sends the maximum likelihood region of I into the maximum likelihood region of A_i, $i = 1, 2, \cdots, \mu$.

2.5 PROOF OF THEOREM 3

That every systematic alphabet is a group alphabet follows trivially from the fact that the sum mod 2 of two letters satisfying parity checks is again a letter satisfying the parity checks. The totality of letters satisfying given parity checks thus constitutes a finite group.

To prove that every group alphabet is a systematic code, consider the letters of a given (n, k)-alphabet listed in a column. One obtains in this way a matrix with 2^k rows and n columns whose entries are zeros and ones. Because the rows are distinct and form a group isomorphic to C_k, there are k linearly independent rows (mod 2) and no set of more than k independent rows. The rank of the matrix is therefore k. The matrix therefore possesses k linearly independent (mod 2) columns and the remaining $n - k$ columns are linear combinations of these k. Maintaining only these k linearly independent columns, we obtain a matrix of k columns and 2^k rows with rank k. This matrix must, therefore, have k linearly independent rows. The rows, however, form a group under mod 2 addition and hence, since k are linearly independent, all 2^k rows must be distinct. The matrix contains only zeros and ones as entries; it has 2^k distinct rows of k entries each. The matrix must be a listing of the numbers from 0 to $2^k - 1$ in binary notation. The other $n - k$ columns of the original matrix considered are linear combinations of the columns of this matrix. This completes the proof of Theorem 3 and Proposition 4.

2.6 PROOF OF THEOREM 4

To prove Theorem 4 we first note that the parity check sequence of the product of two elements of B_n is the mod 2 sum of their separate parity check sequences. It follows then that all elements in a given coset have the same parity check sequence. For, let the coset be S_i, S_iA_2, S_iA_3, $\cdots$ S_iA_μ. Since the elements I, A_2, A_3, $\cdots$, A_μ all have parity check sequence $00 \cdots 0$, all elements of the coset have parity check $R(S_i)$.

In the array (4) there are 2^{n-k} cosets. We observe that there are 2^{n-k} elements of B_n that have zeros in their first k places. These elements have parity check symbols identical with the last $n - k$ places of their symbols. These elements therefore give rise to 2^{n-k} different parity check symbols. The elements must be distributed one per coset. This proves Theorem 4.

2.7 PROOF OF PROPOSITION 5

If

$$n \geqq 2^{n-k} - \binom{n-k}{2} - \binom{n-k}{3} - 1$$

we can explicity exhibit group alphabets having the property mentioned in the paragraph preceding Proposition 5. The notation of the demonstration is cumbersome, but the idea is relatively simple.

We shall use the notation of paragraph 2.1 for elements of B_n, i.e., an element of B_n will be given by a list of integers that specify what places of the sequence for the element contain ones. It will be convenient furthermore to designate the first k places of a sequence by the integers $1, 2, 3, \cdots, k$ and the remaining $n - k$ places by the "integers" $1', 2', 3', \cdots, \ell'$, where $\ell = n - k$. For example, if $n = 8$, $k = 5$, we have

$$10111010 \leftrightarrow 13452'$$
$$10000100 \leftrightarrow 11'$$
$$00000101 \leftrightarrow 1'3'$$

Consider the group generated by the elements $1', 2', 3', \cdots, \ell'$, i.e. the 2^ℓ elements $I, 1', 2', \cdots, \ell', 1'2', 1'3', \cdots, 1'2'3' \cdots \ell'$. Suppose these elements listed according to decreasing weight (say in decreasing order when regarded as numbers in the decimal system) and numbered consecutively. Let B_i be the ith element in the list. Example: if $\ell = 3$, $B_1 = 1'2'3'$, $B_2 = 2'3'$, $B_3 = 1'3'$, $B_4 = 1'2'$, $B_5 = 3'$, $B_6 = 2'$, $B_7 = 1'$.

Consider now the (n, k)-alphabet whose generators are

$$1B_1, 2B_2, 3B_3, \cdots, kB_k$$

We assert that if

$$n \geqq 2^{n-k} - \binom{n-k}{2} - \binom{n-k}{3} - 1$$

this alphabet is as good as any other alphabet of 2^k letters and n places.

In the first place, we observe that every letter of this (n, k)-alphabet (except I) has unprimed numbers in its symbols. It follows that each of the 2^ℓ letters $I, 1', 2', \cdots, \ell', 1'2', \cdots, 1'2' \cdots \ell'$ occurs in a different coset of the given (n, k)-alphabet. For, if two of these letters appeared in the same coset, their product (which contains only primed numbers) would have to be a letter of the (n, k) alphabet. This is impossible since every letter of the (n, k) alphabet has unprimed numbers in its symbol. Since there are precisely 2^ℓ cosets we can designate a coset by the single element of the list $B_1, B_2, \cdots, B_{2^\ell} = I$ which appears in the coset.

We next observe that the condition

$$n \geqq 2^{n-k} - \binom{n-k}{2} - \binom{n-k}{3} - 1$$

guarantees that B_{k+1} is of weight 3 or less. For, the given condition is equivalent to

$$k \geqq 2^\ell - \binom{\ell}{0} - \binom{\ell}{1} - \binom{\ell}{2} - \binom{\ell}{3}$$

We treat several cases depending on the weight of B_{k+1}.

If B_{k+1} is of weight 3, we note that for $i = 1, 2, \cdots, k$, the coset containing B_i also contains an element of weight one, namely the element i obtained as the product of B_i with the letter iB_i of the given (n, k)-alphabet. Of the remaining $(2^\ell - k)$ B's, one is of weight zero, ℓ are of weight one, $\binom{\ell}{2}$ are of weight 2 and the remaining are of weight 3. We have, then $\alpha_0 = 1$, $\alpha_1 = \ell + k = n$. Now every B of weight 4 occurs in the list of generators $1B_1, 2B_2, \cdots, kB_k$. It follows that on multiplying this list of generators by any B of weight 3, at least one element of weight two will result. (E.g., $(1'2'3')(j1'2'3'4') = j4'$) Thus every coset with a B of weight 2 or 3 contains an element of weight 2 and $\alpha_2 = 2^\ell - \alpha_0 - \alpha_1$.

The argument in case B_{k+1} is of weight two or one is similar.

2.8 MODULAR REPRESENTATIONS OF C_n

In order to explain one of the methods used to obtain the best (n, k)-alphabets listed in Tables II and III, it is necessary to digress here to present additional theory.

It has been remarked that every (n, k)-alphabet is isomorphic with C_k. Let us suppose the elements of C_k listed in a column starting with I and proceeding in order I, 1, 2, 3, $\cdots$, k, 12, 13, $\cdots$, $(k - 1)k$, 123, $\cdots \cdots$, $123 \cdots k$. The elements of a given (n, k)-alphabet can be paired off with these abstract elements so as to preserve group multiplication. This can be done in many different ways. The result is a matrix with elements zero and one with n columns and 2^k rows, these latter being labelled by the symbols I, 1, 2, $\cdots$ etc. What can be said about the columns of this matrix? How many different columns are possible when all (n, k)-alphabets and all methods of establishing isomorphism with C_k are considered?

In a given column, once the entries in rows 1, 2, $\cdots$, k are known, the entire column is determined by the group property. There are therefore only 2^k possible different columns for such a matrix. A table showing these 2^k possible columns of zeros and ones will be called a *modular representation* table for C_k. An example of such a table is shown for $k = 4$ in Table VI.

It is clear that the columns of a modular representation table can also be labelled by the elements of C_k, and that group multiplication of these column labels is isomorphic with mod 2 addition of the columns. The table is a symmetric matrix. The element with row label A and column label B is one if the symbols A and B have an odd number of different numerals in common and is zero otherwise.

Every (n, k)-alphabet can be made from a modular representation table by choosing n columns of the table (with possible repetitions) at least k of which form an independent set.

TABLE VI — MODULAR REPRESENTATION TABLE FOR GROUP C_4

	I	1	2	3	4	12	13	14	23	24	34	123	124	134	234	1234
I	0	0	0	0	0	0	0	0	0	0	0	0	0	0	0	0
1	0	1	0	0	0	1	1	1	0	0	0	1	1	1	0	1
2	0	0	1	0	0	1	0	0	1	1	0	1	1	0	1	1
3	0	0	0	1	0	0	1	0	1	0	1	1	0	1	1	1
4	0	0	0	0	1	0	0	1	0	1	1	0	1	1	1	1
12	0	1	1	0	0	0	1	1	1	1	0	0	0	1	1	0
13	0	1	0	1	0	1	0	1	1	0	1	0	1	0	1	0
14	0	1	0	0	1	1	1	0	0	1	1	1	0	0	1	0
23	0	0	1	1	0	1	1	0	0	1	1	0	1	1	0	0
24	0	0	1	0	1	1	0	1	1	0	1	1	0	1	0	0
34	0	0	0	1	1	0	1	1	1	1	0	1	1	0	0	0
123	0	1	1	1	0	0	0	1	0	1	1	1	0	0	0	1
124	0	1	1	0	1	0	1	0	1	0	1	0	1	0	0	1
134	0	1	0	1	1	1	0	0	1	1	0	0	0	1	0	1
234	0	0	1	1	1	1	1	1	0	0	0	0	0	0	1	1
1234	0	1	1	1	1	0	0	0	0	0	0	1	1	1	1	0

We henceforth exclude consideration of the column I of a modular representation table. Its inclusion in an (n, k)-alphabet is clearly a waste of 1 binary digit.

It is easy to show that every column of a modular representation table for C_k contains exactly 2^{k-1} ones. Since an (n, k)-alphabet is made from n such columns the alphabet contains a total of $n2^{k-1}$ ones and we have

Proposition 6. The weights of an (n, k)-alphabet form a partition of $n2^{k-1}$ into $2^k - 1$ non-zero parts, each part being an integer from the set $1, 2, \cdots, n$.

The identity element always has weight zero, of course.

It is readily established that the product of two elements of even weight is again an element of even weight as is the product of two elements of odd weight. The product of an element of even weight with an element of odd weight yields an element of odd weight.

The elements of even weight of an (n, k)-alphabet form a subgroup and the preceding argument shows that this subgroup must be of order 2^k or 2^{k-1}. If the group of even elements is of order 2^{k-1}, then the collection of even elements is a possible $(n, k - 1)$-alphabet. This $(n, k - 1)$ alphabet may, however, contain the column I of the modular representation table of C_{k-1}. We therefore have

Proposition 7. The partition of Proposition 6 must be either into $2^k - 1$ even parts or else into 2^{k-1} odd parts and $2^{k-1} - 1$ even parts. In the latter case, the even parts form a partition of $\alpha 2^{k-2}$ where α is some integer of the set $k - 1, k, \cdots, n$ and each of the parts is an integer from the set $1, 2, \cdots, n$.

2.9 THE CHARACTERS OF C_k

Let us replace the elements of B_n (each of which is a sequence of zeros and ones) by sequences of $+1$'s and -1's by means of the following substitution

$$\begin{aligned} 0 &\leftrightarrow 1 \\ 1 &\leftrightarrow -1. \end{aligned} \tag{13}$$

The multiplicative properties of elements of B_n can be preserved in this new notation if we define the product of two $+1, -1$ symbols to be the symbol whose ith component is the ordinary product of the ith components of the two factors. For example, 1011 and 0110 become respectively $-11\ -1\ -1$ and $1\ -1\ -11$. We have

$$(-11\ -1\ -1)(1\ -1\ -11) = (-1\ -11\ -1)$$

corresponding to the fact that

$$(1011)\ (0110) = (1101)$$

If the $+1$, -1 symbols are regarded as shorthand for diagonal matrices, so that for example

$$-11\ -1\ -1 \leftrightarrow \begin{vmatrix} -1 & 0 & 0 & 0 \\ 0 & 1 & 0 & 0 \\ 0 & 0 & -1 & 0 \\ 0 & 0 & 0 & -1 \end{vmatrix}$$

then group multiplication corresponds to matrix multiplication.

(While much of what follows here can be established in an elementary way for the simple group at hand, it is convenient to fall back upon the established general theory of group representations[8] for several propositions.

The substitution (13) converts a modular representation table (column I included) into a square array of $+1$'s and -1's. Each column (or row) of this array is clearly an irreducible representation of C_k. Since C_k is Abelian it has precisely 2^k irreducible representations each of degree one. These are furnished by the converted modular table. This table also furnishes then the characters of the irreducible representations of C_k and we refer to it henceforth as a *character table.*

Let $\chi^\alpha(A)$ be the entry of the character table in the row labelled A and column labelled α. The orthogonality relationship for characters gives

$$\sum_{A \subset C_k} \chi^\alpha(A)\chi^\beta(A) = 2^k\delta_{\alpha\beta}$$

$$\sum_{\alpha \subset C_k} \chi^\alpha(A)\chi^\alpha(B) = 2^k\delta_{AB}$$

where δ is the usual Kronecker symbol. In particular

$$\sum_{A \subset C_k} \chi^I(A)\chi^\beta(A) = \sum_{A \subset C_k} \chi^\beta(A) = 0, \qquad \beta \neq I$$

Since each $\chi^\beta(A)$ is $+1$ or -1, these must occur in equal numbers in any column $\beta \neq I$. This implies that each column except I of the modular representation table contains 2^{k-1} ones, a fact used earlier.

Every matrix representation of C_k can be reduced to its irreducible components. If the trace of the matrix representing the element A in an arbitrary matrix representation of C_k is $\chi(A)$, then this representation contains the irreducible representation having label β in the character table d_β times where

$$d_\beta = \frac{1}{2^k} \sum_{A \subset C_k} \chi(A)\chi^\beta(A) \tag{14}$$

Every (n, k)-alphabet furnishes us with a matrix representation of C_k by means of (13) and the procedure outlined below (13). The trace $\chi(A)$ of the matrix representing the element A of C_k is related to the weight of the letter by

$$\chi(A) = n - 2w(A) \tag{15}$$

Equations (14) and (15) permit us to compute from the weights of an (n, k)-alphabet what irreducible representations are present in the alphabet and how many times each is contained. It is assumed here that the given alphabet has been made isomorphic to C_k and that the weights are labelled by elements of C_k.

Consider the converse problem. Given a set of numbers $w_1, w_2, \cdots, w_{2^k}$ that satisfy Propositions 6 and 7. From these we can compute quantities $\chi_i = n - 2w_i$ as in (15). It is clear that the given w's will constitute the weights of an (n, k)-alphabet if and only if the 2^k χ_i can be labelled with elements of C_k so that the 2^k sums (14) (β ranges over all elements of C_k) are non-negative integers. The integers d_β tell what representations to choose to construct an (n, k)-alphabet with the given weights w_1.

2.10 CONSTRUCTION OF BEST ALPHABETS

A great many different techniques were used to construct the group alphabets listed in Tables II and III and to show that for each n and k there are no group alphabets with smaller probability of error. Space prohibits the exhibition of proofs for all the alphabets listed. We content ourselves here with a sample argument and treat the case $n = 10$, $k = 4$ in detail.

According to (2) there are $N(10, 4) = 53{,}743{,}987$ different (10, 4)-alphabets. We now show that none is better than the one given in Table III. The letters of this alphabet and weights of the letters are

I	0
1 6 7 8 10	5
2 6 7 9 10	5
3 5 6 8 9 10	6
4 5 7 8 9 10	6
1 2 8 9	4
1 3 5 7 9	5

1 4 5 6 9	5
2 3 5 7 8	5
2 4 5 6 8	5
3 4 6 7	4
1 2 3 5 7 9	6
1 2 4 5 7 10	6
1 3 4 8 10	5
2 3 4 9 10	5
1 2 3 4 6 7 8 9	8

The notation is that of Section 2.1. By actually forming the standard array of this alphabet, it is verified that

$$\alpha_0 = 1, \qquad \alpha_1 = 10, \qquad \alpha_2 = 39, \qquad \alpha_3 = 14.$$

Table II shows $\binom{10}{2} = 45$, whereas $\alpha_2 = 39$, so the given alphabet does not correct all possible double errors. In the standard array for the alphabet, 39 coset leaders are of weight 2. Of these 39 cosets, 33 have only one element of weight 2; the remaining 6 cosets each contain two elements of weight 2. This is due to the two elements of weight 4 in the given group, namely 1289 and 3467. A portion of the standard array that demonstrates these points is

I	1289	3467
·	·	·
·	·	·
12	89	·
18	29	·
19	28	·
34	·	67
36	·	47
37	·	46
·	·	·
·	·	·

In order to have a smaller probability of error than the exhibited alphabet, it is necessary that a (10, 4)-alphabet have an $\alpha_2 > 39$. We proceed to show that this is impossible by consideration of the weights of the letters of possible (10, 4)-alphabets.

We first show that every (10, 4)-alphabet must have at least one element (other than the identity, I) of weight less than 5. By Propositions 6 and 7, Section 2.8, the weights must form a partition of $10 \cdot 8 = 80$ into 15 positive parts. If the weights are all even, at least two must be less than 6 since $14 \cdot 6 = 84 > 80$. If eight of the weights are odd, we see from $8 \cdot 5 + 7 \cdot 6 = 82 > 80$ that at least one weight must be less than 5.

An alphabet with one or more elements of weight 1 must have an $\alpha_2 \leqq 36$, for there are nine elements of weight 2 which cannot possibly be coset leaders. To see this, suppose (without loss of generality) that the alphabet contains the letter 1. The elements 12, 13, 14, $\cdots$ 1 10 cannot possibly be coset leaders since the product of any one of them with the letter 1 yields an element of weight 1.

An alphabet with one or more elements of weight 2 must have an $\alpha_2 \leqq 37$. Suppose for example, the alphabet contained the letter 12. Then 13 and 23 must be in the same coset, 14 and 24 must be in the same coset, $\cdots$, 1 10 and 2 10 must be in the same coset. There are at least eight elements of weight two which are not coset leaders.

Each element of weight 3 in the alphabet prevents three elements of weight 2 from being coset leaders. For example, if the alphabet contains 123, then 12, 13, and 23 cannot be coset leaders. We say that the three elements of weight 2 are "blocked" by the letter of weight 3. Suppose an alphabet contains at least three letters of weight three. There are several cases: (A) if three letters have no numerals in common, e.g., 123, 456, 789, then nine distinct elements of weight 2 are blocked and $\alpha_2 \leqq 36$; (B) if no two of the letters have more than a single numeral in common, e.g., 123, 345, 789, then again nine elements of weight 2 are blocked and $\alpha_2 \leqq 36$; and (C) if two of the letters of weight 3 have two numerals in common, e.g., 123, 234, then their product is a letter of weight 2 and by the preceding paragraph $\alpha_2 \leqq 37$. If an alphabet contains exactly two elements of weight 3 and no elements of weight 2, the elements of weight 3 block six elements of weight 2 and $\alpha_2 \leqq 39$.

The preceding argument shows that to be better than the exhibited alphabet a (10, 4)-alphabet with letters of weight 3 must have just one such letter. A similar argument (omitted here) shows that to be better than the exhibited alphabet, a (10, 4)-alphabet cannot contain more than one element of weight 4. Furthermore, it is easily seen that an alphabet containing one element of weight 3 and one element of weight 4 must have an $\alpha_2 \leqq 39$.

The only new contenders for best (10, 4)-alphabet are, therefore, alphabets with a single letter other than I of weight less than 5, and this letter must have weight 3 or 4. Application of Propositions 6 and 7 show that the only possible weights for alphabets of this sort are: $35^7 6^7$ and $5^8 4 6^6$ where 5^7 means seven letters of weight 5, etc. We next show that there do not exist (10, 4)-alphabets having these weights.

Consider first the suggested alphabet with weights $35^7 6^7$. As explained in Section 2.9, from such an alphabet we can construct a matrix representation of C_4 having the character $\chi(I) = 10$, one matrix of trace 4, seven of trace 0 and seven of trace -2. The latter seven matrices correspond to elements of even weight and together with I must represent a subgroup of order 8. We associate them with the subgroup generated by the elements 2, 3, and 4. We have therefore

$$\chi(I) = 10, \qquad \chi(2) = \chi(3) = \chi(4) = \chi(23)$$
$$= \chi(24) = \chi(34) = \chi(234) = -2.$$

Examination of the symmetries involved shows that it doesn't matter how the remaining χ_i are associated with the remaining group elements. We take, for example

$$\chi(1) = 4, \qquad \chi(12) = \chi(13) = \chi(14) = \chi(123)$$
$$= \chi(124) = \chi(134) = \chi(1234) = 0.$$

Now form the sum shown in equation (14) with $\beta = 1234$ (i.e., with the character χ^{1234} obtained from column 1234 of the Table VI by means of substitution (13). There results $d_{1234} = \frac{1}{2}$ which is impossible. Therefore there does not exist a (10, 4)-alphabet with weights 35^76^7.

The weights 5^846^6 correspond to a representation of C_4 with character $\chi(I) = 10, 0^8, 2, (-2)^6$. We take the subgroup of elements of even weight to be generated by 2, 3, and 4. Except for the identity, it is clearly immaterial to which of these elements we assign the character 2. We make the following assignment: $\chi(I) = 10$, $\chi(2) = 2$, $\chi(3) = \chi(4) = \chi(23) = \chi(24) = \chi(34) = \chi(234) = -2$, $\chi(1) = \chi(12) = \chi(13) = \chi(14) = \chi(123) = \chi(124) = \chi(134) = \chi(1234) = 0$. The use of equation (14) shows that $d_2 = \frac{1}{2}$ which is impossible.

It follows that of the 53,743,987 (10, 4)-alphabets, none is better than the one listed on Table III.

Not all the entries of Table III were established in the manner just demonstrated for the (10, 4)-alphabet. In many cases the search for a best alphabet was narrowed down to a few alphabets by simple arguments. The standard arrays for the alphabets were constructed and the best alphabet chosen. For large n the labor in making such a table can be considerable and the operations involved are highly liable to error when performed by hand.

I am deeply indebted to V. M. Wolontis who programmed the IBM CPC computer to determine the α's of a given alphabet and who patiently ran off many such alphabets in course of the construction of Tables II and III. I am also indebted to Mrs. D. R. Fursdon who evaluated many of the smaller alphabets by hand.

REFERENCES

1. R. W. Hamming, B.S.T.J., **29**, pp. 147–160, 1950.
2. I. S. Reed, Transactions of the Professional Group on Information Theory, PGIT-4, pp. 38–49, 1954.
3. See section 7 of R. W. Hamming's paper, loc. cit.
4. I.R.E. Convention Record, Part 4, pp. 37–45, 1955 National Convention, March, 1955.
5. C. E. Shannon, B.S.T.J., **27**, pp. 379–423 and pp. 623–656, 1948.
6. Birkhoff and MacLane, A Survey of Modern Algebra, Macmillan Co., New York, 1941. Van der Waerden, Modern Algebra, Ungar Co., New York, 1953. Miller, Blichfeldt, and Dickson, Finite Groups, Stechert, New York, 1938.
7. This theorem has been previously noted in the literature by Kiyasu-Zen'iti, Research and Development Data No. 4, Ele. Comm. Lab., Nippon Tele. Corp. Tokyo, Aug., 1953.
8. F. D. Murnaghan, Theory of Group Representations, Johns Hopkins Press, Baltimore, 1938. E. Wigner, Gruppentheorie, Edwards Brothers, Ann Arbor, Michigan, 1944.

Reprinted from *I.R.E. Trans. Inform. Theory*, **IT-2,** 84–86 (1956)

11

A Note on Two Binary Signaling Alphabets

DAVID SLEPIAN†

Summary—A generalization of Hamming's single error correcting codes is given along with a simple maximum likelihood detection scheme. For small redundancy these alphabets are unexcelled. The Reed-Muller alphabets are described as parity check alphabets and a new detection scheme is presented for them.

Introduction

CERTAIN general properties of parity check alphabets for use on the binary symmetric channel were investigated in a recent paper by the author.[1] In particular, it was shown that maximum likelihood detectors for such alphabets assume a particularly simple form. In addition certain optimal small-sized alphabets were presented.

This paper describes two simple families of binary-signaling alphabets and detectors that are easy extensions of the material contained in that earlier paper.[1] Both families consist of parity check alphabets and both contain alphabets of arbitrarily large size. Members of the first are designed for use in situations requiring little redundancy, and for certain ranges of the pertinent parameters these alphabets cannot be excelled. The rate for these alphabets approaches unity with increasing alphabet size, however, so that they cannot be used to approach the Shannon rate in nontrivial cases. It is not known whether members of the second family can be used to approach the Shannon rate.

As in the reference cited,[1] we consider communication over a binary symmetric channel by means of parity check alphabets having k information places, p check places, and $n = p + k$ places all told. That is, if $a_1a_2 \cdots a_kb_1b_2 \cdots b_p$ is a typical letter of the signaling alphabet, where the a's and b's are either zero or one, then $a_1 \cdots a_k$ can be chosen arbitrarily and

$$b_j = \dot{\sum_1^k} c_{ij}a_i, \qquad j = 1, 2, \cdots, p \tag{1}$$

where the dot denotes summation modulo 2. The kp entries of the matrix $C = (c_{ij})$ determine the alphabet in question.

Generalized Hamming Alphabets

If p and k are chosen so that

$$k \geq 2^p - \sum_{i=0}^{3} \binom{p}{i},$$

then a simple parity check alphabet and maximum likelihood detector can be described which cannot be excelled by any other alphabet of 2^k letters using n-digit binary sequences. That is, no other alphabet of the same size has a smaller average probability of error per letter. These small redundancy alphabets are generalizations of Hamming's original single error correcting code.[2]

The C matrix for these alphabets is constructed as follows. The first row contains p ones. Successive rows of the matrix are obtained by writing in any order all the $\binom{p}{1}$ p-place binary sequences with exactly 1 zero, then in any order all the $\binom{p}{2}$ p-place binary sequences with exactly 2 zeros, etc., until k rows have been written. If c_{ij} is the element in the ith row and jth column of this kxp matrix, then the parity check rules are given by (1).

The matrix C just described also serves as a code book for the maximum likelihood detector for this alphabet. For each received (possibly erroneous) letter, form the parity check sequence $f_1f_2 \cdots f_p$ where

$$f_j = e_j \dotplus \dot{\sum_{i=1}^{k}} c_{ij}d_i, \qquad j = 1, 2, \cdots, p$$

and where d_i is the binary symbol in the ith information place of the received letter and e_j is the binary symbol in the jth check position of the received letter. If $f_1f_2 \cdots f_p$ is the same as the rth row of C, the binary digit in the rth information place of the received letter should be altered. If $f_1f_2 \cdots f_p$ is not listed among the rows of C and does not contain exactly three 1's, the binary digits in those check places of the received letter should be changed that correspond to the places of $f_1f_2 \cdots f_p$ that have the binary digit 1. If the parity check sequence $f_1 \cdots f_p$ has exactly three 1's and does not appear as a row of C, a row of C having four 1's is located such that three of these four 1's are in the same places as the three 1's of the parity check sequence. Let this row of C be the ith row and let the 1 in this row that does not correspond to a one in the parity check sequence be located in the jth column. Then in the received letter, the ith information place and the jth check place should be altered.

The alphabets and detectors just described correct $2^p - 1$ single errors if $n > 2^p - 1$. If

$$2^p - \sum_{0}^{3} \binom{p}{i} \leq n \leq 2^p - 1,$$

the alphabets correct all $\binom{n}{i}$ single errors and in addition $2^p - 1 - n$ double errors.

† Bell Telephone Labs., Inc., Murray Hill, N.J.

[1] D. Slepian, "A class of binary signaling alphabets," *Bell Sys. Tech. J.*, vol. 35, pp. 203–234; January, 1956.

[2] R. W. Hamming, "Error detecting and error correcting codes," *Bell Sys. Tech. J.*, vol. 29, pp. 147–160; April, 1950.

Proof of the above statements follows readily from Slepian.[3] The description of the generalized Hamming alphabets presented here is simpler than that given by him. The detection rule has not been given before.

Reed-Muller Alphabets

The Reed-Muller alphabets[4] can also be described as parity check alphabets. Each letter has $n = 2^m$ binary places of which

$$k = \sum_{i=0}^{r} \binom{m}{i}$$

are information places and $p = n - k$ are check places. For each non-negative integer value of m and $r \leq m$ there is one Reed-Muller alphabet.

It is convenient to label the entries in the k information places of a typical letter of an alphabet by the symbols $a_o, a_1, a_2, \cdots, a_m, a_{12}, a_{13}, \cdots, a_{(m-1)m}, a_{123}, \cdots, a_{123\ldots r}, \cdots, a_{(m-r+1)(m-r-2)\ldots m}$. That is, the a's are labelled by all k combinations of the integers 1, 2, $\cdots$, m taken r or fewer at a time. Similarly, the check places, $b_{12\ldots(r+1)}$, $b_{12\ldots r(r+2)}, \cdots, b_{12\ldots m}$, are labelled by all $2^m - k$ combinations of the integers 1, 2, $\cdots$, m taken $r + 1$ or more at a time. It will be convenient to denote a typical a or b with j subscripts by $a_\alpha^{(j)}$ or $b_\alpha^{(j)}$ respectively and to write $\alpha \subset \beta$ for two subscripts α and β if all the numerals in α are also present in β. The special subscript o is to be considered as the empty set and is contained in every β.

Let D_{ij}^r be one or zero accordingly as $\binom{j-i-1}{r-i}$ is odd or even respectively. The parity check rules for the Reed-Muller alphabets can then be written

$$b_\beta^{(j)} = \sum_{i=0}^{r} \cdot D_{ij}^r \sum_{\alpha \subset \beta} \cdot a_\alpha^{(i)} \tag{2}$$

$j = r + 1, r + 2, \cdots m$, all possible β.

Here all sums are modulo 2 and the α sum is over all indexes α that are i-fold combinations of the j integers that comprise β; *e.g.*, if $\beta = 1346$, the α sum is for $i = 2$, $\sum_{\alpha\subset\beta} \cdot a_\alpha^{(2)} = a_{13} + a_{14} + a_{16} + a_{34} + a_{36} + a_{46}$.

A maximum likelihood detector for the Reed-Muller alphabets can be constructed in a straightforward manner.[1] No simplification in the detector has been found that depends on the special nature of the Reed-Muller alphabet. For large m and r the labor involved in constructing the detector is considerable.

An important feature of the Reed-Muller alphabets is that every two letters of an alphabet with parameters m and r differ in at least 2^{m-r} places. This implies that a detector can be built which will correct all errors in a sent letter provided these errors be $l = 2^{m-r-1} - 1$ or fewer in number. Such a detector can be constructed as follows. First a code book is made which associates with each n place binary sequence $z_o, z_1, \cdots, z_{12}, \cdots, z_{12\ldots m}$ having l or fewer ones an $(n - k)$-place parity check sequence $f_{12\ldots r}, \cdots, f_{12\ldots m}$. The f's are given by

$$f_\beta^{(j)} = z_\beta^{(j)} + \sum_{i=0}^{r} \cdot D_{ij}^r \sum_{\alpha \subset \beta} \cdot z_\alpha^{(i)}, \tag{3}$$

$j = r + 1, r + 2, \cdots, m$, all possible β.

The code book contains $\sum_{i=0}^{l} \binom{n}{i}$ pairs of entries. For each received (possibly erroneous) letter, an $(n - k)$-place parity check sequence is formed as in (3) where the digits of the received letter are substituted for the z's. If the parity check sequence is not in the code book, the detector makes no decision as to the sent letter. If the parity check sequence is in the code book, the corresponding z entry having l or fewer ones is found. Those places of the received (possibly erroneous) letter are altered that correspond to places of the z entry that have the digit 1.

It is not difficult to show that if the parity check sequence has 1 or fewer 1's, the detector prescribes that no changes be made in the information places of the received letter. This results in a simplification of the detector when it is only desired to make corrections in the information places. The code book only needs to be constructed for those z sequences that have at least one l in their information places.

The description of the Reed-Muller alphabets given here is new as is the detector described. In some cases this detector is simpler than that given.[3]

The parity check rules (2) for the Reed-Muller alphabets can be derived as follows. Consider the m-row by 2^m column array of the components of the vectors $x_1, x_2, \cdots, x_m$ as given by Reed's.[4] The successive columns of this array are a listing of the integers from 0 to $2^m - 1$ in usual binary notation. The columns of this array can be labelled with sets of integers or α-symbols as used in (2) of this note. In particular, the column having 1's in rows $i_1, i_2, \cdots i_j$ is labelled by the α symbol $i_1 i_2 \cdots i_j$. The column containing no 1's is labelled by the special symbol o.

Permute the columns of the array so that the labels are in the *natural* order o, 1, 2, $\cdots$, m, 12, 13, $\cdots$, 123 $\cdots$ m. Consider the rows of this array as vectors and denote the ith row by y_i. Vectors whose subscripts are α symbols can now be formed from these y's. If α is the collection of integers $i_1 i_2 \cdots i_j$, then the βth component of y_α is the product of the βth components of $y_{i_1}, y_{i_2}, \cdots, y_{i_j}$. The vector y_o has all its components unity.

The vectors $y_o, y_1, \cdots, y_{12}, \cdots, y_{123}$, etc., correspond respectively to Reed's vector $I, x_1, \cdots, x_1x_2, \cdots, x_1x_2x_3$, etc. The y vectors have the distinguishing feature that the component of y_α with label β is unity if and only if $\alpha \subset \beta$.

[3] Slepian, *op. cit.*, paragraph 27.

[4] I. S. Reed, "A class of multiple error-correcting codes and the decoding scheme," Trans. IRE, vol. IT-4, pp. 38–49; September, 1954.

Reed's[4] (5) shows that the Reed-Muller alphabets are obtained by forming all possible linear combinations (modulo 2) of the first k y-vectors. The same alphabet will be obtained if all linear combinations are formed of a set of k vectors w_α that are linearly independent combinations of the k y_α's. The following is such a collection of vectors:

$$w_\alpha = \sum_{\beta \supset \alpha} \cdot y_\beta \tag{5}$$

α composed of r or fewer integers,
β composed of r or fewer integers.

From (4) and (5) it follows that the component of w_α labelled μ will be 1 if and only if there are an odd number of subscript symbols β such that $\alpha \subset \beta$ and $\beta \subset \mu$. If $\alpha \not\subset \mu$, then, the μth component of w_α will be zero.

It is convenient to treat the case when $\alpha \subset \mu$ in two parts.

1) If μ is comprised of $j \leq r$ integers and α is comprised of $i \leq j$ integers, then there are 2^{j-i} β symbols such that $\alpha \subset \beta \subset \mu$. Since 2^{j-i} is even except when $j = i$, this shows that the first k components of w_α are zero with the single exception of the component labelled α which has the value 1. The w_α are therefore linearly independent.

2) If μ is comprised of $j > r$ integers and α is comprised of $i \leq r$ integers, then there are

$$\binom{j-i}{0} + \binom{j-i}{1} + \cdots + \binom{j-i}{r-i}$$

β symbols of r or fewer integers such that $\alpha \subset \beta \subset \mu$. From

$$\binom{j-i}{\nu} = \binom{j-i-1}{\nu} + \binom{j-i-1}{\nu-1}$$

it follows that

$$\sum_0^{r-i} \binom{j-i}{\nu} = \sum_0^{r-i-1} 2\binom{j-i-1}{\nu} + \binom{j-i-1}{r-i}$$

or

$$\sum_0^{r-i} \binom{j-i}{2} \equiv \binom{j-i-1}{r-i} \quad \text{mod } 2.$$

Therefore, when μ is comprised of more than r integers, the component of w_α labelled μ is 1 if and only if $\alpha \subset \mu$ and $\binom{j-i-1}{r-i}$ is odd.

Consider the array of n columns and k rows whose entry in the row labelled α and column labelled μ is the component of w_α labelled μ. From the foregoing it is seen that the first k rows and k columns of this array consist of the $k \times k$ unit matrix. The last $n - k$ columns are linear combinations of the first k columns. Indeed, it is easily seen that the column labelled μ is the mod 2 sum of those columns whose labels are the names of the rows that have a 1 in the column labelled μ. This statement, combined with the preceeding paragraph, establishes the parity check rules (2).

The detector described for the Reed-Muller alphabets follows trivially from the results obtained by Slepian.[1]

Reprinted from *Bell System Tech. J.*, **39**, 1219–1252 (1960)

12

Some Further Theory of Group Codes

By DAVID SLEPIAN

(Manuscript received April 5, 1960)

The notion of equivalence for group codes is explored in some detail. A dual for a code, and the sum and product of two or more codes, are defined. Properties of these constructs are investigated. Indecomposable codes are defined and are shown to be optimal in two different senses. Various classes of codes are enumerated.

INTRODUCTION

This paper is a collection of results on the theory of group error-correcting codes for use on binary channels. It investigates further certain topics introduced in an earlier paper[1] by the author. The reader will be assumed to be familiar with the contents of this earlier paper as well as with the general nature of the coding problem in information theory.

The evident trend to digital transmission systems has given rise in recent years to an increased interest in coding as a possible practical means of error control. Lacking an "explicit solution" to the coding problem in any real sense, many investigators have chosen in an *ad hoc* manner promising special classes of parity-check codes and have examined their properties. A large and useful literature of special codes has resulted.

The approach taken here is different. No special codes are examined; rather, we attempt to shed some additional light on the structure of the class of all group codes. Our original aim was to parametrize in some manner the various equivalence classes of group codes. If such a parametrization could be effected, one could then hope to express the error probability of a code in terms of the parameters, and possibly to see how to choose the parameters to obtain codes of small error probability. We have fallen far short of this goal.

The main results to be found in this paper are as follows. A natural dual for a group code is defined. For any two group codes, a product code and a sum code are defined and certain properties of these operations are investigated. These operations have the important property of maintaining equivalence in the sense that if $\mathcal{A}$ and $\mathcal{A}'$ are equivalent group codes and $\mathcal{B}$ and $\mathcal{B}'$ are equivalent group codes, then $\mathcal{A} + \mathcal{B}$ is equivalent to $\mathcal{A}' + \mathcal{B}'$ and $\mathcal{A}\mathcal{B}$ is equivalent to $\mathcal{A}'\mathcal{B}'$. This result in turn leads to an arithmetic of equivalence classes of codes. The notion of an (additively) indecomposable equivalence class is introduced, and it is shown that an arbitrary equivalence class can be written in a unique manner as a sum of indecomposable equivalence classes. It is then shown that one can limit the search for best codes (with two commonly used meanings for "best") to the indecomposable equivalence classes. Enumeration formulae for the types of equivalence classes are given, and these formulae are evaluated for small values of the pertinent parameters.

In the interest of simplicity of exposition, we have restricted our attention to binary codes, although many of the results obtained hold for codes consisting of sequences of elements drawn from any finite field. Also, in an effort to make the paper available to as wide a class of readers as possible, we have carefully eschewed the specialized vocabulary of modern algebra,* although many of our results could be stated more succinctly in these terms. In addition, as an aid to the casual reader we adopt once more the format of Ref. 1: Part I contains definitions, examples and results; Part II contains additional theory and proofs of the less obvious assertions of Part I. The terminology of Ref. 1 is maintained with one exception: the word "code" is here used as a synonym for "alphabet," as has become accepted practice in the literature.

There is some overlap of material with that found in the paper of Fontaine and Peterson[2] which appeared after much of this work was done. In the interest of making this paper self-contained, we repeat some material that might have been quoted from that paper.

Part I — DEFINITIONS, EXAMPLES AND RESULTS

1.1 *Recall of Previous Paper*[1] *and Some New Definitions*

An (n,k)-alphabet, or (n,k)-code, is an unordered collection of 2^k distinct n-place binary sequences that forms an Abelian group under the operation of mod 2 addition of the sequences term by term. The elements of the group, that is, the n-place binary sequences, are also called "letters." We assume always in this paper that $n \geqq k > 0$.

We denote specific group codes by large script letters, $\mathcal{A}$, $\mathcal{B}$, etc. We denote the letters of $\mathcal{A}$ by A_1, A_2, etc., and the digits of a letter by lower-case Latin letters. Thus, for example, a particular letter of the (n,k)-

* In modern terminology, we are studying properties of subspaces of a finite dimensional linear vector space over a finite field.

code $\mathcal{A}$ is the binary sequence $A_1 = (a_1, a_2, \cdots, a_n)$. It is frequently convenient to regard the letters A_1, A_2, etc. as n-dimensional vectors.

A particular (n,k)-code can be specified by listing its 2^k letters. It can also be specified by listing k of its generators, i.e., any k linearly independent letters of the code. These k generators can be displayed as a binary matrix of rank k, with k rows and n columns. The rows of the matrix are the generators of the code. Such a matrix will be called a *generator matrix* and will be denoted typically by the symbol Ω. When referring to different generator matrices of a specific code $\mathcal{A}$, we shall write $\Omega_1(\mathcal{A})$, $\Omega_2(\mathcal{A})$, etc.

Many generator matrices correspond to the same code. The first generator can be chosen in $2^k - 1$ ways, since the all-zero sequence or identity, I, of the group code cannot serve as a generator. The second generator can be chosen in $2^k - 2^1$ ways. The third can be chosen in $2^k - 2^2$ ways, since the first two generators determine a group of order 2^2. Proceeding in this way, we find

$$\begin{aligned} M_k &= (2^k - 2^0)(2^k - 2^1)(2^k - 2^2) \cdots (2^k - 2^{k-1}) \\ &= 2^{k(k-1)/2}(2^k - 1)(2^{k-1} - 1)(2^{k-2} - 1) \cdots (3)(1) \end{aligned} \tag{1}$$

different generator matrices for a given (n,k)-code. Indeed, if Ω_1 and Ω_2 are generator matrices for the same code, then $\Omega_1 = g\Omega_2$, where g is a nonsingular $k \times k$ binary matrix and all operations implied in the matrix product $g\Omega_2$ are carried out mod 2. The collection of $k \times k$ nonsingular binary matrices forms a group under matrix multiplication (arthimetic mod 2) which we shall denote by G_k. G_k is of order M_k. [G_k is the general linear group of dimension k over a field of two elements, frequently denoted by $GL(k, 2)$.] If Ω is any generator matrix for an (n,k)-code, then, as g runs through G_k, $g\Omega$ gives the M_k distinct generator matrices associated with the code.

In all that follows we shall frequently omit the phrase "all arithmetic mod 2." It will generally be clear from the context whether the field in question is the reals, the complex numbers, or the two element field.

It was shown in Ref. 1 that every group code is a parity-check code and that every parity-check code is a group code. Let Λ be a binary matrix of $n - k = l$ rows and n columns and of rank l. Let λ_{ij} be the entry in the ith row and jth column of Λ, $i = 1, 2, \cdots, l$ and $j = 1, 2, \cdots, n$. The equations

$$\Lambda \tilde{A} = 0 \tag{2}$$

or

$$\sum_{j=1}^{n} \lambda_{ij} a_j = 0, \qquad i = 1, 2, \cdots, l,$$

3

where A is the binary row vector $A = (a_1, a_2, \cdots, a_n)$ and the tilde denotes transpose, have k linearly independent solutions, say $A_1, A_2, \cdots, A_k$. These k vectors can be taken as the generators of an (n,k)-code. Since every linear combination of the vectors $A_1, \cdots, A_k$ also satisfies (2), every generator matrix Ω of this (n,k)-code satisfies

$$\Lambda\tilde{\Omega} = 0.$$

The matrix Λ is called a *parity-check matrix* for the (n,k)-code.

A given (n,k)-code has many parity-check matrices. Indeed, if Λ is one such, so is $g\Lambda$ for every g contained in G_{n-k}. There are therefore M_{n-k} distinct parity-check matrices associated with a given (n,k)-code. We shall denote the different parity-check matrices of a specific (n,k)-code $\mathcal{A}$ by $\Lambda_1(\mathcal{A})$, $\Lambda_2(\mathcal{A})$, etc.

1.2 *Equivalence*

As in Ref. 1, we define two (n,k)-codes to be equivalent if one can be obtained from the other by a fixed permutation of the places of every letter. The concept has been illustrated in Section 1.7 of Ref. 1. Equivalent (n,k)-codes have the same transmission properties over the binary symmetric channel.

We denote the fact that codes $\mathcal{A}$ and $\mathcal{B}$ are equivalent by the symbolism $\mathcal{A} \cong \mathcal{B}$. It is immediately established that this is a true equivalence relation; i.e., that $\mathcal{A} \cong \mathcal{A}$; that $\mathcal{A} \cong \mathcal{B}$ implies $\mathcal{B} \cong \mathcal{A}$; and that if $\mathcal{A} \cong \mathcal{B}$ and $\mathcal{B} \cong \mathcal{C}$, then $\mathcal{A} \cong \mathcal{C}$. The totality of (n,k)-codes can therefore be broken down into disjoint equivalence classes. We denote by $\hat{\mathcal{A}}$ the equivalence class containing $\mathcal{A}$.

This equivalence of codes induces an equivalence relation among the totality of possible generator matrices. Two such matrices, say Ω_1 and Ω_2, will be called equivalent (written $\Omega_1 \cong \Omega_2$) if there exists a g in G_k and an $n \times n$ permutation matrix σ such that $g\Omega_1\sigma = \Omega_2$. That is, two $k \times n$ Ω-matrices are equivalent if one can be obtained from the other by permuting columns and/or forming nonsingular linear combinations of the rows mod 2. Clearly, two equivalent Ω-matrices, when considered as generator matrices, give rise to equivalent codes. Equivalent codes have equivalent generator matrices.

The task of analyzing group codes would be greatly simplified if a canonical form could be found for each equivalence class of Ω-matrices. That is, for a given n and k, we should like to be able to write down one generator matrix from each equivalence class. This would provide a simple means of describing each of the essentially different (n,k)-codes. The number of equivalence classes of (n,k)-codes is very much smaller than the number of distinct (n,k)-codes. They are enumerated in Section 1.9. Here we present further only two results pertaining to equivalence.

Every $k \times n$ Ω-matrix is equivalent to an Ω-matrix whose first k rows and columns are the $k \times k$ unit matrix. That is, Ω is equivalent to the partitioned matrix $\Omega \cong (I_k \vdots M)$, where I_k is the $k \times k$ unit matrix and M is a matrix of k rows and $l = n - k$ columns.

An Ω-matrix with the above structure will be said to be in *M-form*. Unfortunately, two $k \times n$ Ω-matrices in M-form having different M-matrices (even apart from permutations of rows and columns) can be equivalent.

A second result is

Theorem 1: A necessary and sufficient condition for two $k \times n$ Ω-matrices to be equivalent is that their columns can be placed into a one-to-one correspondence that preserves mod 2 addition of the columns.

Examples: Let

$$\Omega_1 = \begin{pmatrix} 1 & 0 & 0 & 1 & 1 \\ 0 & 1 & 0 & 1 & 1 \\ 0 & 0 & 1 & 0 & 1 \end{pmatrix}, \qquad \Omega_2 = \begin{pmatrix} 0 & 1 & 1 & 1 & 1 \\ 1 & 1 & 0 & 0 & 1 \\ 0 & 1 & 0 & 1 & 0 \end{pmatrix}.$$

Then $\Omega_1 \cong \Omega_2$, for if we denote the columns of Ω_1 by $u_1, u_2, \cdots, u_5$ and those of Ω_2 by $v_1, v_2, \cdots, v_5$ and establish the correspondence $u_1 \leftrightarrow v_3$, $u_2 \leftrightarrow v_5$, $u_3 \leftrightarrow v_2$, $u_4 \leftrightarrow v_1$, $u_5 \leftrightarrow v_4$, one sees that u_1, u_2, u_3 are independent as are v_3, v_5, v_2 and that the equations $u_4 = u_1 + u_2$ and $u_5 = u_1 + u_2 + u_3$ have the analogs $v_1 = v_3 + v_5$ and $v_4 = v_3 + v_5 + v_2$. Both Ω_1 and Ω_2 are equivalent to

$$\Omega_3 = \begin{pmatrix} 1 & 0 & 0 & 1 & 0 \\ 0 & 1 & 0 & 1 & 1 \\ 0 & 0 & 1 & 0 & 1 \end{pmatrix}.$$

The matrices Ω_1 and Ω_3 are both in M-form and are equivalent, although they have different M-matrices.

The preceding considerations of equivalence for Ω-matrices have their obvious analogs for parity-check matrices.

1.3 *Duality*

There is a natural duality between (n,k)-codes and (n,l)-codes, where $l = n - k$. In Ref. 1 it was noted that the two sets of codes are equinumerous. We elaborate further on this notion here.

In Section 1.1 it was remarked that every generator matrix $\Omega(\mathcal{A})$ for a given (n,k)-code $\mathcal{A}$ and every parity check matrix $\Lambda(\mathcal{A})$ for this code satisfies

$$\Lambda(\mathcal{A})\tilde{\Omega}(\mathcal{A}) = 0. \tag{3}$$

The transpose of this relation is

$$\Omega(\mathcal{A})\tilde{\Lambda}(\mathcal{A}) = 0.$$

Thus, *every parity check matrix* $\Lambda(\mathcal{A})$ *of an* (n,k)*-code* $\mathcal{A}$ *can be regarded as a generator matrix for a particular* (n,l)*-code hereafter called the dual of* $\mathcal{A}$ *and denoted* $\mathcal{A}^\dagger$. *Every generator matrix* $\Omega(\mathcal{A})$ *is a parity check matrix for* $\mathcal{A}^\dagger$.

The above can be regarded as defining $\mathcal{A}^\dagger$ by the relation

$$\Omega(\mathcal{A}^\dagger) = \Lambda(\mathcal{A}).$$

One immediately finds that

$$(\mathcal{A}^\dagger)^\dagger = \mathcal{A} \tag{4}$$

and that

$$\mathcal{A} \cong \mathcal{B} \quad \text{implies} \quad \mathcal{A}^\dagger \cong \mathcal{B}^\dagger. \tag{5}$$

The equivalence classes of (n,k)-codes can therefore be put in a natural way into one-to-one correspondence with the equivalence classes of (n,l)-codes:

$$\hat{\mathcal{A}} \text{ corresponds to } \widehat{\mathcal{A}^\dagger}.$$

It is convenient to define

$$(\hat{\mathcal{A}})^\dagger = \widehat{\mathcal{A}^\dagger}.$$

There is a simple way of passing from a $k \times n$ generator matrix Ω in M-form for a code in $\hat{\mathcal{A}}$ to a generator matrix Ω' in M-form for a code in $\hat{\mathcal{A}}^\dagger$. If $\Omega = (I_k \vdots M)$ defines a code in $\hat{\mathcal{A}}$, then $\Omega' = (I_l \vdots \tilde{M})$ defines a code in $\hat{\mathcal{A}}^\dagger$. Here $\tilde{M}$ is the transpose of M.

1.4 *The Sum of Two Codes*

Let $\mathcal{A}$ be an (n,k)-code and $\mathcal{B}$ be an (n',k')-code. We define a new code $\mathcal{C}$ by the partitioned generator matrix

$$\Omega(\mathcal{C}) = \begin{pmatrix} \Omega(\mathcal{A}) & \vdots & 0 \\ \cdots\cdots & \vdots & \cdots\cdots \\ 0 & \vdots & \Omega(\mathcal{B}) \end{pmatrix}. \tag{6}$$

The code $\mathcal{C}$ *is an* $(n + n', k + k')$*-code called the sum of* $\mathcal{A}$ *and* $\mathcal{B}$ *and we write* $\mathcal{C} = \mathcal{A} + \mathcal{B}$. It is easy to show that this is a valid definition and does not depend on the particular generator matrices chosen for $\mathcal{A}$ and $\mathcal{B}$.

If $\Lambda(\mathcal{A})$ and $\Lambda(\mathcal{B})$ are parity-check matrices for $\mathcal{A}$ and $\mathcal{B}$ respectively, then

$$\Lambda(\mathcal{C}) = \begin{pmatrix} \Lambda(\mathcal{A}) & \vdots & 0 \\ \cdots\cdots & \vdots & \cdots\cdots \\ 0 & \vdots & \Lambda(\mathcal{B}) \end{pmatrix} \tag{7}$$

is a parity-check matrix for $\mathcal{C} = \mathcal{A} + \mathcal{B}$.

Transmission of a letter from $\mathcal{C}$ amounts to transmitting a letter from $\mathcal{A}$ followed by a letter from $\mathcal{B}$. Because of the independence of the noise on the channel from one transmitted digit to the next,* it follows at once that *if* $Q_1(\mathcal{A})$, $Q_1(\mathcal{B})$ *and* $Q_1(\mathcal{C})$ *(see Section 1.6, Ref. 1) are the probability of no error for codes* $\mathcal{A}$, $\mathcal{B}$ *and* $\mathcal{C} = \mathcal{A} + \mathcal{B}$ *respectively, then* $Q_1(\mathcal{C}) = Q_1(\mathcal{A})Q_1(\mathcal{B})$.

If $\mathcal{C} = \mathcal{A} + \mathcal{B}$, a generator matrix for $\mathcal{C}$ need not appear in the block form (6). A parity-check matrix for $\mathcal{C}$ need not appear in the block form (7). The columns of a generator or parity-check matrix for $\mathcal{C}$, however, separate into two sets. All columns of the first set are linearly independent of all columns of the second set, and vice versa. Furthermore, if a linear combination of the columns sums to zero, the terms of this sum belonging to the first set separately sum to zero. The two sets of columns are said to be independent. (See Section 2.2 of this paper for further detail.) Since column dependences of a matrix are unaffected by premultiplication by a nonsingular matrix, we have that *a code is equivalent to a sum of two codes if and only if the columns of its* Ω*-matrices or* Λ*-matrices separate into independent sets.*

Some readily established properties of the sum just defined follow:

$$\mathcal{A} \cong \mathcal{A}' \text{ and } \mathcal{B} \cong \mathcal{B}' \quad \text{implies} \quad \mathcal{A} + \mathcal{B} \cong \mathcal{A}' + \mathcal{B}'; \tag{8}$$

$$\mathcal{A} + \mathcal{B} \cong \mathcal{B} + \mathcal{A}; \tag{9}$$

$$\mathcal{A} + (\mathcal{B} + \mathcal{C}) = (\mathcal{A} + \mathcal{B}) + \mathcal{C}; \tag{10}$$

if

$$\mathcal{C} = \mathcal{A} + \mathcal{B}, \qquad \mathcal{C}^\dagger = \mathcal{A}^\dagger + \mathcal{B}^\dagger. \tag{11}$$

1.5 *The Product of Two Codes*

We first remind the reader of the definition and elementary properties of the direct or Kronecker product of two matrices. Let $R = (r_{ij})$ be a

---

* Whenever probabilities are discussed in this paper, the usual binary symmetric channel is assumed.

matrix with a rows and b columns. Let $S = (s_{ij})$ be a matrix with c rows and d columns. The Kronecker product $T = R \times S$ of R times S (the order of factors is important) is the matrix of ac rows and bd columns with partitioned structure

$$T = R \times S = \begin{pmatrix} r_{11}S & r_{12}S & \cdots & r_{1b}S \\ r_{21}S & r_{22}S & & r_{2b}S \\ \vdots & & & \vdots \\ r_{a1}S & r_{a2}S & \cdots & r_{ab}S \end{pmatrix}.$$

The rows and columns of T can be labelled by pairs of integers so that a typical element of T is $t_{ij:kl} = r_{ik}s_{jl}$. These indexing pairs are listed in dictionary order, so that ij precedes $i'j'$ if either $i < i'$, or, when $i = i'$, if $j < j'$. For example 14 precedes 23, and 63 precedes 64.

One readily establishes the following properties for the Kronecker product:

$$Q \times (R \times S) = (Q \times R) \times S, \tag{12}$$

$$\widetilde{R \times S} = \tilde{R} \times \tilde{S}, \tag{13}$$

$$(P \times Q)(R \times S) = (PR) \times (QS), \tag{14}$$

$$R \times S = \sigma(S \times R)\mu. \tag{15}$$

In (13), the tilde indicates transpose. In (14), it is assumed that the columns of P are equinumerous with the rows of R and that the columns of Q are equinumerous with the rows of S. The product PR indicates the usual matrix product. In (15), if R has a rows and b columns and S has c rows and d columns, then σ and μ are permutation matrices of dimension ac and bd respectively and these matrices depend only on the numbers a, b, c and d and not the entries of R or S.

Let $\mathcal{A}$ be an (n,k)-code and let $\mathcal{B}$ be an (n',k')-code. We define a new code $\mathcal{C}$ by

$$\Omega(\mathcal{C}) = \Omega(\mathcal{A}) \times \Omega(\mathcal{B}). \tag{16}$$

The code $\mathcal{C}$ so defined is an (nn',kk')-code called the product of $\mathcal{A}$ and $\mathcal{B}$ and we write $\mathcal{C} = \mathcal{A}\mathcal{B}$. It is an easy consequence of the properties of the Kronecker product that $\mathcal{C}$ so defined is an (nn',kk')-code and does not depend on the particular generator matrices used for $\mathcal{A}$ and $\mathcal{B}$ in (16).

From (12) through (15) the following properties of code multiplication are readily established:

$$\mathcal{A} \cong \mathcal{A}' \text{ and } \mathcal{B} \cong \mathcal{B}' \quad \text{implies} \quad \mathcal{A}\mathcal{B} \cong \mathcal{A}'\mathcal{B}', \tag{17}$$

$$\mathcal{A}\mathcal{B} \cong \mathcal{B}\mathcal{A}, \tag{18}$$

$$\mathcal{A}(\mathcal{B}\mathcal{C}) \cong (\mathcal{A}\mathcal{B})\mathcal{C}, \tag{19}$$

$$\mathcal{A}(\mathcal{B} + \mathcal{C}) \cong \mathcal{A}\mathcal{B} + \mathcal{A}\mathcal{C}. \tag{20}$$

We note that $(\mathcal{A}\mathcal{B})^\dagger$ is not equivalent to $\mathcal{A}^\dagger\mathcal{B}^\dagger$ in general.

Let $\mathcal{A}$, $\mathcal{B}$ and $\mathcal{C} = \mathcal{A}\mathcal{B}$ be respectively an (n,k)-, an (n',k')- and an (nn',kk')-code with generator matrices Ω, Ω' and Ω'' and parity-check matrices Λ, Λ' and Λ''. There does not seem to be a simple expression for a parity-check matrix for $\mathcal{C}$ in terms of Λ and Λ'. However, if we confine our examination of codes to equivalences only, the structure of the parity checks for the product of two codes can be described simply.

We may suppose, then, that Ω and Ω' are in M-form. The structure of Ω'' is then given, up to equivalences, by

$$\begin{aligned} \Omega'' &= (I_k \vdots M) \times (I_{k'} \vdots M') \\ &\cong (I_k \times I_{k'} \vdots I_k \times M' \vdots M \times I_{k'} \vdots M \times M'). \end{aligned} \tag{21}$$

Denote the last $nn' - kk'$ columns of this last matrix by N. Then $(I_{nn'-kk'} \vdots \tilde{N})$ is the parity-check matrix for a code equivalent to $\mathcal{C}$.

It is readily seen from (21) that a code $\mathcal{C}'$ equivalent to $\mathcal{C}$ can be described as follows. The k' information places of $\mathcal{B}$ are replaced by letters (n-place binary sequences) of the code $\mathcal{A}$. This accounts for the kk' information places of $\mathcal{C}'$ and for the $k'(n - k)$ check places of $\mathcal{C}'$ described by the block $M \times I_{k'}$ in (21). The $n' - k'$ parity checks of $\mathcal{B}$ are then applied to these k' "information hyperplaces." The block $I_k \times M'$ in (21) describes repeated application of checks of $\mathcal{B}$ over the first k positions of the information hyperplaces of $\mathcal{C}'$ and accounts for $(n' - k')k$ checks. The block $M \times M'$ gives $(n - k)(n' - k')$ additional checks over the information places of $\mathcal{C}'$.

Up to equivalence, the product of two codes can be described in another, perhaps more simple, manner. Let $\mathcal{C} = \mathcal{A}\mathcal{B}$, where $\mathcal{A}$ is an (n,k)-code and $\mathcal{B}$ is an (n',k')-code. Then $\mathcal{C}$ is equivalent to the (nn',kk')-code $\mathcal{C}'$ obtained as follows. $\mathcal{A}$ is equivalent to a code $\mathcal{A}'$ with k information places and $n - k$ check places; $\mathcal{B}$ is equivalent to a code $\mathcal{B}'$ with k' information places and $n' - k'$ check places. In both $\mathcal{A}'$ and $\mathcal{B}'$, the check digits are mod 2 sums only over the information places. Write the kk' information places of $\mathcal{C}'$ in a rectangular array of k' rows and k

columns. Treat each row of the array as the k information places of a letter of $\mathcal{A}'$ and affix the corresponding check digits to obtain k' rows each of n binary digits. Regard each column of the array as the k' information places of a letter of $\mathcal{B}'$ and affix to each column the $n' - k'$ corresponding $\mathcal{B}'$ check digits. The nn' binary digits so obtained, read off in some fixed order, give the corresponding letter of $\mathcal{C}'$. It is to be noted that, in this description of $\mathcal{C}'$, $(n - k)(n' - k')$ of the check digits involve sums over other check digits, whereas in the description given by the last block of (21) these check digits are given as linear sums over the information places only.

1.6 *Arithmetic of Equivalence Classes*

The sum and product of group codes introduced in the preceding two sections provide an arithmetic of equivalence classes of codes. As before, let $\hat{\mathcal{A}}$ denote the equivalence class of codes to which the (n,k)-code $\mathcal{A}$ belongs. We define the sum of two equivalence classes by

$$\hat{\mathcal{A}} + \hat{\mathcal{B}} \equiv \widehat{(\mathcal{A} + \mathcal{B})}.$$

The self-consistency of this definition follows from (8). Similarly we define a product

$$\hat{\mathcal{A}}\hat{\mathcal{B}} \equiv \widehat{\mathcal{A}\mathcal{B}}$$

which is seen to be consistent from (17). Equations (8) through (11) and (17) through (20) give at once

$$\begin{aligned}
\hat{\mathcal{A}} + \hat{\mathcal{B}} &= \hat{\mathcal{B}} + \hat{\mathcal{A}}, \\
\hat{\mathcal{A}} + (\hat{\mathcal{B}} + \hat{\mathcal{C}}) &= (\hat{\mathcal{A}} + \hat{\mathcal{B}}) + \hat{\mathcal{C}}, \\
\hat{\mathcal{A}}\hat{\mathcal{B}} &= \hat{\mathcal{B}}\hat{\mathcal{A}}, \\
\hat{\mathcal{A}}(\hat{\mathcal{B}}\hat{\mathcal{C}}) &= (\hat{\mathcal{A}}\hat{\mathcal{B}})\hat{\mathcal{C}}, \\
\hat{\mathcal{A}}(\hat{\mathcal{B}} + \hat{\mathcal{C}}) &= \hat{\mathcal{A}}\hat{\mathcal{B}} + \hat{\mathcal{A}}\hat{\mathcal{C}}.
\end{aligned}$$

The simple two-letter code, **1**, consisting of the letters 0 and 1 with parameters $n = 1$, $k = 1$ and generator matrix $\Omega = (1)$ has the property

$$\mathbf{1}\hat{\mathcal{A}} = \hat{\mathcal{A}}\mathbf{1} = \hat{\mathcal{A}},$$

for all equivalence classes $\hat{\mathcal{A}}$.

1.7 *Indecomposable Codes*

To avoid repeated cumbersome statements about trivial cases, *in this section and the next we exclude from consideration codes whose generator matrices contain columns of zeros.* Such columns correspond to wasted digits in the code. A new code with smaller n value and the same k value can be obtained by deleting such all-zero columns. This property of possessing no columns of zeros is maintained under equivalence. If $\mathcal{A}$ possesses the property, it is not necessarily true, however, that $\mathcal{A}^\dagger$ has no columns of zeros.

It may happen that an (n,k)-code $\mathcal{A}$ is equivalent to the sum of two or more codes. In this case, we call $\mathcal{A}$ *decomposable.* If $\mathcal{A}$ is not equivalent to the sum of two or more codes, we call $\mathcal{A}$ *indecomposable.*

If $\mathcal{A}$ is decomposable, all codes equivalent to $\mathcal{A}$ are also decomposable; if $\mathcal{A}$ is indecomposable, all codes equivalent to $\mathcal{A}$ are also indecomposable. We can therefore speak of an equivalence class $\hat{\mathcal{A}}$ of codes as being either decomposable or indecomposable according as its members are or are not decomposable.

Theorem 2: Every (n,k)-code $\mathcal{A}$ is equivalent to a sum of indecomposable codes: $\mathcal{A} \cong \mathcal{A}_1 + \mathcal{A}_2 + \cdots + \mathcal{A}_m$, where $\mathcal{A}_1, \mathcal{A}_2, \cdots, \mathcal{A}_m$ are indecomposable. Furthermore, this decomposition is unique in the following sense. If also $\mathcal{A} \cong \mathcal{A}_1' + \mathcal{A}_2' + \cdots + \mathcal{A}_{m'}'$, where $\mathcal{A}_1', \mathcal{A}_2', \cdots, \mathcal{A}_{m'}'$ are indecomposable, then $m = m'$, $\mathcal{A}_1 \cong \mathcal{A}_{i_1}'$, $\mathcal{A}_2 \cong \mathcal{A}_{i_2}'$, $\cdots$, $\mathcal{A}_m \cong \mathcal{A}_{i_m}'$, where $i_1, i_2, \cdots, i_m$ are the integers 1, 2, $\cdots$, m in some order.

Theorem 2 can be stated in terms of equivalence classes as follows: *Every equivalence class $\hat{\mathcal{A}}$ of codes can be expressed as a sum of indecomposable equivalence classes $\hat{\mathcal{A}} = \hat{\mathcal{A}}_1 + \hat{\mathcal{A}}_2 + \cdots + \hat{\mathcal{A}}_m$. The indecomposable summands $\hat{\mathcal{A}}_1, \hat{\mathcal{A}}_2, \cdots, \hat{\mathcal{A}}_m$ are uniquely determined apart from order by $\hat{\mathcal{A}}$.*

A further consequence of Theorem 2 is

Theorem 3 (cancellation law of addition): Let $\hat{\mathcal{A}}$, $\hat{\mathcal{B}}$ and $\hat{\mathcal{C}}$ be any three equivalence classes of group codes. Then, if $\hat{\mathcal{A}} + \hat{\mathcal{B}} = \hat{\mathcal{A}} + \hat{\mathcal{C}}$, it follows that $\hat{\mathcal{B}} = \hat{\mathcal{C}}$. (This theorem holds also when codes with columns of zeros are allowed.)

1.8 *Optimal Properties of Indecomposable Codes*

A useful property of indecomposable codes is stated in the following theorem.

Theorem 4: Let $\mathcal{A}$ be a decomposable (n,k)-code, $k < n$, with probability of no error $Q_1(\mathcal{A})$. There exists an indecomposable (n,k)-code, $\mathcal{P}$, whose probability of no error $Q_1(\mathcal{P})$ satisfies $Q_1(\mathcal{P}) \geqq Q_1(\mathcal{A})$.

In this theorem, $Q_1(\mathcal{A})$ is the probability that a letter of $\mathcal{A}$ be decoded correctly when a maximum likelihood detector is used as the decoder (see Section 1.6, Ref. 1). A similar meaning holds for $Q_1(\mathcal{P})$. The

TABLE I — VALUES

n		k									
		1		2		3		4		5	
	X =	S	R	S	R	S	R	S	R	S	R
1	X	1	1								
	$\bar{X}$	1	1								
2	X	1	1	1							
	$\bar{X}$			1							
3	X	1	1	2	1	1					
	$\bar{X}$			1	1	1					
4	X	1	1	3	1	3	1	1			
	$\bar{X}$					2	1	1			
5	X	1	1	4	2	6	2	4	1	1	
	$\bar{X}$					1	1	3	1	1	
6	X	1	1	6	3	12	5	11	3	5	1
	$\bar{X}$					1	1	4	2	4	1
7	X	1	1	7	4	21	10	27	10	17	4
	$\bar{X}$					1	1	5	4	8	3
8	X	1	1	9	5	34	18	63	28	54	18
	$\bar{X}$							6	5	15	9
9	X	1	1	11	7	54	31	134	71	163	71
	$\bar{X}$							5	5	29	22
10	X	1	1	13	8	82	51	276	164	465	250
	$\bar{X}$							4	4	46	40
11	X	1	1	15	10	120	79	544	361	1283	809
	$\bar{X}$							3	3	64	60
12	X	1	1	18	12	174	121	1048	751	3480	2484
	$\bar{X}$							2	2	89	86
13	X	1	1	20	14	244	177	1956	1503	9256	7240
	$\bar{X}$							1	1	112	110
14	X	1	1	23	16	337	254	3577	2887	24282	20341
	$\bar{X}$							1	1	128	127
15	X	1	1	26	19	453	356	6395	5393	62812	55322
	$\bar{X}$							1	1	144	143
16	X	1	1	29	21	613	490	11217	9763	160106	146237
	$\bar{X}$									145	144
17	X	1	1	32	24	808	661	19307	17273	401824	376725
	$\bar{X}$									129	129
18	X	1	1	36	27	1056	882	32685	29839	992033	947555
	$\bar{X}$									113	113
19	X	1	1	39	30	1361	1157	54413	50557	2.40633	2.32900
	$\bar{X}$									91	91

OF S_{nk}, $\bar{S}_{nk}$, R_{nk}, $\bar{R}_{nk}$

k							
6		7		8		9	
S	R	S	R	S	R	S	R
1 1							
6 5	1 1	1 1					
25 14	5 4	7 6	1 1	1 1			
99 38	31 19	35 22	7 6	8 7	1 1	1 1	
385 105	164 70	170 80	51 35	47 32	8 7	9 8	1 1
1472 273	800 220	847 312	361 190	277 151	79 59	61 44	10 9
5676 700	3749 629	4408 1285	2484 977	1775 821	751 465	436 266	121 96
22101 1794	16749 1700	24297 5632	16749 4875	12616 5098	7240 3689	3557 1948	1503 1041
87404 4579	72783 4463	143270 26792	113662 24920	102445 37191	72783 31227	34942 17934	20341 12476
350097 11635	311233 11505	901491 137493	784390 132811	957357 320663	784390 293070	428260 213773	311233 175114
1.41325 29091	1.31126 28946	5.98528 745413	5.51748 733654	10.1746 3.18608	9.09877 3.04662	6.59254 3.27631	5.51748 2.94948
5.70816 70600	5.44572 70454	41.1752 4.14506	39.2920 4.11584	119.235 34.7994	112.170 34.0492	123.425 61.2716	112.170 58.0573
22.9032 164705	22.2371 164575	287.813 22.9827	280.215 22.9120	1482.30 397.232	1434.04 393.075	2647.03 1296.46	2516.51 1261.52
90.6994 366089	89.0390 365976	2009.86 124.432	1979.34 124.268	18884.5 4558.66	18548.3 4535.64	76284.2 29032.1	59541.8 28634.1

theorem thus states that the search for best codes can be restricted to indecomposable codes when "best" means large values of Q.

Another criterion frequently used to evaluate codes is the nearest neighbor distance, d. This quantity is the smallest nonzero weight of the letters of the code. If $d = 2e + 1$, then the code can correct all combinations of e or fewer digit errors in any transmitted letter. For a given n and k, it is not necessarily true that the code with largest d value has the largest Q_1 value.

The search for codes of largest nearest neighbor distance can also be limited to indecomposable codes as a result of

Theorem 5: Let $\mathcal{A}$ *be an* (n,k)*-code,* $k < n$*, with nearest neighbor distance* $d(\mathcal{A})$*. There exists an indecomposable* (n,k)*-code,* $\mathcal{P}$*, with nearest neighbor distance* $d(\mathcal{P}) \geqq d(\mathcal{A})$.

A convenient test exists for determining whether a given Ω-matrix in M-form is the generator matrix of an indecomposable code. Two elements, m_{rs} and m_{tu}, of M are said to be *connected* if they both have value 1 and lie either in the same column or the same row of M. A *path* in M is a sequence of elements of M each of which is connected to its successor except for the last element of the sequence. In terms of these definitions, we have the following

Test: Let $\mathcal{A}$ *be an* (n,k)*-code with* $k < n$*. Then* $\mathcal{A}$ *is decomposable if and only if* M *contains a path containing elements from every row of* M.

The above test is meaningless for (n,n)-codes. The $(1,1)$-code is indecomposable. For $n \neq 1$, the (n,n)-code is decomposable.

It is easy to show from this test for decomposability that $\mathcal{A}$ is an indecomposable (n,k)-code with no column of zeros if and only if $\mathcal{A}^\dagger$ is indecomposable and has no column of zeros.

The test for decomposability can also be used to establish that $\mathcal{C} = \mathcal{A}\mathcal{B}$ is indecomposable if and only if $\mathcal{A}$ and $\mathcal{B}$ are indecomposable.

1.9 *Enumeration of Equivalence Classes*

Although we have not succeeded in parametrizing the equivalence classes of (n,k)-codes, we can systematically enumerate these classes by a modified Polya scheme.[3] The details of the method are given in Section 2.8. Here we present the results of a computation.

We shall denote by S_{nk} the number of equivalence classes of (n,k)-codes with no columns of zero.

A generator matrix for an (n,k)-code may or may not have repeated columns. The multiplicities of columns in an Ω-matrix are preserved under equivalence. Of interest are the (n,k)-codes whose Ω-matrices have no repeated columns. We denote by $\bar{S}_{nk}$ the number of equivalence classes of (n,k)-codes having no repeated columns and no columns of zeros.

We adopt an analogous notation for the number of indecomposable equivalence classes. The number of equivalence classes of indecomposable (n,k)-codes with no columns of zeros is denoted by R_{nk}. The number of equivalence classes of indecomposable (n,k)-codes with no repeated columns and no columns of zeros is denoted by $\bar{R}_{nk}$.

Table I lists values of S_{nk}, $\bar{S}_{nk}$, R_{nk} and $\bar{R}_{nk}$. The box in row n and column k contains S_{nk} in the upper left corner, $\bar{S}_{nk}$ in the lower left corner, $\bar{R}_{nk}$ in the upper right corner and $\bar{R}_{nk}$ in the lower right corner. All entries are given to six significant figures. Numbers containing a decimal point are to be multiplied by 10^6.

From a table of values of S_{nk}, one can easily construct a table of values of W_{nk}, the number of equivalence classes of (n,k)-codes (zero columns and repetition allowed). Table II is a short table of values of

TABLE II — VALUES OF N_{nk} AND W_{nk}

n		k = 0	1	2	3	4	5
1	N	1	1				
	W	1	1				
2	N	1	3	1			
	W	1	2	1			
3	N	1	7	7	1		
	W	1	3	3	1		
4	N	1	15	35	15	1	
	W	1	4	6	4	1	
5	N	1	31	155	155	31	1
	W	1	5	10	10	5	1
6	N	1	63	651	1395	651	63
	W	1	6	16	22	16	6
7	N	1	127	2667	11811	11811	2667
	W	1	7	23	43	43	23
8	N	1	255	10795	97155	200787	97155
	W	1	8	32	77	106	77
9	N	1	511	43435	788035	3309747	3309747
	W	1	9	43	131	240	240
10	N	1	1023	174251	6347715	53743987	109221651
	W	1	10	56	213	516	705

W_{nk} along with values of N_{nk}, the total number of distinct (n,k)-codes. One has $N_{nk} = N_{nl}$, $W_{nk} = W_{nl}$, $l = n - k$. The familiar appearance of the first five rows of the W_{nk} table provides a good example of the perils of too hasty extrapolation in mathematics.

Part II — ADDITIONAL THEORY AND PROOFS OF THEOREMS OF PART I

2.1 *Proof of Theorem 1*

Theorem 1 asserts that a necessary and sufficient condition for two $k \times n$ Ω-matrices, say Ω and Ω', to be equivalent is that their columns can be placed into a one-to-one correspondence that preserves mod 2 addition of the columns.

The necessity of the condition follows trivially from the fact that equivalence means $g\Omega\sigma = \Omega'$ for some nonsingular g and some permutation matrix σ. For the one-to-one correspondence of the theorem, associate the ith column of $\Omega\sigma$, say c_i, with the ith column of Ω', say c_i', $i = 1, 2, \cdots, n$. Then $gc_i = c_i'$, $i = 1, 2, \cdots, n$. Thus, if $c_i + c_j = c_k$, then $gc_i + gc_j = gc_k$, or $c_i' + c_j' = c_k'$. Since g is nonsingular, it also follows that $c_i' + c_j' = c_k'$ implies $c_i + c_j = c_k$.

To prove the sufficiency of the condition, suppose that the columns of Ω and Ω' can be placed into a one-to-one correspondence that preserves mod 2 addition of columns. Let σ permute the columns of Ω so that the ith column of $\Omega\sigma$ corresponds to the ith column of Ω', $i = 1, 2, \cdots, n$. Let $g \in G_k$ and μ, an $n \times n$ permutation matrix, reduce $\Omega\sigma$ to M-form. Then mod 2 addition of columns is preserved between $g\Omega\sigma\mu$ and $g\Omega'\mu$ when the ith column of the former is associated with the ith column of the latter, $i = 1, 2, \cdots, n$. The first k columns of $g\Omega\sigma\mu$ are independent since the first k columns of $g\Omega\sigma'\mu$ are. Therefore the matrix g_1 formed by the first k rows and k columns of $g\Omega\sigma\mu$ is nonsingular. The matrix $g_1^{-1}g\Omega\sigma\mu$ is in M-form and, when its ith column is associated with the ith column of $g\Omega'\mu$, mod 2 addition of columns is still preserved. But then columns $k + 1, k + 2, \cdots, n$ of these two matrices are identical linear combinations of their identical first k columns, so that $g_1^{-1}g\Omega\sigma\mu = g\Omega'\mu$. It follows then that $\Omega' = g^{-1}g_1^{-1}g\Omega\sigma$, so that Ω' and Ω are equivalent.

2.2 *Decomposition of Sets of Vectors*

In this section we present five lemmas and a theorem concerning linear dependence of vectors. This material is preparatory for the proof of Theorem 2. While it is true that Theorem 2 can be proved much more directly (and abstractly) than is done here, it is felt that the procedure to be followed gives more insight into the nature of the problem at hand than do the shorter more abstract proofs.

Here we shall consider collections of vectors drawn with *possible repetitions* from a finite dimensional vector space over a finite field of scalars. In the application to be made later, the vectors will be columns taken from the generator matrix of a code, and the scalars will as usual be zero or one. The reader may, if he wishes, restrict his considerations to vectors and scalars of this sort. Throughout this section, we agree to exclude the null- or zero-vector from consideration as a member of any of the collections of vectors we may discuss.

Let $S_1, S_2, \cdots, S_m$ be nonempty finite sets of vectors. Denote the vectors of S_i by $\mathbf{v}_{ij}$, $j = 1, 2, \cdots, r_i$, for $i = 1, 2, \cdots, m$. The sets $S_1, S_2, \cdots, S_m$ are then called *independent* if every relation of the form

$$\sum_{i=1}^{m} \sum_{j=1}^{r_i} \alpha_{ij} \mathbf{v}_{ij} = 0$$

implies

$$\sum_{j=1}^{r_i} \alpha_{ij} \mathbf{v}_{ij} = 0, \qquad i = 1, 2, \cdots, m.$$

Clearly, no vector in any one such set can be written as a linear combination of vectors taken only from the other sets. Directly from the definition of independence we also have

Lemma 1: Let the sets S_i be independent and let R_i be a subset of S_i, $i = 1, 2, \cdots, m$. Then the nonempty sets among $R_1, R_2, \cdots, R_m$ are independent.

A set, S, of vectors is called *indecomposable* if S cannot be written as a union of two or more independent subsets of S. Every vector in an indecomposable set containing more than one vector can be written as a linear combination of other vectors in the set. Clearly, a set S that is not indecomposable is the union of independent indecomposable subsets, $S_1, S_2, \cdots, S_m$. In this case we say that S can be *decomposed* into independent indecomposable *components* $S_1, S_2, \cdots, S_m$.

A linear form $l = \alpha_1 \mathbf{v}_1 + \alpha_2 \mathbf{v}_2 + \cdots + \alpha_j \mathbf{v}_j$ is called *irreducible* if no collection of $j - 1$ or fewer of the terms $\alpha_1 \mathbf{v}_1, \alpha_2 \mathbf{v}_2, \cdots, \alpha_j \mathbf{v}_j$ sums to zero; otherwise, the linear form is called *reducible.* Two linear forms are called *disjoint* if the respective sets of vectors with nonzero coefficients in the two forms are disjoint. We have then

Lemma 2: Every reducible linear form that is equal to zero is the sum of disjoint irreducible linear forms each of which is zero.

Proof: Suppose $l = \alpha_1 \mathbf{v}_1 + \alpha_2 \mathbf{v}_2 + \cdots + \alpha_j \mathbf{v}_j$ to be reducible where all the α's are different from zero. Then there are subsets of terms of l that add to zero. Choose such a subset containing a minimal number of terms and call the sum of these terms the linear form l_1. The form l_1 must be irreducible or it would not contain a minimal number of terms. Repeat this procedure for $l - l_1 \equiv l_2 = 0$. After a finite number of steps we obtain an irreducible form l_i and $l = l_1 + l_2 + \cdots + l_i$. The forms so obtained are disjoint by construction.

Let S contain r vectors. One can form $p^r - 1$ linear forms

$$\sum_1^r \alpha_i \mathbf{v}_i$$

of these vectors where not all the α's are zero. Here p is the number of elements in the field of scalars ($p = 2$ in the applications to follow). From this list of linear forms, delete those that do not sum to zero. From the remaining forms, delete those that are reducible. One arrives then at a uniquely determined set $\mathfrak{L}$ of irreducible sums, each one of which is zero. Two vectors of S, say $\mathbf{v}_1$ and $\mathbf{v}_2$, are said to be *directly connected* to each other if they appear together as terms in any one of the irreducible sums of $\mathfrak{L}$. A vector of S not appearing in any of the linear forms of $\mathfrak{L}$ is said to be directly connected to itself. Two vectors of S, $\mathbf{v}_1$ and $\mathbf{v}_2$, are said to be *connected* if there exist vectors

$$\mathbf{v}_{i_1}, \mathbf{v}_{i_2}, \cdots, \mathbf{v}_{i_q}$$

of S such that $\mathbf{v}_1$ is directly connected to $\mathbf{v}_{i_1}$, $\mathbf{v}_{i_q}$ is directly connected to $\mathbf{v}_2$ and $\mathbf{v}_{i_\alpha}$ is directly connected to $\mathbf{v}_{i_{\alpha+1}}$, $\alpha = 1, 2, \cdots, q - 1$. If $\mathbf{v}_1$ is connected to $\mathbf{v}_2$, we write $\mathbf{v}_1 \sim \mathbf{v}_2$. Evidently, for all vectors $\mathbf{v}_1, \mathbf{v}_2, \mathbf{v}_3$ of S we have: (a) $\mathbf{v}_1 \sim \mathbf{v}_1$; (b) $\mathbf{v}_1 \sim \mathbf{v}_2$ implies $\mathbf{v}_2 \sim \mathbf{v}_1$; (c) if $\mathbf{v}_1 \sim \mathbf{v}_2$ and $\mathbf{v}_2 \sim \mathbf{v}_3$, then $\mathbf{v}_1 \sim \mathbf{v}_3$. The vectors of S are therefore uniquely separated into disjoint equivalence classes by the connectedness relation $\sim$.

Lemma 3: The totality of vectors of S belonging to an equivalence class E of connected vectors forms an indecomposable set.

For, suppose E could be written as the union of two independent subsets S_1 and S_2 of E. Since all elements of E are connected, there must be a $\mathbf{v}_1$ in S_1 and a $\mathbf{v}_2$ in S_2 such that $\mathbf{v}_1$ is directly connected to $\mathbf{v}_2$. There is therefore a linear form in $\mathfrak{L}$ of the form

$$\alpha_1 \mathbf{v}_1 + \alpha_2 \mathbf{v}_2 + \sum_3^t \alpha_i \mathbf{v}_i = 0$$

with $\alpha_1 \neq 0$, $\alpha_2 \neq 0$. By the definition of independence, the terms in this sum belonging to S_1 add to zero, as do the terms belonging to S_2. But this contradicts the irreducibility of sums in $\mathfrak{L}$.

Lemma 4: Distinct equivalence classes $S_1, S_2, \cdots, S_m$ *of connected vectors of* S *are independent sets of vectors.*

Proof: Consider any linear form

$$l = \sum\sum \alpha_{ij}\mathbf{v}_{ij}$$

of vectors of S that is zero. Suppose l contains vectors from different equivalence classes with nonzero coefficients. Then, since $l = 0$, l cannot be irreducible, for in this case the vectors in different equivalence classes would be directly connected. Since it is reducible, l can be written by Lemma 2 as the sum of disjoint irreducible forms each of which is zero. But none of these forms can contain vectors from different equivalence classes. Adding together all the irreducible forms containing vectors from any one equivalence class, we get

$$\sum_j \alpha_{ij}\mathbf{v}_{ij} = 0, \qquad i = 1, 2, \cdots, m.$$

Lemma 5: All vectors of an indecomposable subset P *of* S *belong to the same equivalence class of connected vectors.*

For, let R_i be the set of vectors of P that belongs to the equivalence class S_i, $i = 1, 2, \cdots, m$. By Lemmas 1 and 4, the sets R_i are independent and the assumed indecomposable set P is then exhibited as the union of independent subsets. This is a contradiction unless all the R_i but one are empty.

The preceding lemmas and definitions allow us to state finally the following

Theorem 6: A set S *of vectors can be decomposed into independent indecomposable components in only one way.*

Proof: We have seen that S can be separated into equivalence classes of connected vectors in a unique manner. Lemmas 3 and 4 show these equivalence classes to be a decomposition of S into independent indecomposable sets. Suppose now that S could be decomposed in another manner into independent indecomposable sets. Lemma 5 shows that each such indecomposable set is completely contained in an equivalence class. There cannot be more than one such indecomposable set in any equivalence class, for then the equivalence class would be the union of two or more independent subsets which contradicts Lemma 3.

We point out once again in closing this section that the vectors of the set S here considered need not be distinct. S may contain several copies of a single vector of the linear vector space under consideration.

2.3 *Proof of Theorem 2*

Let us regard the columns of a generator matrix $\Omega(\mathcal{A})$ as a collection of vectors. The linear relations satisfied by a set of vectors determine whether or not the set is indecomposable. The linear relations satisfied by the column vectors of generator matrices of equivalent codes are identical (except for possible renumbering of the columns). It follows immediately that a code $\mathcal{A}$ is indecomposable if and only if the columns of any (and hence every) generator matrix $\Omega(\mathcal{A})$ form an indecomposable set of vectors. With this remark, we proceed to the proof of Theorem 2.

That every (n,k)-code $\mathcal{A}$ is equivalent to a sum of indecomposable codes follows readily from the definitions of indecomposable codes and equivalence. Here we show only that if $\mathcal{A} \cong \mathcal{A}_1 + \mathcal{A}_2 + \cdots + \mathcal{A}_m$ and $\mathcal{A} \cong \mathcal{A}_1' + \mathcal{A}_2' + \cdots + \mathcal{A}_{m'}'$ where the $\mathcal{A}_i$ and $\mathcal{A}_i'$ are indecomposable, then $m = m'$ and $\mathcal{A}_j \cong \mathcal{A}_{i_j}'$, $j = 1, 2, \cdots, m$, where $i_1, i_2, \cdots, i_m$ are the integers $1, 2, \cdots, m$ in some order.

If $R, S, \cdots$, are matrices of respective size $r \times r'$, $s \times s'$, $\cdots$, we denote by diag $(R, S, \cdots)$ the $(r + s + \cdots) \times (r' + s' + \cdots)$ partitioned matrix having R in its first r row and r' columns, S in rows $r + 1$ to $r + s$ and columns $r' + 1$ to $r' + s'$, etc., and zeros elsewhere. Set

$$\begin{aligned} \Omega &= \text{diag } [\Omega(\mathcal{A}_1), \Omega(\mathcal{A}_2), \cdots, \Omega(\mathcal{A}_m)], \\ \Omega' &= \text{diag } [\Omega(\mathcal{A}_1'), \Omega(\mathcal{A}_2'), \cdots, \Omega(\mathcal{A}_{m'}')]. \end{aligned} \tag{22}$$

Then, by hypothesis, $\Omega = g\Omega'\sigma$, where $\mathcal{A}_i$ is an indecomposable (n_i, k_i)-code, $i = 1, 2, \cdots, m$; $\mathcal{A}_j'$ is an indecomposable (n_j',k_j')-code, $j = 1, 2, \cdots, m'$; and

$$\sum_{i=1}^{m} k_i = \sum_{j=1}^{m'} k_j' = k,$$

$$\sum_{i=1}^{m} n_i = \sum_{j=1}^{m'} n_j' = n.$$

The columns of Ω decompose into independent indecomposable sets $S_1, S_2, \cdots, S_m$. Here S_1 consists of the first n_1 columns of Ω, S_2 consists of the next n_2 column of Ω, etc. The columns of $\Omega'\sigma$ satisfy linear relations identical with those satisfied by the columns of Ω since $\Omega = g\Omega'\sigma$, and hence, from Theorem 6, the first n_1 columns of $\Omega'\sigma$ are an indecomposable set S_1', the next n_2 columns of $\Omega'\sigma$ are an indecomposable set S_2', etc., and these sets are independent. But the columns of $\Omega'\sigma$ are a reordering of the columns of Ω' and the latter are exhibited as m' independent indecomposable sets in (22). Therefore, $m = m'$ and $n_{i_j}' =$

$n_j, j = 1, 2, \cdots, m$, where $i_1, i_2, \cdots, i_m$ are the integers $1, 2, \cdots, m$ listed in some order. It follows then that S_j' consists entirely of those columns of Ω' that contain $\Omega(\mathcal{A}_{i_j}'), j = 1, 2, \cdots, m$. We can then write $\Omega'\sigma = \mu\Omega''$, where μ is a $k \times k$ permutation matrix,

$$\Omega'' = \text{diag}\,[\Omega(\mathcal{A}_{i_1}')\sigma_1, \Omega(\mathcal{A}_{i_2}')\sigma_2, \cdots, \Omega(\mathcal{A}_{i_m}')\sigma_m],$$

and σ_j is an $n_j \times n_j$ permutation matrix, $j = 1, 2, \cdots, m$. On setting $g'' = g\mu$, we have $g''\Omega'' = \Omega$.

Let T_1 be the matrix of the first n_1 columns of Ω, T_2 be the matrix of the next n_2 columns of Ω, etc. Let T_1'' be the matrix of the first n_1 columns of Ω'', T_2'' be the matrix of the next n_2 columns of Ω'', etc. Then $g''T_j'' = T_j, j = 1, 2, \cdots, m$. But T_j is of rank k_j and g'' is nonsingular, so that $k_{i_j}' \geqq k_j$. From $\sum k_j' = \sum k_j = k$, we find $k_{i_j}' = k_j, j = 1, 2, \cdots, m$.

Now partition g'' in rows according to $k_1, k_2, \cdots, k_m$ and in columns according to $n_1, n_2, \cdots, n_m$. Denote the ith diagonal submatrix of g'' by g_i. Then $g''\Omega'' = \Omega$ yields $g_j\Omega(\mathcal{A}_{i_j}')\sigma_j = \Omega(\mathcal{A}_j), j = 1, 2, \cdots, m$. A comparison of ranks in these equations shows that the g_j are nonsingular. Therefore $\mathcal{A}_j \cong \mathcal{A}_{i_j}', j = 1, 2, \cdots, m$, and the theorem is proved.

2.4 *The Test for Indecomposability*

We have seen that an (n,k)-code $\mathcal{A}$ is indecomposable if and only if the columns of any generator matrix $\Omega(\mathcal{A})$ are an indecomposable collection of vectors. If $\Omega(\mathcal{A})$ is in M-form its first k columns are independent and each contains a single one. The other columns of $\Omega(\mathcal{A})$ can each be expressed as an irreducible sum of these first k columns. From Section 2.2 it follows that the columns of $\Omega(\mathcal{A})$ will form an indecomposable set of vectors if and only if the first k columns of $\Omega(\mathcal{A})$ are connected to each other. The reader can readily translate this statement into the test described in Section 1.8.

2.5 *Proof of Theorem 3*

The hypothesis $\hat{\mathcal{A}} + \hat{\mathcal{B}} = \hat{\mathcal{A}} + \hat{\mathcal{C}}$ means that, for codes $\mathcal{A}$, $\mathcal{B}$ and $\mathcal{C}$ respectively in $\hat{\mathcal{A}}$, $\hat{\mathcal{B}}$ and $\hat{\mathcal{C}}$,

$$\mathcal{A} + \mathcal{B} \cong \mathcal{A} + \mathcal{C}.$$

Then

$$\mathcal{A}_1 + \mathcal{A}_2 + \cdots + \mathcal{A}_\alpha + \mathcal{B}_1 + \mathcal{B}_2 + \cdots + \mathcal{B}_\beta$$
$$\cong \mathcal{A}_1 + \mathcal{A}_2 + \cdots + \mathcal{A}_\alpha + \mathcal{C}_1 + \mathcal{C}_2 + \cdots + \mathcal{C}_\gamma,$$

21

where the $\mathcal{A}_j$, $\mathcal{B}_j$ and $\mathcal{C}_j$ are the (unique) indecomposable code components respectively of $\mathcal{A}$, $\mathcal{B}$ and $\mathcal{C}$. By Theorem 2 we have $\beta = \gamma$, and there is a one-to-one correspondence set up by the equivalence relation $\cong$ between elements of the set $H_1 = \{\mathcal{A}_1, \cdots, \mathcal{A}_\alpha, \mathcal{B}_1, \cdots, \mathcal{B}_\beta\}$ and the set $H_2 = \{\mathcal{A}_1, \cdots, \mathcal{A}_\alpha, \mathcal{C}_1, \cdots, \mathcal{C}_\beta\}$. If all the $\mathcal{B}$'s map into $\mathcal{C}$'s in this correspondence, then $\sum \mathcal{B}_i \cong \sum \mathcal{C}_i$, $\hat{\mathcal{B}} = \hat{\mathcal{C}}$, and the theorem is proved. Suppose then that $\mathcal{B}_1$ maps into $\mathcal{A}_{i_1}$ of H_2. If $\mathcal{A}_{i_1}$ of H_1 maps into a $\mathcal{C}$, say $\mathcal{C}_1$, then $\mathcal{B}_1 \cong \mathcal{A}_{i_1} \cong \mathcal{C}_1$, and we go on to examine another $\mathcal{B}$ of H_1. If, however, $\mathcal{A}_{i_1}$ of H_1 maps into $\mathcal{A}_{i_2}$ of H_2, we then consider $\mathcal{A}_{i_2}$ in H_1. Proceeding in this manner, we must ultimately reach an $\mathcal{A}$ in H_1 that is mapped onto a $\mathcal{C}$, since the $\mathcal{A}$'s in H_1 and H_2 are equinumerous and $\mathcal{B}_1$ of H_1 is mapped onto an $\mathcal{A}$ of H_2. This yields a chain of equivalences starting with $\mathcal{B}_1$ and ending with a $\mathcal{C}$. Each $\mathcal{B}$ then is equivalent to a $\mathcal{C}$ and, by reversing the argument, we find a one-to-one equivalence correspondence among the $\mathcal{B}$'s and $\mathcal{C}$'s. It follows then that $\hat{\mathcal{B}} = \hat{\mathcal{C}}$.

2.6 *Proof of Theorem 4*

Theorem 4 states that if $\mathcal{A}$ is an indecomposable (n,k)-code, $k < n$, with probability of no error $Q_1(\mathcal{A})$, then there exists an indecomposable (n,k)-code, $\mathcal{P}$, with probability of no error $Q_1(\mathcal{P}) \geqq Q_1(\mathcal{A})$.

Proof: The given code $\mathcal{A}$ is equivalent, by Theorem 2, to a code $\mathcal{A}'$ that is the sum of indecomposable codes:

$$\mathcal{A}' = \mathcal{B}_1 + \mathcal{B}_2 + \cdots + \mathcal{B}_m,$$

where $\mathcal{B}_i$ is an indecomposable (n_i, k_i)-code and $\sum k_i = k$, $\sum n_i = n$. Let $\mathcal{B}_i$ have probability of no error $Q_1(\mathcal{B}_i)$ when used with a maximum likelihood detector. Then $\mathcal{A}'$ has probability of no error $Q_1(\mathcal{A}') = Q_1(\mathcal{B}_1)Q_1(\mathcal{B}_2) \cdots Q_1(\mathcal{B}_m)$. [See remark following (7).]

We shall show below that the theorem is true for $m = 2$. The proof for general m then follows readily by induction. For, suppose the theorem to be true for $m = 2, 3, \cdots, r$. If then $\mathcal{A}' = \mathcal{B}_1 + \mathcal{B}_2 + \cdots + \mathcal{B}_r + \mathcal{B}_{r+1}$, by the induction hypothesis there is an indecomposable $(n - n_{r+1}, k - k_{r+1})$-code $\mathcal{B}'$ with $Q_1(\mathcal{B}') \geqq Q_1(\mathcal{B}_1)Q_1(\mathcal{B}_2) \cdots Q_1(\mathcal{B}_r)$. The decomposable code $\mathcal{A}'' = \mathcal{B}' + \mathcal{B}_{r+1}$ has probability of no error $Q_1(\mathcal{A}'') = Q_1(\mathcal{B})Q_1(\mathcal{B}_{r+1})$. Again by the induction hypothesis, there exists an indecomposable (n,k)-code, $\mathcal{P}$, with $Q_1(\mathcal{P}) \geqq Q_1(\mathcal{A}'') = Q_1(\mathcal{B}')Q_1(\mathcal{B}_{r+1}) \geqq Q_1(\mathcal{B}_1)Q_1(\mathcal{B}_2) \cdots Q_1(\mathcal{B}_r)Q_1(\mathcal{B}_{r+1}) = Q_1(\mathcal{A}')$. The theorem is then true also for $m = 2, 3, \cdots, r + 1$.

To prove the theorem for $m = 2$, we distinguish two cases. First suppose $n_2 \neq 1$. We can suppose the generator matrices for $\mathcal{B}_1$ and $\mathcal{B}_2$ written in M-form so that a generator matrix for $\mathcal{A}'$ has the form

$$\Omega(\mathcal{A}') = \left(\begin{array}{c:c:c:c} I_{k_1} & M_1 & 0 & 0 \\ \hdashline 0 & 0 & I_{k_2} & M_2 \end{array}\right). \tag{23}$$

Consider now the (n,k)-code $\mathcal{P}$ with generator matrix

$$\Omega(\mathcal{P}) = \left[\begin{array}{c:c:c:c} & & & 11\cdots1 \\ & & & 00\cdots0 \\ I_{k_1} & M_1 & 0 & \vdots\quad\vdots \\ & & & 00\cdots0 \\ \hdashline 0 & 0 & I_{k_2} & M_2 \end{array}\right], \tag{24}$$

where the upper right section of $\Omega(\mathcal{P})$ has one row of 1's and $k_1 - 1$ rows of zeros. We observe first that $\mathcal{P}$ is indecomposable, since $\mathcal{P}$ is equivalent to a code with generator matrix in M-form with

$$M = \left[\begin{array}{c:c} & 11\cdots1 \\ & 00\cdots0 \\ M_1 & \vdots\quad\vdots \\ & 00\cdots0 \\ \hdashline 0 & M_2 \end{array}\right].$$

Since $\mathcal{B}_1$ and $\mathcal{B}_2$ are indecomposable, both M_1 and M_2 have paths that contain all their rows, by the test of Section 1.8. A single path containing all rows of M is then easily obtained by joining together the paths for M_1 and M_2 by some of the ones of the upper right block of M. The code associated with M is thus indecomposable, and so is $\mathcal{P}$.

The last $k_1 - 1$ rows of $\Omega(\mathcal{B}_1)$ generate an $(n_1, k_1 - 1)$-code. Let the letters of this code be $B_{11}', B_{12}', \cdots, B_{1\sigma}'$, where $\sigma = 2^{k_1-1}$. Let the first row of $\Omega(\mathcal{B}_1)$ be denoted by B_{11}. Then the $\mu_1 = 2^{k_1}$ letters of $\mathcal{B}_1$ are $B_{11}', B_{12}', \cdots, B_{1\sigma}'$ and $B_{11} + B_{11}', B_{11} + B_{12}', \cdots, B_{11} + B_{1\sigma}'$. Let the letters of $\mathcal{B}_2$ be $B_{21}, B_{22}, \cdots, B_{2\mu_2}$ where $\mu_2 = 2^{k_2}$. Then the letters of $\mathcal{A}'$ can be denoted by the $\mu_1\mu_2$ symbols (B_{1i}', B_{2j}) and $(B_{11} + B_{1i}', B_{2j})$, where $i = 1, 2, \cdots, \sigma$ and $j = 1, 2, \cdots, \mu_2$. The notation here is that (B_{1i}', B_{2j}) stands for the sequence B_{1i}' followed by the sequence B_{2j}, for example.

In the notation just introduced, the $\mu_1\mu_2$ letters of $\mathcal{P}$ are (B_{1i}, B_{2j}) and $(B_{11} + B_{1i}', \bar{B}_{2j})$, where $i = 1, 2, \cdots, \sigma$ and $j = 1, 2, \cdots, \mu_2$ and $\bar{B}_{2j}$ denotes the sequence B_{2j} with its last $n_2 - k_2 = l_2$ places complemented.

23

That is, $\bar{B}_{2j}$ is obtained from B_{2j} by changing to zero every one in the last l_2 places of B_{2j} and by changing to one every zero in the last l_2 places of B_{2j}.

Consider now transmitting with $\mathcal{P}$ over a binary symmetric channel using the following decoding rules. Apply the maximum likelihood detector for $\mathcal{B}_1$ to the first n_1 digits of a received sequence R. One thus obtains a letter of $\mathcal{B}_1$, say B_{1i}. If B_{1i} is one of the letters B_{11}', B_{12}', $\cdots$, $B_{1\sigma}'$, apply the maximum likelihood detector for $\mathcal{B}_2$ to the last n_2 places of R to obtain a letter of $\mathcal{B}_2$, say B_{2j}. The pair (B_{1i}, B_{2j}) is taken as the decoded version of R. If,however, B_{1i} is one of the letters $B_{11} + B_{11}'$, $B_{11} + B_{12}'$, $\cdots$, $B_{11} + B_{1\sigma}'$, complement the last l_2 places of R, and then apply the maximum likelihood detector of $\mathcal{B}_2$ to the last n_2 digits of this new sequence derived from R. A letter B_{2j}, say, of $\mathcal{B}_2$ will be obtained. The decoded version of R is taken to $(B_{1i}, \bar{B}_{2j})$.

It is readily seen that on using the indecomposable code $\mathcal{P}$ with this decoding scheme, the probability of no error is $Q_1(\mathcal{B}_1)Q_1(\mathcal{B}_2)$. Since the maximum likelihood detector for $\mathcal{P}$ must do as well, $Q_1(\mathcal{P}) \geqq Q_1(\mathcal{B}_1)\cdot Q_1(\mathcal{B}_2) = Q_1(\mathcal{A}') = Q_1(\mathcal{A})$, and the theorem is proved for this case.

If $n_2 = 1$, but $n_1 \neq 1$, reverse the roles of $\mathcal{B}_1$ and $\mathcal{B}_2$ in the preceding argument. The case $n_1 = n_2 = 1$ has been excluded by the condition $k < n$, for $n_1 = n_2 = 1$ implies $k_1 = k_2 = 1$, or $n = k = 2$.

This completes the proof.

2.7 *Proof of Theorem 5*

The nearest neighbor distance, $d(\mathcal{A})$, of a group code $\mathcal{A}$ is the smallest of the nonzero weights of the letters of $\mathcal{A}$. If $\mathcal{A}$ and $\mathcal{A}'$ are equivalent, $d(\mathcal{A}) = d(\mathcal{A}')$, and indeed the list of weights of letters of $\mathcal{A}$ is the same set of numbers as the list of weights of the letters of $\mathcal{A}'$. It is easy to see that if $\mathcal{A} = \mathcal{B} + \mathcal{C}$ then $d(\mathcal{A}) = \min\,[d(\mathcal{B}), d(\mathcal{C})]$. Thus, if $\mathcal{A} \cong \mathcal{B}_1 + \mathcal{B}_2 + \cdots + \mathcal{B}_m$, $d(\mathcal{A}) = \min\,[d(\mathcal{B}_1), d(\mathcal{B}_2), \cdots, d(\mathcal{B}_m)]$.

The proof of Theorem 5 follows the outline of the proof of Theorem 4. The inductive part of the proof only requires substituting d's for Q's. The pertinent equations are:

$$d(\mathcal{B}') \geqq \min\,[d(\mathcal{B}_1), d(\mathcal{B}_2), \cdots, d(\mathcal{B}_r)],$$

$$d(\mathcal{A}'') = \min\,[d(\mathcal{B}'), d(\mathcal{B}_{r+1})],$$

$$\begin{aligned} d(\mathcal{P}) \geqq d(\mathcal{A}'') &= \min\,[d(\mathcal{B}'), d(\mathcal{B}_{r+1})] \\ &\geqq \min\{\min\,[d(\mathcal{B}_1), \cdots, d(\mathcal{B}_r)], d(\mathcal{B}_{r+1})\} \\ &= \min\,[d(\mathcal{B}_1), \cdots, d(\mathcal{B}_{r+1})] = d(\mathcal{A}') = d(\mathcal{A}). \end{aligned}$$

To prove the theorem for $m = 2$, we again consider a generator matrix for $\mathcal{A}'$ in the form given by (23). Without loss of generality, we suppose $d(\mathcal{A}') = d(\mathcal{B}_1)$, so that $d(\mathcal{B}_1) \leqq d(\mathcal{B}_2)$. Now suppose $l_2 = n_2 - k_2 \geqq 1$. We compare $\mathcal{A}'$ with the indecomposable code $\mathcal{P}$ given by (24). The nonzero letters of $\mathcal{P}$ are the $2^{k_1+k_2} - 1$ nontrivial linear combinations of the rows of $\Omega(\mathcal{P})$. Every such linear combination that contains one or more of the first k_1 rows of $\Omega(\mathcal{P})$ has weight $\geqq d(\mathcal{B}_1)$, since the first n_1 places will be a nonzero letter of $\mathcal{B}_1$ and the last n_2 places have weight $\geqq 0$. Every linear combination of rows of $\Omega(\mathcal{P})$ that does not contain any of the first k_1 rows is just a letter of $\mathcal{B}_2$ preceded by n_1 zeros, and hence has weight $\geqq d(\mathcal{B}_2) \geqq d(\mathcal{B}_1)$. We thus have $d(\mathcal{P}) \geqq d(\mathcal{B}_1) = d(\mathcal{A}')$.

If $l_2 = 0$, then $k_2 = n_2 = 1$, since $\mathcal{B}_2$ is assumed indecomposable. Then $d(\mathcal{B}_2) = 1$ and, since $d(\mathcal{B}_1) \leqq d(\mathcal{B}_2)$, $d(\mathcal{A}') = d(\mathcal{B}_1) = 1$. However, for every indecomposable (n,k)-code $\mathcal{P}$, we have $d(\mathcal{P}) \geqq 1 = d(\mathcal{A}')$, and so the theorem is proved for $m = 2$.

2.8 *Enumeration Formulae*

Let G be a finite group with elements $g_1, g_2, \cdots, g_r$, where r is the order of G. Define $g_i \sim g_j$ if there exists an element $g \in G$ such that $g_i = g g_j g^{-1}$. The equivalence relation $\sim$ partitions G into equivalence classes $C_1, C_2, \cdots, C_p$ called *classes of conjugate elements.* Now suppose that corresponding to each element g_i of G there is a permutation, $\sigma(g_i)$, of m objects $S_1, S_2, \cdots, S_m$ of a set S such that if $g_i g_j = g_k$, then $\sigma(g_i)\sigma(g_j) = \sigma(g_k)$. We define two of the objects of the collection S, say S_i and S_j, to be equivalent if there is a $\sigma(g_l)$, $g_l \in G$, that replaces S_i by S_j. The collection of objects S is then partitioned into equivalence classes. A well-known theorem (p. 231, Ref. 3) gives, for the number of equivalence classes N of S,

$$N = \frac{1}{r} \sum_{i=1}^{p} n(C_i)\chi(C_i). \tag{25}$$

Here $n(C_i)$ is the number of elements of G in the equivalence class C_i and $\chi(C_i)$ is the number of elements of S left invariant by any $\sigma(g_i)$, $g_i \in C_i$. [It is easy to show that if $g_i \sim g_j$, then $\sigma(g_i)$ and $\sigma(g_j)$ leave the same number of elements of S invariant.]

We apply this theorem to the enumeration of (n,k)-codes as follows. For the group G we choose the collection G_k of nonsingular $k \times k$ matrices (mod 2) of order

$$|G_k| = (2^k - 2^0)(2^k - 2^1) \cdots (2^k - 2^{k-1}). \tag{26}$$

Let $\mathbf{v}_1, \mathbf{v}_2, \cdots, \mathbf{v}_{2^k-1}$ be the nonzero k-place binary column vectors. For the sets $S_1, S_2, \cdots, S_m$ we choose the $m = (2^k - 1)^n$ possible collections of the $\mathbf{v}$'s taken n at a time (repetitions of $\mathbf{v}$'s within any S allowed). The elements of G_k permute the $2^k - 1$ vectors $\mathbf{v}$ among themselves by ordinary matrix multiplication. That is, if $g_i\mathbf{v}_j = \mathbf{v}_l$, we say that g_i induces a permutation $\mu(g_i)$ that replaces $\mathbf{v}_j$ by $\mathbf{v}_l$. The permutation $\mu(g_i)$ of the $\mathbf{v}$'s in turn induces a permutation $\sigma(g_i)$ of the sets $S_1, S_2, \cdots, S_m$. We note that if $n \leqq 2^k - 1$, then

$$\bar{m} = \binom{2^k - 1}{n}$$

of the m S's have the property of containing only distinct vectors (no repetitions), and these $\bar{m}$ special S's are permuted among themselves under $\sigma(g_i)$. We denote by $\bar{\sigma}(g_i)$ the permutation of these $\bar{m}$ special S's induced by g_i.

We now define two $k \times n$ binary matrices Ω and Ω', regardless of their rank, to be equivalent if there exists a $g \in G_k$ and an $n \times n$ permutation matrix ν such that $\Omega' = g\Omega\nu$. The number of equivalence classes of $k \times n$-matrices none of which has columns of zeros is then clearly the same as the number of equivalence classes of the sets $S_1, \cdots, S_m$. Applying (25), we write

$$T_{nk} = \frac{1}{|G_k|} \sum_i n(C_i)\chi(C_i), \tag{27}$$

$$\bar{T}_{nk} = \frac{1}{|G_k|} \sum_i n(C_i)\bar{\chi}(C_i), \tag{28}$$

where $|G_k|$ is given by (26), $n(C_i)$ is the number of elements of G_k in class C_i, and $\chi(C_i)$ and $\bar{\chi}(C_i)$ are the number of objects left invariant respectively by $\sigma(g_i)$ and $\bar{\sigma}(g_i)$, $g_i \in C_i$. The quantities T_{nk} and $\bar{T}_{nk}$ are, respectively, the number of equivalence classes of $k \times n$ matrices with no columns of zeros and the number of equivalence classes of $k \times n$ matrices with no columns of zeros and no repeated columns.

The matrices Ω in the above enumeration may have rank less than k. It is easy to show, however, that

$$S_{nk} = T_{n,k} - T_{n,k-1}, \tag{30}$$

$$\bar{S}_{nk} = \bar{T}_{n,k} - \bar{T}_{n,k-1}, \tag{31}$$

$k = 2, \cdots, n$, $n = 1, 2, \cdots$, where, as in Section 1.9, S_{nk} and $\bar{S}_{nk}$ are, respectively, the number of equivalence classes of (n,k)-codes with no column of zeros and the number with neither repeated columns nor columns of zeros. We also have $S_{n1} = 1$ for $n = 1, 2, \cdots$ and $\bar{S}_{11} = 1$, $\bar{S}_{n1} = 0$ for $n > 1$.

The group G_k has been well studied, and the detail needed to evaluate (27) and (28) can be taken from the literature. Here we omit all derivations and only present such definitions and formulae as needed for our purpose. The structure of G_k is given in detail by Dickson;[4] a recipe for getting the cycle structure of the permutations of the **v**'s induced by elements of G_k is given by Elspas.[5]

A polynomial of degree $d > 0$,

$$P(x) = x^d + a_1 x^{d-1} + a_2 x^{d-2} + \cdots + a_d ,$$

where the a's are zero or one, is said to be irreducible if it cannot be written as the product of two or more polynomials with coefficients zero or one, where each factor is of degree greater than zero. (All addition of coefficients is to be done mod 2.) For each d there are a finite number of irreducible polynomials. In what follows, we shall exclude from consideration the irreducible polynomial $P(x) = x$. The first few irreducible polynomials are $x + 1$, $x^2 + x + 1$, $x^3 + x + 1$, $x^3 + x^2 + 1$. A more comprehensive table of irreducible polynomials is given by Church,[6] where, for each irreducible polynomial, P, there is also listed the smallest integer e such that P divides $x^e - 1$. We suppose the irreducible polynomials to be numbered, and denote them by $P_1, P_2, P_3, \cdots$. We let d_i denote the degree of P_i and e_i denote the smallest integer e such that P_i divides $x^e - 1$. We further let t_d be the number of irreducible polynomials of degree d or less.

A partition of an integer α into positive integral parts $\lambda_1, \lambda_2, \cdots$, say $\alpha = \lambda_1 + \lambda_2 + \cdots + \lambda_p$, can also be written in the form

$$\alpha = 1\alpha_1 + 2\alpha_2 + \cdots + \alpha\alpha_\alpha = \sum_1^\alpha i\alpha_i .$$

Here α_i designates how many parts have the value i. We shall use boldface Greek letters to denote partitions. The absolute value sign will denote the value of the integer being partitioned. For example, $\boldsymbol{\alpha}$ will denote a particular partition,

$$\sum_1^\alpha i\alpha_i ,$$

of the integer $\alpha = |\boldsymbol{\alpha}|$. When dealing with many partitions $\boldsymbol{\alpha}_1, \boldsymbol{\alpha}_2, \boldsymbol{\alpha}_3$, etc., we shall denote the numbers of parts of various size of $\boldsymbol{\alpha}_i$ by α_{i1}, α_{i2}, etc., so that

$$|\boldsymbol{\alpha}_i| = \sum_{j=1}^{|\boldsymbol{\alpha}_i|} j\alpha_{ij} .$$

We admit the single partition of zero, **0**, into one part. For this partition, all α's are zero.

The classes of conjugate elements of G_k can be specified conveniently by t_k-place symbols. The ith place in such a class symbol corresponds to the ith of the irreducible polynomials of degree $\leq k$. Each place in such a class symbol is occupied by a partition. If the symbol for a class of G_k is

$$(\boldsymbol{\alpha}_1, \boldsymbol{\alpha}_2, \cdots, \boldsymbol{\alpha}_{t_k}), \tag{32}$$

we require

$$\sum_{i=1}^{t_k} | \boldsymbol{\alpha}_i | d_i = k. \tag{33}$$

The various classes of G_k are given by all the distinct symbols (32) that can be formed subject to (33). The sums in (27) and (28) are over such class symbols.

We now give a recipe for the integers $n(C)$ of (27) and (28). (See p. 235, Ref. 4.) We first write

$$n(C) = \frac{| G_k |}{D(C)}.$$

Then, if C is specified by (32),

$$D(C) = \prod_{j=1}^{t_k} f(\boldsymbol{\alpha}_j, d_j).$$

Here

$$f(\boldsymbol{\alpha}_i, j) = 2^{j\theta(\boldsymbol{\alpha}_i)} \prod_{l=1}^{|\boldsymbol{\alpha}_i|} \Omega(\alpha_{il}, j),$$

where

$$\Omega(r,j) = (2^{rj} - 2^{0j})(2^{rj} - 2^{1j}) \cdots (2^{rj} - 2^{(r-1)j})$$

and

$$\theta(\boldsymbol{\alpha}_i) = \sum_{j=1}^{|\boldsymbol{\alpha}_i|} \alpha_{ij}^{2}(j-1) + 2 \sum_{j=1}^{|\boldsymbol{\alpha}_i|-1} j\alpha_{ij} \sum_{l=j+1}^{|\boldsymbol{\alpha}_i|} \alpha_{il}.$$

To compute the quantities $\chi(C_i)$ and $\bar{\chi}(C_i)$ of (27) and (28), we need to know the cycle structure of the permutation of the **v**'s induced by an element of class C_i of G_k. Let an element of C_i, as given by (32), permute the **v**'s into ν_i cycles of length i, where $i = 1, 2, \cdots, 2^k - 1$. An algorithm for finding the ν's is given by Elspas.[5] Introduce indeter-

28

minates $z_1, z_2, \cdots$, and define the product of two z's by the rule

$$z_a z_b = c z_d ,$$

where c is the greatest common divisor of a and b and d is the least common multiple of a and b. Then the ν's may be obtained from

$$z_1 + \sum_{l=1}^{2^k-1} \nu_l(C) z_l = \prod_{i=1}^{t_k} \prod_{j=1}^{|\alpha_i|} H(i,j)\alpha_{ij} ,$$

where the linear forms $H(i,j)$ in the z's are obtained recursively by

$$H(i,j) = H(i,j-1) + \frac{2^{d_i(j-1)}(2^{d_i} - 1)}{q_{ij}} z_{q_{ij}} ,$$

$$i = 1, 2, \cdots ,$$

$$q_{ij} = e_i 2^{b_j},$$

where b_j is the smallest integer such that $2^{b_j} \geqq j$, and $H(i,0) = z_1$, $i = 1, 2, \cdots$.

An element of G_k permutes the **v**'s in cycles. A collection S_j of n **v**'s will remain invariant under this permutation only if S_j is composed of complete sets of the **v**'s that are permuted in cycles. It is not hard to determine the number of S_j that remain fixed when the cycle structure of the permutation of the **v**'s is given. We write only the final result:

$$\sum_0^\infty T_{nk} t^n = \frac{1}{|G_k|} \sum_i n(C_i) \prod_{j=1}^{2^k-1} (1 - t^j)^{-\nu_j(C_i)},$$

$$\sum_0^\infty \bar{T}_{nk} t^n = \frac{1}{|G_k|} \sum_i n(C_i) \prod_{j=1}^{2^k-1} (1 + t^j)^{\nu_j(C_i)}.$$

The utterly formidable series of formulae and algorithms from (32) on were used, along with (30) and (31), to compute the S_{nk} and $\bar{S}_{nk}$ given on Table I. The R_{nk} were found from the S_{nk} by a generating function scheme which will not be described in detail here. When the R_{nk} are known for $k = 1, 2, \cdots, k_0$ and $n = 1, 2, \cdots, n_0$, these numbers can be used to find the number of equivalence classes of decomposable $(n_0 + 1, k_0)$-codes, $(n_0, k_0 + 1)$-codes and $(n_0 + 1, k_0 + 1)$-codes. By subtracting the number of decomposable equivalence classes from the appropriate S_{nk}, new values of R_{nk} are found.

The programming of these formulae for the IBM 704 presented a number of interesting problems. All quantities involved are integers. In the program, they were maintained as integers. The division indicated in (27) then provides a check as to the accuracy of the sum. Unfortunately, the integers involved are frequently enormous. Modest answers in Table I of magnitude 10^1 to 10^2 were obtained as the result of computations involving integers of magnitude 10^{30}. The total machine time needed to compute the results presented was about 45 minutes.

2.9 *An Alternate Approach to Enumeration*

In Ref. 1 we regarded any subgroup of order 2^k of the group B_n of n-place binary sequences under mod 2 addition as an (n,k)-code. Thus codes with columns of zeros were admitted. It was also pointed out that G_n is the group of automorphisms of B_n. If we regard the elements of B_n as column vectors, then multiplication of each element of B_n by an $n \times n$ matrix $g \in G_n$ sends the element into a new element of B_n and this defines the automorphism associated with g.

In an automorphism of B_n, subgroups of B_n are sent into subgroups. We denote by $g\mathcal{a}$ the subgroup into which the (n,k)-code $\mathcal{a}$ is sent under the automorphism g. As g runs through G_n, $g\mathcal{a}$ runs through all N_{nk} (n,k)-codes.

Now let H be the subgroup of G_n that leaves $\mathcal{a}$ invariant, i.e., H consists of all those elements $g \in G_n$ for which $g\mathcal{a} = \mathcal{a}$. Let S_n be the subgroup of G_n consisting of all $n!$ $n \times n$ permutation matrices. Then the elements S_nH (the collection of distinct elements of G_n obtained by multiplying every element of S_n on the right by every element of H) send $\mathcal{a}$ into an equivalent code, and it is easy to show that S_nH contains all elements of G_n that send $\mathcal{a}$ into an equivalent code. Let $g_2 \in G_n$ send $\mathcal{a}$ into a nonequivalent code $\mathcal{a}_2$. Then $g_2 \notin S_nH$. Every element of the collection S_ng_2H (i.e., all elements sg_2h with $s \in S_n$, $h \in H$) then sends $\mathcal{a}$ into a code equivalent to $\mathcal{a}_2$, and again it is easily shown that every element of G_n that sends $\mathcal{a}$ into a code equivalent to $\mathcal{a}_2$ is contained in S_ng_2H.

A collection of the form S_ngH is called a *double coset* of G_n with respect to S_n and H. Two double cosets of G_n with respect to S_n and H, say S_ng_1H and S_ng_2H, are either disjoint or identical. The group G_n can thus be decomposed into disjoint double cosets $S_ng_1H, S_ng_2H, \cdots, S_ng_pH$. The argument of the preceding paragraph can be continued to show that p, the number of double cosets of G_n with respect to S_n and H, is the number, W_{nk}, of equivalence classes of (n,k)-codes (zero columns permitted).

The following formula[7] for the number, p, of double cosets of a finite group G of order $|G|$ with respect to the subgroups H_1 and H_2 respectively of order $|H_1|$ and $|H_2|$,

$$p = \frac{|G|}{|H_1||H_2|} \sum_i \frac{n_1(C_i)n_2(C_i)}{n(C_i)}, \tag{34}$$

could then be applied to the case at hand to compute W_{nk}. In (34) the sum is over the classes C_i of conjugate elements of G, $n(C_i)$ is the number of elements of G in class C_i, and $n_j(C_i)$ is the number of elements of C_i that lie in H_j, $j = 1, 2$. An appropriate choice for $\mathcal{Q}$ in the enumeration in question would be the (n,k)-code whose last $n - k$ columns are zero. The set of all matrices of G_n whose last $n - k$ rows contain only zero in their first k columns then makes up the subgroup H. We do not carry out the details of the enumeration by this method further here.

2.10 *Equivalence for M-forms*

We have commented in Section 1.2 that two equivalent Ω-matrices both in M-form may have different M-matrices. It is natural to inquire into the different M-forms possible for Ω-matrices within an equivalence class.*

The M-forms of all matrices equivalent to Ω can be obtained as follows. Make any permutation of the columns of Ω that causes the resultant matrix, Ω', to have its first k columns linearly independent. Premultiply Ω' by the inverse of the matrix formed by its first k columns.

Now let

$$\Omega = \begin{pmatrix} 100\cdots0 & m_{11}m_{12}\cdots m_{1l} \\ 010\cdots0 & m_{21}m_{22}\cdots m_{2l} \\ \vdots & \vdots \\ 000\cdots1 & m_{k1}m_{k2}\cdots m_{kl} \end{pmatrix} = (I_k \vdots M),$$

where $l = n - k$. The permutations of the columns of Ω that replace its first k columns by independent columns can be generated by repeated applications of three types of elementary permutations: (a) interchange of position of two among the last l columns of Ω; (b) interchange of position of two among the first k columns of Ω; (c) interchanging one of the first k columns with one of the last l columns. A type (a) transposition is a column transposition of M and Ω is still in M-form. A type (b) transposition involving columns i and j yields a matrix that can be brought into M-form by premultiplication by the permutation matrix that interchanges rows i and j. The new M differs from the old only by interchange of rows i and j. A type (c) transposition, which interchanges column j of M with column i of I_k, is valid only if $m_{ij} = 1$ (otherwise the first k columns of the new Ω would not be independent). Let such a transposition send Ω into Ω'. Let column j of M have ones in rows i, p_1, p_2, $\cdots$, p_r and zeros elsewhere. Then Ω' can be brought into M-form

* The equivalence described here has been investigated independently and in a more general setting by Tucker.[8]

by premultiplication by a matrix that adds row i of Ω' to rows $p_1, p_2, \cdots, p_r$. The new M-matrix is then obtained from the original M-matrix by these operations: leave column j unchanged; except in column j, add row i to rows $p_1, p_2, \cdots, p_r$. We call this a *pivotal operation on M about the position m_{ij}*, provided $m_{ij} = 1$.

Define two M-matrices to be equivalent if one can be obtained from the other by repeated applications in any order of permutations of rows or columns or by pivotal operations. Then two Ω-matrices are equivalent if and only if when reduced to M-form their M-matrices are equivalent. Equivalent M-matrices, when prefixed by a unit matrix, yield equivalent Ω-matrices. We have not been able to find a systematic method of reducing a given $k \times l$ binary matrix to a canonical form by means of pivotal operations and permutations of rows and columns.

2.11 *Miscellaneous Comments and Problems*

The Q for the sum of two codes is the product of the Q's for the summands. What is the relationship for the Q of a product in terms of the Q's of the factors? What is the relationship between the Q of a code and the Q of its dual? Answers to both of these questions probably require some detailed knowledge of the structure of the codes involved beyond a mere statement of their Q's. What detail must be known?

Decomposition of codes with respect to addition has been explored. Certain optimal properties of indecomposable codes and a unique decomposition theorem have been proved. Decomposition with respect to multiplication can be defined in a similar manner. Do analogous theorems hold in this case?

When $n < 2^k - 1$, an Ω-matrix need not have repeated columns. If an indecomposable Ω-matrix does have repeated columns, the corresponding code can be viewed as having several check digits that are identical linear combinations of the information places. Intuitively, this seems like a wasteful use of the check digits. Is it possible to prove a theorem to the effect that if $n < 2^k - 1$, there is an (n,k)-code with no repeated columns with a Q as great as that for any (n,k)-code with repeated columns? All cases of known best group codes with $n < 2^k - 1$ have no repeated columns.

A strong statement about group codes with no repeated columns that might be conjectured is the following: "Let $\mathcal{A}$ be an (n,k)-code with $n < 2^k - 2$. Let $\mathcal{B}$ be any $(n + 1, k)$-code formed from $\mathcal{A}$ by adjoining to $\Omega(\mathcal{A})$ any one of the columns already present in $\Omega(\mathcal{A})$. Let $\mathcal{C}$ be an $(n + 1, k)$-code formed by adjoining to $\Omega(\mathcal{A})$ a column $\mathbf{c}$ not already present in $\Omega(\mathcal{A})$. Then $\mathbf{c}$ can be chosen so that $Q(\mathcal{C}) \geqq Q(\mathcal{B})$ for all $\mathcal{B}$."

This conjecture has been shown not to be true for all $\mathcal{A}$. E. F. Moore of Bell Telephone Laboratories has constructed a code $\mathcal{A}$ such that the new code formed by repeating a parity check of $\mathcal{A}$ is strictly better than any code formed from $\mathcal{A}$ by adding a new type parity check. The falsity of this conjecture does not preclude the possibility of a thoerem of the sort mentioned in the previous paragraph. One should not expect to pass from a good (n,k)-code to a good $(n + 1, k)$-code in any simple manner: the structure of a best $(n + 1, k)$-code may be quite different from the structure of a best (n,k)-code.

In this connection, we point out that there are many (n,k)-codes that cannot be improved by the addition of a single parity check. This situation obtains whenever the coset leaders of the given code are unique (or, in geometrical terms, when there are no vertices of the n-cube on the boundaries of the maximum-likelihood regions). Adding a single parity check to such a code to form an $(n + 1, k)$-code leaves the value of Q unaltered.

The notions of addition and multiplication for group codes can be easily generalized to hold for block codes. How much of the theory developed remains in this case?

The foregoing are but a few of the many questions that arise naturally from this work. Most of them have not yet been investigated in any detail. We have, it is clear, raised more questions than we have answered. Perhaps this is inherent in the nature of research.

ACKNOWLEDGMENTS

Much of the work reported here was done during the Spring of 1959 while the author was a visiting professor at the University of California in Berkeley. He is indebted to his many friends in the Electrical Engineering Department there for providing a stimulating atmosphere in which to work, and is particularly indebted to Prof. A. J. Thomasian, with whom he discussed many parts of this work.

The author extends his thanks and admiration to Mrs. W. Mammel of Bell Telephone Laboratories, who by ingenious and unusual programs converted the formulae of Section 2.8 into the tables of Section 1.9 (with some aid from an IBM 704).

REFERENCES

1. Slepian, D., A Class of Binary Signalin8 Alphabets, B.S.T.J., **35**, 1956, pp. 203–234.
2. Fontaine, A. B., and Peterson, W. W., Group Code Equivalence and Optimum Codes, I.R.E. Trans., **IT-5**, 1959, pp. 60–70.

3. Riordan, J., The Combinatorial Significance of a Theorem of Pólya, J. Soc. Ind. & Appl. Math., **5**, 1957, p. 225–237.
4. Dickson, L. E., *Linear Groups*, Dover Publications, New York, 1958.
5. Elspas, B., Autonomous Linear Sequential Networks, I.R.E. Trans., **CT-6**, 1959, pp. 45–60.
6. Church, R., Tables of Irreducible Polynomials, Ann. Math., **36**, 1935, pp. 198–209.
7. Littlewood, D. E., *Theory of Group Characters and Matrix Representations of Groups*, Clarendon Press, Oxford, 1950, pp. 166–167.
8. Tucker, A. W., Combinatorial Equivalence of Matrices, mimeographed notes, Princeton Univ., Princeton, N. J.

BCH and Reed–Solomon Codes

III

Editor's Comments on Papers 13 Through 18

13 Hocquenghem: *Codes correcteurs d'erreurs*

14 Bose and Ray-Chaudhuri: *On a Class of Error Correcting Binary Group Codes*

15 Bose and Ray-Chaudhuri: *Further Results on Error Correcting Binary Group Codes*

16 Reed and Solomon: *Polynomial Codes over Certain Finite Fields*

17 Gorenstein and Zierler: *A Class of Error-Correcting Codes in p^m Symbols*

18 Mattson and Solomon: *A New Treatment of Bose–Chaudhuri Codes*

Although multiple-error-correcting codes had been discussed prior to 1959, there existed no nontrivial method of a general nature for the construction of such codes other than the Reed–Muller and maximum-length codes. In that year Hocquenghem [13] and Bose and Chaudhuri [14] independently obtained such a construction for binary codes by generalizing Hamming codes. Interestingly, the work of Bose and Chaudhuri [14] uses an earlier result of Bose (1947) on symmetrical factorial designs. These codes are now known as the Bose–Chaudhuri–Hocquenghem (BCH) codes. That the BCH codes are cyclic, an important notion introduced by Prange (1957), was shown by Peterson [19], whose approach in describing the codes is the more familiar one used today. The nonprimitive BCH codes received more attention in the second paper by Bose and Chaudhuri [15]. These codes were generalized to codes over GF(q) by Gorenstein and Zierler [17], who also describe a decoding algorithm for them, as discussed in the next section.

At about the same time, and quite independently of the Bose and Chaudhuri [14] and Hocquenghem [13] papers, Reed and Solomon [16] introduced a class of codes that were later established as a subclass of BCH codes. They are optimal in the sense that it is impossible for any linear code over GF(q) with the same length and number of codewords to have a distance greater than the Reed–Solomon codes. As such, they form an important and interesting subclass of BCH codes. The Reed–Solomon codes were used extensively by Forney (1966) in his concatenation scheme. Reed and Solomon [16] also give a decoding scheme for their codes which is a variant of the majority-logic-decoding algorithm of Reed [5].

In the final paper of the section, Mattson and Solomon [18] introduced what would now be termed a polynomial approach to coding. It is attractive from several points of view and has been used periodically over the last decade for several purposes. One very significant use to which it has been put is its extension to include multivariable polynomials for defining polynomial codes, a very general class of cyclic codes introduced by Kasami, Lin, and Peterson [29]. Mattson and Solomon [18] also consider an interesting subclass of quadratic residue codes which were introduced by Prange (1958).

Reprinted from *Chiffres,* **2,** 147–156 (1959)

13

Codes correcteurs d'erreurs

par A. Hocquenghem,

Professeur au Conservatoire des Arts et Métiers,
Ingénieur conseil à la S.E.A.

Généralisant un travail de Hamming, l'auteur construit des codes permettant de corriger k erreurs dans une transmission de digits binaires.

The paper is a generalization of Hamming's work. The author gives a coding system available to correct k errors in a transmission of binary digits.

Eine Arbeit von Hamming verallgemeinernd, entwickelt der Autor Kodes die es ermöglichen bei Übertragung binärer bits k Fehler zu korrigieren.

Обобщая работу Хамминга, автор предлагает коды, которые дают возможность исправлять k ошибок в передаче двоичных цифр.

1. Introduction.

Introduisons dans un système de transmission un mot, constitué par un nombre de n chiffres binaires :

$$a_1 \; a_2 \; \ldots\ldots\ldots \; a_n$$

Le mot reçu peut différer du mot initial par un certain nombre d'erreurs (certains chiffres a_i étant altérés en $1 - a_i$). Pour essayer de détecter et de corriger ces erreurs, on n'utilise que m chiffres du mot comme support de l'information, les chiffres restant appelés chiffres de test devant servir à la vérification du mot après la transmission. Donner une loi de détermination de ces chiffres de test en fonction des m chiffres d'information de façon à pouvoir

1a

détecter — ou corriger — un nombre maximum k d'erreurs, c'est former un code détecteur — ou correcteur — de k erreurs.

L'exemple le plus simple est le code détecteur d'une erreur. Dans ce cas $m = n - 1$, et on choisit le chiffre de test de façon que le nombre total de chiffres 1 du mot soit pair. La vérification du mot consiste alors en un test de parité.

Hamming (*Bell System Technical Journal,* 1950) a donné la loi de formation d'un code correcteur d'une erreur. Le nombre de chiffres de test est l'entier N déterminé par les inégalités

$$\text{(11)} \qquad \mathrm{Log}_2(1 + n) \leqq \mathrm{N} < 1 + \mathrm{Log}_2(1 + n)$$

Dans le cas général d'un code correcteur de k erreurs, le nombre de configurations d'erreurs possibles est :

$$\mathrm{H} = 1 + \mathrm{C}_n^1 + \mathrm{C}_n^2 + \ldots + \mathrm{C}_n^k$$

Par suite le code le plus économique utiliserait un nombre de chiffres de test égal à l'entier immédiatement supérieur à $\mathrm{Log}_2\mathrm{H}$. A part le code de Hamming, on n'a pu construire de tels codes. Ceux que nous proposons utilisent un nombre de chiffres de test égal à

$$n - m = kN$$

La différence

$$k\mathrm{N} - \mathrm{Log}_2\mathrm{H}$$

est de l'ordre de $\mathrm{Log}_2(k!)$, donc assez faible pour que ces codes soient satisfaisants.

Après avoir défini un anneau dans lequel nous ferons nos calculs, nous exposerons le code de Hamming sous cette optique, puis les principes de formation des codes qui nous conduiront à une détermination quasi-expérimentale et à une détermination systématique de ces codes. Nous terminerons par un exemple de code correcteur de 2 erreurs.

2. Définition de l'anneau $\mathcal{A}$

Les éléments de l'anneau $\mathcal{A}$ sont les nombres entiers écrits en numération binaire.

A chaque élément de l'anneau $\mathcal{A}$ nous faisons correspondre un polynôme ayant comme coefficients les chiffres de l'élément. Le polynôme est alors défini sur le corps de caractéristique 2.

Toute opération sur les éléments de $\mathcal{A}$ sera faite sur les polynômes correspondants — au cours de ces opérations tout coefficient pair sera remplacé par O, tout coefficient impair par 1.

Le résultat sera un polynôme auquel correspondra un élément de l'anneau $\mathcal{A}$.

On a donc toutes les opérations habituelles sur les nombres entiers — afin d'éviter toute ambiguïté, toutes les expressions calculées selon ces règles seront suivies de l'indication $(\mathcal{A})$.

Exemples :

Addition : $101 + 111 = 10 \quad (\mathcal{A})$
Multiplication : $101 \times 111 = 11.011 \quad (\mathcal{A})$
Puissance : $101^2 = 10.001 \quad (\mathcal{A})$
Division : $1.101 = 111 \times 10 + 11 \quad (\mathcal{A})$
En particulier : $p + p = 0,\ (p + q)^2 = p^2 + q^2 \quad (\mathcal{A})$

Lorsque le polynôme sera irréductible sur le corps de caractéristique 2, nous dirons que le nombre correspondant est irréductible (il n'admet pas, dans l'anneau $\mathcal{A}$, d'autre diviseur que lui-même et l'unité).

On peut classer évidemment les nombres dans l'anneau $\mathcal{A}$ par ordre de grandeur, mais beaucoup plus important est le nombre de chiffres. On démontre que parmi les nombres ayant un nombre de chiffres donné, il existe toujours un nombre irréductible.

Etant donné un mot écrit en binaire

$$a_1\, a_2 \ldots a_n$$

nous attacherons à chaque indice i un nombre p_i de l'anneau $\mathcal{A}$) et au mot lui-même nous attacherons le nombre

$$T = a_1\, p_1 + a_2\, p_2 + \ldots + a_n\, p_n \qquad (\mathcal{A})$$

C'est la considération du nombre T qui, grâce à un choix convenable des nombres p_i nous permettra de corriger les erreurs éventuelles.

3. Code de Hamming.

Nous retrouvons le code de Hamming en faisant

$$p_i = i$$

Les chiffres de test sont les chiffres du mot d'indices

$1, 2, 2^2, \ldots, 2^{N-1}$ (N défini par les inégalités 11).

L'information sera portée par les chiffres

$$a_3\, a_5\, a_6\, a_7\, a_9 \ldots a_n$$

On détermine les chiffres de test par la condition

$$T = \Sigma\, p_i\, a_i = 0$$

condition qui s'écrit ici

$$a_1 + 2a_2 + 4a_4 + \ldots + 2^{N-1}\, a_{2^{N-1}} = 3a_3 + 5a_5 + 6a_6 + \ldots + na_n \qquad (\mathcal{A})$$

Le second membre est un nombre binaire connu d'au plus N chiffres. L'égalité détermine donc parfaitement les valeurs des chiffres de test.

Si, après transmission, il n'y a pas d'erreur, on retrouvera $T = 0$.

S'il y a une erreur, portant par exemple sur le chiffre a_α remplacé par $(1 - a_\alpha)$, le nombre T' prendra la valeur :

$$T = a_1 + 2a_2 + \ldots + \alpha(1 - \alpha_\alpha) + \ldots + na_n = \alpha \qquad (\mathcal{A})$$

La valeur de T sera l'indice du chiffre erronné.

S'il y a deux erreurs, portant sur les chiffres d'indice α et β, T prendra la valeur :

$$T = \alpha + \beta \neq 0 \qquad (\mathcal{A})$$

S'il y a plus de deux erreurs, T pourrait être nul. Le code obtenu est donc correcteur d'une erreur, détecteur de deux erreurs.

Il est commode, pour automatiser le contrôle, de supposer les nombres p disposés en matrice. Par exemple pour $n = 7$, on aura la matrice

$$\begin{vmatrix} 0\,0\,0\,1\,1\,1\,1 \\ 0\,1\,1\,0\,0\,1\,1 \\ 1\,0\,1\,0\,1\,0\,1 \end{vmatrix}$$

Aux nombres

$$0\,0\,1\,0\,1\,1\,0 \qquad \text{et} \qquad 0\,0\,1\,0\,0\,1\,0$$

correspondront les matrices

$$\begin{vmatrix} 0\,0\,0\,0\,1\,1\,0 \\ 0\,0\,1\,0\,0\,1\,0 \\ 0\,0\,1\,0\,1\,0\,0 \end{vmatrix} \qquad \text{et} \qquad \begin{vmatrix} 0\,0\,0\,0\,0\,1\,0 \\ 0\,0\,1\,0\,0\,1\,0 \\ 0\,0\,1\,0\,0\,0\,0 \end{vmatrix}$$

Le nombre T s'obtient en faisant suivre chaque ligne de la matrice de son chiffre de parité (§ 1).

On obtient ici :

$$\begin{vmatrix} 0 \\ 0 \\ 0 \end{vmatrix} \qquad \text{et} \qquad \begin{vmatrix} 1 \\ 0 \\ 1 \end{vmatrix} = 5$$

Le premier nombre est correct, le 5e chiffre du second nombre est faux.

4. Principe d'un code correcteur de k erreurs.

Voyons maintenant à quelles conditions doivent satisfaire les nombres p pour que le calcul de T permette de corriger k erreurs.

Nous supposerons que les chiffres de test sont en nombre suffisant pour que, connaissant les chiffres d'information, on puisse réaliser la condition

(41) $$T = \Sigma a_i p_i = 0 \qquad (\mathcal{A})$$

Si après transmission, les chiffres de rang

$$\alpha_1, \alpha_2 \ldots, \alpha_j \qquad (j \leqq k)$$

sont erronés, le nombre T calculé sur le mot déformé prendra la valeur :

$$T = p_{\alpha_1} + p_{\alpha_2} + \ldots + p_{\alpha_j} \qquad (\mathcal{A})$$

Il faut que le nombre ainsi trouvé soit caractéristique des rangs $\alpha_1, \alpha_2, \ldots, \alpha_j$, c'est-à-dire que :

$$p_{\alpha_1} + p_{\alpha_2} + \ldots + p_{\alpha_j} \neq p_{\alpha'_1} + p_{\alpha'_2} + \ldots + p_{\alpha'_{j'}} \qquad (\mathcal{A})$$

lorsque

$$j \leqq k, \quad j' \leqq k$$

et les deux ensembles

$$(\alpha_1, \alpha_2, \ldots, \alpha_j), \qquad (\alpha'_1, \alpha'_2, \ldots, \alpha'_j),$$

non identiques.

Cette condition peut encore s'écrire :

(42) $$p\lambda_1 + p\lambda_2 + \ldots + p\lambda_l \neq 0 \qquad (\mathcal{A})$$

lorsque $l \leqq 2k$

et les $\lambda_1, \lambda_2, \ldots, \lambda_l$ étant tous différents.

On devra donc choisir les nombres p tels que l'addition, dans l'anneau $\mathcal{A}$, d'au plus $2k$ de ces nombres donne un résultat non nul.

Une fois déterminé un ensemble de n nombres p, il faudra choisir les chiffres de test. Il est commode pour cela de remplacer l'ensemble obtenu par un autre ensemble de n nombres mais contenant les puissances successives de 2 :

$$1, 2, 2^2, \ldots, 2^{K-1}$$

K désignant le nombre de chiffres du plus grand nombre p obtenu.

Disposons pour cela les nombres p en une matrice M de n colonnes et K lignes (K < n), chaque nombre p étant donc représenté par une colonne

$$p_i = \begin{vmatrix} \varpi_i^K \\ \\ \varpi_i^2 \\ \varpi_i^1 \end{vmatrix}$$

Si le rang de cette matrice (dans l'anneau $\mathcal{A}$) est $K' < K$, c'est que $K - K'$ lignes de cette matrice sont des combinaisons linéaires des K' lignes restantes. Si l'on supprime ces $K - K'$ lignes, on obtiendra une matrice M' de nombres p' qui vérifieront encore la condition (42).

Ceci étant, nous pourrons extraire de la matrice M' une matrice carrée Δ de K' lignes dont le déterminant calculé dans l'anneau $\mathcal{A}$ ne sera pas nul. On aura donc

$$\det \Delta = 1$$

puisque les seules valeurs possibles sont 0 ou 1. En multipliant la matrice M' par Δ^{-1}, on obtiendra la matrice M'' formée de nombres p'' tels que

$$p''_i = \begin{vmatrix} \varpi''^{K'}_i \\ \\ \varpi''^{1}_i \end{vmatrix} = \Delta^{-1} \begin{vmatrix} \varpi'^{K'}_i \\ \\ \varpi'^{1}_i \end{vmatrix}$$

et par suite les nombres p''_i vérifieront encore la condition (42). De plus, la matrice M'' contiendra à ce moment la matrice $\Delta^{-1} \times \Delta$, c'est-à-dire la matrice unité, donc l'ensemble des p'' contiendra les puissances successives de 2 :

$$1,\ 2,\ 2^2,\ \ldots,\ 2^{K'-1}$$

Les indices correspondants seront pris comme chiffres de test et la condition (41) déterminera ces chiffres en fonction des chiffres d'information par égalité de deux nombres binaires de K' chiffres.

Tout le problème se ramène donc à construire des ensembles de nombres p satisfaisant à la condition (42).

5. Formation de proche en proche d'une suite de nombres p.

Prenons d'abord :

$$p_1 = 1,\ p_2 = 2,\ p_3 = 2^2,\ \ldots,\ p_{2k} = 2^{2k-1}$$

puis :

$$p_{2k+1} = 2^{2k} - 1$$

$$p_{2k+2} = 2^{2k}$$

Ces nombres satisfont déjà aux conditions (42). Pour prolonger cette suite dans l'ordre des p croissants, supposons être arrivé au nombre p_i de l chiffres. Considérons l'ensemble des nombres p_1 à p_i et de leurs sommes dans l'anneau $\mathcal{A}$ par groupes de 2, 3, ... $(2k - 1)$. Tous les nombres obtenus ont au plus l chiffres.

S'il existe un nombre non contenu dans l'ensemble ainsi formé et compris entre p_i et 2^l, ce nombre sera pris pour valeur de

p_{i+1} (s'il y a plusieurs nombres on choisira évidemment le plus petit). Sinon on prendra $p_{i+1} = 2^l$.

On peut ainsi continuer pas à pas jusqu'à l'obtention des n nombres p. Si p_n a K chiffres, les nombres

$$1,\ 2,\ 2^2,\ \ldots,\ 2^{K-1}$$

seront inclus dans la suite des p. La suite sera donc directement utilisable pour former un code. Il restera à établir le tableau de correspondance entre les H valeurs de la somme

$$p_{\alpha_1} + p_{\alpha_2} + \ldots + p_{\alpha_j} \qquad (\mathcal{A}) \qquad (j \leq k)$$

et la valeur des indices $\alpha_1, \alpha_2, \ldots, \alpha_j$.

Le procédé ainsi défini est assez long à exploiter. Cependant, pour des valeurs raisonnables de n et k, il ne dépasse pas les possibilités d'une calculatrice de moyenne puissance.

La détermination à priori du nombre K de chiffres de test paraît assez difficile. Aussi allons-nous exposer un procédé plus systématique de recherche des nombres p.

6. Formation systématique des nombres p.

La théorie des congruences, si utilisée dans les preuves des opérations arithmétiques, va nous fournir un mode de calcul des nombres p. Désignons par ρ un nombre irréductible de N + 1 chiffres et par q' le reste de la division dans l'anneau $\mathcal{A}$ d'un nombre q par ρ. Le nombre q' aura au maximum N chiffres.

Nous poserons alors :

$$p_i = i + 2^{N}(i^3)' + 2^{2N}(i^5)' + \ \ldots \ + 2^{(k-1)N}(i^{2k-1})' \qquad (\mathcal{A})$$
$$(i = 1,\ 2,\ \ldots,\ n)$$

c'est-à-dire que le nombre p_i est formé de la juxtaposition des restes successifs de la division par ρ des puissances impaires dans l'anneau $\mathcal{A}$ du nombre i. Nous allons montrer que ces nombres p_i satisfont à la condition (42).

En effet, supposons :

$$(61) \qquad p_{\lambda_1} + p_{\lambda_2} + \ldots + p_{\lambda_l} = 0 \qquad (\mathcal{A}) \qquad (l \leq 2k)$$

Cela entraînerait :

$$(62) \qquad S_1 = S'_3 = S'_5 = \ \ldots \ = S'_{2k-1} = 0$$

en posant :

$$S_i = (\lambda_1)^i + (\lambda_2)^i + \ldots + (\lambda_l)^i \qquad (\mathcal{A})$$

et

$$S'_i \equiv S_i \pmod{\rho} \qquad (\mathcal{A})$$

Or, si nous considérons le produit

$$\Pi(\lambda_i + \lambda_j) \quad \begin{matrix} i = 2, 3, \ldots, l \\ j = 1, 2, \ldots, l-1 \end{matrix} \quad i > j \qquad (\mathfrak{A})$$

ce produit peut s'écrire sous forme d'un déterminant de Van der Monde dont le carré contiendra la ligne

$$S_1 \; S_2 \; S_3 \; \ldots\ldots \; S_l$$

Comme $S_{2i} = S_i^2$, les conditions (62) entraînent

$$\Pi(\lambda_i + \lambda_j) \equiv 0 \pmod{\rho} \qquad (\mathfrak{A})$$

Donc un des facteurs, par exemple $\lambda_i + \lambda_j$, serait divisible par ρ. Comme la somme dans $\mathfrak{A}$ des nombres $\lambda_i + \lambda_j$ a moins de chiffres que le nombre ρ, il en résulterait

$$\lambda_i + \lambda_j = 0 \qquad \lambda_i = \lambda_j$$

Par suite l'hypothèse (61) ne peut être réalisée que si au moins deux indices étaient égaux.

Donc les nombres p que nous avons formés remplissent la condition (42) et peuvent servir à former un code correcteur de k erreurs. Naturellement on les transformera comme il est indiqué au § 4 pour former une suite contenant des puissances de 2. Dans le cas général, p_i comprenant kN chiffres, il y aura lieu d'utiliser kN chiffres de test.

7. Exemple.

Nous avons formé un code de 15 chiffres correcteur pour 2 erreurs. Ici $N = \text{Log}_2 16 = 4$, il y aura 8 chiffres de test.

En prenant $\rho = 19 = 10.011$, on calcule aisément les nombres p et la matrice M :

0	1	1	1	1	0	0	1	1	1	1	1	1	1	1
0	0	1	1	0	0	0	0	1	1	1	0	0	0	1
0	0	1	0	1	0	0	1	1	1	0	0	1	0	0
1	0	1	0	0	1	1	0	1	1	0	0	0	0	0
0	0	0	0	0	0	0	1	1	1	1	1	1	1	1
0	0	0	1	1	1	1	0	0	0	0	1	1	1	1
0	1	1	0	0	1	1	0	0	1	1	0	0	1	1
1	0	1	0	1	0	1	0	1	0	1	0	1	0	1
1	2	3	4	5	6	7	8	9	10	11	12	13	14	15

Cette matrice M est de rang 8, car le déterminant formé avec les colonnes 1, 2, 4, 8, 6, 12, 7, 14 (choisies parce qu'elles présentent le plus de zéros) vaut 1.

La matrice Δ (§ 4) sera formée avec ces colonnes. En l'inversant on trouve la matrice :

$$\Delta^{-1} = \left| \begin{array}{cccccccc} 0 & 1 & 1 & 1 & 1 & 1 & 0 & 0 \\ 1 & 1 & 0 & 0 & 1 & 0 & 0 & 0 \\ 0 & 1 & 0 & 0 & 0 & 0 & 0 & 0 \\ 0 & 0 & 1 & 0 & 0 & 0 & 0 & 0 \\ 0 & 0 & 0 & 1 & 0 & 0 & 0 & 1 \\ 1 & 0 & 0 & 0 & 1 & 1 & 1 & 0 \\ 0 & 1 & 1 & 1 & 1 & 1 & 0 & 1 \\ 1 & 0 & 1 & 0 & 0 & 1 & 1 & 0 \end{array} \right|$$

Le produit $\Delta^{-1}M$ (dans l'anneau $\mathcal{A}$) donne la matrice définitive :

$$\left| \begin{array}{ccccccccccccccc} 1 & 0 & 1 & 0 & 0 & 0 & 0 & 0 & 0 & 0 & 0 & 0 & 1 & 0 & 1 \\ 0 & 1 & 0 & 0 & 1 & 0 & 0 & 0 & 1 & 1 & 1 & 0 & 0 & 0 & 1 \\ 0 & 0 & 1 & 1 & 0 & 0 & 0 & 0 & 1 & 1 & 1 & 0 & 0 & 0 & 1 \\ 0 & 0 & 1 & 0 & 1 & 0 & 0 & 1 & 1 & 1 & 0 & 0 & 1 & 0 & 0 \\ 0 & 0 & 0 & 0 & 1 & 1 & 0 & 0 & 0 & 1 & 1 & 0 & 1 & 0 & 1 \\ 0 & 0 & 0 & 0 & 0 & 0 & 1 & 0 & 0 & 1 & 1 & 0 & 1 & 0 & 0 \\ 0 & 0 & 0 & 0 & 1 & 0 & 0 & 0 & 1 & 0 & 1 & 1 & 0 & 0 & 0 \\ 0 & 0 & 1 & 0 & 1 & 0 & 0 & 0 & 0 & 1 & 0 & 0 & 1 & 1 & 1 \end{array} \right|$$

$$\begin{array}{ccccccccccccccc} 1 & 2 & 3 & 4 & 5 & 6 & 7 & 8 & 9 & 10 & 11 & 12 & 13 & 14 & 15 \end{array}$$

Les chiffres de rang

1, 2, 4, 6, 7, 8, 12, 14

serviront de chiffres de test, les 7 autres chiffres seront les supports de l'information.

Pour vérifier et corriger un mot on fera la somme

$$T = \Sigma p_i a_i \qquad (\mathcal{A})$$

Si elle est nulle, il n'y aura pas eu d'altération du mot (ou plus de 4 erreurs). Si T n'est pas nulle, on pourra retrouver les chiffres faux (en admettant qu'il n'y en ait pas plus de 2) en utilisant la table suivante qui donne les valeurs possibles de T suivies entre parenthèses des rangs des chiffres faux.

1 (14) — 2 (12) — 3 (12,14) — 4 (7) — 5 (7,14)
6 (7,12) — 8 (6) — 9 (6,14) — 10 (6,12) — 12 (6,7)
15 (9,10) — 16 (8) — 17 (8,14) — 18 (8,12) — 19 (10,11)
20 (7,8) — 24 (6,8) — 27 (2,5) — 28 (9,11) — 29 (1,13)
32 (4) — 33 (4,14) — 34 (4,12) — 36 (4,7) — 40 (4,6)
41 (5,9) — 42 (5,10) — 44 (3,13) — 46 (2,11) — 48 (4,8)
49 (1,3) — 50 (2,9) — 53 (5,11) — 61 (2,10) — 64 (2)
65 (2,14) — 66 (2,12) — 68 (2,7) — 72 (2,6) — 75 (5,8)
77 (8,13) — 78 (4,11) — 80 (2,8) — 82 (4,9) — 83 (5,6)
85 (6,13) — 88 (3,15) — 89 (5,12) — 90 (5,14) — 91 (5)
93 (4,10) — 96 (2,4) — 98 (8,9) — 102 (6,11) — 103 (5,7)
105 (1,15) — 106 (7,11) — 108 (11,12) — 110 (11) — 111 (11,14)
112 (9,12) — 113 (8,10) — 114 (9) — 115 (9,14) — 116 (13,15)
117 (6,10) — 118 (7,9) — 121 (7,10) — 122 (6,9) — 123 (4,5)
124 (10,14) — 125 (10) — 126 (8,11) — 127 (10,12) — 128 (1)
129 (1,14) — 130 (1,12) — 132 (1,7) — 135 (11,15) — 136 (1,6)
144 (1,8) — 145 (3,4) — 152 (10,15) — 153 (7,13) — 155 (9,15)
156 (13,14) — 157 (13) — 159 (12,13) — 160 (1,4) — 161 (3,8)
169 (2,15) — 176 (3,14) — 177 (3) — 178 (5,15) — 179 (3,12)
181 (3,7) — 185 (3,6) — 189 (4,13) — 192 (1,2) — 195 (3,9)
198 (5,13) — 201 (4,15) — 204 (3,10) — 219 (1,5) — 221 (2,13)
223 (3,11) — 224 (10,13) — 225 (6,15) — 232 (14,15) — 233 (15)
234 (3,5) — 235 (12,15) — 237 (7,15) — 238 (1,11) — 239 (9,13)
241 (2,3) — 242 (1,9) — 243 (11,13) — 249 (8,15) — 253 (1,10)

On remarquera que le nombre T prend 121 valeurs possibles $\left(1 + C_{15}^{1} + C_{15}^{2}\right)$ et qu'on utilise un nombre de 8 chiffres pour l'écrire. Le code utilise un chiffre de test de plus qu'il n'est théoriquement indispensable, mais il n'est pas sûr qu'on puisse construire des codes n'ayant qu'un nombre de chiffres de test strictement égal à l'entier par excès de Log_2 H.

Reprinted from *Inform. Cont.*, **3**, 68–79 (1960)

14

On A Class of Error Correcting Binary Group Codes*

R. C. Bose and D. K. Ray-Chaudhuri

University of North Carolina and Case Institute of Technology

A general method of constructing error correcting binary group codes is obtained. A binary group code with n places, k of which are information places is called an (n,k) code. An explicit method of constructing t-error correcting (n,k) codes is given for $n = 2^m-1$ and $k = 2^m-1-R(m,t) \geqq 2^m-1-mt$ where $R(m,t)$ is a function of m and t which cannot exceed mt. An example is worked out to illustrate the method of construction.

SECTION 1

Consider a binary channel which can transmit either of two symbols 0 or 1. However, due to the presence of "noise" a transmitted zero may sometimes be received as 1, and a transmitted 1 may sometimes be received as 0. When this happens we say that there is an error in transmitting the symbol. The symbols successively presented to the channel for transmission constitute the "input" and the symbols received constitute the "output."

A v-letter n-place binary signalling alphabet A_n may be defined as a set of v distinct sequences $\alpha_0, \alpha_1, \cdots, \alpha_{v-1}$ of n binary digits. The individual sequences may be called the letters of the alphabet. Given a set of v distinct messages, we get an encoder $E_{n,v}$ by setting up a $(1,1)$ correspondence between the messages and the letters of the alphabet. To transmit a message over the channel the n individual symbols of the corresponding letter of the alphabet are presented to the channel in succession. The output is then an n-place binary sequence belonging to the set B_n of all possible binary sequences. A decoder $D_{n,v}$ is obtained by partitioning B_n into v disjoint sets $S_1, S_2, \cdots, S_v$ and setting up a

* This research was supported by the United States Air Force through the Air Force Office of Scientific Research of the Air Research and Development Command, under Contract No. AF 49(638)-213. Reproduction in whole or in part is permitted for any purpose of the United States Government.

correspondence between these subsets and the letters of the alphabet so that if a sequence belonging to S_i is received as an output, it is read as the letter α_i and interpreted as the corresponding message. The encoder $E_{n,v}$ together with the decoder $D_{n,v}$ constitute a binary n-place code.

Each sequence of B_n can be regarded as an n-vector with elements from the Galois field $GF(2)$. The addition of these vectors may then be defined in the usual manner, the sum of two vectors being obtained by adding the corresponding elements (mod 2). For example, if $n = 6$ and $\gamma_1 = (110011)$ and $\gamma_2 = (101001)$ then $\gamma_1 + \gamma_2 = (011010)$. Clearly the set B_n of all binary n-place sequences forms a group under vector addition. The weight $w(\gamma)$ of any sequence is defined as the number of unities in the sequence. Thus in the example considered $w(\gamma_1) = 4$, $w(\gamma_2) = 3$. The Hamming distance $d(\gamma_1, \gamma_2)$ between two sequences γ_1 and γ_2 is defined as the number of places in which γ_1 and γ_2 do not match (Hamming, 1950). Clearly $d(\gamma_1, \gamma_2) = w(\gamma_1 + \gamma_2)$. In the example $d(\gamma_1, \gamma_2) = 3 = w(\gamma_1 + \gamma_2)$. The Hamming distance satisfies the three conditions for a metric, namely

$$(a)\ d(\gamma_1, \gamma_2) = 0 \text{ if and only if } \gamma_1 = \gamma_2 .$$

$$(b)\ d(\gamma_1, \gamma_2) = d(\gamma_2, \gamma_1),$$

$$(c)\ d(\gamma_1, \gamma_2) + d(\gamma_2, \gamma_3) \geqq d(\gamma_1, \gamma_3).$$

Let the letter α_i of the alphabet A_n be transmitted over the channel. Let ϵ_i be the vector which has unities in those places, where an error occurs in transmitting a symbol of α_i. Then ϵ_i is the noise vector. The output received is the sequence $\alpha_i + \epsilon_i$, and the number of errors is $w(\epsilon_i)$. The code is said to be t-error correcting if $\alpha_i + \epsilon_i$ belongs to S_i whenever $w(\epsilon_i) \leqq t (i = 0, 1 \cdots, v - 1)$. It is clear that under these circumstances if there are t or a lesser number of errors in transmitting a letter α_i, the received message will be correctly interpreted.

A particularly important class of codes has been studied by Slepian (1956). For this class $v = 2^k$ and the letters of the alphabet A_n form a subgroup of B_n. The null sequence is the unit element of B_n, and must also belong to A_n. We shall suppose without loss of generality that $\alpha_0 = (0, 0, \cdots, 0)$. Slepian's decoder may be described as follows: If $r = n - k$, then the group B_n can be partitioned into 2^r cosets with respect to the subgroup A_n. The coset containing a particular sequence β consists of the sequences

$$\alpha_0 + \beta, \alpha_1 + \beta, \cdots, \alpha_{v-1} + \beta.$$

In the jth coset we can choose a sequence β_j whose weight does not exceed the weight of any other sequence in the coset, and call it the coset leader. Let $\beta_0, \beta_1, \cdots, \beta_{u-1}$, $(u = 2^r)$ be the coset leaders, where $\beta_0 = \alpha_0$ is the null sequence and leader of the 0th coset A_n. Let S_j be the set of sequences

$$\alpha_j + \beta_0, \alpha_j + \beta_1, \cdots, \alpha_j + \beta_{u-1} \qquad j = 0,1, \cdots, v - 1.$$

Then the decoder is obtained by partitioning B_n into $S_0, S_1, \cdots, S_{v-1}$ and setting up the rule that if the sequence received as an output belongs to S_j, it is read as the letter α_j. The code thus obtained may be called an (n,k) binary group code. It is clear that a transmitted message will be correctly interpreted if and only if the error vector happens to be a coset leader. Hence a necessary and sufficient condition for the code to be t-error correcting is that if β is any n-place binary sequence for which $w(\beta) \leq t$, then β is a coset leader. The following lemma, due to Hamming (1950), is then easy to deduce.

LEMMA 1. The necessary and sufficient condition for an (n,k) binary group code to be t-error correcting is that each letter of the alphabet except the null letter has weight $2t + 1$ or more.

Since the $v = 2^k$ messages can be transmitted by a k-place binary code if there is no possibility of error, the number $r = n - k$ is called the redundancy for an (n,k) binary group code. In constructing a t-error correcting (n,k) binary group code for given n and t one would like to maximize k (that is, maximize the number of different messages that it is possible to transmit). Varšamov (1957) has shown that if k satisfies the inequality

$$S_r^{2t-1} + \binom{k-1}{1} S_r^{2t-2} + \cdots + \binom{k-1}{2t-2} S_r^{1} + \binom{k-1}{2t-1} < 2^r \qquad (1)$$

where

$$S_r^{q} = 1 + \binom{r}{1} + \binom{r}{2} + \cdots + \binom{r}{q} \qquad (2)$$

then a t-error correcting (n,k) group code exists.

The main result of the present paper is the following: If $n = 2^m - 1$, then there exists a t-error correcting (n,k) binary group code with $k \geqq 2^m - 1 - mt$.

The method of proof is constructive and is illustrated by considering the case $n = 15$, $t = 3$, for which a 3-error correcting (15,5) binary group code is explicitly obtained.

As an example of comparison between Varšamov's result and our theorem consider the case $n = 31$. Varšamov's result then shows that a 2-error correcting binary group code can be obtained with $k = 18$, and a 3-error correcting binary group code can be obtained with $k = 13$ but is inconclusive for larger values of k. Our method, however, gives an explicit construction for a 2-error correcting binary group code with $k = 21$, and a 3-error correcting binary group code with $k = 16$.

The following table gives some of the values of n, k and t for which a t-error correcting (n,k) binary group code can be constructed by our method. The transmission rate $R = k/n$ is also given.

TABLE I

t	n	k	R
1	15	11	0.73
2	15	7	0.47
2	31	21	0.68
2	63	51	0.81
2	127	113	0.89
3	15	5	0.33
3	31	16	0.52
3	63	45	0.71
3	127	106	0.83
4	63	39	0.64
4	127	99	0.78
5	127	92	0.72

SECTION 2

We shall now prove a theorem which gives a necessary and sufficient condition for the existence of a t-error correcting (n,k) group code. This theorem appears in a different form in an earlier paper by Bose (1947) but is given here for the sake of completeness.

THEOREM 1. The necessary and sufficient condition for the existence of a t-error correcting (n,k) binary group code is the existence of a matrix A of order $n \times r$ and rank $r = n - k$ with elements from $GF(2)$, such that any set of $2t$ row vectors from A are independent.

PROOF OF SUFFICIENCY. The matrix A has the property (P_{2t}) that any

$2t$ row vectors of A are independent. Clearly $r \geqq 2t$. The property (P_{2t}) is invariant under the following operations: (1) interchange of two rows or columns and (2) replacement of the ith column by the sum of ith and jth column, $i \neq j$. By these operations A can be transformed to the matrix.

$$A^* = \left\| \begin{matrix} I_r \\ C \end{matrix} \right\| \tag{3}$$

where A^* has the property (P_{2t}), I_r is the unit matrix of order r, and C is a matrix of order $k \times r$. Consider the matrix

$$C^* = \|C, I_k\|. \tag{4}$$

Then C^* is of order $k \times n$. We shall show that the k rows of C^* (under vector addition (mod 2)) are generators of a group G of order 2^k such that if α is any arbitrary (nonnull) element of G, then $w(\alpha) \geqq 2t + 1$. Let α be the sum of any d row vectors of C^*, $d \leqq k$. We can write $\alpha = (\gamma, \epsilon)$, where γ is the part coming from C and ϵ the part coming from I_k. Now

$$w(\alpha) = w(\gamma) + w(\epsilon) = w(\gamma) + d.$$

Hence

$$w(\alpha) \geqq 2t + 1 \text{ if } d > 2t.$$

Suppose $d \leqq 2t$. If $w(\alpha) < 2t + 1$, then $w(\gamma) \leq 2t - d$. Let $w(\gamma) = c$. There are exactly c positions in γ which are occupied by unity. Corresponding to each such position we can find a row vector of I_r which has unity in this position (and zero in all other positions). Then these c vectors of I_r together with the d row vectors of C whose sum is γ, constitute a set of $c + d$ vectors which are dependent. Since $c + d \leqq 2t$, this contradicts the fact that A^* has the property (P_{2t}). Thus the weight of any nonnull element of G is greater than or equal to $2t + 1$. It follows from Lemma 1 that the sequences of the subgroup generated by the k rows of C^* form the alphabet of a t-error correcting (n,k) group code.

Proof of Necessity. Suppose there exists a t-error correcting (n,k) binary group code. We can then find a set of k n-place binary sequences, or n-vectors with elements from $GF(2)$, which under addition generate the group of sequences which constitute the letters of the alphabet. By Lemma 1 if α is a sequence of this group $w(\alpha) \geq 2t + 1$. Consider the $k \times n$ matrix C^* whose row vectors are given by these sequences. If we interchange any two rows or columns of C^*, or replace the ith row of C^* by the sum of the ith and the jth row $(i \neq j)$, the transformed matrix still retains the property that its rows generate under addition a group, each sequence of which has weight $2t + 1$ or more. Hence we can without loss of generality take C^* in the canonical form (4) where C is of order $r \times k$ and I_k is the unit matrix of order k. By retracing the arguments used in proving the first part of the theorem, we see that the matrix A^* of order $n \times r$, given by (3), has the property that any two $2t$ row vectors are independent. This proves that the condition of the theorem is necessary.

COROLLARY 1. The existence of a t-error correcting (n,k) binary group code implies the existence of a t-error correcting $(n - c, k - c)$ binary group code, $0 < c < k$.

If in the matrix C^* given by (4) we delete the last c rows and the last c columns, we get a matrix

$$C_1^* = \|C_1, I_{k-c}\|$$

of order $(k - c) \times (n - c)$, the rows of which generate a group for which each nonnull element is of weight $2t + 1$ or more. The rows of C_1^* generate the alphabet of the required code.

Let V_r denote the vector space of all r-vectors whose elements belong to GF(2). One may then ask the following question. What is the maximum number of vectors in a set Σ chosen from V_r, such that any $2t$ distinct vectors from Σ are independent. This number may be denoted $n_{2t}(r)$, and the problem of finding the set Σ may be called the packing problem (of order $2t$) for V_r. For a given t, $n_{2t}(r)$ is a monotonically increasing function of r.

Let $k = k_t(n)$ denote the maximum value of k such that a t-error correcting (n,k) binary group code for given t and n exists. We can then state the following.

THEOREM 2. If $n_{2t}(r) \geq n > n_{2t}(r - 1)$, then $k_t(n) = n - r$.

From Theorem 1 there exists a t-error correcting $[n_{2t}(r), n_{2t}(r) - r]$ binary group code. Taking $c = n_{2t}(r) - n$ in Corollary 1, there exists a t-error correcting $(n, n - r)$ group code. But a t-error correcting $(n, n - r + 1)$ binary group code cannot exist, since from Theorem 1 its existence would imply that $n_{2t}(r - 1) \geq n$. Hence $k_t(n) = n - r$ is the maximum value of k for which a t-error correcting (n,k) binary group code exists.

Thus the problem of finding a t-error correcting n-place binary group code, with the maximum transmission rate k/n, is equivalent to determining the smallest r for which there exists a set of n or more distinct vectors of V_r, such that any $2t$ distinct vectors from the set are independent.

SECTION 3

The theorem to be proved in the next section depends upon the following lemma.

LEMMA 2. If $x_1, x_2, \cdots, x_l$ are different nonzero elements of the Galois field $GF(2^m)$, then the equations

$$x_1^{2i-1} + x_2^{2i-1} + \cdots + x_l^{2i-1} = 0, \; i = 1,2, \cdots, t \qquad (5)$$

cannot simultaneously hold if $l \leqq 2t$.

Suppose, if possible, the Eqs. (5) hold simultaneously. Let

$$x^l + p_1 x^{l-1} + p_2 x^{l-2} + \cdots + p_l = 0 \qquad (6)$$

be the algebraic equation whose roots are $x_1, x_2, \cdots, x_l$. Then p_j belongs to $GF(2^m)$ and is the sum of the products of the roots taken j at a time ($j = 1,2, \cdots, l$). We define s_j as the sum of the jth powers of the roots. For a field of characteristic 2 the well-known relations between the symmetric functions s_j and p_j become (Levi, 1942),

$$\begin{aligned} & s_1 + \delta_1 p_1 = 0 \\ & s_2 + p_1 s_1 + \delta_2 p_2 = 0 \\ & s_3 + p_1 s_2 + p_2 s_1 + \delta_3 p_3 = 0 \\ & \vdots \\ & s_l + p_1 s_{l-1} + p_2 s_{l-2} + \cdots + \delta_l p_l = 0 \end{aligned} \qquad (7)$$

where $\delta_i = 0$ or 1 according as i is even or odd. From Eqs. (5) $s_j = 0$ when j is odd ($j < 2t$). It then follows from (7) that $s_j = 0$ if j is even ($j \leqq l$) and $p_j = 0$ if j is odd ($j \leqq l$).

CASE I. If l is odd then $p_l = x_1 x_2 \cdots x_l \neq 0$, since $x_1, x_2, \cdots, x_l$ are nonzero. This is a contradiction.

CASE II. If l is even, say $l = 2c$, the Eq. (6) becomes

$$x^{2c} + p_2 x^{2c-2} + \cdots + p_{2c} = 0 \tag{8}$$

therefore

$$(x^c + q_1 x^{c-1} + \cdots + q_c)^2 = 0 \tag{9}$$

where q_j is the unique square root of p_{2j} in $GF(2^m)$. Hence (6) cannot have more than c distinct roots, which again is a contradiction, since $x_1, x_2, \cdots, x_l$ are distinct by hypothesis.

Hence the lemma is true whether l is odd or even.

SECTION 4

Let V_m be the vector space of m-vectors with elements from $GF(2)$. We can institute a correspondence between the vector $\alpha = (a_0, a_1, \cdots, a_{m-1})$ of $\mathbf{V}_m$ and the element $a_0 + a_1 x + \cdots + a_{m-1} x^{m-1}$ of $GF(2^m)$, where x is a given primitive element of the field. This is a (1,1) correspondence in which the null vector α_0 of V_m corresponds to the null element of $GF(2^m)$, and the sum of any two vectors of V_m corresponds to the sum of the corresponding elements of $GF(2^m)$. We can therefore identify the vector α of V_m and the corresponding element of $GF(2^m)$. This in effect defines a multiplication of the vectors of V_m and converts it into a field. In particular we can speak of powers of any vector.

Let V_{mt} be the vector space of all mt-vectors with elements from $GF(2)$. To any vector α_i of V_m there corresponds a unique vector α_i^* of V_{mt} defined by

$$\alpha_i^* = (\alpha_i, \alpha_i^3, \cdots, \alpha_i^{2t-1}) \tag{10}$$

though the converse is not true.

There are $n = 2^m - 1$ distinct nonnull vectors in V_m. Let

$$M^* = \begin{Vmatrix} \alpha_1, \alpha_1^3, \cdots, \alpha_1^{2t-1} \\ \alpha_2, \alpha_2^3, \cdots, \alpha_2^{2t-1} \\ \vdots \\ \alpha_n, \alpha_n^3, \cdots, \alpha_n^{2t-1} \end{Vmatrix} \tag{11}$$

be the $n \times mt$ matrix, which has for row vectors the corresponding vectors $\alpha_1^*, \alpha_2^*, \cdots, \alpha_n^*$. We shall show that M^* has the property

(P_{2t}) that any of $2t$ distinct row vectors belonging to M^* are independent. For this it is sufficient that the sum of any l row vectors of M^*, $l \leqq 2t$, is nonnull. This is ensured by Lemma 2, since α_i can also be regarded as elements of $GF(2^m)$.

Now rank $(M^*) \leqq mt$. Since there is essentially only one Galois field $GF(2^m)$, this rank is a definite function of m and t and will be denoted by $R(m,t)$. When $R(m,t) < mt$, we can choose $R(m,t)$ independent columns of M^*, and delete the other columns dependent on them. The matrix A so obtained has still the property (P_{2t}). Using Theorem 1 we have

THEOREM 3. If $n = 2^m - 1$, we can obtain a t-error correcting (n,k) binary group code where

$$k = 2^m - 1 - R(m,t) \geq 2^m - 1 - mt.$$

When n is not of the form $2^m - 1$ t-error correcting (n,k) binary group codes can be deduced from those obtainable from Theorem 3, by using Corollary 1 of Theorem 1. Stronger results than those which can be obtained in this way will be given in a subsequent communication.

SECTION 5

The proofs of the theorems in Sections 2 and 4 are constructive in the sense that they give an actual procedure for obtaining the required codes. We shall illustrate the procedure to be followed by taking the case $m = 4$, $t = 3$. Then $n = 15$ and the rank $R(m,t)$ turns out to be 10. We thus obtain a 3-error correcting (15,5) group code. The roots of the equation

$$x^4 = x + 1 \tag{12}$$

are primitive elements of $GF(2^4)$, that is all the nonzero elements of the field can be expressed as the powers of a root x (Carmichael, 1937, p. 262). Using (12) the powers of n can be expressed alternatively as polynomials in x of degree 3 or less.The 15 nonzero elements or vectors are listed in the following table in two equivalent forms, (1) as powers of the primitive element, and (2) as polynomials of degree 3 or less in the primitive element. We thus have the following table of the 15 nonzero elements or vectors.

$$
\begin{array}{llllll}
x^0 & = 1 & & = (1,0,0,0) & = \alpha_1 \\
x & = & x & = (0,1,0,0) & = \alpha_2 \\
x^2 & = & x^2 & = (0,0,1,0) & = \alpha_3 \\
x^3 & = & x^3 & = (0,0,0,1) & = \alpha_4 \\
x^4 & = 1 + x & & = (1,1,0,0) & = \alpha_5 \\
x^5 & = & x + x^2 & = (0,1,1,0) & = \alpha_6 \\
x^6 & = & x^2 + x^3 & = (0,0,1,1) & = \alpha_7 \\
x^7 & = 1 + x & + x^3 & = (1,1,0,1) & = \alpha_8 \\
x^8 & = 1 & + x^2 & = (1,0,1,0) & = \alpha_9 \\
x^9 & = & x \quad + x^3 & = (0,1,0,1) & = \alpha_{10} \\
x^{10} & = 1 + x + x^2 & & = (1,1,1,0) & = \alpha_{11} \\
x^{11} & = & x + x^2 + x^3 & = (0,1,1,1) & = \alpha_{12} \\
x^{12} & = 1 + x + x^2 + x^3 & & = (1,1,1,1) & = \alpha_{13} \\
x^{13} & = 1 & + x^2 + x^3 & = (1,0,1,1) & = \alpha_{14} \\
x^{14} & = 1 & + x^3 & = (1,0,0,1) & = \alpha_{15}
\end{array}
$$

In obtaining the powers of the elements it should be remembered that each nonzero element of $GF(2^m)$ satisfies $x^{2^m-1} = 1$. Since $m = 4$ we have

$$x^{15} = 1.$$

Thus for example

$$\alpha_7^3 = (x^6)^3 = x^{18} = x^3 = (0,0,0,1) = \alpha_4$$

$$\alpha_7^5 = (x^6)^5 = x^{30} = x^0 = (1,0,0,0) = \alpha_1 .$$

It is now easy to calculate the matrix M^* given by (11). For example, the seventh row is $(\alpha_7, \alpha_7^3, \alpha_7^5)$ or (0 0 1 1, 0 0 0 1, 1 0 0 0). Thus

$$
M^* = \left\| \begin{array}{cccc} 1\,0\,0\,0 & 1\,0\,0\,0 & 1\,0\,0\,0 \\
0\,1\,0\,0 & 0\,0\,0\,1 & 0\,1\,1\,0 \\
0\,0\,1\,0 & 0\,0\,1\,1 & 1\,1\,1\,0 \\
0\,0\,0\,1 & 0\,1\,0\,1 & 1\,0\,0\,0 \\
1\,1\,0\,0 & 1\,1\,1\,1 & 0\,1\,1\,0 \\
0\,1\,1\,0 & 1\,0\,0\,0 & 1\,1\,1\,0 \\
0\,0\,1\,1 & 0\,0\,0\,1 & 1\,0\,0\,0 \\
1\,1\,0\,1 & 0\,0\,1\,1 & 0\,1\,1\,0 \\
1\,0\,1\,0 & 0\,1\,0\,1 & 1\,1\,1\,0 \\
0\,1\,0\,1 & 1\,1\,1\,1 & 1\,0\,0\,0 \\
1\,1\,1\,0 & 1\,0\,0\,0 & 0\,1\,1\,0 \\
0\,1\,1\,1 & 0\,0\,0\,1 & 1\,1\,1\,0 \\
1\,1\,1\,1 & 0\,0\,1\,1 & 1\,0\,0\,0 \\
1\,0\,1\,1 & 0\,1\,0\,1 & 0\,1\,1\,0 \\
1\,0\,0\,1 & 1\,1\,1\,1 & 1\,1\,1\,0
\end{array} \right\|
$$

where the vertical divisions separate the parts coming from α, α^3 and α^5. From M^* we can drop the last null column and the 11th column which is identical with the 10th. The 10×15 matrix of rank 10 so obtained we can take as the matrix A of Theorem 1. From what has been shown in Section 4, this matrix has the property $P(6)$ that any 6 row vectors are independent. Using operations (1) and (2) of Section 2, we can then transform A to A^* where

$$A^* = \left\| \begin{matrix} I_{10} \\ C \end{matrix} \right\|$$

and I_{10} is the unit matrix of order 10, and C is the 5×10 matrix given by

$$C = \left\| \begin{matrix} 1 & 1 & 1 & 0 & 1 & 1 & 0 & 0 & 1 & 0 \\ 0 & 1 & 1 & 1 & 1 & 0 & 0 & 1 & 0 & 1 \\ 1 & 1 & 0 & 1 & 1 & 0 & 1 & 1 & 1 & 0 \\ 0 & 1 & 1 & 0 & 0 & 1 & 1 & 1 & 1 & 1 \\ 1 & 1 & 0 & 1 & 0 & 1 & 1 & 0 & 0 & 1 \end{matrix} \right\|$$

Taking

$$C^* = \|C, I_5\|$$

we have a matrix of order 5×15 whose rows generate under vector addition (mod 2), the group of 32 sequences which constitute the letters of the alphabet of the required 3-error correcting (15,5) binary group alphabet. It is easy to verify that of the 31 nonnull sequences 15 have weight 7, 15 have weight 8 and one has weight 15, which checks with Lemma 1.

RECEIVED: September 24, 1959; revised October 20, 1959.

REFERENCES

BOSE, R. C. (1947). Mathematical theory of the symmetrical factorial design. *Sankhya* **8**, 107–166.

CARMICHAEL, R. D. (1937). "Introduction to the Theory of Groups of Finite Order." Gin and Co., New York.

DWORK, B. M. AND HELLER, R. M. (1959). Results of a geometric approach to the theory and construction of non-binary, multiple error and failure correcting codes. IRE *Convention Record*, Pt. 4, pp. 123–192.

HAMMING, R. W. (1950). Error detecting and error correcting codes. *Bell System Tech. J.* **29**, 147–160.

LEVI, F. W. (1942). "Algebra," Vol. I p. 147. University of Calcutta.

SACKS, G. E. (1958). Multiple error correction by means of parity checks. IRE *Trans. on Information Theory,* **IT-4,** pp. 145–147.

SLEPIAN, D. (1956). A class of binary signalling alphabets. *Bell System Tech. J.* **35,** 203–234.

VARSAMOV, R. R. (1957). The evaluation of signals in codes with correction of errors. *Doklady Akad. Nauk SSSR* [*N.S.*] **117,** 739–741 (Russian).

Copyright © 1960 by Academic Press, Inc.
Reprinted from *Inform. Contr.*, **3**, 279–290 (1960)

15

Further Results on Error Correcting Binary Group Codes*

R. C. Bose and D. K. Ray-Chaudhuri

University of North Carolina and Case Institute of Technology

The present paper is a sequel to the paper "On a class of error-correcting binary group codes", by R. C. Base and D. K. Ray-Chaudhuri, appearing in *Information and Control* in which an explicit method of constructing a t-error correcting binary group code with $n = 2^m - 1$ places and $k = 2^m - 1 - R(m,t) \geqq 2^m - 1 - mt$ information places is given. The present paper generalizes the methods of the earlier paper and gives a method of constructing a t-error correcting code with n places for any arbitrary n and $k = n - R(m,t) \geqq [(2^m - 1)/c] - mt$ information places where m is the least integer such that $cn = 2^m - 1$ for some integer c. A second method of constructing t-error correcting codes for n places when n is not of the form $2^m - 1$ is also given.

SECTION I

This paper is a continuation of our previous paper, Bose and Ray-Chaudhuri (1960), "On a class of error correcting binary group codes." The notation used there will be followed throughout, with the minimum of explanation.

It was shown that we can obtain a t-error correcting n-place binary group code (n,k) with k information places, if $n = 2^m - 1$ and $k = n - R(m,t)$ where $R(m,t) \leqq mt$ is the rank of a certain matrix whose properties have been investigated. Peterson (1960) has investigated certain interesting properties of these codes, and given the exact value of $R(m,t)$. In Section II, we have generalized our results to the case when $n = (2^m - 1)/c$ where c is the smallest integer for which $cn + 1$ is a power of 2. This generalization enables us to obtain as a special case certain

* This research was supported in part by the United States Air Force Office of Scientific Research of the Air Research and Development Command, under Contract No. AF(638)-213. Reproduction in whole or in part is permitted for any purpose of the United States Government.

codes with the same values of n and k as those investigated by Prange (1958–1959).

Let V_r denote the vector space of r-vectors with coordinates from $GF(2)$. Following the notation of Bose and Chaudhuri (1960) we shall denote by $n_{2t}(r)$ the maximum number of vectors that it is possible to choose in V_r such that no $2t$ are dependent. A matrix with elements from $GF(2)$ is said to possess the property (P_{2t}) if no set of $2t$ rows are dedependent. It was shown in our earlier paper that the problem of finding a t-error correcting n-place binary group code (n,k) with k information places, and the maximum transmission rate k/n can be completely solved if we can determine $n_{2t}(r)$ for every value of r and can construct a matrix with r columns and $n_{2t}(r)$ rows, possessing the property (P_{2t}). We constructed a matrix with mt columns and $2^m - 1$ rows, possessing the property (P_{2t}), which establishes the inequality

$$n_{2t}(mt) \geqq 2^m - 1 \tag{1}$$

In Section III, of the present paper we shall find lower bounds for $n_{2t}(r)$ for values of r which are not multiples of t, and construct the corresponding matrix with the property (P_{2t}). This enables us in certain instances to obtain t-error correcting (n,k) binary group codes for which the transmission rate k/n is better than for codes obtainable by using corollary 1, Theorem 1 of our earlier paper.

In Section IV we have given a table which, for given $n \leqq 100$ and $t \leqq 6$, enables us to calculate the best corresponding value of k obtainable by our methods.

SECTION II

For a given positive integer n, let $c = c(n)$ be the smallest integer such that $1 + cn$ is a power of 2. Let this power be denoted by $m = m(n)$. Thus,

$$n = (2^m - 1)/c \tag{2}$$

For example, if $n = 21$, then $c = 3$, $m = 6$; if $n = 31$, $c = 1$, $m = 5$. Again, if $n = 73$, $c = 7$, $m = 9$.

Let x be a primitive element of the Galois field $GF(2^m)$. Then

$$1, x, x^2, \cdots, x^{nc-1}$$

are all the distinct nonzero elements of the field and

$$x^{nc} = 1.$$

Each element of $GF(2^m)$ can be expressed as a polynomial of degree $m - 1$ or less with coefficients from $GF(2)$. Let V_m be the vector-space of m-vectors, with elements from $GF(2)$. Then, as explained in Bose and Ray-Chaudhuri (1960) we can institute a (1,1) correspondence between the vector $\alpha = (a_0, a_1, \cdots, a_{m-1})$ of V_m, and the element

$$a_0 + a_1x + \cdots + a_{m-1}x^{m-1}$$

of $GF(2^m)$. Then the null vector α_0 of V_m corresponds to the null element of $GF(2^m)$, and the sum of any two vectors of V_m corresponds to the sum of the corresponding elements of $GF(2^m)$. We can then identify the vector α of V_m and the corresponding element of $GF(2^m)$. This in effect defines a multiplication of the vectors of V_m and converts it into a field. In particular, we can speak of the powers of any vector. Let us set

$$\begin{aligned} \alpha_i = x^{ci} &= a_{i0} + a_{i1}x + \cdots + a_{i,m-1}x^{m-1} \\ &= (a_{i0}, a_{i1}, \cdots, a_{i,m-1}) \end{aligned} \tag{3}$$

where $i = 1, 2, \cdots, n$. Then $\alpha_1, \alpha_2, \cdots, \alpha_n$ are all the distinct elements of $GF(2^m)$ which are powers of x^c, that is, α_1. In particular, $\alpha_n = x^{cn} = 1$. Let

$$\alpha_i^* = (\alpha_i, \quad \alpha_i^3, \quad \cdots, \quad \alpha_i^{2t-1}) \tag{4}$$

and

$$M^* = \begin{Vmatrix} \alpha_1, & \alpha_1^3, & \cdots & \alpha_1^{2t-1} \\ \alpha_2, & \alpha_2^3, & \cdots & \alpha_2^{2t-1} \\ \cdots & \cdots & \cdots & \cdots \\ \alpha_n, & \alpha_n^3, & \cdots & \alpha_n^{2t-1} \end{Vmatrix} \tag{5}$$

When the α_i's are regarded as m-vectors over $GF(2)$, M^* is a matrix of order $n \times mt$ with elements from $GF(2)$. We shall now prove

LEMMA 1. Any $2t$ row vectors belonging to M^* are independent, i.e., M^* possesses the property (P_{2t}).

This result was proved in our earlier paper by using the properties of power sums. It is possible to generalize this proof. However, we shall give the following alternative proof based on considerations suggested by W. W. Peterson in a private communication.

Let $\beta_1, \beta_2, \cdots, \beta_{2t}$ be any $2t$ elements of $GF(2^m)$ chosen out of $\alpha_1, \alpha_2, \cdots, \alpha_n$. We then have to show that the matrix

$$D = \left\| \begin{matrix} \beta_1 & \beta_1^3 & \cdots & \beta_1^{2t-1} \\ \beta_2 & \beta_2^3 & \cdots & \beta_2^{2t-1} \\ \vdots & \vdots & & \vdots \\ \beta_{2t} & \beta_{2t}^3 & \cdots & \beta_{2t}^{2t-1} \end{matrix} \right\|$$

has rank $2t$. Since $x \rightarrow x^2$ is an automorphism of $GF(2^m)$, any linear relation between $\beta_1^u, \beta_2^u, \cdots, \beta_{2t}^u$ implies a corresponding linear relation among $\beta_1^{2u}, \beta_2^{2u}, \cdots, \beta_{2t}^{2u}$ and vice versa. Hence, the rank of

$$D_1 = \left\| \begin{matrix} \beta_1 & \beta_1^2 & \cdots & \beta_1^{2t-1} & \beta_1^{2t} \\ \beta_2 & \beta_2^2 & \cdots & \beta_2^{2t-1} & \beta_2^{2t} \\ \vdots & \vdots & & \vdots & \vdots \\ \beta_{2t} & \beta_{2t}^2 & \cdots & \beta_{2t}^{2t-1} & \beta_{2t}^{2t} \end{matrix} \right\|$$

is the same as the rank of D. However.

$$\det D_1 = \beta_1\beta_2 \cdots \beta_{2t} \prod_{j<i}^{2t} (\beta_i - \beta_j) \neq 0$$

since $\beta_1, \beta_2, \cdots, \beta_{2t}$ are all distinct and nonzero. This shows that rank $(D_1) = 2t$ and completes the proof of the Lemma.

Let $mt < n$. The columns of M^* are not always independent as is clear from the example for the case $n = 15$, $c = 1$, $m = 4$, $t = 3$, discussed in Section 5 of Bose and Chaudhuri (1960). As before, we shall denote the rank of M^* by $R(m,t)$. If $R(m,t) < mt$, then we can choose $R(m,t)$ independent columns of M^* and drop the remaining columns of M^* and thus get a matrix of order $n \times R(m,t)$ with the property (P_{2t}).

Lemma 2. The rank $R(m,t)$ is the number of distinct residue classes (mod n) among the integers $2^j u (u = 1, 3, \cdots, 2t - 1; j \geqq 0)$.

This Lemma has been proved by Peterson (1960) for the special case $c = 1$, and his proof can be easily extended to the general case. We shall make a few remarks useful for application of the Lemma.

Denote by $(2^j u)$ the residue class corresponding to the integer $2^j u$. Since

$$\begin{aligned} 2^m u &= (2^m - 1)u + u \\ &= ncu + u \\ &= u \pmod{n} \end{aligned}$$

there cannot be more than m distinct residue classes among $(2^j u)$ with a fixed value of u, and in counting the number of residue classes it is sufficient to confine ourselves to values of j in the range $0 \leqq j \leqq m - 1$. Hence,

$$R(m,t) \leqq mt.$$

If we arrange the integers $2^j u$ reduced (mod n) in a rectangular scheme, each row corresponding to one value of u, then

(i) If k is the least nonzero positive integer such that

$$u = 2^k u (\text{mod } n)$$

then $k \leqq m$. If $k = m$ the residue classes in the corresponding row are all distinct. If $k < m$, then k is a factor of m, and there are k distinct residue classes in the corresponding row.

(ii) If any two rows have one element in common they coincide entirely.

(a) To u we can associate the set of m columns of the submatrix

$$M_u^* = \begin{Vmatrix} \alpha_u \\ \alpha_u^3 \\ \vdots \\ \alpha_u^n \end{Vmatrix} \tag{6}$$

of M. The number of independent columns in M_u^* is exactly k. We can therefore delete $m - k$ suitable columns from M_u^* without changing the rank of M^*, or the property (P_{2t}).

(b) When two rows of the scheme corresponding to say u_1 and $u_2 (u_1 < u_2 \leqq 2t - 1)$ are identical we can delete the submatrix $M^*_{u_2}$ without changing the rank of M^* or the property (P_{2t}).

After the operations (a) and (b) we get from M^* a matrix of order $n \times R(m,t)$ with rank $R(m,t)$ and possessing the property (P_{2t}). Let the matrix so obtained be called A^* which is of order $n \times R(m,t)$ and possesses the property (P_{2t}).

The matrix A^* can serve as the parity check matrix.

Using Theorem 1 of Bose and Chaudhuri (1960) we now get the following results:

THEOREM 1. If n is any integer, and c is the least integer such that $1 + cn = 2^m$, then there exists a t-error correcting binary group code (n,k) for which the number of information places is

$$k = n - R(m,t)$$

where $R(m,t)$ is given by Lemma 2. The letters of the code are binary

n-vectors orthogonal to the columns of A^*, i.e., form the left null-space of A^*, where A^* is the matrix defined in the remarks following Lemma 2.

Every n-place binary sequence $(a_0, a_1, \cdots, a_{n-1})$ may be regarded as a polynomial $a_0 + a_1y + \cdots + a_{n-1}y^{n-1}$ in an indeterminate y. Let R_n denote the set of all such polynomials of degree less than n with coefficients 0 and 1. The addition of polynomials in R_n can be defined in the usual way, i.e., by adding the coefficients mod 2. Let the multiplication be defined mod 2 and mod $(y^n - 1)$. With these operations R_n becomes a ring. Let

$$A^* = \begin{Vmatrix} b_{10} & b_{20} & \cdots & b_{r0} \\ b_{11} & b_{22} & \cdots & b_{r1} \\ \vdots & & & \\ b_{1,n-1} & b_{2,n-1} & \cdots & b_{r,n-1} \end{Vmatrix}$$

and let

$$\beta_i' = (b_{i0}, b_{i1}, \cdots, b_{i,n-1}), \qquad i = 1, 2, \cdots, r = R(m,t)$$

Let $\bar{V}(A^*)$ denote the vector space generated by $\beta_1', \beta_2', \cdots, \beta_r'$, and $V(A^*)$ denote the vector space orthogonal to $\bar{V}(A^*)$. If now the vectors of $V(A^*)$ are regarded as polynomials, then for the case $c = 1$, Peterson (1960) has proved that $V(A^*)$ is an ideal in R_n, generated by a certain polynomial $f(y)$ of degree $r = R(m,t)$. Peterson's arguments at once extend to the general case and we have the following:

Let $f_j(x)$ be the minimum polynomial of x^{cj} over $GF(2)$, where x is a primitive element of $GF(2^m)$. Then $V(A^*)$ is the ideal generated by

$$f(y) = \underset{j=1,3,\cdots,(2t-1)}{\text{L.C.M.}} f_j(y)$$

The polynomial $f(y)$ can also be expressed in an alternative form. Let $(p_1, p_2, \cdots, p_r)$ be a set of $r = R(m,t)$ integers containing one integer for each of the $R(m,t)$ distinct residue classes considered in Lemma 2. Then

$$f(y) = (y - x^{cp_1})(y - x^{cp_2}) \cdots (y - x^{cp_r}).$$

For a polynomial $f(y) = a_0 + a_1y + \cdots a_{n-1}y^{n-1}$ we shall call $(a_{n-1}, a_{n-2}, \cdots, a_0)$ the reversed vector corresponding to $f(y)$. Let

$$\bar{f}(y) = (y^n - 1)/f(y)$$

and let $I[\bar{f}(y)]$ denote the set of 2^r reversed vectors corresponding to the

2^r polynomials of the ideal generated by $\bar{f}(y)$. Peterson's arguments then show that

$$\bar{V}(A^*) = I[\bar{f}(y)]$$

It follows that we can take

$$A^* = [\beta_1, \beta_2, \cdots, \beta_r]$$

where β_i' is the reversed vector corresponding to the polynomial

$$y^{i-1}\bar{f}(y), \qquad (i = 1, 2, \cdots, r),$$

and β_i is the transpose β_i'.

Example 1. Let $n = 21$. Then $c = 3$, $m = 6$. Let $t = 2$. To determine $R(m,t)$, we write the integers $2^j u(u = 1, 3; j = 0,1,2,3,4,5)$ in the following scheme, each row corresponding to one value of u:

$$\begin{array}{cccccc} 1, & 2, & 4, & 8, & 16, & 11 \\ 3, & 6, & 12, & 3, & 6, & 12 \end{array}$$

We thus get nine distinct residue classes and $R(m,t) = 9$. The number of information places is $k = 21 - 9 = 12$ and we get a 2 error-correcting binary group code (21, 12). To actually construct the code, we have to compute

$$f(y) = (y - x^3)(y - x^6)(y - x^{12})(y - x^{24})$$
$$(y - x^{48})(y - x^{33})(y - x^9)(y - x^{18})(y - x^{36})$$

where x is a primitive element of $GF(2^6)$. A minimum function of $GF(2^6)$ is $x^6 + x + 1$. Hence, using the relation $x^6 + x + 1 = 0$, the coefficients of the polynomial $f(y)$ will be all reduced to 0 and 1. The 2^{12} message sequences will be the 21-place binary vectors corresponding to the elements of the ideal generated by $f(y)$ in R_{21}. $\bar{V}(M^*)$ is the ideal generated by

$$\bar{f}(y) = (y^{21} - 1)/\bar{f}(y)$$
$$= y^{12} + y^{11} + y^9 + y^7 + y^3 + y^2 + y + 1.$$

Hence the parity check matrix A^* can be taken as

$$A^* = \|\beta_1 \vdots \beta_2 \vdots \cdots \beta_9\|$$

where β_i' is the (1×21) reversed vector corresponding to $y^{i-1}\bar{f}(y)$,

$i = 1, 2, \cdots, 9$. A 2 error-correcting (21, 12) code has also been studied by Prange (1958).

Example 2. Let $n = 73$. Then $c = 7$, $m = 9$. Let $t = 4$. The residue classes corresponding to the integer $2^j u(u = 1,3,5,7; 0 \leqq j \leqq 8)$ can be exhibited as

1,	2,	4,	8,	16,	32,	64,	55,	37
3,	6,	12,	24,	48,	23,	46,	19,	38
5,	10,	20,	40,	7,	14,	28,	56,	39
7,	14,	28,	56,	39,	5,	10,	20,	40.

The third and the fourth rows in this scheme are identical. Hence $R(m,t) = 27$ and $k = 46$. We thus get a 4 error-correcting binary group code (73, 46). This 4 error-correcting (73, 46) group code has also been obtained by Prange (1959).

Example 3. Let $n = 85$. Then $c = 3$, $m = 8$. Let $t = 6$. The residue classes corresponding to the integers $2^j u(u = 1,3,5,7,9,11; 0 \leqq j \leqq 7)$ can be exhibited as

1	2	4	8	16	32	64	43
3	6	12	24	48	11	22	44
5	10	20	40	80	75	65	45
7	14	28	56	27	54	23	46
9	18	36	72	59	33	66	47
11	22	44	3	6	12	24	48

The rows corresponding to $u = 3$ and 11 coincide. Hence $R(m,t) = 40$ and $k = 45$. We thus get a 6 error-correcting binary group code (85, 45).

SECTION III

We shall now discuss a method which enables us to get matrices possessing the property (P_{2t}) by adjoining other matrices. For the purpose of this section the subscripts carried by a matrix will denote the number of rows and columns of the matrix. Thus, $A_{n,r}$ denotes a matrix with n rows and r columns. $O_{n,r}$ will denote a matrix with n rows and r columns, each of whose elements is zero. Also $O_{n,1}$ will denote a column vector with n zero elements, and $O_{1,r}$ a row vector with r zero elements. Finally, $j_{r,1}$ will denote a column vector with r unities as elements. The elements of all the matrices considered belong to $GF(2)$.

LEMMA 3. If $A_{n,r}$ possesses the property (P_{2t}) then the matrix

$$\| A_{n,r}, \quad B_{n,s} \| \tag{7}$$

obtained from it by adjoining s new columns ($s > 0$) also possesses the property (P_{2t}).

Proof is obvious.

LEMMA 4. If the matrix $F_{n,r}$ possesses the property (P_{2t}), then the matrix

$$G_{n+1,r+1} = \left\| \begin{array}{c:c} F_{n,r} & j_{n,1} \\ \hdashline O_{1,r} & 1 \end{array} \right\| \tag{8}$$

possesses the property (P_{2t}).

Denote the matrix $\|F_{n,r} \vdots j_{n,1}\|$ formed by the first n rows of $G_{n+1,r+1}$ by $\bar{F}_{n,r+1}$. From Lemma 3, no $2t$ rows of $\bar{F}_{n,r+1}$ can be dependent. Again, consider the $2t$ rows obtained by choosing $2t - 1$ rows from $\bar{F}_{n,r+1}$ and adjoining the last row of $G_{n+1,r+1}$. These cannot be dependent. Otherwise the corresponding $2t - 1$ rows of $F_{n,r}$ which possess the property (P_{2t}) would be dependent. This completes the proof of the Lemma.

THEOREM 2. If the matrix $A_{n,r-r_0}$, $r > r_0$, possesses the property (P_{2t-2}) and the matrices

$$\| A_{n,r-r_0} \vdots T_{n,r_0} \| \qquad \text{and} \qquad F_{n',r_0+d-1} \tag{9}$$

$d \geqq 1$, possess the property (P_{2t}), then the matrix

$$M_{n+n'+1,r+d} = \left\| \begin{array}{c:c:c:c} A_{n,r-r_0} & T_{n,r_0} & O_{n,d-1} & O_{n,1} \\ \hdashline O_{n',r-r_0} & \multicolumn{2}{c:}{F_{n',r_0+d-1}} & j_{n',1} \\ \hdashline O_{1,r-r_0} & \multicolumn{2}{c:}{O_{1,r_0+d-1}} & 1 \end{array} \right\| \tag{10}$$

also possesses the property (P_{2t}).

Clearly the matrix $\bar{A}_{n,r+d}$ consisting of the first n rows of $M_{n+n'+1,r+d}$ has the property (P_{2t}). Also from Lemma 4, the matrix $\bar{G}_{n'+1,r+d}$ formed by the last $n' + 1$ rows of $M_{n+n'+1,r+d}$ has the property (P_{2t}). To prove the theorem we have to show that the $2t$ rows obtained by choosing any c rows of $\bar{A}_{n,r+d}$ and any $2t - c$ rows of $\bar{G}_{n'+1,r+d}$, $0 \leqq c \leqq 2t$, cannot be dependent. From what has been said this is true for $c = 2t$ or 0. If $c = 2t - 1$, then the last coordinate of the chosen rows adds up to unity. Hence they cannot be dependent because the matrix $A_{n,r-r_0}$ has the property (P_{2t-2}). This completes the proof of the theorem.

As in Section II, let $c = c(n)$ be the smallest integer such that $1 + cn$ is a power of 2, this power being $m = m(n)$. Let $R(m,t)$ be defined as in Lemma 2. We then have

THEOREM 3.

$$n_{2t}[R(m,t) + d] \geqq 1 + n + n_{2t}[R(m,t) - R(m,t-1) + d - 1]$$

where $n_{2t}(r)$ has been defined in the introduction, and

$$1 \leqq d < R(m+1,t) - R(m,t).$$

Let M^* be the matrix given by (2.3). We can then write

$$M^* = \| M_1^*, M_3^*, \cdots, M_u^*, \cdots, M_{2t-1}^* \|$$

where M_u^* is defined by (2.4).

Using the operations (a) and (b) described under Lemma 2, we can drop redundant columns from M^* and arrive at a matrix with n rows and $R(m,t)$ columns. Let the number of columns in the block coming from M_{2t-1}^* be r_0 and the submatrix of these columns be T_{n,r_0}. Let the number of columns coming from the part $\|M_1^*, M_3^*, \cdots, M_{2t-3}^*\|$ be $r - r_0$ and the submatrix of these columns be $A_{n,r-r_0}$. Then $r = R(m,t)$, $r - r_0 = R(m,t-1)$, and the matrices $\|A_{n,r-r_0} \vdots T_{n,r_0}\|$ and $\|A_{n,r-r_0}\|$ possess the properties (P_{2t}) and (P_{2t-2}) respectively. Let

$$r_0 + d - 1 = R(m,t) - R(m,t-1) + d - 1$$

and let

$$n' = n_{2t}(r_0 + d - 1)$$

Then there exists a matrix F_{n',r_0+d-1} with elements from $GF(2)$ and possessing the property (P_{2t}). We can now construct the matrix

$$M_{n+n'+1,r+d}$$

given by (3.4). The required result then follows from Theorem 2.

The most useful case of Theorem 3 is when $c = 1$, $n = 2^m - 1$. For this case we have

COROLLARY (1). $n_{2t}[R(m,t) + d] \geqq 2^m + n_{2t}[R(m,t) - R(m,t-1) + d - 1]$ A less powerful but simpler result is

COROLLARY (2). $n_{2t}(mt + d) \geqq 2^m + n_{2t}(m + d - 1)$

This follows by applying our reasoning to M^* without dropping any redundant column.

Example 4. Let us consider the case $t = 2$, $c = 1$, so that $n = 2^m - 1$. Then $R(m,2) = 2m$ and corollary (2) gives the same result as corollary (1). We know that $n_4(2m) \geqq 2^m - 1$. But one may want to get a bound on $n_4(2m + 1)$. From corollary (2) we have

$$n_4(2m + 1) \geqq 2^m + n_4(m)$$

For example,

(i)
$$n_4(21) \geqq 2^{10} + n_4(10)$$
$$\geqq 2^{10} + 2^5 - 1$$

(ii)
$$n_4(15) \geqq 2^7 + n_4(7)$$
$$\geqq 2^7 + 2^3 + n_4(3)$$
$$\geqq 2^7 + 2^3 + 3$$

SECTION IV

It is easy to see by exhaustive trial that $n_4(m) = m$ for $m = 1,2,3$; $n_4(4) = 5$, $n_4(5) = 6$, and $n_4(6) = 8$. Similarly, we can easily see that $n_{2t}(m) = m$ for $m = 1, 2, \cdots 2t$; $n_6(7) = 8$ and $n_{12}(13) = 14$. Using these facts and the results we have obtained, we can construct the following table where $L_{2t}(r)$ denotes the number of vectors in V_r that we can actually obtain such that no $2t$ are dependent. Thus $L_{2t}(r)$ is a lower bound for $n_{2t}(r)$.

The three asterisks in Table I indicate those cases corresponding to

TABLE I

$t = 1$		$t = 2$		$t = 3$		$t = 4$		$t = 5$		$t = 6$	
r	$L_2(r)$	r	$L_4(r)$	r	$L_6(r)$	r	$L_8(r)$	r	$L_{10}(r)$	r	$L_{12}(r)$
2	3	6	7	6	7	14	15	25	31	30	31
3	7	7	11	7	8	15	20	26	37	31	37
4	15	8	15	8	9	16	21	27	63	32	38
5	31	9	21*	9	10	17	22	28	67	33	63
6	63	10	31	10	15	18	23	29	68	34	70
7	127	11	36	11	18	19	24	30	69	35	71
		12	63	12	19	20	31	31	70	36	72
		13	71	13	20	21	37	32	71	37	73
		14	127	14	21	22	38	33	72	38	74
				15	31	23	39	34	73	39	75
				16	37	24	63	35	127	40	85*
				17	38	25	70			41	86
				18	63	26	71			42	127
				19	70	27	73*				
				20	72	28	127				
				21	127						

the three examples given after Theorem 1. Given n and t, $n \leqq 100$, $t \leqq 6$, we can find out from Table I the maximum possible k for which we can obtain by our methods a t-error correcting (n,k) group code. For this purpose we need to use the fact that if $n_{2t}(r) = n$, then for any positive integer c we have a t-error correcting $(n - c, n - r - c)$ group code. Thus, for instance, if we are seeking the largest value of k for $n = 90$, $t = 4$, we shall note that $L_8(27) < 90 < L_8(28)$ and decide that the required value of k is $90 - 28 = 62$.

RECEIVED: February 23, 1960.

REFERENCES

BOSE, R. C., AND RAY-CHAUDHURI, D. K. (1960). On a class of error correcting binary group codes. *Information and Control* **3**, 68.

PETERSON, W. W. (1960) Encoding and error-correction procedures for Bose-Chaudhuri codes. To appear in *IRE Trans. on Inform. Theory*.

PRANGE, E. (1958). Some cyclic error-correcting codes with simple decoding algorithms. Tech. note AFCRC-TN-58-156, Air Force Cambridge Research Center, Bedford, Massachusetts.

PRANGE, E. (1959). The use of coset equivalence in the analysis and decoding of group codes. Tech. Rept. AFCRC-TR-59-164, Air Force Cambridge Research Center, Bedford, Massachusetts.

16

POLYNOMIAL CODES OVER CERTAIN FINITE FIELDS*†

I. S. REED AND G. SOLOMON‡

Introduction. A code is a mapping from a vector space of dimension m over a finite field K (denoted by $V_m(K)$) into a vector space of higher dimension $n > m$ over the same field ($V_n(K)$). K is usually taken to be the field of two elements Z_2, in which case it is a mapping of m-tuples of binary digits (bits) into n-tuples of binary digits. If one transmits n bits, the additional $n - m$ bits are "redundant" and allow one to recover the original message in the event that noise corrupts the signal during transmission and causes some bits of the code to be in error. A multiple-error-correcting code of order s consists of a code which maps m-tuples of zeros and ones into n-tuples of zeros and ones, where m and n both depend on s, and a decoding procedure which recovers the message completely, assuming no more than s errors occur during transmission in the vector of n bits. The Hamming code [1] is an example of a systematic one bit error-correcting code. We present here a new class of redundant codes along with a decoding procedure.

Let K be a field of degree n over the field of two elements Z_2. K contains 2^n elements. Its multiplicative group is cyclic and is generated by powers of α where α is the root of a suitable irreducible polynomial over Z_2. We discuss here a code E which maps m-tuples of K into 2^n-tuples of K.

Consider the polynomial $P(x)$ of degree $m - 1$

$$P(x) = a_0 + a_1x + \cdots + a_{m-1}x^{m-1},$$

where $a_i \in K$ and $m < 2^n$. Code E is the mapping of the m-tuple $(a_0, a_1, \cdots, a_{m-1})$ into the 2^n-tuple $(P(0), P(\alpha), P(\alpha^2), \cdots, P(1))$; this m-tuple might be some encoded message and the corresponding 2^n-tuple is to be transmitted. This mapping of m symbols into 2^n symbols will be shown to be $(2^n - m)/2$ or $(2^n - m - 1)/2$ symbol correcting, depending on whether m is even or odd.

A natural correspondence is established between the field elements of K and certain binary sequences of length n. Under this correspondence, code E may be regarded as a mapping of binary sequences of mn bits into binary sequences of $n2^n$ bits. Thus code E can be interpreted to be a systematic multiple-error-correcting code of binary sequences.

* Received by the editors January 21, 1959 and in revised form August 26, 1959.

† The work reported here was performed at Lincoln Laboratory, a technical center operated by Massachusetts Institute of Technology with the joint support of the Army, Navy and Air Force, under contract.

‡ Staff members, Lincoln Laboratory, Massachusetts Institute of Technology, Lexington 73, Massachusetts.

One should note that the binary representation of code E allows in general for the correction of more than $(2^n - m - 1)/2$ bits since each symbol of the code is represented by n consecutive bits. Hence when the binary errors are strongly correlated or occur in "bursts," this code may be more desirable than other more "efficient" multiple-error-correction codes.

Finally, it should be mentioned that code E may be generalized to polynomials of the mth degree in several variables over K. Evidently, for $K = Z_2$, such codes reduce to Reed-Muller codes [2].

The code E. Consider the field $K = Z_2(\alpha)$. This is the vector space over Z_2 with basis $1, \alpha, \alpha^2, \cdots, \alpha^{n-1}$, where α is the root of a suitable irreducible polynomial over Z_2. The nonzero elements of K form a multiplicative cyclic group. Thus we may represent the elements of K in the order

$$0, \beta, \beta^2, \cdots, \beta^{2^n-2}, \beta^{2^n-1} = 1$$

where β is a generator of the multiplicative cyclic group.

Let $P(x) = a_0 + a_1x + a_2x^2 + \cdots + a_{m-1}x^{m-1}$. The code E sends

$$(a_0, a_1, \cdots, a_{m-2}, a_{m-1}) \rightarrow (P(0), P(\beta), P(\beta^2), \cdots, P(\beta^{2^n-2}), P(1)).$$

Upon receiving the message $(P(0), P(\beta), \cdots, P(1))$, we may decode the message by solving simultaneously any m of the 2^n equations,

$$\begin{aligned} P(0) &= a_0 \\ P(\beta) &= a_0 + a_1\beta + a_2\beta^2 + \cdots + a_{m-1}\beta^{m-1} \\ P(\beta^2) &= a_0 + a_1\beta^2 + a_2\beta^4 + \cdots + a_{m-1}\beta^{2m-2} \\ &\cdot \quad \cdot \quad \cdot \quad \cdot \quad \cdot \quad \cdot \quad \cdot \\ P(1) &= a_0 + a_1 + a_2 + \cdots + a_{m-1}. \end{aligned}$$

We note that any m of these equations are linearly independent since the coefficient determinant for, say, $P(\alpha_1), \cdots, P(\alpha_m)$, is

$$\begin{vmatrix} 1 & \alpha_1 & \alpha_1^2 & \cdots & \alpha_1^{m-1} \\ 1 & \alpha_2 & \alpha_2^2 & \cdots & \alpha_2^{m-1} \\ & \cdot & \cdot & \cdot & \cdot \\ 1 & \alpha_m & \alpha_m^2 & \cdots & \alpha_m^{m-1} \end{vmatrix}$$

which is a Vandermonde determinant whose value is

$$= \prod_{j<i} (\alpha_1 + \alpha_j) \neq 0.$$

Thus in the case of no errors in the received values of $P(\cdot)$, we obtain $\binom{2^n}{m}$ determinations of $(a_0, \cdots, a_{m-1})$.

Any errors occurring in the values of $P(\cdot)$ will immediately disturb the unanimity of the values obtained for the a_n's. Indeed, for sufficiently small numbers of errors, by looking at the largest number of determinations for any $(a_0, \cdots, a_{m-1})$ (the plurality of votes received by any m-tuple) we may detect the order of error made and correct it. We prove the following statement.

Lemma. For s errors we can get at most $\binom{s+m-1}{m}$ determinations for a wrong m-tuple.

Proof. We look upon the simultaneous solution of m equations as the intersection of m hyperplanes. The linear independence guarantees that they meet at only one point. To obtain more than one solution for any m-tuple, we would need more than m hyperplanes meeting at that point. For a wrong m-tuple, we can have at most $s + m - 1$ hyperplanes intersecting at a single wrong point, where s is the number of mistaken equations and where the remaining $m - 1$ equations are chosen from the $2^n - s$ correct ones. Any more correct hyperplanes would determine the correct solution, i.e., a different point of intersection from the assumed wrong one. Therefore, there are at most $\binom{s+m-1}{m}$ determinations for any wrong value. Note that we get $\binom{2^n-s}{m}$ determinations for the correct one, and a total of $\binom{2^n}{m} - \binom{2^n-s}{m}$ wrong determinations.

Thus, by examining the vote received by the individual candidates $(a_0, \cdots, a_{m-1})$, we may determine the correct message and the number s.

Note that this is valid only when

$$\binom{2^n-s}{m} > \binom{s+m-1}{m}$$

or

$$2^n - s > s + m - 1$$

or

$$s < \frac{2^n - m + 1}{2}.$$

The code will thus correct errors of order less than $(2^n - m + 1)/2$. For m odd, we get corrections up to $s = (2^n - m - 1)/2$, and detection at $s = (2^n - m + 1)/2$. For m even, we can correct up to $s = (2^n - m)/2$ and not detect any further errors.

Translation of K into a binary alphabet. We represent the elements of K by n-tuples of zeros and ones, $V_n(Z_2)$, and define a multiplication on $V_n(Z_2)$ corresponding to the multiplication of K. We again note that the multiplicative group of K is generated by powers of β. Let us consider an irreducible polynomial f which generates K over Z_2. Suppose $f(x) = x^n + c_1x^{n-1} + \cdots + c_{n-1}x + c_n = 0$, $c_i \in Z_2$. Following N. Zierler [3], we associate the following finite difference equation

$$a_{n+k} + c_1a_{n-1+k} + c_2a_{n-2+k} + \cdots + c_na_{0+k} = 0$$

where $a_i \in Z_2$.

Thus for any fixed f (giving rise to $(c_1, \cdots, c_n)$) and arbitrary $(a_0, \cdots, a_{n-1})(a_i \neq 0$ for $i = 0, 1, \cdots, n-1)$ we have a sequence

$$a_0, a_1, \cdots, a_{n-r}, a_n, a_{n+1}, a_{n+2}, \cdots$$

where the values of a_i for $i \geqq n$ are determined by the above difference equation. Zierler has shown that for suitable irreducible f, the sequence (a_n) is periodic of period $2^n - 1$, i.e., $a_{2^n-1} = a_0$, $a_{2^n+m-1} = a_m$ and the $2^n - 1$ sequences of length n obtained by translating the n-tuple $(a_0, a_1, \cdots, a_{n-1})$ along the derived sequence are all distinct.

Thus if we define

$$\begin{aligned} \beta &= (a_0, \cdots, a_{n-1}) \\ \beta^2 &= (a_1, \cdots, a_n) \\ &\cdot \quad \cdot \quad \cdot \quad \cdot \quad \cdot \\ \beta^m &= (a_{m-1}, \cdots, a_{n+m-2}) \end{aligned}$$

we have a multiplication table for the n-tuples. In other words, multiplication of the elements is simply translation along this periodic sequence generated by f. Note too that the elements β satisfy the algebraic equations satisfied by corresponding elements in K. We have thus defined multiplication on $V_n(Z_2)$ to make this correspond with the multiplication on K.

We remark that the initial choice of $\beta = (a_0, \cdots, a_{n-1})$ is arbitrary and there are $2^n - 1$ such representations. There are of course many other ways of associating vectors with powers of β. The referee has suggested another natural algebraic association of $V_n(Z_2)$ with K.

We identify K with the ring of polynomials in x with coefficients in Z_2, (i.e., $Z_2[x]$) modulo the prime ideal generated by the irreducible $f(x)$. Let $\beta = (a_0, a_1, \cdots, a_{n-1})$ be a nonzero vector of $V_n(Z_2)$. We associate with β the polynomial $\beta(x) = a_0 + a_1x + a_2x^2 + \cdots a_{n-1}x^{n-1} \bmod f(x)$. Consider $\beta(x)^k \bmod f(x)$. This again is a polynomial of formal degree $(n-1)$. Let β^k be the vector whose components are the n coefficients of this $(n-1)$-degree polynomial. This establishes a one-one correspondence of $V_n(Z_2)$

with K(if $(0, 0, \cdots, 0)$ is added to correspond to zero in K). While this may be a more natural choice, we prefer our first representation as the more suitable for computability.

Example. Let $n = 3$, $m = 3$. $K = Z_2(\alpha)$ where α is root of $x^3 + x + 1 = 0$. $P(x) = b_0 + b_1x + b_2x^2$.

Code E: $(b_0, b_1, b_2) \rightarrow (P(0), P(\alpha), P(\alpha^2), \cdots, P(\alpha^6), P(1))$.

Binary translation of this code. To $f(x) = x^3 + x + 1$ we associate the difference equation

$$a_n = a_{n-2} + a_{n-3} \qquad (\text{for } n = 3, 4, 5, \cdots).$$

Choose $a_0 = 1$, $a_1 = 1$, $a_2 = 0$. Then

$$\{a_n\} = (1, 1, 0, 0, 1, 0, 1, 1, 1, 0, 0, 1, 0, 1, \cdots).$$

$\{a_n\}$ has period 7, i.e., $a_7 = a_0$, $a_8 = a_1$.

$$\begin{aligned} 0 &= (0, 0, 0) \\ \alpha &= (1, 1, 0) \\ \alpha^2 &= (1, 0, 0) \\ \alpha^3 &= (0, 0, 1) \\ \alpha^4 &= (0, 1, 0) \\ \alpha^5 &= (1, 0, 1) \\ \alpha^6 &= (0, 1, 1) \\ 1 = \alpha^7 &= (1, 1, 1). \end{aligned}$$

The message $(0, \alpha, \alpha^3) \rightarrow (P(0), P(\alpha), P(\alpha^2), \cdots, P(\alpha^6), P(1))$ translates into (via $P(x) = \alpha x + \alpha^3 x^2$)

$$(0\ 0\ 0\ 1\ 1\ 0\ 0\ 0\ 1)$$

$$\rightarrow (0\ 0\ 0, 0\ 0\ 1, 1\ 1\ 0, 1\ 1\ 0, 1\ 1\ 1, 0\ 0\ 0, 0\ 0\ 1, 1\ 1\ 1).$$

This code is error correcting up to $(2^3 - 3 - 1)/2 = 2$ symbols.

REFERENCES

1. R. W. HAMMING, *Error detecting and error correcting codes*, Bell System Tech. J., 26 (1950), pp. 147–160.
2. I. S. REED, *A class of multiple-error-correcting codes and the decoding scheme*, Trans. I.R.E., Prof. Group on Information Theory No. 4 (1954), pp. 38–49.
3. N. ZIERLER, *Linear recurring sequences*, this Journal, 7 (1959), pp. 31–48.

Reprinted from *J. Soc. Ind. Appl. Math.*, **9**, 207–214 (1961)

17

A CLASS OF ERROR-CORRECTING CODES IN p^m SYMBOLS*

DANIEL GORENSTEIN† AND NEAL ZIERLER‡

1. Intıoduction. Bose and Chaudhuri [7] have introduced a class of binary linear error-correcting codes of block length of the form $2^m - 1$ for which W. W. Peterson [8] has devised an economical decoding procedure. The principal purpose of this note is to construct an efficient decoding procedure, in the spirit of Peterson's for the binary case, for the class of codes in p^m symbols and of block length prime to p (where p is any prime and m is any positive integer) obtained in an obvious generalization of the Bose-Chaudhuri codes. It turns out that the "polynomial codes" discussed by Reed and Solomon [3] belong to the general class, and hence may be decoded by our procedure.

There are two areas (at least) of application for codes in $q > 2$ symbols. First, data to be transmitted may appear in such a form and second, although the *B-C* codes tend to be highly efficient for the correction of independent errors, still greater efficiency may be obtained with the general codes when the errors occur in bursts. A code C of the general class is essentially uniquely determined by a triple (q, n, e) where q is the number of symbols, n is the block length (number of symbols per block), and C can correct every instance of e or fewer errors in a block. The parameter k, the number of information symbols per block, is computed as a function of q, n and e (see below). Two examples of the second application are as follows:

Example 1. Suppose we wish to transmit binary data, that errors occur in bursts of length 5 (binary symbols) or less and that acceptable reception results when we are able to correct 2 bursts in a block of length about 60. A *B-C* code for the job has $n = 63$, $e = 10$, $k = 18$ (and, of course, $q = 2$), and so gives a transmission rate $R = k/n = \frac{18}{63} = \frac{2}{7}$. On the other hand, we can take the general code with $q = 2^4$ (so a burst of length 5 causes errors in exactly two symbols, each symbol being encoded in a natural way as a binary 4-tuple) $n = 15$, $e = 4$ and $k = 7$. Then the binary block length is $4 \times 15 = 60$ and the transmission rate is $\frac{7}{15}$.

---

* Received by the editors May 24, 1960 and in revised form November 15, 1960.

† Consultant to Lincoln Laboratory,§ Massachusetts Institute of Technology, Lexington, Massachusetts and Clark University, Worcester, Massachusetts.

‡ Lincoln Laboratory,§ Massachusetts Institute of Technology, Lexington, Massachusetts. Now at Arcon Corporation, Lexington, Massachusetts.

§ Operated with support from the U. S. Army, the U. S. Navy and the U. S. Air Force.

Example 2. Suppose again that binary data are to be transmitted, that errors occur in bursts of length 9 or less and that we must be able to correct 4 such bursts in a binary block of length around 2050. An appropriate *B-C* code has $n = 2047$, $e = 36$ and k very near 1670, so $R \sim \frac{4}{5}$. A suitable general code has $q = 2^8$, $n = 255$, $e = 8$, $k = 239$ and binary block length $8 \times 255 = 2040$; its rate is then $\frac{239}{255} \sim \frac{15}{16}$.

In §2 some basic facts concerning linear codes over general finite fields are established. In §3 we assemble the necessary machinery from the theory of linear recurring sequences (including several results which do not appear in [6]). The codes are constructed in §4, and the decoding procedure is obtained in §5.

2. Linear codes. Let p be a prime, m and n positive integers, K the field with $q = p^m$ elements and $V = V_n$ the vector space (over K) of n-tuples of elements of K. For u, v in V define the inner product $u \cdot v = \sum_{i=1}^{n} u_i v_i$ and let $\| u \|$, the norm of n, denote the number of nonzero components of u. If A is a subset of V, define $A^{\perp} = \{v \in V\colon u \cdot v = 0$ for all $u \in A\}$. Suppose now that A is a subspace of V and let $\bar{v}$ denote the A-coset to which $v \in V$ belongs. Define $W(\bar{v})$, the weight of $\bar{v}$, to be min $\| u \|$: $u \in \bar{v}$ and a *leader* of $\bar{v}$ to be an element u of $\bar{v}$ with $\| u \| = W(\bar{v})$. Evidently, the function W may be defined on V by $W(v) = W(\bar{v})$. Taking A as code alphabet, consider the following idealized decoding procedure: assign $v - u$ as (estimate of) transmitted message when v is received where u is an arbitrarily chosen leader of $\bar{v}$. A is said to be *e-error correcting*, e a positive integer, if the foregoing procedure decodes correctly whenever e or fewer errors occur in the transmission of a member of A, i.e., if every element of V of norm $\leqq e$ is the unique leader of the coset to which it belongs. It is convenient to characterize the property of being e-error correcting in the following two ways.

LEMMA 2.1. *A is e-error correcting if and only if $\| u \| > 2e$ for every $u \neq 0 \in A$.*

LEMMA 2.2. *Let L be any n-rowed matrix with coefficients in K whose columns, regarded as members of V, belong to and linearly span $A^{\perp}$. Then A is e-error correcting if and only if every set of $2e$ rows of L is linearly independent.*

Proofs. Suppose $\| u \| > 2e$ for every $u \neq 0$ in A, $v \in A$ is sent and $v + w$ is received where $\| w \| \leqq e$. Then for $u \in A$, $\| v + w - u \| > e$ if $u \neq v$ since $v - u \neq 0 \in A$ and so w is the unique leader of $\overline{v + w} = \bar{w}$. On the other hand, if A contains a nonzero vector u with $\| u \| = t \leqq 2e$, let $s = \min(t, e)$, let $i_1, \cdots, i_s$ be distinct indices of nonzero components of u and let $w \in V$ such that $w_{i_j} = -u_{i_j}$, $j = 1, \cdots, s$ and $w_i = 0$ for $i \notin \{i_1, \cdots, i_s\}$. Then $\| w \| = s \leqq e$; but $\| u + w \| = t - s \leqq s$ (for if

$t \geqq e$, then $s = e$ and $t - s = t - e \leqq 2e - e = e = s$ while if $t < e$, $s = t$ and $t - s = 0$) so either w is not a leader of $\bar{w}$ or, if it is a leader, it is not unique.

Lemma 2.2 follows at once from Lemma 2.1 and the fact that $u \in A$ if and only if $uL = 0$.

REMARK. If A is merely an additive subgroup of V, the foregoing remains valid except for Lemma 2.2, and it would be natural to call A a "group code" in this case, as does Slepian [4] for the case $q = 2$. Of course, such an A is always linear over the prime subfield of K.

Define two codes (= subspaces of V) to be *equivalent* if one is mapped on the other by a coordinate permutation. If the k-dimensional subspace A of V has the property that every k-tuple of elements of K appears as the first k components of some member of A, A is said to be a *check code*. As in [4], we note that every linear code is equivalent to a check code; indeed, if A is any k-dimensional subspace, let M be a $k \times n$ matrix whose rows are a basis of A over K. Since M has rank k, we can find k columns, say columns $i_1, \cdots, i_k$, which are linearly independent, and we let π be any coordinate permutation such that $\pi(i_j) = j$, $j = 1, \cdots, k$. Evidently $\pi(A)$ is a check code equivalent to A.

3. Linear recurring sequences. Let K be the q-element field, let d be a positive integer and let $f \in K[x]$: $f(x) = c_d x^d + \cdots + c_0$ with $c_0 c_d \neq 0$. In [6] we study families $G(f)$ = the set of all linear recurring sequences generated by f = the set of all sequences $u_0, u_1, \cdots$ of elements of K satisfying $c_0 u_i + c_1 u_{i-1} + \cdots + c_d u_{i-d} = 0$ for $i = d, d+1, \cdots$. We show there that if we identify the sequence u with the member $\sum_{i=0}^{\infty} u_i x^i$ of $K(x)$, then $G(f) = \{f_1/f : \deg f_1 < d\}$. For present purposes, however, we wish to consider, rather than sequences of arbitrarily large period, only sequences whose periods divide some fixed $n > 0$; thus, we consider only $G(f)$ where f divides $1 - x^n$ and treat the members of $G(f)$ as n-tuples of elements of K. The embedding of $G(f)$ in $K(x)$ of [6] now goes over into an embedding in the quotient ring $R = K[x]/(1 - x^n)$ as follows. Identify an element $(u_0, \cdots, u_{n-1})$ of V, the space of n-tuples of elements of K, with (the coset modulo $1 - x^n$ in $K[x]$ which contains) $u_0 + \cdots + u_{n-1} x^{n-1}$. Then $G(f) = \{f_1(1 - x^n)/f : \deg f_1 < d\}$ = the ideal in R generated by $(1 - x^n)/f$.

LEMMA 3.1. $G(f) = ((1 - x^n)/f)$.

COROLLARY 3.1. *If* $fg \mid 1 - x^n$ *and* $h \in G(f)$ *then* $g \mid h$.

Proof. Since $(1 - x^n)/fg$ is a polynomial g_1 and $h \in G(f)$, there exists a polynomial f_1 (of degree $< \deg f$) such that $h = f_1(1 - x^n)/f = f_1 g_1 g$.

If $f(x) = c_d x^d + \cdots + c_0$, let $f'(x) = c_0 x^d + \cdots + c_d$. Clearly, $f \mid 1 - x^n$ implies $f' \mid 1 - x^n$.

COROLLARY 3.2. *If* $fg \mid 1 - x^n$ *then* $G(f) \perp G(g')$.

Proof. Let $h \in G(f)$. Since $g \mid h$ by Corollary 3.1, every member v of $G(g)$ satisfies the recurrence relation associated with h by [6, Theorem 1], i.e., $h_0 v_{n-1} + \cdots + h_{n-1} v_0 = 0$. Since $(v_0, \cdots, v_{n-1}) \in G(g)$ if and only if $(v_{n-1}, \cdots, v_0) \in G(g')$, the result follows.

REMARK. It is interesting to note the following special case of Corollary 3.2. Let $q = 2$, $g = x + 1$, f any polynomial which is not a multiple of $x + 1$. Then one period of any sequence generated by f contains an even number of 1's.

By a count of dimensions we have, finally,

COROLLARY 3.3. $G(f)^{\perp} = G((1 - x^n)/f')$. $= (f')$

4. A class of codes in p^m symbols. Let K be the $q = p^m$ element field and let n and e be positive integers with $(p, n) = 1$ and $2e \leqq n$. Then the equation $x^n - 1 = 0$ is separable over K and has a primitive root a in an extension $F = K(a)$ of K whose degree r satisfies: $r = \min\{j : n \mid q^j - 1\}$. Let B be the $n \times 2e$ matrix with $B_{ij} = a^{j(i-1)}$, $i = 1, \cdots, n$, $j = 1, \cdots, 2e$; i.e.,

$$B = \begin{bmatrix} 1 & 1 & \cdots & 1 \\ a & a^2 & & a^{2e} \\ a^2 & a^4 & & a^{4e} \\ \cdot & \cdot & & \cdot \\ \cdot & \cdot & & \cdot \\ \cdot & \cdot & & \cdot \\ a^{n-1} & a^{2(n-1)} & \cdots & a^{2e(n-1)} \end{bmatrix}.$$

It follows at once from the fact that a is primitive and $2e \leqq n$ that the $2e$-square matrix formed from any $2e$ rows of B is a Vandermonde matrix in distinct arguments and so is nonsingular.[1] Now choose a basis of F over K and replace each element a^i of B by its r-tuple of coefficients relative to this basis. Then B becomes an $n \times 2re$ matrix L with coefficients in K which clearly also has the property that every set of $2e$ rows of L is linearly independent over K. Let $f_i(x)$ be the minimum polynomial of a^i over K, $i = 1, \cdots, 2e$. Then the ith column of B evidently belongs to $G(f_i')$ over F and it follows at once that column $r(i - 1) + j$ of L belongs to $G(f_i')$ over K for $i = 1, \cdots, 2e; j = 1, \cdots, r$.

LEMMA 4.1. *Let $1 \leqq i \leqq 2e$. Then columns $r(i - 1) + j$, $j = 1, \cdots, r$ of L span $G(f_i')$.*

Proof. Consider the $n \times r$ matrix $L^{(i)}$ consisting of the r columns of L in question. As members of F relative to the chosen basis, its rows are $1, a^i, a^{2i}, \cdots, a^{i(n-1)}$. Hence any d_i consecutive rows are linearly independent

[1] Cf. [1, Ch. X, §5, ex. 6] and [2].

over K where d_i = degree f_i so the row rank, and hence the column rank, of $L^{(i)}$ is d_i.

Let f = least common multiple $\{f_1', \cdots, f_{2e}'\}$. Then, from Lemma 4.1 and [6, Theorem 2] we have

LEMMA 4.2. *The linear span over K of the columns of L is $G(f)$.*

The orthogonal complement A of $G(f)$ is an e-error correcting code by Lemma 2.2, and $A = G(g)$ where $g = (1 - x^n)/f'$ by Corollary 3.3.

DEFINITION. *Let q, n and e be as in the first paragraph of this section. A (q, n, e) code is a subspace $G(g)$ of the space V of n-tuples of elements of K $(= GF(q))$ where $g = (1 - x^n)/\text{l.c.m.}\ \{f_1, \cdots, f_{2e}\}$, f_i the minimum polynomial over K of a^i, where a is a primitive nth root of unity.*

We have seen that a (q, n, e) code is an e-error correcting linear code of length n in q symbols containing q^k n-tuples where k = degree g. According to [6, Theorem 3], every (q, n, e) code is closed under coordinate translation (i.e., under $v \to u$ with $u_i = v_{i+j \bmod n}$, $i = 1, \cdots, n$, any fixed j). Furthermore, such a code is essentially unique in the following sense.

LEMMA 4.3. *Two (q, n, e) codes are equivalent under a coordinate permutation.*

Proof. Let A and B be (q, n, e) codes corresponding to primitive nth roots a and b respectively. Then $b = a^t$ for some t prime to n and the coordinate permutation $i \to ti$ modulo n maps A on B.

The "polynomial codes" of Reed and Solomon [3] may be described as follows. Choose parameters q and k where q is a prime power and $0 < k < q - 1$. Let V_{q-1} be the space of $(q - 1)$-tuples of elements of the q-element field K, let a be a primitive $(q - 1)$th root of unity in K and let T be the mapping from V_k to V_{q-1} defined by:

$$T(u_0, u_1, \cdots, u_{k-1}) = (h(1), h(a), h(a^2), \cdots, h(a^{q-2}))$$

where $h(x) = u_0 + u_1 x + \cdots + u_{k-1}x^{k-1}$. T is obviously a linear mapping of V_k in V_{q-1} and the image of the ith natural basis vector of V_k is $(1, a^i, a^{2i}, \cdots, a^{(q-2)i})$, $i = 0, 1, \cdots, k - 1$. But this is a linear recurring sequence generated by the polynomial $f_i' = x - a^i$ so $TV_k = G(g)$ where $g = (x - 1)(x - a^{-1}) \cdots (x - a^{-(k-1)})$, which is exactly the $\left(q, q - 1, \left[\frac{q - k - 1}{2}\right]\right)$ code corresponding to a.

5. Decoding (q, n, e) **codes.** Let A be the (q, n, e) code corresponding to the primitive nth root of unity a and let B denote the associated $n \times 2e$ matrix of §4. Suppose the element $u = (u_0, \cdots, u_{n-1})$ of A is transmitted and $u + v$ is received where $v = (v_0, \cdots, v_{n-1})$ is the error n-tuple. Then $(u + v)B = uB + vB = vB$ is a $2e$-tuple $(s_1, \cdots, s_{2e})$ of members of $F = K(a)$ satisfying: $s_j = v_0 + v_1 a^j + v_2 a^{2j} + \cdots + v_{n-1}a^{(n-1)j}$, $j =$

$1, \cdots, 2e$. Let $0 \leqq t \leqq e$ and suppose that $\| v \| \leqq t$, i.e., at most t errors have occurred. The decoding procedure consists essentially of computing the s_j (i.e., $(u + v)B$) and then applying the following theorem (which will be proved below).

Let $t_0 = \| v \|$ and let $i_1, \cdots, i_{t_0}$ denote the indices of the nonzero components of v. Let $\sigma_1, \cdots, \sigma_t$ denote the elementary symmetric functions of $a^{i_1}, \cdots, a^{i_{t_0}}$:

$$\sigma_1 = a^{i_1} + \cdots + a^{i_{t_0}},$$

$$\sigma_2 = a^{i_1+i_2} + a^{i_1+i_3} + \cdots + a^{i_1+i_{t_0}} + a^{i_2+i_3} + \cdots + a^{i_{t_0-1}+i_{t_0}},$$

$$\vdots$$

$$\sigma_{t_0} = a^{i_1+i_2+\cdots+i_{t_0}},$$

and if $t_0 < t$,

$$\sigma_{t_0+1} = \cdots = \sigma_t = 0.$$[2]

THEOREM 5.1. *Let $0 \leqq t \leqq e$, suppose $\| v \| = t_0 \leqq t$ and let $i_1, \cdots, i_{t_0}$ and $\sigma_1, \cdots, \sigma_t$ be as in the preceding paragraph. Then the following matrix equation holds:*

$$\begin{bmatrix} s_1 & s_2 & \cdots & s_t \\ s_2 & s_3 & \cdots & s_{t+1} \\ s_3 & s_4 & \cdots & s_{t+2} \\ \vdots & \vdots & & \vdots \\ s_t & s_{t+1} & \cdots & s_{2t-1} \end{bmatrix} \begin{bmatrix} (-1)^{t+1}\sigma_t \\ (-1)^{t}\sigma_{t-1} \\ (-1)^{t-1}\sigma_{t-2} \\ \vdots \\ \sigma_1 \end{bmatrix} = \begin{bmatrix} s_{t+1} \\ s_{t+2} \\ s_{t+3} \\ \vdots \\ s_{2t} \end{bmatrix}.$$

Furthermore, letting M_t denote the foregoing matrix, $| M_t |$ (the determinant of M_t) $= v_{i_1} \cdots v_{i_{t_0}} \sigma_t \prod_{y>z} (a^{i_y} - a^{i_z})^2$. In particular, M_t is nonsingular if and only if $t_0 = t$.

Assuming, as we do, that $\| v \| \leqq e$, we reduce M_e to the form $\begin{bmatrix} I_{t_0} & 0 \\ 0 & 0 \end{bmatrix}$ by a sequence of elementary row operations,[3] where I_j denotes the $j \times j$ identity matrix, simultaneously subjecting the matrix I_e to the same sequence of operations, at the end of which it has the form $\begin{bmatrix} M_{t_0}^{-1} & - \\ - & - \end{bmatrix}$. Having found t_0 and $M_{t_0}^{-1}$ simultaneously in this way, we compute the σ_i: $M_{t_0}^{-1} \begin{bmatrix} s_{t_0+1} \\ \vdots \\ s_{2t_0} \end{bmatrix} = \begin{bmatrix} (-1)^{t_0+1}\sigma_{t_0} \\ \vdots \\ \sigma_1 \end{bmatrix}$. Then the i_j, $j = 1, \cdots, t_0$ are found by substituting successively $1, a, a^2, \cdots$ in $h(x) = x^{t_0} - \sigma_1 x^{t_0-1} + \cdots +$

[2] See [5, §26].

[3] [1, Ch. X, § 3, Theorem 8]

$(-1)^{t_0}\sigma_{t_0} = (x - a^{i_1}) \cdots (x - a^{i_{t_0}})$. The v_{i_j} are then determined by solving the t_0 simultaneous linear equations

$$v_{i_1}a^{ji_1} + \cdots + v_{i_{t_0}}a^{ji_{t_0}} = s_j \qquad (j = 1, \cdots, t_0)\,.$$

In particular, the coefficient matrix

$$\begin{bmatrix} a^{i_1} & a^{2i_1} & \cdots & a^{t_0 i_1} \\ \vdots & \vdots & & \vdots \\ a^{i_{t_0}} & a^{2i_{t_0}} & \cdots & a^{t_0 i_{t_0}} \end{bmatrix}$$

may be inverted by a sequence of elementary row operations, or explicit formulae may be used.[4]

It remains to prove the assertions of Theorem 5.1, which we do in the following general form. Let F be an arbitrary field, let t be a positive integer and let $y_1, \cdots, y_t, x_1, \cdots, x_t$ be elements of F. Let $s_j = y_1x_1^j + \cdots + y_tx_t^j$, $j = 1, \cdots, 2t$, let $\sigma_1, \cdots, \sigma_t$ be the elementary symmetric functions of $x_1, \cdots, x_t$ and let $h(x) = x^t - \sigma_1 x^{t-1} + \cdots + (-1)^t\sigma_t$. Let W denote the Vandermonde matrix,

$$W = \begin{bmatrix} 1 & \cdots & 1 \\ x_1 & & x_t \\ x_1^2 & & x_t^2 \\ \vdots & & \vdots \\ x_1^{t-1} & \cdots & x_t^{t-1} \end{bmatrix}$$

and let Λ denote the diagonal matrix:

$$\Lambda = \begin{bmatrix} y_1x_1 & & & \\ & y_2x_2 & & \\ & & \ddots & \\ & & & y_tx_t \end{bmatrix}.$$

Letting W' denote the transpose of W, direct computation gives

$$W'\begin{bmatrix} (-1)^{t+1}\sigma_t \\ \vdots \\ \sigma_1 \end{bmatrix} = \begin{bmatrix} (-1)^{t+1}\sigma_t + (-1)^t\sigma_{t-1}x_1 + \cdots + \sigma_1 x_1^{t-1} \\ \vdots \\ (-1)^{t+1}\sigma_t + (-1)^t\sigma_{t-1}x_t + \cdots + \sigma_1 x_t^{t-1} \end{bmatrix} = \begin{bmatrix} x_1^t \\ \vdots \\ x_t^t \end{bmatrix}.$$

Multiplying this vector by Λ gives $\begin{bmatrix} y_1x_1^{t+1} \\ \vdots \\ y_tx_t^{t+1} \end{bmatrix}$ and multiplying this by W

[4] See, e.g., [2].

gives $\begin{bmatrix} s_{t+1} \\ \vdots \\ s_{2t} \end{bmatrix}$. On the other hand, $W\Lambda W'$ is readily seen to be

$$M_t = \begin{bmatrix} s_1 & s_2 & \cdots & s_t \\ s_2 & s_3 & \cdots & s_{t+1} \\ \vdots & & & \\ s_t & s_{t+1} & \cdots & s_{2t-1} \end{bmatrix}$$

(which proves the matrix equation of Theorem 5.1). Finally $|M_t| = |W\Lambda W'| = |W||\Lambda||W'| = |\Lambda||W|^2 = y_1 \cdots y_t \sigma_t \prod_{y>z}(x_y - x_z)^2$.

NOTE. It should be pointed out that (q, n, e) codes come in compatible families in the following sense. A decoder for a (q, n, e) code A will also decode every (q, n, e') code with $e' < e$ provided only that A and A' correspond to the same primitive nth root of unity. This property could be exploited in a "variable rate" system in which error-correcting capability is matched to time-varying channel statistics.

REFERENCES

1. G. Birkhoff and S. MacLane, *A Survey of Modern Algebra,* Macmillan, New York, 1944.
2. N. Macon and A. Spitzbart, *Inverses of Vandermonde Matrices,* Amer. Math. Monthly, 65 (1958), pp. 95–100.
3. I. S. Reed and G. Solomon, *Polynomial codes,* this Journal, 8 (1960), pp. 300–304.
4. D. Slepian, *A class of binary signaling alphabets,* Bell System Tech. J., 35 (1956), pp. 203–234.
5. B. van der Waerden, *Moderne Algebra,* Ungar, New York, 1943. English Transl., Ungar, 1949.
6. N. Zierler, *Linear recurring sequences,* this Journal, 7 (1959), pp. 31–48.
7. R. C. Bose and D. K. Ray-Chaudhuri, *On a class of error-correcting binary group codes,* Information and Control, 3 (1960), pp. 68–79.
8. W. W. Peterson, *Encoding and error correction procedures for the Bose-Chaudhuri Codes,* IRE Transactions on Information Theory, vol. IT-6 (1960), pp. 459–470.

Reprinted from *J. Soc. Appl. Math.*, **9**, 654–669 (1961)

18

A NEW TREATMENT OF BOSE-CHAUDHURI CODES†

H. F. MATTSON* AND G. SOLOMON**

Abstract. Letting A be any (k, n) Bose-Chaudhuri code, we first attach to each a in A (via difference equations over $GF(2)$) a polynomial $g_a(x)$ such that the coordinates of a are the values of $g_a(x)$ on the nth roots of unity. The degree of these polynomials is such that the minimum nonzero weight d of vectors in A is immediately seen to be at least d_0, the usual Bose-Chaudhuri lower bound. This lower bound d_0 is improved over a class of $(h + 1, p)$ codes, where $p = 2h + 1$ has certain prime values, in a number of general theorems. In particular, the (12, 23) Golay code is proved very simply to have $d = 7$; and a (24, 47) code is shown to have $d \geqq 9$, thus improving by 4 the usual lower bound $d_0 = 5$ for that code.

1. Introduction. In constructing an automatic, high-speed communication system, it is usual to take as a "message" an ordered n-tuple in two arbitrary symbols, say 0, 1. Thus a message would consist of $a = (a_0, \cdots, a_{n-1})$ where each a_i is 0 or 1. Since the usual addition and multiplication modulo 2 are convenient to perform on an electronic machine, one is led to consider the set V of all these n-tuples in 0, 1 as a vector space over $F = GF(2)$, the field of two elements, with addition defined component-wise.

In actual systems, errors may occur in transmission of the a_i. In order to cope with this effect, one can decide that only those a in some subset A of V will be transmitted and then try to choose A in such a way that if $a \in A$ is sent and $a^* \in V$ is received, it will then be possible to recover a from a^* provided not too many errors are present. The subset A is usually called a *code*, or *error-correcting code*. Elements of A will be called code-vectors.

In particular, if we choose A as a (linear) subspace of V, it is very simple to describe the number of errors in a single message which one can "correct". For if we define the *weight* $w(a)$ of $a \in V$ as the total number of a_i which equal 1, then the function $\rho(a, b) = w(a + b)$ is a metric on V—the Hamming distance—and is invariant under translation. Therefore, the distance

$$d(a) = \min \{\rho(a, b); b \in A, b \neq a\}$$

is the same for all $a \in A$; in particular $d(a) = d(0) = \min \{w(b); b \in A,$

† Received by the editors March 27, 1961 and in revised form July 24, 1961.

* Sylvania Applied Research Laboratory, Waltham, Massachusetts. This work was supported in part by the Air Force Cambridge Research Laboratory under Contract No. AF 19(604)-6639.

** Staff Member, M.I.T. Lincoln Laboratory (operated with support from the U. S. Army, Navy, and Air Force).

$b \neq 0\} = d$ is this value. Thus, taking[1] $e = [(d - 1)/2]$, we easily see that if a is sent and a^* received, then the nearest element of A to a^* is a itself whenever the total number of errors present in a^* is at most e.

How to determine an efficient procedure for finding the element of A nearest to a given element of V—the so-called decoding problem—is an important problem in this field. Another important problem is to determine d for a given (k, n) *group code* A, i.e., a k-dimensional subspace of V.

The group codes to which we confine ourselves here are due mainly to Bose and Ray-Chaudhuri [1], who gave a constructive procedure for obtaining a large class of cyclic codes having preassigned error-correcting properties. Their codes can, in many cases, be shown to possess greater error-correcting ability than they were able to demonstrate. These codes have been studied recently by several authors [3, 4, 8, 9]. Their estimates of the error-correcting properties of these codes are based on methods using matrices, linear recursive sequences, and rings of polynomials. We have reproduced their estimates in general and, in a sub-class of Bose-Chaudhuri codes, have improved these estimates. Our methods are based on the treatment of linear recursions as finite-difference equations [7] and illuminate the subject in a particularly simple way. In particular, we change the difficult combinatorial problem of determining the number of 1's in a code-vector into the more tractable algebraic problem of determining the number of zeros of a certain polynomial on a given finite set.

This paper is reasonably self-contained. We do assume, however, certain basic algebraic notions, such as the elementary properties of polynomials and the existence of extension fields. Nevertheless, the reader will be able to understand all of the paper without much trouble even if he has only a nodding acquaintance with algebra.

We give one particular case ($p = 23$) in detail in §5, so as to illustrate the general theorems of the report. For the casual reader this example is probably a good abstract of our work.

Summary. In §2 we present that part of [7] which we need. For completeness, we give the proof (which is quite simple).

In §3 we define group codes obtained from linear recursions and derive a basic result (Lemma 2), which rests on §2. Applying Lemma 2 to the Bose-Chaudhuri codes as defined in [8], we obtain a simple proof of the previously known result (Theorem 1) giving a lower bound on the minimum nonzero weight of code-vectors.

The rest of this report is restricted to a certain class of $(h + 1, p)$ Bose-Chaudhuri codes, defined for certain primes $p = 2h + 1$. In §4 we present general results on these codes, including an apparently strong improvement

[1] The square brackets here (only) denote the usual greatest-integer function.

(Theorem 2) in the lower bound on the above-mentioned minimum weight for the (h, p) subcode of even-weight vectors in the case $p \equiv 1(\text{mod } 8)$. In Theorem 3 we prove that the previously known lower bound can always be improved by 1, unless that lower bound is quite good already (namely, as big as h). In Theorem 5 we give relations between p and odd weights of code-vectors.

In §5 we give particular results for $p = 23$ (the Golay case [2]) and $p = 47$; our methods give a short and simple proof that the Golay code corrects three errors, and without too much labor, we raise from 5 to 9 the above-mentioned lower bound in the case $p = 47$.

2. Difference equations over $GF(2)$. Throughout this paper, let $F = GF(2)$ denote the field of two elements 0, 1. Let $a_0, a_1, \cdots$ be a linear recursive sequence in F with the a_j determined by the recursion

$$(1)\qquad a_{k+i} + b_1 a_{k+i-1} + b_2 a_{k+i-2} + \cdots + b_k a_i = 0 \qquad (i = 0, 1, 2, \cdots),$$

in which the coefficients $b_1, \cdots, b_k \in F$ are independent of i and the values $a_0, a_1, \cdots, a_{k-1}$ are preassigned.

Equation (1) is simply a linear difference equation with constant coefficients. In order to find a "general solution" of (1), we set $a_j = \beta^j$ for all j in (1), as in the classical case, obtaining

$$\beta^i(\beta^k + b_1\beta^{k-1} + \cdots + b_{k-1}\beta + b_k) = 0,$$

which is satisfied if β is a root of the polynomial $f(x) = x^k + b_1 x^{k-1} + \cdots + b_k$. To obtain a finite field K which contains F and all the roots of $f(x)$ is a standard algebraic procedure, which we assume done without further ado.

We restrict our attention to the case in which the roots $\beta_1, \cdots, \beta_k$ of $f(x)$ are distinct. For any $c_1, \cdots, c_k \in K$, let

$$(2)\qquad a_j = c_1\beta_1^{\,j} + \cdots + c_k\beta_k^{\,j} \qquad (j = 0, 1, 2, \cdots).$$

Then this sequence $\{a_j\}$ satisfies (1), except it may not take the preassigned values for $j = 0, 1, \cdots, k - 1$. But there is a unique set of c's in K, obtained as the solution of k linear simultaneous equations having a van der Monde coefficient matrix such that the sequence defined in (2) satisfies (1) for all $j = 0, 1, 2, \cdots$.

Thus every linear recursive sequence $\{a_i\}$ for (1) is obtainable in the form (2). Conversely, every sequence in the form (2) is linear recursive for (1) over K; it lies in F provided only that $a_0, a_1, \cdots, a_{k-1}$ are in F. We state these results in

LEMMA 1. *If $a_0, a_1, \cdots$ is a sequence in F given by the recursion* (1), *and if the polynomial $f(x) = x^k + b_1x^{k-1} + \cdots + b_k$ has no repeated roots,* *then there exist uniquely determined elements* $c_1, \cdots, c_k$ *in* K *such that*

$$a_i = c_1\beta_1^i + \cdots + c_k\beta_k^i \qquad (i = 0, 1, 2, \cdots),$$

where $\beta_1, \cdots, \beta_k$ *are the roots of* $f(x)$.

A detailed treatment of this subject, with additional results on linear recursive sequences not needed here, appears in [7].

3. The codes. Let n be an odd integer, and let $f(x) = x^k + b_1x^{k-1} + \cdots + b_k$ be a polynomial with coefficients in F (the set of all such polynomials is denoted by $F[x]$); suppose also that $f(x)$ divides $x^n + 1$. Let K denote the smallest field containing F and all the roots of $x^n + 1$. Let V denote the vector space of all ordered n-tuples $a = (a_0, a_1, \cdots, a_{n-1})$ with each a_i in F, with addition of vectors a and $b = (b_0, \cdots, b_{n-1})$ defined by $a + b = (a_0 + b_0, \cdots, a_{n-1} + b_{n-1})$. We define a certain (k, n) group code A as the following subspace of[2] V:

$$A = \{a; a = (a_0, \cdots, a_{n-1}) \in V, a_{i+k} + b_1a_{i+k-1} + \cdots + b_ka_i = 0, \quad i = 0, 1, \cdots, n-k-1\}. \tag{3}$$

Thus, to form a vector of A, we choose coordinates $a_0, \cdots, a_{k-1}$ arbitrarily from F and determine succeeding coordinates by the linear recursion in (3). A is a cyclic code of dimension k over[3] F. A criterion for the above cyclic code is that if $a(x)$ denotes the polynomial $a_0x^{n-1} + a_1x^{n-2} + \cdots + a_{n-1}$ in $F[x]$ and if $f^{\#}(x) = (x^n + 1)/f(x)$, then the vector a in V belongs to the cyclic code A if and only if the associated polynomial $a(x)$ is a multiple of $f^{\#}(x)$. This result is proved in [5] for all n; we essentially prove it in Corollary 2 by our methods for odd n.

Recall that the *weight* $w(a)$ of a vector $a = (a_0, \cdots, a_{n-1}) \in V$ is the total number of a_i which equal 1. Throughout this paper we shall use the notation d for the minimum nonzero weight of vectors a in A:

$$d = \min\{w(a); a \neq (0, 0, \cdots, 0), a \in A\}.$$

We now present the definition of Bose-Chaudhuri codes given in [8]. Let β be any primitive nth root of unity over F, choose $s < n$, and let $f^{\#}(x)$ be the polynomial over F of least degree which divides $x^n + 1$ and has $\beta, \beta^2, \beta^3, \cdots, \beta^s$ among its roots. Define $f(x) = (x^n + 1)/f^{\#}(x)$.

DEFINITION. *The Bose-Chaudhuri code for* β *and* s *is defined as the code* A *attached to* $f(x)$ *by our definition* (3).

It is known that for the above-defined code A, we have[4] $d \geq d_0 = s + 1$. We shall later prove this result (as Theorem 1) by our methods.

[2] The reader may note the inessential distinction between our definition and that of some authors, who "reverse" the code, i.e., who for our $f(x)$ would use $x^kf(1/x)$.

[3] See Corollary 1, (iii), for definition.

[4] See [8]. A is the code W^* of [8, p. 22], with our $f(x)$ equal to the h there.

If $\phi(x)$ is any polynomial over F and z any root of $\phi(x)$, then z^2 is also a root of $\phi(x)$; this fact follows from the property that if L is any field containing F, then $(\alpha + \beta)^2 = \alpha^2 + \beta^2$ for all $\alpha, \beta \in L$, which implies $\phi(z)^2 = \phi(z^2)$. If $\phi(x)$ is irreducible over F, then all the roots of $\phi(x)$ are obtainable on repeated squaring of any one of them: if $\phi(z) = 0$, then all the roots of $\phi(x)$ are $z, z^2, z^4, \cdots, z^{2^{j-1}}$, where j is the degree of $\phi(x)$. Also, $z^{2^j} = z$.

In passing, we note that choosing s above to be even is free; for if s is odd, then β^{s+1} is automatically a root of $f^{\#}(x)$, since $\beta^{(s+1)/2}$ is by definition a root of $f^{\#}(x)$. Therefore, we shall always take s to be even in the above definition.

We now set up certain definitions basic to this paper. Let n be odd and let $f(x) \in F[x]$ divide $x^n + 1$, as before. Let ζ be a primitive nth root of unity, i.e., an element ζ of K such that $\zeta^n = 1$ and no lower power of ζ is 1. We define

$$E(\zeta) = \{e; 0 \leqq e < n, f(\zeta^e) = 0\}. \tag{4}$$

In other words, if $f(x)$ has degree k, $E(\zeta)$ consists of integers $e_1, \cdots, e_k$ with $0 \leqq e_i < n$ such that $\zeta^{e_1}, \cdots, \zeta^{e_k}$ are all the roots of $f(x)$. For example, when $n = 7$, we have $x^7 + 1 = (x + 1)(x^3 + x + 1)(x^3 + x^2 + 1) = (x + 1)f_0(x)f_1(x)$. Suppose we take $f(x) = (x + 1)f_0(x)$. Since $n = 7$ is prime, any root of $x^7 + 1$ other than 1 is a primitive 7th root of 1. If we take ζ as a root of $f_0(x)$, then $E(\zeta) = \{0, 1, 2, 4\}$. But if we take ζ to be a root of $f_1(x)$, then, since the roots of $f_1(x)$ are ζ, ζ^2, ζ^4, the roots of $f_0(x)$ are now $\zeta^3, \zeta^5, \zeta^6$; thus $E(\zeta) = \{0, 3, 5, 6\}$.

The basis for all the work of this paper is the following result. As above let n be odd, let ζ be a primitive nth root of unity, let $f(x)$ divide $x^n + 1$, and let A be the code attached to $f(x)$ by (3). Then we have

LEMMA 2. *For each $a = (a_0, \cdots, a_{n-1}) \in A$ there is a polynomial $g_a(x)$ with coefficients in K such that $a_i = g_a(\zeta^i)$ for $i = 0, 1, \cdots, n - 1$. ζ being fixed, this polynomial is uniquely determined by a. The degree of $g_a(x)$ is at most m, the largest integer in $E(\zeta)$.*

Proof. The roots of $f(x)$ are $\zeta^{e_1}, \cdots, \zeta^{e_k}$, $e_i \in E(\zeta)$, and $f(x)$ has no repeated roots. We may therefore apply Lemma 1 to our vector $a = (a_0, \cdots, a_{n-1}) \in A$, since the recursion (3) defining $a \in A$ is of the type (1). We obtain $c_1, \cdots, c_k \in K$, the c_j's depending on a, of course, such that

$$a_i = c_1(\zeta^{e_1})^i + \cdots + c_k(\zeta^{e_k})^i \tag{5}$$

for $i = 0, 1, \cdots, n - 1$. But we can write (5) as

$$a_i = c_1(\zeta^i)^{e_1} + \cdots + c_k(\zeta^i)^{e_k} \qquad (i = 0, 1, \cdots, n - 1), \tag{6}$$

and this observation leads us to define the polynomial

$$g_a(x) = c_1 x^{e_1} + \cdots + c_k x^{e_k},$$

which, by Lemma 1, is uniquely determined by the vector $a \in A$ (and the choice of ζ). Equation (6) is the statement $a_i = g_a(\zeta^i)$.

The mapping $a \to g_a(x)$ of A into $K[x]$ is linear over F. That is, if $a, b \in A$ then $g_{a+b}(x) = g_a(x) + g_b(x)$.

We emphasize that $g_a(x)$ must take the value 0 or 1 on the group of nth roots of unity $Z = \{1, \zeta, \cdots, \zeta^{n-1}\}$; the weight of a, therefore, is n less the number of zeros of $g_a(x)$ on Z. In this way the combinatorial problem of determining the weight of a vector is transformed into the more algebraic problem of determining the number of zeros of a polynomial on a given set. In particular, we shall next reproduce the Bose-Chaudhuri lower bound d_0 for d simply by proving that we can choose ζ so that the degree of each nonzero $g_a(x)$ is at most $n - d_0$. (The number of zeros of $g_a(x)$ on Z is then at most $n - d_0$, so the weight of a is at least $n - (n - d_0) = d_0$.)

With d_0 as defined by Bose-Chaudhuri via [8], we shall now prove that the minimum nonzero weight d in the code A is at least d_0. We slightly reword the definition of d_0, by singling out $f(x)$ instead of s. As before, let n be odd and let $f(x) \in F[x]$ divide $x^n + 1$ in such a way that there is a primitive nth root β of 1 which is not a root of $f(x)$. Let A be the (Bose-Chaudhuri) code attached by (3) to $f(x)$. Then we have

THEOREM 1. *Let β^{d_0} be the least positive power of β which is a root of $f(x)$. Then d_0 is necessarily odd, and $d \geqq d_0$.*

Proof. As earlier noted, it suffices to prove that for some primitive nth root of unity, ζ, the set $E(\zeta)$ has $n - d_0$ as maximum.

We are given that $\beta, \beta^2, \cdots, \beta^{d_0-1}$ are not roots of $f(x)$ and that β^{d_0} is a root of $f(x)$. It follows immediately that $E(\zeta)$ for $\zeta = \beta^{-1}$ does not contain $n - 1, n - 2, \cdots, n - (d_0 - 1)$ but does contain $n - d_0$.

4. The case where n takes certain prime values—general results. For the rest of the paper we restrict our attention to the case where n is an odd prime $p = 2h + 1$ such that $x^p + 1 = (x + 1)f_0(x)f_1(x)$, where $f_0(x)$ and $f_1(x)$ are irreducible over[5] F.

We immediately observe the following three simple consequences:

(i) $x^p + 1$ factors over F as above if and only if 2 has multiplicative order h modulo p. For if n_i is the degree of $f_i(x) (i = 0, 1)$, and if ζ is a root of $f_i(x)$, then we must have $\zeta^{2^{n_i}} = \zeta$, or $2^{n_i} \equiv 1 (\mathrm{mod}\, p)$. No lower power of 2 can satisfy this congruence, however, or we would not find enough roots of $f_i(x)$. Therefore $n_0 = n_1$, and since $n_0 + n_1 = 2h$, we have $n_0 = n_1 = h$. Conversely, if 2 has multiplicative order h modulo p, then $x^p + 1$ splits over F into three irreducible factors $x + 1, f_0(x)$, and $f_1(x)$. Thus in this case we may, and do, take K as $GF(2^h)$.

(ii) We may, and do, take $f_0(x) = x^h + 0x^{h-1} + \cdots + 1$ and $f_1(x)$

[5] We are grateful to Eugene Prange for suggesting investigation of the Bose-Chaudhuri codes attached to such primes p.

$= x^h + x^{h-1} + \cdots + 1$, since each $f_i(x)$ has degree h and $f_0(x)f_1(x) = x^{p-1} + x^{p-2} + \cdots + x + 1$, the latter implying that the coefficients of x^{h-1} in $f_0(x)$ and $f_1(x)$ are not the same.

(iii) The powers of 2 are precisely the quadratic residues modulo p and $p \equiv \pm 1 (\text{mod } 8)$. For a cyclic group has at most one subgroup of a given order and the subgroup of quadratic residues mod p also has order h. Since, in particular, 2 is a quadratic residue mod p, the law of quadratic reciprocity[6] tells us that $p \equiv \pm 1 (\text{mod } 8.)$[7]. We shall occasionally modify our treatment to deal with one or the other of these two possibilities. Two examples of such primes are $p = 7$ and $p = 17$.

Let us denote by R the set of least positive quadratic residues mod p:

$$R = \{r_i\,; 0 < r_i < p, r_i \equiv 2^{i-1} (\text{mod } p), i = 1, \cdots, h\}.$$

Let $R' = \{s_1, \cdots, s_h\}$ denote the set of least positive quadratic non-residues mod p; $R \cup R' = \{1, 2, \cdots, p - 1\}$. Let s_1 denote the least member of R' (note that s_1 is thus odd) and choose the notation so that $s_2 \equiv 2s_1, s_3 \equiv 2s_2, \cdots, s_1 \equiv 2s_h (\text{mod } p)$, as we may do since $R' \equiv s_1 R$ (mod p). Thus we have chosen $r_1 = 1$; and, as we shall see in a moment, the most advantageous choice of ζ leads to $d_0 = s_1$.

We define our code A as the $(h + 1, p)$ code attached by (3) to $f(x) = (x + 1) f_0(x)$. In this section we present some general results on the minimum nonzero weight d of vectors a in A, and in the next section we shall present further results on d for particular values of p. The code attached by (3) to $(x + 1)f_1(x)$ is equivalent to A, as we shall prove below, after Corollary 2, so we may confine our attention to A. (Two codes are called equivalent if one can be obtained from the other by a permutation of coordinates.)

We now carry out some of the procedures of the previous section for this code A. The distinction between the two cases $p \equiv \pm 1 (\text{mod } 8)$ arises from the fact that for odd primes q, -1 is a quadratic residue (mod q) if and only if $q \equiv 1 (\text{mod } 4)$. In other words

$$p - 1 \in R, \quad \text{if} \quad p \equiv 1 (\text{mod } 8);$$

$$p - 1 \in R', \quad \text{if} \quad p \equiv -1 (\text{mod } 8).$$

Let ζ be any primitive pth root of unity. Any choice of ζ leads either to $\{0\} \cup R$ or $\{0\} \cup R'$ as $E(\zeta)$; we always choose ζ so that $p - 1$ is not in

[6] See any book on number-theory, e.g., [10, p. 127].

[7] As a side point which may be of some interest, we note that for primes less than 3,000,000, those congruent to 1 mod 8 are scarce compared with those congruent to -1 mod 8. See Shanks, Daniel, *Quadratic residues and the distribution of primes*, Math. Tables Aids Comput., 13 (1959), pp. 272–284 (Math. Rev., 21 (1960), Rev. No. 7186, p. 1325).

$E(\zeta)$. That is,

$$\text{Choose } \zeta \text{ to be a root of } f_1(x) \quad \text{if} \quad p \equiv 1 \pmod 8;$$
$$\text{Choose } \zeta \text{ to be a root of } f_0(x) \quad \text{if} \quad p \equiv -1 \pmod 8.$$

(Recall that in (4) $E(\zeta)$ is defined so that $(\zeta^e + 1)f_0(\zeta^e) = 0$ when e runs through $E(\zeta)$.) In Theorem 1, d_0 is defined as the smallest odd exponent such that β^{d_0} is a root of $f_0(x)$, where β is a root of $f_1(x)$. In other words, since the roots of $f_1(x)$ are $\beta, \beta^2, \beta^4, \cdots, \beta^{2^{h-1}}$, i.e., $\beta^{r_1}, \beta^{r_2}, \cdots, \beta^{r_h}$, we find immediately that $d_0 = s_1$. Thus we have

THEOREM 1′. *Under the restrictions of this section, and letting* s_1 *be the smallest quadratic nonresidue* $(\bmod\, p)$ *(*s_1 *is necessarily odd), we have* $d \geqq s_1 = d_0$.

We summarize the preceding remarks in

LEMMA 3. *The primitive pth root of unity* ζ *is chosen and the polynomial* $g_a(x)$ *of Lemma 2 is given as follows: If* $a \in A$, *then*

$$g_a(x) = \begin{cases} c_0 + c_1 x^{r_1} + c_2 x^{r_2} + \cdots + c_h x^{r_h}, & (f_0(\zeta) = 0) \\ & p \equiv -1 \pmod 8) \\ c_0 + c_1 x^{s_1} + c_2 x^{s_2} + \cdots + c_h x^{s_h}, & (f_1(\zeta) = 0) \\ & p \equiv +1 \pmod 8). \end{cases}$$

The degree of $g_a(x)$ is at most $p - d_0 = p - s_1$ in both cases.

For convenience we shall need a generic set of exponents for $g_a(x)$. Let us say then that $g_a(x) = c_0 + c_1x^{e_1} + c_2x^{e_2} + \cdots + c_hx^{e_h}$, where $e_i = r_i$ for all $i = 1, \cdots, h$ when $p \equiv -1 \pmod 8$ and $e_i = s_i$ for all i when $p = +1 \pmod 8$. By our choice of notation we have $e_i \equiv 2^{i-1}e_1 \pmod p$ for all i; and the roots of $f_0(x)$ are $\zeta^{e_1}, \cdots, \zeta^{e_h}$. Notice that e_h is not necessarily the degree of $g_a(x)$ (even when $c_h \neq 0$).

We now investigate the coefficients c_i of $g_a(x)$. For this we need the following lemma, due to Reed [6].

LEMMA 4. *Let* K_0 *be any field containing* F *and* $h(x) = \sum c_j'x^j$ *any polynomial over* K_0. *Let* β *be a primitive mth root of unity in* K_0, *with* m *odd, and suppose degree* $h(x) < m$. *Then we have*

$$c_j' = \textstyle\sum_{i=0}^{m-1} h(\beta^i)\beta^{-ji} \qquad (j = 0, 1, 2, \cdots).$$

Proof. The sum in question is

$$\textstyle\sum_{i=0}^{m-1} \sum_{k \neq j} c_k'\beta^{i(k-j)} + c_j' \sum_{i=0}^{m-1} 1,$$

in which the second part is c_j', since m is odd; the rest is 0 because

$$\sum_{i=0}^{m-1} \beta^{i(k-j)} = \frac{x^m + 1}{x + 1}$$

with $x = \beta^{k-j} \neq 1$.

When we apply Lemma 4 to our polynomial $g_a(x)$ we obtain, recalling Lemma 2,

LEMMA 5. *Let a be in A. The coefficients* $c_0, c_1, \cdots, c_h$ *of* $g_a(x)$ *are given by the formulas*

$$c_0 = \sum_{i=0}^{p-1} a_i, \qquad (c_0 \in F)$$

$$c_j = \sum_{i=0}^{p-1} a_i \zeta^{-ie_j}, \qquad (c_j \in K) \qquad (j = 1, \cdots, h).$$

In particular, $c_0^2 = c_0, c_1^2 = c_2, c_2^2 = c_3, \cdots, c_h^2 = c_1$.

COROLLARY 1.

(i) *The linear mapping from* A *to* $F \times K$ *given by* $a \to (c_0, c_1)$ *(where* $a \in A$ *and* $g_a(x) = c_0 + c_1x^{e_1} + \cdots + c_hx^{e_h}$*) is one-one and onto (by equality of dimensions over* F*).*

(ii) *Furthermore, the polynomial* $g_a(x)$ *satisfies*

$$g_a(\zeta^i) = c_0 + T(c_1\zeta^{ie_1}), \qquad a \in A,$$

where T *denotes the trace from* K *to* F *(if* $z \in K$*, then* $T(z)$ *is defined as* $z + z^2 + z^4 + \cdots + z^{2^{h-1}}$*).*

(iii) *The code* A *is cyclic; i.e., for each* $a = (a_0, a_1, \cdots, a_{p-1})$ *in* A*, the vector* $a' = (a_1, a_2, \cdots, a_{p-1}, a_0)$ *is also in* A*. (For* $a_i' = g_a(\zeta^{i+1})$*; thus* $g_a(\zeta x)$ *is the polynomial for the code-vector corresponding to* $(c_0, \zeta c_1)$ *under the mapping in* (i) *above. We could have proved this result earlier, of course.)*

(iv) *The polynomial* $g_a(x)$*,* $a \in A$*, has degree* $p - d_0 = p - s_1$ *unless it is constant (since* $c_1, \cdots, c_h$ *are all* 0 *or all not* 0*).*

COROLLARY 2. *Let* β *be a root of* $f_1(x)$*; if* $a \in A$*, then*

$$\sum_{i=0}^{p-1} a_i\beta^{-i} = 0.$$

Proof. Since $\beta = \zeta^e$ for some $e \not\equiv 0, e_1, \cdots, e_h \pmod p$, the quantity in question is 0 as the coefficient of x^e in $g_a(x)$ when $a \in A$.

Let A_1 be the code attached by (3) to $f(x) = (x+1)f_1(x)$. We shall show that A and A_1 are equivalent. Let S be any quadratic nonresidue mod p, and consider the permutation of coordinates sending Si to i for each $i = 0, 1, \cdots, p-1 \pmod p$. If ζ is chosen according to Lemma 3, then $a \in A$ is mapped by our permutation to $b = (g_a(1), g_a(\beta), g_a(\beta^2), \cdots, g_a(\beta^{p-1}))$, where $\beta = \zeta^S$. That is, $b_i = g_a(\zeta^{Si}) = a_{Si}$. By Corollary 1, (i), the set of $g_a(x)$, $a \in A$, coincides with the corresponding set of polynomials for the code A_1; and by Lemma 3, β is a proper choice of pth root of unity for A_1, since β and ζ are not conjugate. In the case $p \equiv -1 \pmod 8$, we may choose $S = -1$ and simply obtain A_1 as the "reverse" of A (followed by one cyclic shift).

We now prove a simple result on the even weights of A when $p \equiv 1 \pmod 8$. That is, we investigate d for the code belonging to $f(x) = f_0(x)$

according to the definition in (3). The result is analogous to Theorem 1 in that the argument rests entirely on the degree of the polynomial $g_a(x)$.

THEOREM 2. *Let d' denote the minimum nonzero even weight attained on A. If $p \equiv 1 (\bmod 8)$ then $d' \geqq 2\, d_0$.*

Proof. Let a be in A, with $w(a)$ even. By Lemmas 3 and 5,

$$g_a(x) = x^{d_0}(c_1 + \cdots + c_h x^{s_h - d_0});$$

and the polynomial in parentheses has degree $p - 2\, d_0$. Therefore $w(a) \geqq 2\, d_0$.

We now prove

THEOREM 3. *If $d_0 < h$, then $d \geqq d_0 + 1$.*

Proof. By Theorem 1, we know $d \geqq d_0$; and if $d = d_0$, then for some $a \in A$ all roots of $g_a(x)$ are pth roots of unity. Since d_0 is odd, $g_a(0) = c_0 = 1$; therefore, if c denotes the leading coefficient of $g_a(x)$, the constant term $1/c$ (of $(1/c)g_a(x)$) is a pth root of unity. Thus $c^p = 1$, and by Lemma 5 all coefficients c_j in $g_a(x)$ satisfy $c_j^{\,p} = 1$. Thus for all i we have $a_i = g_a(\zeta^i) = 1 + T(c\zeta^{e_1 i})$; here $c_1\zeta^{e_1 i}$ runs through all pth roots of 1 as i goes from 0 to $p - 1$. Thus a_i assumes the value 0 h times (when $c_1\zeta^{e_1 i}$ is a root of $f_0(x)$) and the value 1 h times, and also the value $1 + T(1) = 1 + h \cdot 1$. Thus if h is odd, a has weight h, contradicting $w(a) = d_0 < h$; if h is even, then a has weight $h + 1$, also a contradiction.

Note that there is nothing gained in splitting the above hypothesis into "$d_0 < h(h$ odd) and $d_0 < h + 1(h$ even)" since d_0 is odd. Note also that we have proved that for this class of primes the Bose-Chaudhuri lower bound is never best possible, except perhaps in the case when d_0 is already as big as h, e.g., $p = 7$ (which indeed may be the only such case).

THEOREM 4. *If $d_0 < h$ and $p \equiv 1(\bmod 8)$, then $d \geqq d_0 + 2$.*

Proof. From Theorem 3 we have $d \geqq d_0 + 1$, and if $d = d_0 + 1$, an even number, then Theorem 2 implies $d' = d_0 + 1 \geqq 2\, d_0$, or $d_0 \leqq 1$. But $d_0 \geqq 3$ in general. Therefore, $d > d_0 + 1$.

A possible way of showing that $d \geqq d_0 + 2$ when $p \equiv -1(\bmod 8)$ is given by

COROLLARY 3. *Let $p \equiv -1(\bmod 8)$ and let $c_{j+1} = c_1^{2^j}$ be the leading coefficient of $g_a(x)$. If $2^j - 1$ is prime to $2^h - 1$ and if $d_0 < h$, then $d \geqq d_0 + 2$.*

Proof. From Theorem 3 we know that $d \geqq d_0 + 1$. If $a \in A$ has weight $d_0 + 1$, then $c_0 = 0$, by Lemma 5; thus $(1/x)g_a(x) = c_1 + \cdots + c_h x^{r_h - 1}$, which has degree $p - d_0 - 1$ by Theorem 1. Thus $w(a) = d_0 + 1$ implies that all roots of $(1/x)g_a(x)$ are pth roots of unity. But $c_{j+1}/c_1 = c_1^{2^j - 1}$, by hypothesis, has the same order as c_1; and therefore c_1 is a pth root of unity. This means a has weight $h + 1$ (since $T(1) = 1$), as in the proof of Theorem 3.

The corresponding result for the case $p \equiv 1(\mathrm{mod}\, 8)$ is vacuous, since here $j = h/2$.

Our next result is a consequence of the cyclic property of A. As before, let the vector $a = (a_0, \cdots, a_{p-1})$ correspond to the polynomial $a(x) = a_0x^{p-1} + a_1x^{p-2} + \cdots + a_{p-1}$. Then the cyclic shift $(a_1, a_2, \cdots, a_{p-1}, a_0)$ of a corresponds to the polynomial $xa(x)$ reduced modulo $x^p + 1$. If a is in A, then $a(x)$ is a multiple of $f_1(x)$, and conversely, as we remarked in §3. Identifying vectors a in V with their associated polynomials $a(x)$, we shall prove

THEOREM 5. *Let $a \in A$ have odd weight m. Then*

$$m^2 - m + 1 \geqq p \qquad \text{if } p \equiv -1(\mathrm{mod}\, 8),$$

and

$$m^2 \geqq p \qquad \text{if } p \equiv +1(\mathrm{mod}\, 8).$$

Proof. Let A_1 be the code attached to $(x + 1)f_1(x)$. Let $a \in A$ have odd weight m; since A and A_1 are equivalent, there must be a vector $b \in A_1$ of weight m. Now consider the polynomial $a(x)b(x)$. Since b has weight m, $a(x)b(x)$ is a sum of m cyclic shifts of a, and therefore is an odd-weight vector in A. By the same argument it is in A_1. Therefore $a(x)b(x)$ is a multiple of both $f_0(x)$ and $f_1(x)$, and hence of the product $f_0(x)f_1(x) = x^{p-1} + \cdots + 1$. Since the weight is odd, $a(x)b(x) \equiv x^{p-1} + \cdots + 1 \;(\mathrm{mod}\, x^p + 1)$. But there are at most m^2 terms in $a(x)b(x)$; therefore $m^2 \geqq p$.

In order to refine this estimate in the case $p = -1(\mathrm{mod}\, 8)$, we use the particular coordinate permutation $i \to -i(\mathrm{mod}\, p)$ known to yield an equivalence between A and A_1 in this case. Then for each coordinate $a_i = 1$ in a, the term x^{p-1-i} in $a(x)$ is matched by the term x^{p-1+i} in $b(x)$; of the m^2 products in $a(x)b(x)$, m are now the same, namely, $x^{p-1-i}x^{p-1+i} = x^{2p-2} (\equiv x^{p-2}(\mathrm{mod}\, x^p + 1))$. Thus m of the 1's collapse into a single 1, so that $p \leqq m^2 - (m - 1)$.

5. Particular values of p (23 and 47). We apply our methods to the Golay (12, 23) code $A[2]$, which is known to be three-error-correcting, i.e., $d = \min \{w(a); a \neq 0, a \in A\} = 7$ [5, pp. 7 ff.].

We factor $x^{23} + 1 = (x + 1)f_0(x)f_1(x)$ into its irreducible parts,[8] where

$$f_0(x) = x^{11} + x^9 + x^7 + x^6 + x^5 + x + 1$$

and

$$f_1(x) = x^{11} + x^{10} + x^6 + x^5 + x^4 + x^2 + 1.$$

[8] A table of irreducible factors of $x^n + 1$ over F, for odd $n \leqq 35$, appears in [5, pp. 22–23].

If β is a root of f_0, then the roots of f_0 are

$$\beta, \beta^2, \beta^4, \beta^8, \beta^{16}, \beta^9, \beta^{18}, \beta^{13}, \beta^3, \beta^6, \beta^{12},$$

while the roots of f_1 are

$$\beta^5, \beta^{10}, \beta^{20}, \beta^{17}, \beta^{11}, \beta^{22}, \beta^{21}, \beta^{19}, \beta^{15}, \beta^7, \beta^{14}.$$

The code A is the set of all linear recursive sequences in F (of length 23) generated by the difference equation associated with $f_0(x)(x + 1)$, as in §2. For $a = (a_0, a_1, a_2, \cdots, a_{11}, a_{12}, \cdots, a_{22})$ the general term a_k is given by Lemma 1 as

$$(9) \qquad a_k = \sum_{i=0}^{11} c_i(\beta^{r_i k}), \qquad r_i \equiv 2^{i-1} (\text{mod } 23), \qquad 0 < r_i < 23,$$

(and $r_0 = 0$) when the c_i, $i = 0, 1, \cdots, 11$ are determined by the first 12 values $(a_0, a_1, \cdots, a_{11})$. The c_i are in $K = GF(2^{11})$, the smallest field over F containing the 23rd roots of unity.

The code A has dimension 12 in $V_{23}(F)$ and is clearly cyclic.

We may look upon (9) as the value of a polynomial $g_a(x)$ when x runs through the 23rd roots of unity, namely

$$g_a(x) = \sum_{i=0}^{11} c_i x^{r_i}.$$

Then

$$a_k = \sum_{i=0}^{11} c_i(\beta^{r_i k}) = \sum_{i=0}^{11} c_i(\beta^k)^{r_i}$$

or,

$$a_k = g_a(\beta^k) \qquad (k = 0, \cdots, 22).$$

The coefficients c_i—using Lemma 5—are given very simply by

$$(10) \qquad \begin{aligned} c_i &= (c_1)^{2^{i-1}} \qquad (i = 1, \cdots, 11), \\ c_0 &= \sum_0^{22} a_i . \end{aligned}$$

The exponents r_i of (9) are

$$1, 2, 4, 8, 16, 9, 18, 13, 3, 6, 12.$$

We note that to each code word a is assigned a pair (c_0, c_1). c_0 is either zero or one according as the weight of a is even or odd. c_1 is an element of $K = GF(2^{11})$ and is given very simply by the formula

$$c_1 = \sum_{i=0} a_i \beta^{-i},$$

where β is the fixed root of $f_0(x)$ chosen before.

For every c in K we have two code words, one of even and one of odd weight. We thus have a natural mapping of our code A onto $F \times K$ as in Corollary 1.

We note that x^{18} is the highest-degree term of our polynomial, since $\max r_i = 18$, and we see clearly that $g_a(x)$ can have at most 18 roots. Therefore the minimum weight of a in A is the minimum number of 1's that $g_a(x)$ takes on as values over the 23rd roots of unity. We get the Bose-Chaudhuri lower bound immediately, namely $23 - 18 = 5$.

We can do better, however, by examining the coefficients of $g_a(x)$. We do so now. If $w(a) = 5$, then $c_0 = 1$; and our polynomial is (let $c = c_1$ in (10))

$$g_a(x) = c^{2^6}x^{18} + c^{2^4}x^{16} + c^{2^7}x^{13} + \cdots + 1.$$

The product of the roots of $g_a(x)$ is of course $1/c^{2^6}$.

If all the 18 roots of $g_a(x)$ are 23rd roots of unity, then their product is a 23rd root of unity—so $(c^{2^6})^{23} = 1$; therefore, $(c^{2^6})^{2^5} = c$ is a 23rd root of 1. In this case, each exponent on c can be reduced mod 23 and we have

$$g_a(x) = c^{18}x^{18} + c^{16}x^{16} + c^{13}x^{13} + \cdots + 1.$$

If we let $y = cx$, we obtain

$$g_a(x) = y^{18} + y^{16} + y^{13} + y^{12} + \cdots y^2 + y + 1.$$

This is a polynomial over F, which if it contains a root β must contain all its conjugates (obtained by squaring). Therefore, $g_a(x)$ can have only 12 roots of unity as zeros; i.e., $w(a)$ is at least 11. We therefore consider only those c's which are not 23rd roots of 1. We conclude $w(a) > 5$.

We now eliminate $w(a) = 6$ very simply. For if $w(a) = 6$, then $c_0 = 0$, and by (10), and

$$\begin{aligned} g_a(x) &= cx + c^2x^2 + \cdots + c^{2^4}x^{16} + c^{2^6}x^{18} \\ &= x(c + c^2x + \cdots + c^{2^6}x^{17}). \end{aligned}$$

The products of the 17 nonzero roots is clearly

$$c/c^{2^6} = (c^{2^6-1})^{-1}.$$

If all these 17 roots are 23rd roots of unity, then

$$(c^{2^6-1})^{23} = 1;$$

but $2^6 - 1$ is prime to $2^{11} - 1$, so c itself must be a 23rd root of unity—again this brings $w(a)$ up to 11.

We get immediately that $w(a) \geqq 7$. We know however of the existence of a vector a with $w(a) = 7$ (see [5]), so that ends the investigation.

Alternatively, we would use our general results to prove $d \geqq 7$ for this code as follows:

We have $d_0 = s_1 = 5$ and $h = 11$; and $c_1^{2^6}$ is the leading coefficient of $g_a(x)$. Now $2^6 - 1$ is prime to $2^{11} - 1$; therefore $d \geqq 7 = d_0 + 2$, by Corollary 3.

Perhaps we should observe that the (12, 23) code A of our definition is not precisely Golay's code but is obtainable from his code on permuting the coordinates suitably.

Some Results for $p = 47$. In this example we rely on the general results of §4. We again have $d_0 = 5$, since $R = \{1, 2, 4, 8, 16, 32, 17, 34, 21, 42, 37, 27, 7, 14, 28, 9, 18, 36, 25, 3, 6, 12, 24\}$. Here $g_a(x)$ has degree $r_{10} = 42$, so that the leading coefficient $c_{10} = c_1^{2^9}$. Now $2^9 - 1$ is prime to $2^{23} - 1$, because $(2^{23} - 1) - 2^5(2^9 + 1)(2^9 - 1) = 31$; and thus the g.c.d. $(2^{23} - 1, 2^9 - 1)$ is either 1 or 31. It cannot be 31, because $2^5 \equiv 1 (\text{mod } 31)$, implying $2^9 \equiv 2^4 \not\equiv 1 (\text{mod } 31)$. Thus by Corollary 3, $d \geqq 7$.

Theorem 5 eliminates vectors of weight 7. We shall eliminate vectors of weight 8 by the following procedure (which also works for vectors of weight 7): Let $a \in A$ have weight 8; then $g_a(x)$ has precisely 39 zeros $\beta_1, \cdots, \beta_{39}$ on the set $Z = \{1, \zeta, \zeta^2, \cdots, \zeta^{46}\}$. Thus we factor $g_a(x)$ (in a possibly larger field than K) as

$$g_a(x) = c_{10}x(x + \gamma_1)(x + \gamma_2)(x + \beta_1) \cdots (x + \beta_{39})$$

where γ_1 and γ_2 are not both 47th roots of 1 (and are not 0).

We define $\alpha_1, \cdots, \alpha_8$ as the 47th roots of 1 other than the β_i's; thus $(x + \alpha_1) \cdots (x + \alpha_8)(x + \beta_1) \cdots (x + \beta_{39}) = x^{47} + 1$.

Our first observation is that $\gamma_1 = \gamma_2 = \gamma$. For $\gamma_1 + \gamma_2 + \beta_1 + \cdots + \beta_{39} = 0$ is the coefficient of x^{41} in $g_a(x)$, and $\alpha_1 + \cdots + \alpha_8 + \beta_1 + \cdots + \beta_{39} = 0$; thus $\gamma_1 + \gamma_2 = \alpha_1 + \cdots + \alpha_8$. Now each $\alpha_i = \zeta^{j_i}$ and the j_ith coordinates of a are precisely the "1-positions" of a. Since $p \equiv -1 (\text{mod } 8)$, we have chosen ζ as a root of $f_0(x)$; thus ζ^{-1} is a root β of $f_1(x)$. We apply Corollary 2 to conclude $\alpha_1 + \cdots + \alpha_8 = 0 (= \beta_1 + \cdots + \beta_{39})$.

Now define

$$\textstyle\sum_{i=0}^{39} b_i x^{39-i} = (x + \beta_1) \cdots (x + \beta_{39}).$$

Then $b_0 = 1$ and $b_1 = \beta_1 + \cdots + \beta_{39} = 0$. We now have

$$g_a(x) = c_{10}x(x^2 + \gamma^2)(x^{39} + b_2x^{37} + b_3x^{36} + \cdots + b_{39}).$$

It follows immediately that the coefficient of x^{40-i} in $g_a(x)$ is

$$c_{10}(\gamma^2 b_i + b_{i+2}) \qquad (i = 0, 1, \cdots, 37). \tag{11}$$

Furthermore,

$$c_{10}\gamma^2\beta_1 \cdots \beta_{39} = c_1;$$

or

$$(\gamma^2 c_{10}/c_1)^{47} = 1, \tag{12}$$

incidentally proving that γ^2 and hence all conjugates, including γ, are in K.

We finally define

$$\textstyle\sum_{i=0}^{8} d_i x^{8-i} = (x + \alpha_1) \cdots (x + \alpha_8),$$

in which $d_0 = 1$ and $d_1 = 0$.

We shall now derive various relations between the b_i's, d_i's and γ, which, together with (12), will lead to a contradiction. We first exploit (11) and our list for R. Since $40, 39, 38 \notin R$, we have

$$b_2 = \gamma^2 \qquad (13)$$
$$b_3 = b_1 = 0$$
$$b_4 = \gamma^4 .$$

Continuing this process, and noting that $37 = r_{11}$ and $36 = r_{18}$, etc., we obtain

$$b_5 = c_{11}/c_{10} = c_{10} \qquad (13)$$
$$b_6 = c_{18}/c_{10} + \gamma^6$$
$$b_7 = \gamma^2 c_{10} .$$

We have

$$(x^8 + d_2 x^6 + d_3 x^5 + \cdots + d_8)(x^{39} + b_2 x^{37} + b_4 x^{35} + \cdots + b_{39}) = x^{47} + 1; \qquad (14)$$

we examine the coefficients of $x^{46}, x^{45}, \cdots$ obtained by multiplying out the left-hand side of (14), first obtaining $b_2 + d_2 = 0$ as the coefficient of x^{45}; thus $b_2 = d_2 = \gamma^2$, from (13). Expressing the next few coefficients and using $d_2 = \gamma^2$ and (11) we find $d_3 = 0$

$$0 = b_4 + b_2 d_2 + d_4 = d_4$$
$$0 = b_5 + b_3 d_2 + d_5 = c_{10} + d_5$$
$$0 = b_6 + b_4 d_2 + b_2 d_4 + d_6 = c_{18}/c_{10} + d_6 \qquad (15)$$
$$0 = b_7 + b_5 d_2 + b_2 d_5 + d_7 = \gamma^2 d_5 + d_7$$
$$0 = b_8 + b_6 d_2 + b_2 d_6 + d_8 = c_8/c_{10} + \gamma^2 d_6 + d_8 .$$

Having now determined all the d_i's in (15), we look for a contradiction in the succeeding equations. We write down the coefficient of x^{35} on the left-hand side of (14), namely

$$0 = b_{12} + b_{10} d_2 + b_7 d_5 + b_6 d_6 + b_5 d_7 + b_4 d_8 \qquad (16)$$
$$= c_{10}(b_7 + d_7) + \gamma^4 c_8/c_{10} + \gamma^6 d_6 + b_6 d_6$$

where we have used (13) and (15), which also give $b_7 = d_7$, etc., so that (16) becomes $\gamma^4 c_8/c_{10} + (c_{18}/c_{10})^2 = 0$. By Lemma 5, this equation is equivalent to

$$\gamma^2 c_7/c_9 = c_{18}/c_{10} \qquad (17)$$

We now multiply (17) by c_9c_{10}/c_1c_7 to obtain $\gamma^2 c_{10}/c_1 = c_9c_{18}/c_1c_7$, which, by (12), must be a 47th root of 1. Using Lemma 5 again, we write

$$c_9c_{18}/c_1c_7 = c_1^{2^{17}+2^8-2^6-1};$$

we now prove that $c_1^{47} = 1$ by showing $2^{17} + 2^8 - 2^6 - 1$ is prime to $2^{23} - 1$. Let δ denote the greatest common divisor of these two numbers. Then δ divides

$$2^6(2^{17} + 2^8 - 2^6 - 1) - (2^{23} - 1) = 3(2^{12} - 21),$$

and therefore δ divides

$$3\cdot 2(2^{23} - 1) - (2^{12} + 21)3(2^{12} - 21) = 3\cdot 439.$$

Now, if q is a prime dividing δ, then $2^{23} \equiv 1 \pmod q$, so that $q \equiv 1 \pmod{46}$. Thus $3 \nmid \delta$, and since 439 is prime, we need only observe that $439 \equiv -21 \pmod{46}$, so that $439 \nmid \delta$ either. Therefore $\delta = 1$, which implies $c_1^{47} = 1$, or $w(a) = h = 23$ (by the proof of Theorem 3), a contradiction. Thus $d \geqq 9$ for this code.

Since Eugene Prange has found many vectors of weight 11 in A, the results to date are $9 \leqq d \leqq 11$ for this (24, 47) code.

REFERENCES

1. R. C. Bose and D. K. Ray-Chaudhuri, *On a class of error correcting binary group codes*, Information and Control, 3 (1960), pp. 68–79.
2. Marcel J. E. Golay, *Notes on digital coding*, Proc. I.R.E., 37 (1949), p. 657.
3. Daniel Gorenstein and Neal Zierler, *A class of cyclic, linear, error-correcting codes in p^m symbols*, Group Report 55–19, Lincoln Laboratory, (1960); *A class of error-correcting codes in p^m symbols*, this Journal, 9 (1961), pp. 207–214.
4. W. W. Peterson, *Error-Correcting Codes*, Mass. Inst. of Technology and Wiley, New York, 1961.
5. Eugene Prange, *Cyclic error-correcting codes in two symbols*, Report No. AFCRC-TN-57-103, USAF Cambridge Research Laboratory, Bedford, Mass., 1957.
6. Irving Reed and Gustave Solomon, *A decoding procedure for a polynomial code*, Group Report 47-24, Lincoln Laboratory, 1959.
7. Gustave Solomon, *Linear recursive sequences as finite difference equations*, Group Report 47-37, Lincoln Laboratory 1960.
8. Edwin Weiss, *Some connections between linear recursive sequences and error-correcting codes: informal lectures*, Group Report 55-22, Lincoln Laboratory, 1960.
9. ———, *Residue class rings and linear recursive sequences*, Group Report 55-24, Lincoln Laboratory, 1960.
10. Hermann Weyl, *Algebraic Theory of Numbers*, Ann. of Math. Studies No. 1, Princeton Univ. Press, 1940.

Decoding

IV

Editor's Comments on Papers 19 and 20

A major concern of many coding theorists is the practical implementation of coding and decoding schemes. For most coding schemes, and certainly for any systematic coding scheme, the coding operation is simple and inexpensive in terms of digital circuitry. Unfortunately, decoding is generally quite the opposite and is the single most important obstacle preventing more widespread application of error-correcting codes. Perhaps the continuing development of technology in the area of digital circuits will make existing error-correction schemes economically feasible.

The first error-correction procedure for binary BCH codes is given in the paper by Peterson [19]. Extension to the nonbinary case is nontrivial and is given in the paper by Gorenstein and Zierler [17] in the previous section. The procedure in the nonbinary case consists of essentially four steps (Peterson and Weldon, 1972). The syndrome digits, S_j, are first calculated as

$$S_j = \sum_{i=1}^{t} Y_i X_i^j = r(\alpha^j) = e(\alpha^j), \quad m_0 \leqslant j \leqslant m_0 + 2t_0 - 1,$$

where $r(x)$ and $e(x)$ are the received and error polynomials, respectively; X_i the error-location variables; and Y_i the error-magnitude variables. The code-generator polynomial has α^j, $m_0 \leqslant j \leqslant m_0 + 2t_0 - 1$ as roots. The elementary symmetric functions, σ_j, are defined by

$$\sigma(x) = \prod_{i=1}^{t} (X + X_i) = \sum_{i=0}^{t} \sigma_i X^{t-i}, \quad \sigma_0 = 1,$$

where it is assumed that $t \leqslant t_0$ errors are actually made.

A very significant contribution to the decoding operation is the development by Berlekamp (1966) of an efficient recursive algorithm to find the polynomial $\sigma(x)$ from the known syndrome digits. As mentioned in the introduction, this work unfortunately appeared in a form inappropriate for inclusion in this volume. The paper by Massey [20], however, apart from giving interesting results on the synthesis of shift registers for generating prescribed sequences, also contains an excellent discussion of a tutorial nature on the Berlekamp algorithm. Once the polynomial $\sigma(x)$ is obtained, a search in the appropriate finite field for its roots (i.e., the error locators) is conducted. The most efficient search method to date is that of Chien (1964). The error magnitudes may then be determined with the help of the method given in Forney (1965).

19

Encoding and Error-Correction Procedures for the Bose-Chaudhuri Codes*

W. W. PETERSON†, MEMBER, IRE

Summary Bose and Ray-Chaudhuri have recently described a class of binary codes which for arbitrary m and t are t-error correcting and have length $2^m - 1$ of which no more than mt digits are redundancy. This paper describes a simple error-correction procedure for these codes. Their cyclic structure is demonstrated and methods of exploiting it to implement the coding and correction procedure using shift registers are outlined. Closer bounds on the number of redundancy digits are derived.

INTRODUCTION

BOSE and Chaudhuri[1] have recently discovered a new class of codes with some remarkable properties. For any positive integers m and t, there is a code in this class that consists of blocks of length $2^m - 1$, that corrects t errors, and that requires no more than mt parity check digits. Thus, the codes cover a wide range in rate and error-correcting ability, unlike most other known classes of codes.[2] These codes are a generalization of the Hamming codes;[3] the case $t = 1$ gives the Hamming code in each case.

In this paper two important properties of these codes are described. First, a method for error correction is described which is a generalization of the simple error-correction procedure that can be used with Hamming codes. The procedure requires a number of operations which increases only as a small power of the length of the codes.

Second, it is shown that these are cyclic codes[4] and,

* Received by the PGIT, December 6, 1959. Part of this work was supported by the U. S. Army Signal Corps, the U. S. Air Force Office of Scientific Research, Air Research and Development Command, and the U. S. Navy Office of Naval Research at the Research Laboratory of Electronics, Mass. Inst. Tech., Cambridge, Mass.; and part of the work was done at the IBM Research Lab., Yorktown, N. Y.

† On leave from the University of Florida, Gainesville. Presently at the Dept. of Elec. Engrg. and Res. Lab. of Electronics, Mass. Inst. Tech., Cambridge, Mass.

[1] R. C. Bose and D. K. Ray-Chaudhuri, "On a class of error-correcting binary group codes," to be published in *Information and Control.*

[2] The only others of which I am aware are I. S. Reed, "A class of multiple-error-correcting codes and decoding scheme," IRE TRANS. ON INFORMATION THEORY, vol. IT-4, pp. 38–49, September, 1954; P. Elias, "Error free coding," IRE TRANS. ON INFORMATION THEORY, vol. IT-4, pp. 29–37, September, 1954; and I. S. Reed and G. Solomon, "Polynomial code," to be published in *J. Soc. Ind. Appl. Math.*

[3] R. W. Hamming, "Error detecting and error correcting codes," *Bell Sys. Tech. J.*, vol. 29, pp. 147–160; April, 1950.

[4] E. Prange, "Some Cyclic Error-Correcting Codes with Simple Decoding Algorithms," Air Force Cambridge Research Center, Bedford, Mass., Tech. Note AFCRC-TN-58-156, April, 1958; "Cyclic Error-Correcting Codes in Two Symbols," Air Force Cambridge Research Center, Bedford, Mass., Tech. Note ARCRC-TN-57-103, September, 1957; "The Use of Coset Equivalence in the Analysis and Decoding of Group Codes," Air Force Cambridge Research Center, Bedford, Mass., Tech. Rept. AFCRC-TR-59-164, June, 1959.

therefore, the encoding can be accomplished very efficiently with a shift register. The theory of the cyclic structure also provides a closer bound on the number of parity checks required to correct a given number of errors.

Construction of the Bose-Chaudhuri Codes

Given an irreducible polynomial $p(X)$ of degree m with 1 and 0 as coefficients, a representation of the Galois Field with 2^m elements $GF(2^m)$ can be formed. It consists of all polynomials of degree $m - 1$ or less. They can be added (modulo 2) term by term in the ordinary way. The rule for multiplication is to multiply in the ordinary way, reducing the answer modulo 2 and modulo $p(X)$ to a polynomial of degree $m - 1$ or less. (That is, consider $p(X) = 0$, and use this equation to eliminate terms of power greater than $m - 1$.) It can be shown then that certain of these polynomials, called primitive elements, have the property that the first $2^m - 1$ powers of such an element are exactly all the $2^m - 1$ nonzero field elements. Also, every nonzero field element is a root of the equation

$$X^{2^m-1} = 1$$

and conversely. Thus if α is any element of the field, $\alpha^{-1} = \alpha^{2^m-2}$.

The field elements can also be thought of as vectors whose components are the coefficients of the polynomials. The sum of two vectors corresponds to the sum of the corresponding polynomials.

The Bose-Chaudhuri codes are described by giving the matrix of parity check rules, which is the matrix

$$M = \begin{bmatrix} 1 & 1 & \cdots & 1 \\ \alpha & \alpha^3 & \cdots & \alpha^{2t-1} \\ \alpha^2 & (\alpha^3)^2 & \cdots & (\alpha^{2t-1})^2 \\ \vdots & \vdots & \vdots & \vdots \\ \alpha^{2^m-2} & (\alpha^3)^{2^m-2} & \cdots & (\alpha^{2t-1})^{2^m-2} \end{bmatrix} \quad (1)$$

where α is a primitive element of the field.

This is a $2^m - 1 \times t$ matrix of $GF(2^m)$ elements, but thinking of each field element as a vector of m binary digits, this is a $2^m - 1 \times mt$ matrix of binary digits. A vector of $2^m - 1$ binary digits is considered a code word if it satisfies the parity check described by each column; *i.e.*, if the product of this vector with the matrix is zero. In other words the set of all code words is the (left) null space of this matrix.

The code that Bose and Ray-Chaudhuri use as an example will be used to illustrate the ideas discussed in this paper. Let α denote a root of the equation $X^4 = X + 1$. This happens to be a primitive element of the field. Then the 15 nonzero field elements are given in Table I.

Taking $t = 3$, the following matrix of parity check rules results:

TABLE I
REPRESENTATION OF $GF(2^4)$

α^0	$= 1$	$= (1\,0\,0\,0)$
α^1	$= \alpha$	$= (0\,1\,0\,0)$
α^2	$= \alpha^2$	$= (0\,0\,1\,0)$
α^3	$= \alpha^3$	$= (0\,0\,0\,1)$
α^4	$= 1 + \alpha$	$= (1\,1\,0\,0)$
α^5	$= \alpha + \alpha^2$	$= (0\,1\,1\,0)$
α^6	$= \alpha^2 + \alpha^3$	$= (0\,0\,1\,1)$
α^7	$= 1 + \alpha + \alpha^3$	$= (1\,1\,0\,1)$
α^8	$= 1 + \alpha^2$	$= (1\,0\,1\,0)$
α^9	$= \alpha + \alpha^3$	$= (0\,1\,0\,1)$
α^{10}	$= 1 + \alpha + \alpha^2$	$= (1\,1\,1\,0)$
α^{11}	$= \alpha + \alpha^2 + \alpha^3$	$= (0\,1\,1\,1)$
α^{12}	$= 1 + \alpha + \alpha^2 + \alpha^3$	$= (1\,1\,1\,1)$
α^{13}	$= 1 + \alpha^2 + \alpha^3$	$= (1\,0\,1\,1)$
α^{14}	$= 1 + \alpha^3$	$= (1\,0\,0\,1)$
α^{15}	$= 1 = \alpha^0$	

$$M = \begin{bmatrix}
1\,0\,0\,0 & 1\,0\,0\,0 & 1\,0\,0\,0 \\
0\,1\,0\,0 & 0\,0\,0\,1 & 0\,1\,1\,0 \\
0\,0\,1\,0 & 0\,0\,1\,1 & 1\,1\,1\,0 \\
0\,0\,0\,1 & 0\,1\,0\,1 & 1\,0\,0\,0 \\
1\,1\,0\,0 & 1\,1\,1\,1 & 0\,1\,1\,0 \\
0\,1\,1\,0 & 1\,0\,0\,0 & 1\,1\,1\,0 \\
0\,0\,1\,1 & 0\,0\,0\,1 & 1\,0\,0\,0 \\
1\,1\,0\,1 & 0\,0\,1\,1 & 0\,1\,1\,0 \\
1\,0\,1\,0 & 0\,1\,0\,1 & 1\,1\,1\,0 \\
0\,1\,0\,1 & 1\,1\,1\,1 & 1\,0\,0\,0 \\
1\,1\,1\,0 & 1\,0\,0\,0 & 0\,1\,1\,0 \\
0\,1\,1\,1 & 0\,0\,0\,1 & 1\,1\,1\,0 \\
1\,1\,1\,1 & 0\,0\,1\,1 & 1\,0\,0\,0 \\
1\,0\,1\,1 & 0\,1\,0\,1 & 0\,1\,1\,0 \\
1\,0\,0\,1 & 1\,1\,1\,1 & 1\,1\,1\,0
\end{bmatrix} \quad (2)$$

Of these twelve columns, the last one is trivial and the next to last is a duplicate; these two can be dropped. The rest are independent, and the result is a code with fifteen digit code words of which ten are parity checks and five are information places. The code corrects all triple errors.

An Error-Correction Procedure

Consider the result of multiplying a vector $(r_0, r_1, r_2, \cdots, r_{n-1})$ of $n = 2^m - 1$ components by the matrix M in (1). The result is a vector of t Galois field elements. The first component is

$$r_0 + r_1\alpha + r_2\alpha^2 + \cdots + r_{n-1}\alpha^{n-1} = r(\alpha)$$

where

$$r(X) = r_0 + r_1X + \cdots + r_{n-1}X^{n-1}$$

is the polynomial which corresponds naturally to the given vector. (In what follows no distinction will be made between a vector and the corresponding polynomial.) The other components are clearly $r(\alpha^3), r(\alpha^5), \cdots, r(\alpha^{2t-1})$.

In these terms an equivalent definition of the Bose-Chandhuri codes can be given. A vector is a code word if it is in the left null space of M, *i.e.*, if the parity checks $r(\alpha), r(\alpha^3), r(\alpha^5), \cdots, r(\alpha^{2t-1}0)$ are zero. This can be restated as follows:

Definition: A polynomial $s(X)$ is a code vector for a t-error correcting Bose-Chaudhuri code if, and only if, $\alpha, \alpha^3, \cdots, \alpha^{2t-1}$ are roots of $s(X)$.

The first step in devising a decoding method is to characterize the information contained in the parity check calculation for a received vector which may contain errors. Let $e = (e_0, e_1, \cdots, e_{n-1})$ be the vector of errors, *i.e.*, if the errors occur in the positions $i_1, i_2, \cdots, i_\nu$, then

$$e_i = 1 \text{ for } i = i_1, i_2, \cdots, i_\nu$$

$$e_i = 0 \text{ otherwise.}$$

There is a one to one correspondence between the elements of the error vector and the elements of $GF(2^m)$ which constitute the first column of the parity check matrix M given by (1), e_i corresponding to the element α^i occurring in the i-th position in the first column of M. The elements $X_1, X_2, \cdots, X_\nu$ of $GF(2^m)$ which correspond in this way to $e_{i_1}, e_{i_2}, \cdots, e_{i_\nu}$ may be called the error position numbers. Thus $X_j = \alpha^{i_j}$ $(j = 1, 2, \cdots, \nu)$.

Lemma 1: If a received vector r has errors in digits numbered $X_1, X_2, \cdots, X_\nu$, then the parity check vector $r \times M$ is of the form $(S_1, S_3, S_5, \cdots, S_{2t-1})$ where

$$S_j = \sum_{i=1}^{\nu} X_i^j. \tag{3}$$

Proof: Assume that the vector s was transmitted, and $r = s + e$ received, where e has ones in the positions $i_1, i_2, \cdots, i_\nu$ and zeros in all other positions. In terms of corresponding polynomials,

$$r(X) = s(X) + e(X)$$

and the result of the parity check calculation is

$$[r(\alpha), r(\alpha^3), \cdots, r(\alpha^{2t-1})].$$

But $s(\alpha) = s(\alpha^3) = \cdots = s(\alpha^{2t-1}) = 0$, so that $r(\alpha) = s(\alpha) + e(\alpha) = e(\alpha)$, $r(\alpha^3) = e(\alpha^3)$, etc. Thus, the result of the parity check calculation is $[e(\alpha), e(\alpha^3), \cdots, e(\alpha^{2t-1})]$. But

$$e(\alpha^j) = e_0 + e_1\alpha^j + e_2\alpha^{2j} + \cdots + e_{n-1}\alpha^{(n-1)j}$$

$$= \sum_{i=1}^{\nu} \alpha^{i_\nu j} = \sum_{i=1}^{\nu} X_i^j \quad \text{Q.E.D.}$$

It is interesting to note that for $t = 1$, if the error occurs, for example, in the component numbered X_1, then the result of the parity check calculation is exactly $S_1 = X_1$ which is the Galois field binary code for the error position number. This is exactly analogous to the method of error-correction for Hamming codes in which the parity check calculation gives the ordinary binary code for the position of the error. In this sense the Bose-Chaudhuri codes for $t = 1$ are equivalent to the Hamming single-error correcting code.

The S_j are the power sum symmetric functions.[5] Thus the parity checks give the first t odd power sum symmetric functions. The first t even ones can be found from the fact that modulo 2, $(a + b)^2 = a^2 + b^2$, and hence

$$S_1^2 = \left[\sum_{i=1}^{\nu} X_i\right]^2 = \sum_{i=1}^{\nu} X_i^2 = S_2. \tag{4}$$

Similarly, $S_4 = S_1^4$, $S_6 = S_3^2$, etc.

Suppose that there are t errors. Then the error position numbers $X_1 \cdots X_t$ satisfy the equations

$$S_j = \sum_{i=1}^{t} X_i^j \qquad j = 1, 3, \cdots 2t - 1.$$

This is a set of t equations in t unknowns, the X_i. The solution would tell the positions of the errors. It appears impossible to solve the equations by any direct method, and trying all combinations of t of the $2^m - 1$ field elements would require too many computations. There is, however, an interesting compromise.

The elementary symmetric functions σ_i are related to the power sum symmetric functions S_i by Newton's identities:[5]

$$\left.\begin{array}{l} S_1 - \sigma_1 = 0 \\ S_2 - S_1\sigma_1 + 2\sigma_2 = 0 \\ S_3 - S_2\sigma_1 + S_1\sigma_2 - 3\sigma_3 = 0 \\ S_4 - S_3\sigma_1 + S_2\sigma_2 - S_1\sigma_3 + 4\sigma_4 = 0 \\ S_5 - S_4\sigma_1 + S_3\sigma_2 - S_2\sigma_3 + S_1\sigma_4 - 5\sigma_5 = 0 \\ \cdots \text{ etc.} \end{array}\right\} \tag{5}$$

If it is possible to solve Newton's identities for the elementary symmetric functions σ_i, the error position numbers must satisfy the equation

$$X^t - \sigma_1 X^{t-1} + \sigma_2 X^{t-2} \cdots \pm \sigma_t$$

$$= (X - X_1)(X - X_2) \cdots (X - X_t) = 0. \tag{6}$$

Eq. (6) can be solved effectively by merely substituting each of the $n = 2^m - 1$ field elements into the equation. For each digit in the received vector, the corresponding $GF(2^m)$ element is substituted in the equation. If the equation is satisfied, this bit is wrong and must be changed. If the equation is not satisfied, the bit is correct.

[5] See, for example, van der Waerden, footnote 8; J. Riordan, "An Introduction to Combinatorial Analysis," John Wiley and Sons, Inc., New York, N. Y., 1958; T. Muir and W. H. Metzler, "A Treatise on the Theory of Determinants," ch. 21, 1930; or any book on the Theory of Equations.

The proof that it is indeed possible to solve for the ordinary symmetric functions from the power sum symmetric functions is given by the following theorem:[6]

Theorem 1: The $k \times k$ matrix

$$M_k = \begin{bmatrix} 1 & 0 & 0 & 0 & \cdots & 0 \\ S_2 & S_1 & 1 & 0 & \cdots & 0 \\ S_4 & S_3 & S_2 & S_1 & \cdots & 0 \\ \vdots & \vdots & \vdots & \vdots & & \\ S_{2k-4} & S_{2k-5} & S_{2k-6} & S_{2k-7} & & S_{k-3} \\ S_{2k-2} & S_{2k-3} & S_{2k-4} & S_{2k-5} & \cdots & S_{k-1} \end{bmatrix}$$

is nonsingular if power sum symmetric functions S_i are power sums of k or $k - 1$ distinct field elements, and is singular if the S_i are power sums of fewer than $k - 1$ distinct field elements.

The proof requires the following two lemmas:

Lemma 2: If the S_i are power sums of $v \leq k - 2$ distinct field elements, M_k is singular.

Proof:

$$M_k \begin{bmatrix} 0 \\ 1 \\ \sigma_1 \\ \cdot \\ \cdot \\ \cdot \\ \sigma_{k-2} \end{bmatrix} = \begin{bmatrix} 0 \\ 0 \\ 0 \\ 0 \\ 0 \\ 0 \\ 0 \end{bmatrix}$$

by Newton's identities, (5), and thus M_k has a nontrivial null space and must be singular. Q.E.D.

Lemma 3: If the S_i are power sums of k indeterminants $X_1, \cdots, X_k$, then the determinant

$$|M_k| = \prod_{i<i} (X_i + X_j).$$

Proof: If $X_i = X_j$, all of the power sums contain two identical terms, which cancel because the field has characteristic 2 (*i.e.*, $2 = 0$). Then it is just as if there were no more than $k - 2$ distinct elements used in forming the power sums, and, by Lemma 2, the determinant is zero. Therefore, $X_i + X_j$ is a factor of the determinant, for all i and j, and the left-hand side must be divisible by the right-hand side. It is easy to check that the left-hand side is homogeneous of degree $k(k - 1)/2$, the same as the right-hand side, and therefore they must differ at most by a constant factor.

To determine the constant factor, a single special case suffices. If k is odd, let the X_i be the roots of the equation

$$X^k - 1 = 0.$$

Then

$$\sum X_i^j = S_j = 0 \quad \text{if} \quad j \not\equiv 0 \mod t,$$
$$= 1 \quad \text{if} \quad j \equiv 0 \mod t.$$

There will be exactly one 1 in each row and each column and it follows that $|M_k| = 1$ in this case. For k even, letting the X_i be all of the roots of the equation

$$X^k - X = 0$$

gives the same result. The constant factor, which could be only 0 or 1, must be 1.

Now Theorem 1 follows from the fact that if the determinant $|M_k|$ is zero it must be that some $X_i = X_j$. Since all of the nonzero X_i are distinct, $X_i = X_j = 0$, and there were fewer than $k - 1$ errors. Q.E.D.

If there are actually $t - 1$ errors, it can be seen from Newton's identities, Cramer's Rule and Theorem 1 that the solution for the σ's will yield $\sigma_t = 0$. The corresponding polynomial equation will have zero as one root.

Now let us review the error-correcting procedure. The t-error correcting Bose-Chaudhuri codes give, as the parity checks on received sequences, the odd power-sum symmetric functions up to S_{2t-1} and the intermediate even functions can be calculated simply from these. If it is assumed that no more than t errors occur, then by Theorem 1, with $k = t$, it is either possible to solve for the error position numbers, or there are $t - 2$ or fewer errors. In the latter case, $\sigma_{t-1} = \sigma_t = 0$, and two equations can be dropped, giving a set of $t - 2$ equations in $t - 2$ unknowns to which Theorem 1 can be applied again. Eventually, if there were any errors at all, a set of equations that can be solved for the elementary symmetric functions of the error-position numbers will be found.

The correction procedure consists of three phases:

1) calculate the parity checks and the even numbered S_i;

2) from these, calculate the elementary symmetric functions σ_i; and

3) finally, substitute each field element into the equation

$$X^t + \sigma_1 X^{t-1} + \sigma_2 X^{t-2} \cdots + \sigma_t = 0. \tag{7}$$

Those field elements which satisfy this equation correspond to error positions.

The second step involves a certain amount of trial and error because it is possible to solve the equations and obtain correct solutions only when the number of equations used equals or exceeds by one the number of errors that actually occur. This step might be carried out, as an alternative to the procedure described in the preceding paragraph, by starting with the assumption that two errors occurred, solving, and checking the solution. If the solution doesn't check, four errors would be assumed, and so forth. When a set of answers that checks occurs, it must be the correct solution.

[6] Similar results for a real field appear, for example, in H. O. Faulkes, "Theorems of Kakeya and Polya on Power sums," *Math. Z.*, vol. 65, pp. 345–352; 1956.

If it is assumed that the length n of the code approaches infinity and that the number of errors corrected t is a fixed fraction of n, the number of operations required for error correction can be crudely estimated as follows. The first phase, calculating parity checks, requires a number of operations proportional to the number of digits multiplied by the number of parity checks, or no more than nmt operations. This quantity nmt is proportional to $n^2 \log n$. The second phase requires solving a $t \times t$ set of equations. The number of operations for this task is typically proportional to t^3, but it may have to be done $t/2$ times. This will increase in the limit no faster than n^4. Finally, substituting in a t-degree polynomial requires t multiplications and t additions of m digit numbers, and must be done n times, so that 2 tmn is a rough estimate of the number of operations. This again would vary as $n^2 \log n$. Thus, the total number of operations certainly would increase as a small power of n.

Consider, as an example, the code corresponding to the matrix in (2), which corrects triple errors. The appropriate equations are

$$S_1 + \sigma_1 = 0,$$
$$S_3 + S_2\sigma_1 + S_1\sigma_2 + \sigma_3 = 0, \quad \text{and} \qquad (8)$$
$$S_5 + S_4\sigma_1 + S_3\sigma_2 + S_2\sigma_3 = 0.$$

The parity checks for the received vectors give S_1, S_3, and S_5. $S_2 = S_1^2$, and $S_4 = S_1^4$. Solving for the σ's gives

$$\sigma_1 = S_1, \quad \sigma_2 = (S_1^2S_3 + S_5)/(S_1^3 + S_3) \quad \text{and}$$
$$\sigma_3 = (S_1S_5 + S_3^2 + S_1^3S_3 + S_1^6)/(S_1^3 + S_3), \qquad (9)$$

provided that $S_1^3 + S_3 \neq 0$. If there is only one error $S_1^3 + S_3 = 0$. Furthermore, if $S_1^3 + S_3 = 0$, the Newton's identities yield $\sigma_3 = \sigma_1\sigma_2$, and the equation

$$X^3 + \sigma_1X^2 + \sigma_2X + \sigma_3$$
$$= X^3 + \sigma_1X^2 + \sigma_2X + \sigma_1\sigma_2$$
$$= (X + \sigma_1)(X^2 + \sigma_2) = (X + \sigma_1)(X + \sqrt{\sigma_2})^2 = 0$$

has two equal roots, which must be zero, and therefore there is only one error.

As a numerical example, suppose that the vector of all zeros is transmitted, and that errors occur in the 2nd, 5th, and 7th positions. Then

$$r = (0\ 1\ 0\ 0\ 1\ 0\ 1\ 0\ 0\ 0\ 0\ 0\ 0\ 0\ 0)$$
$$r \times M = (1\ 0\ 1\ 1 \quad 1\ 1\ 1\ 1 \quad 1\ 0\ 0\ 0)$$
$$S_1 = (1\ 0\ 1\ 1) \qquad S_3 = (1\ 1\ 1\ 1) \qquad S_5 = (1\ 0\ 0\ 0).$$

Referring to Table I, one finds

$$S_2 = S_1^2 = (1\ 0\ 1\ 1)^2 = (\alpha^{13})^2 = \alpha^{26} = \alpha^{11} = (0\ 1\ 1\ 1)$$

and

$$S_4 = S_1^4 = \alpha^{52} = \alpha^7 = (1\ 1\ 0\ 1).$$

Then,

$$S_3 + S_1^3 = (1\ 0\ 1\ 0) \neq 0,$$
$$\sigma_1 = S_1 = (1\ 0\ 1\ 1) = \alpha^{13},$$
$$\sigma_2 = (S_1^2S_3 + S_5)/S_3 + S_1^3 = (0\ 0\ 1\ 0)/(1\ 0\ 1\ 0)$$
$$= \alpha^2/\alpha^8 = \alpha^9/\alpha^{15} = \alpha^9.$$

Similarly,

$$\sigma_3 = \alpha^{11}.$$

It is then easy to verify that the equation,

$$X^3 + \alpha^{13}X^2 + \alpha^9X + \alpha^{11} = 0,$$

is satisfied by the three values $X = \alpha$, α^4, and α^6, and only these. These correspond to the errors in r.

Some Properties of Cyclic Codes and Shift Register Generators

Codes for which the code points comprise a cyclic subspace of vectors of zeros and ones have been studied recently by Prange,[4] and, along with theoretical results, he found several efficient codes that can be decoded easily. He has noted that the codes can be coded with the use of a shift-register generator.[7] In this section, some of the theory of cyclic codes and linear recurrent sequences is reviewed briefly from a point of view that is especially well adapted to the study of the Bose-Chaudhuri codes.

A subset C of vectors of n binary digits is called a *cyclic subspace* if it has the following two properties:

1) If v_1 and v_2 are in C, their sum modulo 2 is also in C; that is, C is a subspace, or subgroup; and
2) If $v = (a_0, a_1, \cdots, a_{n-1})$ is in C, the vector $v^1 = (a_{n-1}, a_0, a_1, \cdots, a_{n-2})$ obtained by shifting v cyclically one place is also in C.

Let R_n denote the set of all polynomials

$$a_0 + a_1X \cdots + a_{n-1}X^{n-1}$$

of degree less than n with coefficients 1 and 0. They form a group under modulo 2 addition. Multiplication can be defined modulo $X^n - 1$; that is, these polynomials can be multiplied in the ordinary way, modulo 2, and then reduced again to polynomials of degree less than n by the use of the equation $X^n = 1$. Then R_n is a ring in the mathematical sense. A subset I of R_n is called an *ideal*[8] if it satisfies the following two properties:

1) I is a subgroup of R_n; and
2) if $p(X)$ is in I and $a(X)$ is R_n, then the product $p(X)\ a(X)$ is in I.

[7] N. Zierler, "Linear recurring sequences," *J. Soc. Ind. Appl. Math.*, vol. 7, pp. 31–48; March, 1959.

[8] Galois fields and other aspects of algebra used in this paper are treated in many books on modern algebra. See, for example, A. A. Albert, "Fundamental Concepts of Modern Algebra," University of Chicago Press, Chicago, Ill., 1956; G. Birkoff and S. MacLane, "A Survey of Modern Algebra," The Macmillan Co., New York, N. Y., 1953; B. L. van der Waerden, "Modern Algebra," F. Ungar Publishing Co., New York, N. Y., vol. 1 and 2, 1949, 1950.

Considering polynomials $p(X) = a_0 + a_1X \cdots + a_{n-1}X^{n-1}$ to be vectors $(a_0, a_1, \cdots, a_{n-1})$, a cyclic shift is the same as multiplication by X modulo $X^n - 1$. Therefore, *every ideal is a cyclic subspace*. Conversely, if $p(X)$ is in a cyclic subspace C, so is $Xp(X)$. It follows that $X^i p(X)$ must also be in C, and since C is a subspace,

$$\sum_i c_i X^i p(X) = p(X) \sum_i c_i X^i$$

must also be in C. Thus, if $p(X)$ is in C, so is the product of $p(X)$ and any polynomial. Therefore, *every cyclic subspace is an ideal*.

The important but well-known properties of ideals given in the following three lemmas and two theorems are proved here to make the paper self-contained.

Lemma 4: If $p(X)$ and $q(X)$ are in an ideal I, the greatest common divisor (GCD), $d(X)$, of $p(X)$ and $q(X)$ is in I.

This follows directly from the fact that it is always possible to express the $d(X)$ in the form

$$d(X) = a(X)p(X) + b(X)q(X)$$

where $a(X)$ and $b(X)$ are polynomials.

Lemma 5: All polynomials in an ideal I are multiples of the unique polynomial of least degree in I. (That is, every ideal is a principal ideal.)

Proof: Let $p(X)$ be a polynomial of least degree in I. Then, if $q(X)$ is any other polynomial in I, the greatest common divisor of $p(X)$ and $q(X)$ is in I. If $p(X)$ does not divide $q(X)$, then the greatest common divisor of $p(X)$ and $q(X)$ would have lower degree than $p(X)$, which is a contradiction. Therefore, every polynomial in I is divisible by $p(X)$. If $p_1(X)$ and $p_2(X)$ both have minimum degree, each must be divisible by the other, and hence they are equal.

The ideal consisting of all multiples of $p(X)$ is denoted $[p(X)]$. The polynomial of least degree in an ideal is called its generator.

Lemma 6: The generator $p(X)$ of an ideal is a factor of $X^n - 1$.

Proof: The GCD $d(X)$ of $p(X)$ and $X^n - 1$ can be expressed in the form

$$d(X) = a(X)p(X) + b(X)(X^n - 1)$$
$$\equiv a(X)p(X) \mod X^n - 1;$$

hence, $d(X)$ is in the ideal. But $p(X)$ is divisible by $d(X)$, and since $d(X)$ is in the ideal, $d(X)$ is divisible by $p(X)$. Hence, $p(X) = d(X)$.

These results can be summarized as follows:

Theorem 2: A set of polynomials is an ideal in the ring of polynomials modulo $X^n - 1$ if and only if it consists of all multiples of degree less than n of a factor of $X^n - 1$.

Corollary: If $p(X)$ is a polynomial of degree k which divides into $X^n - 1$, $[p(X)]$ is a vector space of dimension $n - k$.

Proof: The elements of $[p(X)]$ are of the form $c(X)p(X)$ where $c(X)$ is an arbitrary polynomial of degree less than $n - k$. Then the $n - k$ coefficients of $c(X)$ are arbitrary.

Theorem 3: If $p(X)\, q(X) = X^n - 1$, the ideals $[p(X)]$ and $[q(X)]$ are null spaces of each other. That is, a polynomial $p_1(X)$ is in $[p(X)]$ if, and only if, $p_1(X)\, q_1(X) = 0$ modulo $(X^n - 1)$ for every polynomial $q_1(X)$ in $[q(X)]$.

Proof: Since $p_1(X)$ is in $[p(X)]$, $p_1(X)$ is a multiple of $p(X)$, for example, $a(X)\, p(X)$. Similarly, $q_1(X) = b(X)\, q(X)$. Then $p_1(X)\, q_1(X) = a(X)\, b(X)\, (X^n - 1) = 0$. Conversely, if $p_1(X)\, q(X) = 0$, then $p_1(X)\, q(X)$ must be a multiple of $X^n - 1$, and $p_1(X)$ must be a multiple of $(X^n - 1)/q(X) = p(X)$.

Note that the fact that the product of two polynomials is zero implies that the dot product of the corresponding two vectors is zero, if in one of them the order of the components is reversed. That is, if

$$(a_0 + a_1X \cdots + a_{n-1}X^{n-1})(b_0 + b_1X \cdots + b_{n-1}X^{n-1}) = 0$$

then

$$(a_0, a_1 \cdots a_{n-1}) \cdot (b_{n-1}, b_{n-2}, \cdots b_1, b_0)$$
$$= a_0b_{n-1} + a_1b_{n-2} \cdots + a_{n-1}b_0 = 0,$$

since this is the coefficient of X^{n-1} in the product of the polynomials. Hence, if $[p(x)]$ and $[q(x)]$ are null spaces of each other, the corresponding vector-spaces are null-spaces of each other provided that the order of components in the vectors of one of these is reversed.

Now let us consider a recursion relation (or difference equation) of the form

$$\sum_{j=0}^{k} a_j R_{i-j} = 0, \tag{10a}$$

or

$$R_i = \sum_{j=1}^{k} a_j R_{i-j} \qquad a_0 = a_k = 1. \tag{10b}$$

The solution of these equations for given coefficients a_n will be a sequence of binary digits, $\{R_i\}$. Given the digits $R_0, \cdots, R_{k-1}$, (10) is the rule for calculations R_k, then R_{k+1}, and so forth. Also, the sum of two solutions is again a solution because the equation is linear. Therefore, the solutions form a vector space of dimension k. The solutions are characterized in the following theorem.

Theorem 4: Let $p(X) = \sum_{j=0}^{k} a_j X^j$, $a_0 = a_k = 1$, and let n be the smallest integer for which $X^n - 1$ is divisible by $p(X)$. Let $q(X) = (X^n - 1)/p(X)$. Then the solutions of the difference equation

$$R_i = \sum_{j=1}^{k} a_j R_{i-j}$$

are periodic of period n, and the set made up of the first period of each possible solution, considered as polynomials, is the ideal $[q(X)]$.

Proof: That any vector taken from $[q(X)]$ is a solution can be seen by multiplying a polynomial from $[q(X)]$, for example, $q_1(X)$, by $p(X)$. The digits in the product are formed by the summation in (10a), and, since the product is zero, (10a) is satisfied. Therefore, any sequence formed by repetition of a vector taken from $[q(X)]$ is a solution of (10). Since $q(X) = X^n - 1/p(X)$ has degree $n - k$, then $[q(X)]$ has dimension k, by the corollary to Theorem 2. This is the same as the dimension of the space of solutions, and therefore $[q(X)]$ must include all solutions.

The Cyclic Structure of the Bose-Chaudhuri Codes

It is shown in this section that the Bose-Chaudhuri codes are examples of cyclic codes as studied by Prange.[4] As such they can be generated with very simple equipment, as is illustrated for the (15,5) code in the next section. Out of this theory also comes a better estimate of the number of parity check digits required to correct a given number of errors.

By the alternative definition of the Bose-Chaudhuri codes given in the second section of this paper, a code consists of all polynomials $f(X)$ which have $\alpha, \alpha^3, \cdots, \alpha^{2t-1}$ as roots. Each element α^i of the field is a root of a unique irreducible polynomial $p_i(X)$ of minimum degree. Then $f(X)$ must be divisible by each of the polynomials $p_1(X), p_3(X), \cdots, p_{2t-1}(X)$ and, hence, by their least common multiple:[9]

$$f(X) = \underset{i=1,3,\cdots,2t-1}{\mathrm{LCM}} [p_i(X)]. \qquad (11)$$

Since each of the factors $p_i(X)$ is irreducible, the least common multiple of the $p_i(X)$ is simply the product of the polynomials $p_i(X)$, with the duplicates omitted. Duplications are quite possible; they will occur, in fact, for any α^i and α^j that are roots of the same polynomial $p_i(X)$. In other words, should α^i and α^j happen to be roots of the same irreducible polynomial, the columns in the parity check matrix will be dependent, although not necessarily identical. The parity checks produced by the column of powers of α^j will be satisfied if and only if the parity checks produced by the column of powers of α^i are satisfied, and thus one set or the other is unnecessary.

Finally, the set of all sequences that comprise the code can, by Theorem 4, be generated by a recursion relation defined by the polynomial $X^n - 1/f(X)$, and hence by a shift register generator.

At this point it is interesting to study the limiting cases of the minimum and maximum numbers of parity checks. It has already been noted that the nontrivial minimum is the Hamming code. On the other extreme, the last two columns which might be included in the parity check matrix are powers of $\alpha^{2^m-2} = \alpha^{-1}$ and $\alpha^{2^m-1} = 1$. The last one is a root of the irreducible polynomial $1 + x$ and the resulting code would be the ideal generated by $(1 + x^n)/(1 + x)$. This ideal consists of the zero vector and the vector of all ones, so the code is the trivial repetition of a single information digit $n = 2^m - 1$ times. If α is a primitive element, so is α^{-1}, and therefore the irreducible polynomial of which α^{-1} is a root is primitive. It can be shown then that when only the last two columns, corresponding to α^{-1} and 1, are omitted from the parity check matrix, the resulting code consists of a maximal length sequence, all its shifts, all complements, and a sequence of all 1's, which is then the code studied by San Soucie and Green.[10] This code can also be shown to be equivalent to the Reed-Muller first-order code with any one digit dropped.[11]

It is possible to predict easily which powers of α are roots of the same polynomial, and thus, incidentally, find the degree of the polynomial of which α^i is a root. The method is based on the fact that if a is a root of $f(X)$, then a^2 is also, since $f(a^2) = [f(a)]^2 = 0$. It turns out that $a, a^2, a^4, a^8, \cdots$ are, in fact, all of the roots. In Table II information is given for $m = 4$ and 5. Note that in the first case, $\alpha^{15} = 1$; and in the second, $\alpha^{31} = 1$.

The code for $m = 4$, $t = 3$ has for its generator, by (11),

$$f(X) = p(X)p_3(X)p_5(X)$$

and therefore has $4 + 4 + 2 = 10$ parity checks, and 5 information places. The code for $m = 5$, $t = 5$ has

$$f(X) = p(X)p_3(X)p_5(X)p_7(X)$$

for its generator, and therefore has 20 parity checks. All codes for $m = 4$ and 5 are listed in Table III.

TABLE II
Roots of Polynomials $p_j(X)$

	Polynomial	Roots
$m = 4$	$p(X)$	$\alpha, \alpha^2, \alpha^4, \alpha^8$
	$p_3(X)$	$\alpha^3, \alpha^6, \alpha^{12}, \alpha^9$
	$p_5(X)$	α^5, α^{10} $(\alpha^{20} = \alpha^5)$
	$p_7(X)$	$\alpha^7, \alpha^{14}, \alpha^{13}, \alpha^{11}$
$m = 5$	$p(X)$	$\alpha, \alpha^2, \alpha^4, \alpha^8, \alpha^{16}$
	$p_3(X)$	$\alpha^3, \alpha^6, \alpha^{12}, \alpha^{24}, \alpha^{17}$
	$p_5(X)$	$\alpha^5, \alpha^{10}, \alpha^{20}, \alpha^9, \alpha^{18}$
	$p_7(X)$	$\alpha^7, \alpha^{14}, \alpha^{28}, \alpha^{25}, \alpha^{19}$
	$p_9(X) = p_5(X)$	
	$p_{11}(X)$	$\alpha^{11}, \alpha^{22}, \alpha^{13}, \alpha^{26}, \alpha^{21}$
	$p_{13}(X) = p_{11}(X)$	
	$p_{15}(X)$	$\alpha^{15}, \alpha^{30}, \alpha^{29}, \alpha^{27}, \alpha^{23}$

TABLE III
Rate and Error Correction Ability of Bose-Chaudhuri Codes for $m = 4$ and 5

Length of Code Words	Number of Parity Checks	Number of Information Places	Number of Errors Corrected
n	$n - k$	k	t
15	4	11	1
15	8	7	2
15	10	5	3
31	5	26	1
31	10	21	2
31	15	16	3
31	20	11	5
31	25	6	7

[9] See, for example, Birkhoff and MacLane, *op. cit.*, p. 396.

[10] J. H. Green, Jr. and R. L. San Soucie, "An error-correcting encoder and decoder of high efficiency," Proc. IRE, vol. 46, pp. 1741–1744; October, 1958.

[11] N. Zierler, "On a variation of the first-order Reed-Muller Codes," Lincoln Laboratory Group Rept. 34-80; October, 1958.

Code parameters for some larger codes were calculated on the IBM 704 computer. The results are plotted in Fig. 1. The vertical axis represents rate (percentage of all digits available for information), and the horizontal axis represents the number of errors correctable as a percentage of the total number of digits. The dashed curve represents asymptotic values of a lower bound on the rate of the best code that corrects errors in a given percentage of the digits.[12] The curves drawn for the Bose-Chaudhuri codes for large n fall below the bound for the best code. In fact, it is shown in the Appendix that they approach zero as the length of the code increases indefinitely. This may mean that these codes are truly not optimum, or it may mean that the number of errors correctable by the procedure given in this paper is not the total number of errors correctable by Bose-Chaudhuri codes in the case of very long codes.[13]

The polynomial $p(X)$ can be any primitive polynomial of degree m. The other polynomials $p_i(X)$ are determined by the particular choice of $p(X)$, and the question arises as to how they may be calculated. One simple method is based on the fact that every element of $GF(2^m)$ is a root of the polynomial $X^{2^m-1} - 1$. Therefore, each element is a root of one of the factors of $X^{2^m-1} - 1$. One needs only to factor this polynomial and test to see which factor has X^i as a root. The following alternative method is useful. It has been noted that the degree m_i of $p_i(X)$ can be easily determined. Then if

$$p_i(X) = a_0 + a_1X + \cdots + a_{m-1}X^{m_i-1} + X^{m_i},$$

since α^i is a root of $p_i(X)$,

$$0 = a_0\alpha^0 + a_1\alpha^1 + \cdots + a_{m-1}\alpha^{m_i-1} + \alpha^{m_i},$$

and if α^i is written as a vector with m components, the resulting set of linear equations can be solved for the coefficients a_i of $p_i(X)$.

There is also an explicit formula

$$p(X^{1/j})p(\alpha X^{1/j}) \cdots p(\alpha^{j-1}X^{1/j})$$

where α is a primitive jth root of unity. It can be shown that when the multiplication is carried out only integral powers of X remain, and these have only ones or zeros as coefficients.

Consider again the sample code discussed by Bose and Chaudhuri. The irreducible factors of $X^{15} - 1$ are

$$X^{15} - 1 = (X - 1)(X^2 + X + 1)(X^4 + X^3 + X^2 + X + 1)(X^4 + X^3 + 1)(X^4 + X + 1).$$

A root of the last factor was taken as α; and thus

$$p(X) = X^4 + X + 1.$$

Then α^3 satisfies the equation $X^5 - 1 = 0$, since $\alpha^{15} = 1$. But $X^5 - 1 = (X - 1)(X^4 + X^3 + X^2 + X + 1)$, and since α^3 is not a root of the first factor, it must be a root of the second. Similarly, α^5 satisfies $X^3 - 1 = 0 = (X - 1)(X^2 + X + 1)$, and so α^5 is a root of $X^2 + X + 1$. The fact that this has degree 2 ties in with the observation that the column of powers of α^5 contained only two independent parity checks.

All code points must be multiples, then, of

$$\begin{aligned} f(X) &= p(X)p_3(X)p_5(X) \\ &= (1 + X + X^4)(1 + X + X^2 + X^3 + X^4) \\ &\quad \cdot(1 + X + X^2) \\ &= 1 + X + X^2 + X^4 + X^5 + X^8 + X^{10} \\ &= (1\ 1\ 1\ 0\ 1\ 1\ 0\ 0\ 1\ 0\ 1\ 0\ 0\ 0\ 0), \end{aligned} \tag{12}$$

and it can easily be checked that this vector, any cyclic permutation of it, and any sum of permutations, actually do satisfy the parity checks defined by the matrix M in (2).

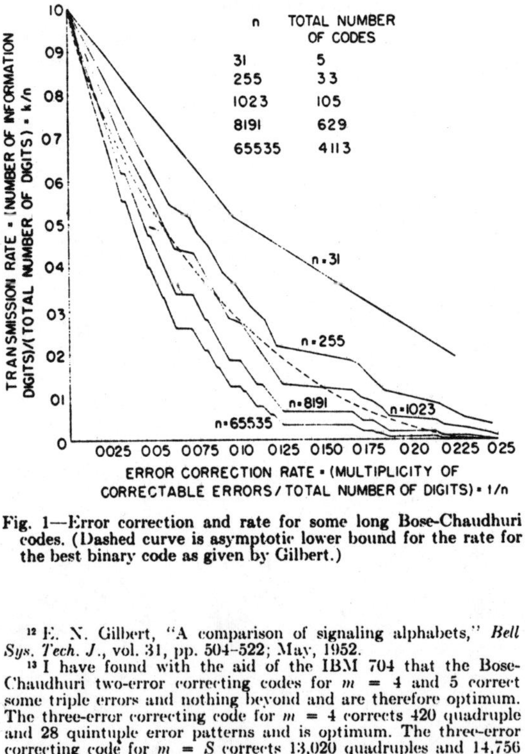

Fig. 1—Error correction and rate for some long Bose-Chaudhuri codes. (Dashed curve is asymptotic lower bound for the rate for the best binary code as given by Gilbert.)

[12] E. N. Gilbert, "A comparison of signaling alphabets," *Bell Sys. Tech. J.*, vol. 31, pp. 504–522; May, 1952.

[13] I have found with the aid of the IBM 704 that the Bose-Chaudhuri two-error correcting codes for $m = 4$ and 5 correct some triple errors and nothing beyond and are therefore optimum. The three-error correcting code for $m = 4$ corrects 420 quadruple and 28 quintuple error patterns and is optimum. The three-error correcting code for $m = 8$ corrects 13,020 quadruples and 14,756 quintuples and nothing beyond—this seems good but has not been proved optimum. (See A. B. Fontaine and W. W. Peterson, "Group code equivalence and optimum codes," IRE TRANS. ON INFORMATION THEORY, vol. IT-5, pp. 60–70; May, 1959.) Thus, any non-optimum behavior of these codes occurs only in codes so large that they are difficult to analyze by looking at code words themselves or searching for coset leaders even with the aid of a computer.

Mechanizing the Coding and Error-Correction

Shift registers with feedback corrections can be used in a number of ways in mechanizing coding and error-correction procedures. The following uses will be discussed in this section:

1) coding using a shift register with one stage for each information digit in the code,
2) coding using a shift register with one stage for each parity check digit in the code,
3) counting in the Galois field code,
4) multiplying and dividing Galois field elements, and
5) calculating parity checks on received vectors.

Both the methods of coding apply to any cyclic code. The methods will be illustrated using the Bose-Chaudhuri (15, 5) code described by the matrix M in (2).

Every cyclic code is an ideal generated by some polynomial $f(X)$, *i.e.*, a polynomial is a code vector if and only if it is divisible by $f(X)$. This means that, by Theorem 4, a vector is a code vector if and only if it satisfies the recursion relation corresponding to the polynomial $(X^n - 1)/f(X)$. For the code used as an example, by (12),

$$f(X) = 1 + X + X^2 + X^4 + X^5 + X^8 + X^{10}$$

$$(1 - X^{15})/f(X) = 1 + X + X^3 + X^5.$$

Then every sequence satisfying the recursion relation

$$R_i = R_{i-1} + R_{i-3} + R_{i-5}$$

is a code point, and conversely. Such sequences can be generated by putting information digits in the shift register generator shown in Fig. 2 and shifting 15 times. The first five digits coming out will be information digits, and the next ten digits will be a set of parity checks which make the whole sequence a code point. The symbols come out of this encoder low order digits first. The order can be reversed by reversing the order of the shift register feedback connections.

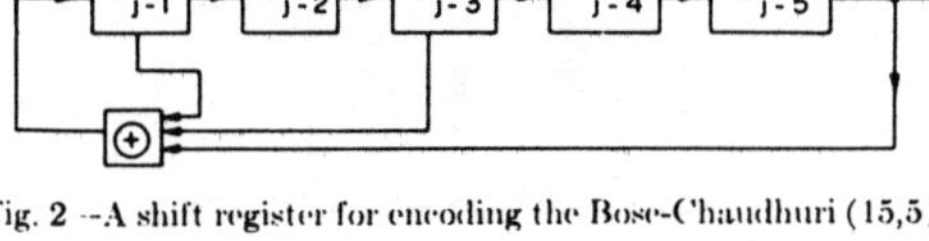

Fig. 2—A shift register for encoding the Bose-Chaudhuri (15,5) code.

A second method of coding is based again on the fact that the coded vector must be, considered as a polynomial, a multiple of $f(X)$. Let $t_0(X)$ be a polynomial in which the k coefficients of the terms involving X^{n-1} through X^{n-k} are arbitrary information digits, and the coefficients of lower order terms are zero. This corresponds to a vector in which the first $n - k$ components are zero, the last k digits arbitrary information digits. Then $t_0(X)$ can be divided by $f(X)$ to produce a quotient and a remainder

$$t_0(X) = f(X)q(X) + r(X),$$

where $r(X)$ has degree less than $(n - k)$, which is the degree of $f(X)$. Then

$$t_0(X) + r(X) = f(X)q(X)$$

and, hence, $t_0(X) + r(X)$ is a code point. But $r(X)$ corresponds to a vector in which all components except the first $n - k$ are zero, since $r(X)$ has degree less than $n - k$. Thus, the sum consists of $n - k$ check digits, the coefficients of $r(X)$, and k information digits, the coefficients of $t_0(X)$.

The next problem is to calculate $r(X)$. In general, the calculation of the remainder after division by a polynomial can be accomplished with a shift register. The method is illustrated in Fig. 3(a). Assuming the divisor is the $f(X)$ for the code used in the example, *i.e.*, $1 + X + X^2 + X^4 + X^5 + X^8 + X^{10}$, the operation of the circuit can be understood as follows: The answer is the same as results from reducing the dividend modulo $f(X)$. This means that the dividend polynomial should be reduced to a polynomial of degree less than 10 using the relation

$$X^{10} = 1 + X + X^2 + X^4 + X^5 + X^8. \qquad (13)$$

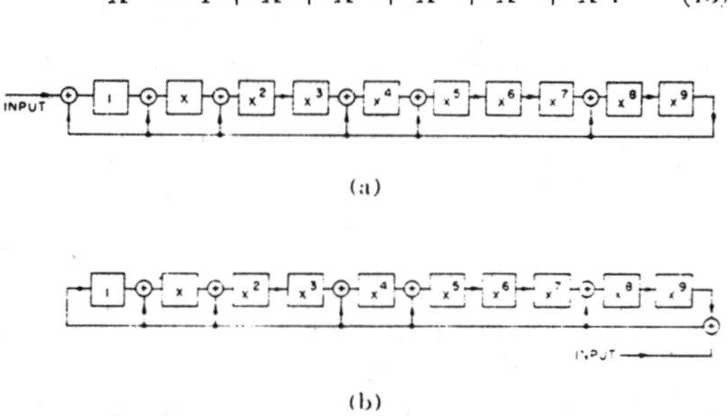

Fig. 3—Shift register for calculating residues modulo $f(X) = 1 + X + X^2 + X^4 + X^5 + X^8 + X^{10}$. (a) Basic circuit; (b) basic circuit with automatic premultiplication by X^{10}.

Now assume that a single one is shifted into the low-order position and then shifted right a number of times. Thinking of the contents of the register as a polynomial with low order digits at the left, each shift corresponds to multiplying by X, at least until a shift out of the high-order position. A one in the high-order position corresponds to X^9, and shifting it out makes it X^{10}. This results in the circuit in adding into the lower order positions the equivalent of X^{10} given in (13), and, hence, in this case the shift still corresponds to multiplying by X and modulo $f(X)$. Thus, successive shifts give successive powers of X modulo $f(X)$.

Now this is a linear device, and a polynomial (which is the sum of powers of X) can be reduced modulo $f(X)$ by shifting it into the device, high power terms first, until the constant term is shifted into the low-order position.

In using this device for calculating the $r(X)$ in (17), the modification shown in Fig. 3(b) can be made to avoid the last $n - k$ shifts which would add $n - k$ zeros into the low-order positions. It amounts to multiplying the input digits by $X^{n-k} = X^{10}$ before adding.

The procedure for coding is then to shift all the information digits into the device in Fig. 3(a) or 3(b). If the device in Fig. 3(a) is used, $n - k$ more shifts must be made with no input. Then the correct check digits remain in the register and should simply follow the information digits, high order digits first, to make a complete code vector. Note that the number of stages in this shift register is $n - k$, while the shift register shown in Fig. 2 has k stages.

A counter which counts in terms of Galois field elements is shown in Fig. 4(a). It works on the same principle as the device shown in Fig. 3(a), but using the primitive polynomial $p(X) = X^4 + X + 1$ of which α is a root. If a 1 is placed in the low-order position, successive shifts give successive powers of α using the relation $\alpha^4 = \alpha + 1$, and these are exactly the representations of $GF(2^4)$ elements given in Table I.

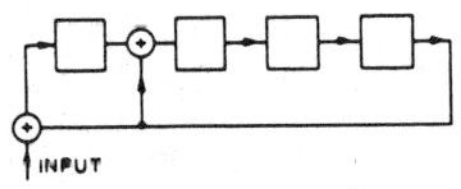

(a)

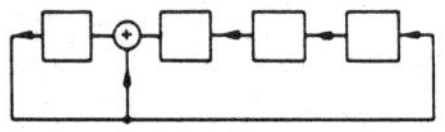

(b)

Fig. 4—Galois field counters for $GF(2^4)$. (a) Increasing powers of α; and (b) decreasing powers of α.

In the device shown in Fig. 3(b), a left shift corresponds to division by α and a 1 shifted out of the low order end α^{-1} is replaced by its equivalent $1 + \alpha^3$. Thus, this device can count down, or give Galois field elements in reverse order. A multiplier can be mechanized by putting one factor in a device A like that shown in Fig. 3(a), the other in a device B like that shown in Fig. 3(b). Then both devices are shifted until the code for 1 appears in device B. The product then appears in A. Division can be done in an anologous manner. Multiplication can also be done in a manner analogous to that used in digital computers with a shift register such as that shown in Fig. 3(a) used in place of an accumulator.

The parity checks corresponding to the first column of Galois field elements in the matrix M of (2) correspond to the Galois field representation of

$$r(\alpha) = r_0 + r_1\alpha + r_2\alpha^2 \cdots r_{2^m-2}\alpha^{2^m-2}.$$

This can be calculated by using the relation $\alpha^4 + \alpha + 1 = 0$ to eliminate terms of degree higher than 3 in α. This, in turn, is exactly what will result if the vector $(r_0, r_1, \cdots, r_{2^m-2})$ is shifted into the shift register shown in Fig. 3(a) high-order digits first. Note that shifting fifteen times multiplies by α^{15}, but $\alpha^{15} = 1$. Similarly, the device in Fig. 3(b) could be used with the low-order digits entering first.

Calculation of the other parity checks is slightly more complicated. It requires calculating $r(\alpha^j)$ for the first odd values of j. The first step is to devise a shift register which automatically multiplies by α^j. The example $j = 5$ should make the principles clear. Note that

$$1 \cdot \alpha^5 = \alpha^5 = \alpha + \alpha^2$$
$$\alpha \cdot \alpha^5 = \alpha^6 = \alpha^2 + \alpha^3$$
$$\alpha^2 \cdot \alpha^5 = \alpha^7 = 1 + \alpha + \alpha^3$$
$$\alpha^3 \cdot \alpha^5 = \alpha^8 = 1 + \alpha^2,$$

so that

$$\begin{aligned}\alpha^5(a_0 + a_1\alpha + a_2\alpha^2 + a_3\alpha^3) &= a_0(\alpha + \alpha^2) + a_1(\alpha^2 + \alpha^3) \\ &\quad + a_2(1 + \alpha + \alpha^3) + a_3(1 + \alpha^2) \\ &= (a_2 + a_3) + (a_0 + a_2)\alpha \\ &\quad + (a_0 + a_1 + a_3)\alpha^2 + (a_1 + a_2)\alpha^3.\end{aligned}$$

Thus, the new value of a_0 is the old $a_2 + a_3$, the new a_1 is the old $a_0 + a_2$, etc. A shift register with feedback connections shown in Fig. 5 will give this result. Then, if the received vector $(r_0, r_1, \cdots, r_{2^m-2})$ is shifted into this device, after fifteen shifts the result $r(\alpha^5)$ will remain in the register.

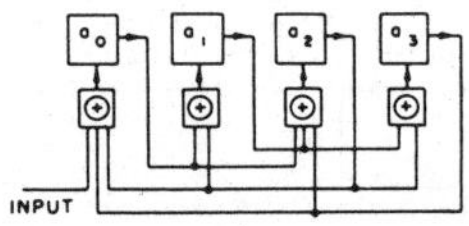

Fig. 5—A circuit for calculating the parity checks $r(\alpha^5)$.

Conclusion

Relatively simple coding and error-correcting methods have been described for the Bose-Chaudhuri codes. The study of coding and error-correction methods for these codes gives additional insight into the remarkable structure of the codes.

Appendix

A bound on the rate of Bose-Chaudhuri codes which correct $t = 2^\lambda$ errors is derived in this Appendix, and it is shown on the basis of this bound that if t is made a fixed fraction of n, the number of digits in the code, the rate must approach zero as n increases indefinitely.

This problem is purely number-theoretic, and can be formulated as follows: The quantity to be studied is the rate, which is the quotient of the number k of information digits and $n = 2^m - 1$, the total number of digits. Since there is one independent parity check for each distinct residue of $j2^i$ for $1 \leq j \leq 2t$, $0 \leq i < m$, the number of such residues in $n - k$. Since $2^m = 2^0$ modulo $2^m - 1$, the condition $0 \leq i < m$ can be replaced by $1 \leq i \leq m$. For convenience in what follows, j will be allowed to take on the value zero also; this adds one distinct residue.

Let $N(s)$ be the number of distinct residues of $j2^i$ for $0 \leq j < 2t = 2^{\lambda+1}$ and $m - s < i \leq m$. Then

$$n - k = N(m) - 1 \geq N(s) - 1 \quad \text{if} \quad s \leq m$$

and

$$k = n - N(m) + 1$$

$$= 2^m - N(m) \leq 2^m - N(s) \quad \text{if} \quad s \leq m. \qquad (14)$$

An equation for $N(s)$, valid only for $s \leq \lambda$, will be derived but this will give an upper bound on k by (14).

Consider first the residues for a particular value of i, $m - \lambda \leq i \leq m$. They can be arranged as follows:

$$\begin{array}{ccccc}
0\cdot 2^i, & 1\cdot 2^i, & 2\cdot 2^i, & \cdots, & (2^{m-i} - 1)2^i \\
(2^{m-i} + 0)2^i, & (2^{m-i} + 1)2^i, & (2^{m-i} + 2)2^i, & \cdots, & (2\cdot 2^{m-i} - 1)2^i \\
(2\cdot 2^{m-i} + 0)2^i, & (2\cdot 2^{m-i} + 1)2^i, & (2\cdot 2^{m-i} + 2)2^i, & \cdots, & (3\cdot 2^{m-i} - 1)2^i \\
\vdots & \vdots & \vdots & & \vdots \\
(2^{\lambda+1} - 2^{m-i} + 0)2^i, & (2^{\lambda+1} - 2^{m-i} + 1)2^i, & (2^{\lambda+1} - 2^{m-i} + 2)2^i, & \cdots, & (2^{\lambda+1} - 1)2^i.
\end{array}$$

In this array there are $2^{\lambda+1-m+i}$ rows. Since $2^m \equiv 1$, the array can be rewritten

$$\begin{array}{ccccc}
0, & 1\cdot 2^i, & 2\cdot 2^i, & \cdots, & (2^{m-i} - 1)2^i \\
1, & 1 + 1\cdot 2^i, & 1 + 2\cdot 2^i, & \cdots, & 1 + (2^{m-i} - 1)2^i \\
2, & 2 + 1\cdot 2^i, & 2 + 2\cdot 2^i, & \cdots, & 2 + (2^{m-i} - 1)2^i \\
\vdots & \vdots & \vdots & & \vdots \\
2^{\lambda+1+i-m} - 1, & (2^{\lambda+1+i-m} - 1) + 1\cdot 2^i, & (2^{\lambda+1+i-m} - 1) + 2\cdot 2^i, & \cdots, & 2^{\lambda+1+i-m} - 1 + (2^{m-i} - 1)2^i.
\end{array}$$

This consists exactly of 2^{m-i} sets of $2^{\lambda+1+i-m}$ successive numbers starting at each multiple of 2^i. The arrangement is shown graphically in Fig. 6.

The important facts can be seen clearly in Fig. 6 but are tedious to prove formally. For each i there are $2^{\lambda+1}$ residues and therefore, in particular, $N(1) = 2^{\lambda+1}$. Two adjacent columns in Fig. 6 have half their residues in common. In particular, $N(2) = 2^{\lambda+1} + 2^\lambda$. Now in adding the contributions to $N(s)$ for larger values of s it is necessary to determine exactly how many residues have occurred in all previous columns combined. There is one other case which must be considered besides the previous adjacent column. Note that the residues and nonresidues of $j\cdot 2^i$ for a particular value of i fall in blocks of $2^{\lambda+1-i-m}$ successive numbers. In determining which residues for a particular value of i, for example, i_0, have occurred before, each block of 2^{i_0+1} successive numbers is treated the same. Each will have two blocks of $2^{\lambda+1+i_0-m}$ residues. The first will already have been counted in the $i_0 + 1^{st}$ column. The fraction of the others to be omitted is the same as the fraction of blocks of length 2^{i_0+1} which were counted as residues for $i \geq i_0 + m - \lambda$, which is the same as $N(\lambda - i_0)/2^m$. Then, since $s = m - i_0$,

$$N(s) = N(s - 1) + 2^\lambda\cdot[1 - 2^{-m}N(\lambda - m + s)] \qquad (15)$$

for $0 < s \leq \lambda$. [$N(s)$ should be considered zero for $s \leq 0$.]

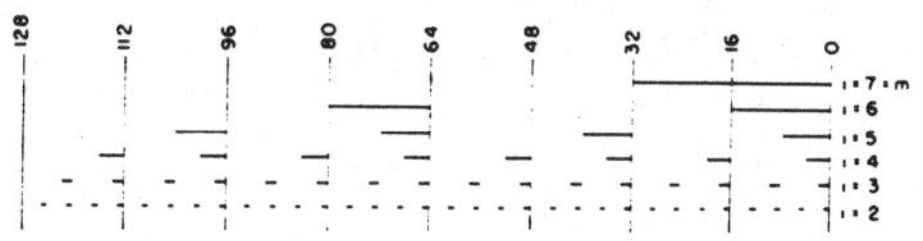

Fig. 6—Distribution of residues of $j2^i (m = 7, \lambda = 4)$.

Now let

$$R(s) = 1 - N(s)\cdot 2^{-m}.$$

Since $N(s)$ includes the zero residue, the actual number of parity digits is at least $N(s) - 1$. The actual number of information digits is at most $2^m - 1 - N(s) + 1 = 2^m - N(s)$. The actual rate would be at most $[2^m - N(s)]/(2^m - 1)$, but for large m, this is approximately $R(s)$. Then

$$N(s) = 2^m[1 - R(s)],$$

and substitution in (15) results in a difference equation for $R(s)$:

$$R(s) = R(s-1) - 2^{\lambda - m}R(s - m + \lambda) \qquad (16)$$

for $0 < s$. [$R(s)$ should be considered to be 1 for $s \leq 0$.] Clearly,

$$1 \geq R(s) \geq 0 \quad \text{for all} \quad s. \qquad (17)$$

It follows at once from (16) and (17) that $R(s)$ is nonincreasing. Now if there exists $\epsilon > 0$ such that $R(s) > \epsilon$ for all s, choose any $s_0 > m - \lambda + (2^{m-\lambda}/\epsilon)$. Then $R(s_0) = [R(s_0) - R(s_0 - 1)] + [R(s_0 - 1) - R(s_0 - 2)] + \cdots + [R(m - \lambda + 1) - R(m - \lambda)] + R(m - \lambda)$ trivially $= R(m - \lambda) - 2^{\lambda - m}[R(s_0 - m + \lambda) + R(s_0 - m + \lambda -) + \cdots + R(1)]$ by (16) $< R(m - \lambda) - 2^{\lambda - m}(s_0 - m + \lambda)\epsilon$ by hyoothesis $< R(m - \lambda) - 1$ by choice of $s_0 < 0$ by half of (17), contradicting the other half, and proving that $R(S) \to 0$ as $s \to \infty$ must hold.

Now suppose that it is required that errors be corrected in a fraction 2^{-v} of the number of digits in a code word Then

$$2^{-v} = 2^{\lambda}/2^m - 1 \approx 2^{\lambda - m},$$

so $v \approx m - \lambda$. Then, taking $s = \lambda$, $R(\lambda) = R(m - v$ is an upper bound on the rate for a code with $2^m -$ digits. As m increases this approaches zero. Since rate is a monotone nonincreasing function of the number of errors correctable and the rate approaches zero for arbitrarily small fractions $t/n = 2^{-v}$, it must approach zero for any fraction $t/n > 0$.

Acknowledgment

I have benefited greatly from discussions with many people at the IBM Research Laboratory and at the Research Laboratory of Electronics at Massachusetts Institute of Technology. E. Prange, of the Air Force Cambridge Research Center, J. Griesmer and J. Selfridge of IBM, and M. P. Schutzenberger and S. Golomb at the Research Laboratory of Electronics were especially helpful.

Most of all, I am indebted to R. C. Bose of the University of North Carolina, for lecturing on his and Chaudhuri's fine work so soon after it was done and for the very stimulating discussion we had during his visit to the IBM Research Laboratory in August 1959.

Part of the computation work was done at the M.I.T. Computation Center.

Reprinted from *I.E.E.E. Trans. Inform. Theory*, **IT-15**, 122–127

20

Shift-Register Synthesis and BCH Decoding

JAMES L. MASSEY, MEMBER, IEEE

***Abstract*—It is shown in this paper that the iterative algorithm introduced by Berlekamp for decoding BCH codes actually provides a general solution to the problem of synthesizing the shortest linear feedback shift register capable of generating a prescribed finite sequence of digits. The shift-register approach leads to a simple proof of the validity of the algorithm as well as providing additional insight into its properties. The equivalence of the decoding problem for BCH codes to a shift-register synthesis problem is demonstrated, and other applications for the algorithm are suggested.**

I. Introduction

In the following section, the problem of finding the shortest linear feedback shift register that can generate a given finite sequence of digits is studied. In Section III, an algorithm is developed that yields a simple recursive solution for this problem by synthesizing for $n = 1, 2, \cdots$ the shortest register that can generate the first n digits of this sequence. Sections IV and V provide a review of certain properties of shift-register sequences and of Bose–Chaudhuri–Hocquenghem (BCH) codes, and culminate in a demonstration that the major decoding problem for BCH codes is a shift-register synthesis problem of the type above. The shift-register synthesis algorithm of Section III is then seen to coincide with the iterative algorithm introduced recently by Berlekamp [1] for decoding the BCH codes. Finally, some additional applications for the algorithm are suggested.

II. Length Properties of LFSR's

A general *linear feedback shift register* (LFSR) of length L is shown in Fig. 1 and consists of a cascade of L unit delay cells, or stages, with provision to form a linear combination of the cell contents, which then serves as the input to the first stage. The output of the LFSR is assumed to be taken from the last stage. The initial contents $s_0, s_1, \cdots, s_{L-1}$ of the L stages coincide with the first L output digits, and the remaining output digits are uniquely determined by the recurson

$$s_j = -\sum_{i=1}^{L} c_i s_{j-i}, \qquad j = L, L+1, L+2, \cdots. \tag{1}$$

The output digits and the feedback coefficients $c_1, c_2, \cdots, c_L$ are assumed to lie in the same field F, which can be either a finite field $GF(q)$, or an infinite field, such as the real number field. There is no requirement that $c_L \neq 0$ (i.e., the last stage of the LFSR need not be tapped).

An LFSR is said to *generate* a finite sequence $s_0, s_1, \cdots, s_{N-1}$ when this sequence coincides with the first N output digits of the LFSR for some initial loading. If $L \geq N$, the LFSR always generates the sequence. If $L < N$, it follows from (1) that the LFSR generates the sequence if and only if

$$s_j + \sum_{i=1}^{L} c_i s_{j-i} = 0, \qquad j = L, L+1, \cdots, N-1. \tag{2}$$

The following simple theorem will play a key role in the subsequent development

Theorem 1

If some LFSR of length L generates the sequence $s_0, s_1, \cdots, s_{N-1}$ but not the sequence $s_0, s_1, \cdots, s_{N-1}, s_N$, then any LFSR that generates the latter sequence has length L', satisfying

$$L' \geq N + 1 - L. \tag{3}$$

Proof: For $L \geq N$, the theorem is trivially true so we may suppose that $L < N$. Let $c_1, c_2, \cdots, c_L$ and $c_1', c_2', \cdots, c_{L'}'$ denote the connection coefficients of the two LFSR's in question and assume that $L' \leq N - L$, in violation of (3). By hypothesis

$$-\sum_{i=1}^{L} c_i s_{j-i} \begin{cases} = s_j, & j = L, L+1, \cdots, N-1 \\ \neq s_N, & j = N, \end{cases} \tag{4}$$

and

$$-\sum_{k=1}^{L'} c_k' s_{j-k} = s_j, \qquad j = L', L'+1, \cdots, N. \tag{5}$$

Therefore, it follows that

$$-\sum_{i=1}^{L} c_i s_{N-i} = +\sum_{i=1}^{L} c_i \sum_{k=1}^{L'} c_k' s_{N-i-k} \tag{6}$$

where the use of (5) in rewriting the left-hand side of (6) is justified by the fact that $\{s_{N-L}, s_{N-L+1}, \cdots, s_{N-1}\}$ is a subset of $\{s_{L'}, s_{L'+1}, \cdots, s_{N-1}\}$. Upon interchange of the order of summation, (6) becomes

$$\begin{aligned} -\sum_{i=1}^{L} c_i s_{N-i} &= +\sum_{k=1}^{L'} c_k' \sum_{i=1}^{L} c_i s_{N-k-i} \\ &= -\sum_{k=1}^{L'} c_k' s_{N-k} \\ &= s_N \end{aligned} \tag{7}$$

where use has been made of (4) and (5), respectively. The use of (4) is justified by the fact that $\{s_{N-L'}, s_{N-L'+1}, \cdots, s_{N-1}\}$ is a subset of $\{s_L, s_{L+1}, \cdots, s_{N-1}\}$. But (7)

Manuscript received March 7, 1967; revised June 24, 1968. This work was supported in part by the National Aeronautics and Space Administration (Grant NsG-334 at the M. I. T. Research Laboratory of Electronics and Grant NGR 15-004-026 at the University of Notre Dame, Notre Dame, Ind.) and by the Joint Services Electronics Program, Contract DA 208-043-AMC-02536)E at Massachusetts Institute of Technology, Cambridge.
The author is with the University of Notre Dame, Notre Dame, Ind. 46556

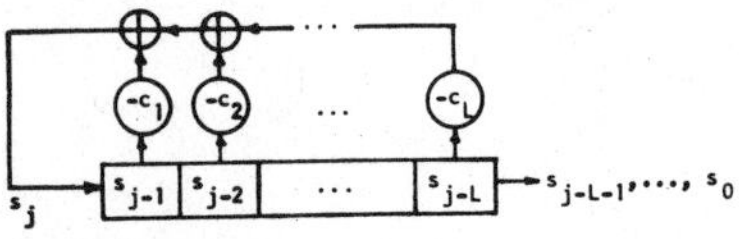

Fig. 1. General L-stage linear feedback shift-register (LFSR).

contradicts (4) proving that the assumption $L' \leq N - L$ is untenable. We conclude that $L' \geq N + 1 - L$ as was to be shown.

Now let s denote an infinite sequence $s_0, s_1, s_2, \cdots$ so that $s_0, s_1, \cdots, s_{N-1}$ forms the first N digits of s. We define $L_N(\mathbf{s})$ as the minimum of the lengths of all the LFSR's that generate $s_0, s_1, \cdots, s_{N-1}$. By our earlier remarks, $L_N(\mathbf{s}) \leq N$. Moreover, $L_N(\mathbf{s})$ must be monotonically nondecreasing with increasing N. By way of convention, we shall say that the all-zero sequence is generated by the LFSR with length $L = 0$, and therefore that $L_N(\mathbf{s}) = 0$ if and only if $s_0, s_1, \cdots, s_{N-1}$ are all zeros.

Lemma 1

If some LFSR of length $L_N(\mathbf{s})$ generates $s_0, s_1, \cdots, s_{N-1}$, but not $s_0, s_1, \cdots, s_{N-1}, s_N$, then

$$L_{N+1}(\mathbf{s}) \geq \max [L_N(\mathbf{s}), N + 1 - L_N(\mathbf{s})].$$

Proof: From the monotonicity of $L_N(\mathbf{s})$, we have $L_{N+1}(\mathbf{s}) \geq L_N(\mathbf{s})$. Under the hypothesis of the lemma, Theorem 1 implies that $L_{N+1}(\mathbf{s}) \geq N + 1 - L_N(\mathbf{s})$. Therefore the lemma follows.

Lemma 1 will be used in the next section to demonstrate the minimality of the length of a shift register found by a synthesis algorithm for LFSR's. A consequence of the resulting development will be a proof that the inequality in Lemma 1 can be replaced by an equality.

III. The LFSR Synthesis Algorithm

In this section, a recursive algorithm is derived for producing one of the LFSR's of length $L_N(\mathbf{s})$, which generates $s_0, s_1, \cdots, s_{N-1}$ for $N = 1, 2, 3, \cdots$. The discussion will be facilitated by defining the *connection polynomial* of the LFSR of Fig. 1 as the polynomial

$$C(D) = 1 + c_1D + c_2D^2 + \cdots + c_LD^L \tag{8}$$

which has degree at most L in the indeterminate D. By way of convention, we take $C(D) = 1$ for the LFSR of length $L = 0$.

When $s_0, s_1, \cdots, s_{N-1}$ are all zeros but $s_N \neq 0$, then $L_{N+1}(\mathbf{s}) = N + 1$ since any shorter LFSR must be initially loaded with all zeros and thus could generate only further zeros. Moreover, any LFSR with $L = N + 1$ suffices to generate $s_0, s_1, \cdots, s_{N-1}, s_N$ in this case. Note further that Lemma 1 holds with equality in this circumstance.

For a given s, let

$$C^{(N)}(D) = 1 + c_1^{(N)}D + \cdots + c_{L_N(\mathbf{s})}^{(N)}D^{L_N(\mathbf{s})} \tag{9}$$

denote the connection polynomial of a minimal-length $L_N(\mathbf{s})$ LFSR that generates $s_0, s_1, \cdots, s_{N-1}$. As an inductive hypothesis, assume that $L_N(\mathbf{s})$ and some $C^{(N)}(D)$ have been found for $N = 1, 2, \cdots, n$ with equality obtaining in Lemma 1 for $N = 1, 2, \cdots, n - 1$. We seek then to find $L_{n+1}(\mathbf{s})$ and some $C^{(n+1)}(D)$, and to show that equality obtains in Lemma 1 for $N = n$.

By the induction hypothesis, we have from (2) that

$$s_j + \sum_{i=1}^{L_n(\mathbf{s})} c_i^{(n)}s_{j-i} = \begin{cases} 0, & j = L_n(\mathbf{s}), \cdots, n-1 \\ d_n & j = n, \end{cases} \tag{10}$$

where d_n, which we call the *next discrepancy*, is the difference between s_n and the $(n + 1)-st$ digit generated by the minimal-length LFSR, which we have found to generate the first n digits of s. If $d_n = 0$, then this LFSR also generates the first $n + 1$ digits of s so that $L_{n+1}(\mathbf{s}) = L_n(\mathbf{s})$, and we may now take $C^{(n+1)}(D) = C^{(n)}(D)$.

If $d_n \neq 0$, a new LFSR must be found to generate the first $n + 1$ digits of s. In this latter case, let m be the sequence length before the last *length change* in the minimal-length registers, i.e.,

$$\begin{aligned} L_m(\mathbf{s}) &< L_n(\mathbf{s}) \\ L_{m+1}(\mathbf{s}) &= L_n(\mathbf{s}). \end{aligned} \tag{11}$$

Since a length change was required, the LFSR with connection polynomial $C^{(m)}(D)$ and length $L_m(\mathbf{s})$ could not have generated $s_0, s_1, \cdots, s_{m-1}, s_m$. Therefore, from (2) we have

$$s_j + \sum_{i=1}^{L_m(\mathbf{s})} c_i^{(m)}s_{j-i} = \begin{cases} 0, & j = L_m(\mathbf{s}), \cdots, m-1 \\ d_m \neq 0, & j = m. \end{cases} \tag{12}$$

By the induction hypothesis, Lemma 1 holds with equality for $N = m$ so that

$$L_{m+1}(\mathbf{s}) = L_n(\mathbf{s}) = \max [L_m(\mathbf{s}), m + 1 - L_m(\mathbf{s})]$$

and in particular, because of (11), this gives

$$L_n(\mathbf{s}) = m + 1 - L_m(\mathbf{s}). \tag{13}$$

We now claim that the connection polynomial

$$C(D) = C^{(n)}(D) - d_nd_m^{-1}D^{n-m}C^{(m)}(D) \tag{14}$$

is a valid next choice for $C^{(n+1)}(D)$. Note first from (14) that the degree of $C(D)$ is at most

$$\max [L_n(\mathbf{s}), n - m + L_m(\mathbf{s})] = \max [L_n(\mathbf{s}), n + 1 - L_n(\mathbf{s})]$$

where the equality follows from (13). Hence $C(D)$ is an allowable connection polynomial for a LFSR of length L where

$$L = \max [L_n(\mathbf{s}), n + 1 - L_n(\mathbf{s})]. \tag{15}$$

Moreover, it follows from (14) that

$$\begin{aligned} s_j + \sum_{i=1}^{L} c_is_{j-i} &= s_j + \sum_{i=1}^{L_n(\mathbf{s})} c_i^{(n)}s_{j-i} - d_nd_m^{-1} \\ &\quad \cdot \left[s_{j-n+m} + \sum_{i=1}^{L_m(\mathbf{s})} c_i^{(m)}s_{j-n+m-i}\right] \\ &= \begin{cases} 0 & j = L, L+1, \cdots, n-1 \\ d_n - d_nd_m^{-1}d_m = 0, & j = n \end{cases} \end{aligned}$$

where the last equalities result from the use of (10) and (12). Therefore, it follows from (2) that the LFSR of length L with connection polynomial $C(D)$ generates the $n + 1$ digits $s_0, s_1, \cdots, s_n$. Since L in (15) satisfies Lemma 1 with equality, we conclude that $L = L_n(\mathbf{s})$, and therefore that equality in Lemma 1 is always obtained. Thus we have proved Theorem 2.

Theorem 2

If some LFSR of length $L_N(\mathbf{s})$, which generates $s_0, s_1, \cdots, s_{N-1}$, also generates $s_0, s_1, \cdots, s_{N-1}, s_N$, then $L_{N+1}(\mathbf{s}) = L_N(\mathbf{s})$. Conversely, if some LFSR of length $L_N(\mathbf{s})$ that generates $s_0, s_1, \cdots, s_{N-1}$ fails to generate $s_0, s_1, \cdots, s_{N-1}, s_N$, then $L_{N+1}(\mathbf{s}) = \max [L_N(\mathbf{s}), N + 1 - L_N(\mathbf{s})]$.

Moreover, our proof of Theorem 2 was a constructive proof, which establishes the validity of the following algorithm for synthesizing a shortest LFSR to generate the sequence $s_0, s_1, \cdots, s_{n-1}$.

LFSR Synthesis Algorithm (Berlekamp Iterative Algorithm):

1) $1 \to C(D)$ $\quad 1 \to B(D)$ $\quad 1 \to x$
   $0 \to L$ $\quad 1 \to b$ $\quad 0 \to N$
2) If $N = n$, stop. Otherwise compute

$$d = s_N + \sum_{i=1}^{L} c_i s_{N-i}.$$

3) If $d = 0$, then $x + 1 \to x$, and go to 6).
4) If $d \neq 0$ and $2L > N$, then
   $C(D) - d\, b^{-1}\, D^x\, B(D) \to C(D)$
   $x + 1 \to x$
   and go to 6).
5) If $d \neq 0$ and $2L \leqq N$, then
   $C(D) \to T(D)$ [temporary storage of $C(D)$]
   $C(D) - d\, b^{-1}\, D^x\, B(D) \to C(D)$
   $N + 1 - L \to L$
   $T(D) \to B(D)$
   $d \to b$
   $1 \to x$.
6) $N + 1 \to N$ and return to 2).

For every n, when $N = n$ and step 2) has just been reached, then the quantities produced by the algorithm bear the following relations to the quantities appearing in the development preceding Theorem 2:

$$C(D) = C^{(n)}(D)$$
$$L = L_n(\mathbf{s})$$
$$x = n - m$$
$$d = d_n \text{ (assuming the computation in step 2) is performed)}$$
$$B(D) = C^{(m)}(D)$$
$$b = d_m.$$

That the algorithm implements the procedure derived preceding Theorem 2 should be evident except for the following two points. First, step 5) is carried out only when, according to Theorem 2, a length change is needed. In this case, the present $C(D)$ for subsequent iterations will be the last connection polynomial before the latest length change and therefore becomes the new $B(D) = C^{(m)}(D)$. Second, suppose that the first nonzero d occurs in step 2) with $N = k$. This implies $s_0 = s_1 = \cdots = s_{k-1} = 0$ and $s_k \neq 0$. At this time, $L = L_k(\mathbf{s}) = 0$ and, therefore, the sequence length before the last length change is undefined, since no LFSR can have length less than zero. Thus the rule of (14) for computing the next connection polynomial is not applicable. However, in this case, the initialization in step 1) has the effect of causing step 5) to be applied, which then results in $C(D) = C^{(k+1)}(D) = 1 - dD^{k+1}$ and $L = L_{k+1}(\mathbf{s}) = k + 1$. We have already pointed out that any length $k + 1$ LFSR is a valid solution for this case.

In Fig. 2 the results are shown for the application of the algorithm to the binary [$F = GF(2)$] sequence $s_0, s_1, \cdots, s_4 = 1, 0, 1, 0, 0$. Note that the resulting LFSR is singular (i.e., $c_3 = 0$) and the last stage is not tapped.

A logical circuit for implementing the algorithm is shown in Fig. 3 and is seen to require $3L_o + 1$ memory cells, where each cell can store a digit in the field F, and where L_o is the maximum length of an LFSR that can be produced with this circuitry.

Up to this point we have considered only the problem of finding one of the minimal-length registers that generate a specified sequence, but the set of *all* minimal-length $L_n(\mathbf{s})$ LSFR's that generate $s_0, s_1, \cdots, s_{n-1}$ can also readily be found from the LFSR synthesis algorithm. From Theorem 2, we observe that when some LFSR of length $L_N(\mathbf{s})$ that generates $s_0, s_1, \cdots, s_{N-1}$ fails to generate $s_0, s_1, \cdots, s_{N-1}, s_N$, there will then be a length change [$L_{N+1}(\mathbf{s}) > L_N(\mathbf{s})$] if and only if $2L_N(\mathbf{s}) \leqq N$. It follows that the minimal-length LFSR is *unique* if and only if $2L_N(\mathbf{s}) \leqq N$. Therefore, when the algorithm terminates with $2L > n$, the resulting minimal-length LFSR is not unique. In this case, however, the resulting LFSR would be the unique solution if the additional digits $s_n, s_{n+1}, \cdots, s_{2L-1}$ were to be specified in agreement with the output sequence of this LFSR. Moreover, for any assignment of these $2L - n$ additional digits, only steps 3) or 4) of the algorithm would be used to produce new connection polynomials, i.e., the pattern of the $2L - n$ next discrepancies d serve only to determine a polynomial multiple of the unchanging $B(D)$, which will be added to produce the final $C(D)$, and some choice of this pattern must result in every possible LFSR of length $L_n(\mathbf{s})$ that generates $s_0, s_1, \cdots, s_{n-1}$. These remarks are summarized in the following theorem.

Theorem 3

Suppose the LFSR synthesis algorithm is applied to the sequence $s_0, s_1, \cdots, s_{n-1}$ and let L, $C(D)$, x, and $B(D)$ denote the values when the algorithm terminates. If $2L \leqq n$, then $C(D)$ is the connection polynomial of the unique minimal-length L LFSR that generates the sequence. If $2L > n$, then the set of polynomials

N	L	C(D)	LFSR	x	B(D)	b	s_N	d
0	0	1	→	1	1	1	1	1
1	1	1 + D		1	1	1	0	1
2	1	1		2	1	1	1	1
3	2	$1 + D^2$		1	1	1	0	0
4	2	$1 + D^2$		2	1	1	0	1
5	3	1		1	$1 + D^2$	1		

Fig. 2. Example of application of the LFSR synthesis algorithm to the binary sequence $s_0, s_1, s_2, s_3, s_4 = 1, 0, 1, 0, 0$.

$$\{C(D) + Q(D)\, D^x B(D):$$
$$\text{degree of } Q(D) \text{ less than } 2L - n\},$$

is the set of connection polynomials for all of the minimal-length-L LFSR's that generate the sequence.

For instance, in the example shown in Fig. 2, Theorem 3 gives the allowable $Q(D)$ to be either 0 or 1. Hence the set of connection polynomials $\{1, 1 + D + D^3\}$ specifies both $L = 3$ LFSR's that generate the given $n = 5$ sequence. The following is an immediate consequence of Theorem 3

Corollary

If $2L_n(\mathbf{s}) < n$, then the LFSR synthesis algorithm will already have produced the unique minimal-length solution, i.e., $L = L_n(\mathbf{s})$ and $C(D) = C^{(n)}(D)$, when $N = 2L_n(\mathbf{s})$ in (2), i.e., after only the first $2L_n(\mathbf{s})$ digits have been processed by the algorithm.

For instance, if the sequence $s_0, s_1, \cdots, s_{n-1}$ is a non-zero cycle of length $n = 2^{100}$-1 from a 100 stage maximal-length LFSR, then the algorithm has necessarily found the unique generating LFSR after the first $2L = 200$ digits have been processed.

The LFSR synthesis algorithm given in this section is (practically) identical to the iterative algorithm developed by Berlekamp [1] for decoding the BCH codes, as will be seen in Section V. It should be noted that when $2L = N + 1$ and $d \neq 0$, it is then permissible to modify step 4) of the algorithm so that $B(D)$ is replaced by the old $C(D)$. The reason for this is that it can be shown that rather than taking $C^{(m)}(D)$ as the last connection polynomial before a length change, it suffices more generally to choose $C^{(m)}(D)$ as any of the previous connection polynomials for which $d_m \neq 0$ and $m - L_m(\mathbf{s})$ is maximized. When $d_n \neq 0$ and $2L_n(\mathbf{s}) = n + 1$, then $n - L_n(\mathbf{s}) = m - L_m(\mathbf{s})$ so that $C^{(n)}(D)$ is an allowable replacement for $C^{(m)}(D)$. Berlekamp's algorithm contains an additional test for deciding whether to replace $C^{(m)}(D)$ in this case, but there seems to be no advantage deriving from it so that we have excluded such a test from the LFSR synthesis algorithm.

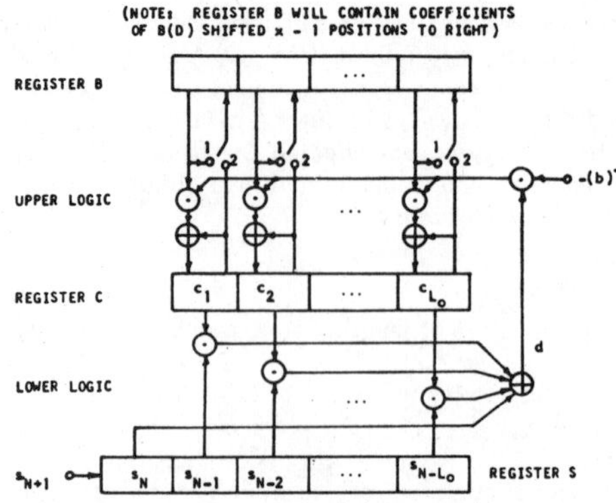

RULES OF OPERATION: ACTIVATE LOWER LOGIC.
IF d = 0, SHIFT B AND S REGISTERS ONE POSITION.
IF d ≠ 0 AND 2L > N, MOVE SWITCHES TO POLE 1 AND ACTIVATE UPPER LOGIC, THEN SHIFT B AND S REGISTERS ONE POSITION.
IF d ≠ 0 AND 2L ≤ N, MOVE SWITCHES TO POLE 2 AND ACTIVATE UPPER LOGIC, REPLACE b BY d AND REPLACE L BY N + 1 - L, THEN SHIFT B AND S REGISTERS ONE POSITION AND LOAD A 1 INTO THE FIRST STAGE OF REGISTER B.

Fig. 3. A logical circuit for implementing the LFSR synthesis algorithm.

IV. Classical Description of LFSR Sequences

In this section, we review some properties of LFSR-generated sequences with a view toward applying this material to BCH codes in the sequel.

It will prove convenient to describe the sequence $\mathbf{s} = s_0, s_1, \cdots$ by its Huffman D-transform

$$S(D) = s_0 + s_1 D + s_2 D^2 + \cdots. \tag{16}$$

From (8) and (16), we see that (2) simply specifies that the degree j term in the product $C(D)\ S(D)$ vanishes for $j = L, L + 1, L + 2, \cdots$. Hence, (2) may be rewritten as

$$C(D)\ S(D) = P(D)$$

or

$$S(D) = \frac{P(D)}{C(D)} \tag{17}$$

where

$$P(D) = p_0 + p_1 D + \cdots + p_{L-1} D^{L-1} \tag{18}$$

is a polynomial of degree less than L. Moreover, from (17) and (18), we find the matrix equation

$$\begin{bmatrix} p_0 \\ p_1 \\ \vdots \\ p_{L-1} \end{bmatrix} = \begin{bmatrix} 1 & 0 & \cdots & 0 & 0 \\ c_1 & 1 & \cdots & 0 & 0 \\ & & \vdots & & \\ c_{L-1} & c_{L-2} & \cdots & c_1 & 1 \end{bmatrix} \begin{bmatrix} s_0 \\ s_1 \\ \vdots \\ s_{L-1} \end{bmatrix} \tag{19}$$

which relates the coefficients of $P(D)$ to the connection coefficients and the initial contents of the LFSR. Since the matrix in (19) is nonsingular, it follows that for every

$P(D)$ as in (18) there will be a unique corresponding assignment of initial conditions. We may summarize in Theorem 4.

Theorem 4

The output sequences generated by an L-stage LFSR with connection polynomial $C(D)$ is the set $\{s\}$ of sequences corresponding to the set of transforms

$$\left\{S(D) = \frac{P(D)}{C(D)}, \text{ degree of } P(D) \text{ less than } L\right\}.$$

Theorem 4 shows that **s** is an output sequence of some LFSR if and only if its transform $S(D)$ is a rational function, i.e., a ratio of polynomials $A(D)/B(D)$, with $B(0) \neq 0$. Moreover, if $A(D)$ and $B(D)$ are relatively prime polynomials (i.e., have no common factor of degree one or greater), then it follows directly from Theorem 4 that $B(D)$, within a constant factor required to make $B(0) = 1$, is the unique connection polynomial of the shortest LFSR that generates **s**, and the length of this LFSR is the maximum of the degree of $B(D)$ and the degree of $A(D)$ plus one. Restating these remarks, we have the following.

Corollary

If $S(D) = P(D)/C(D)$ where $P(D)$ and $C(D)$ are relatively prime polynomials and $C(0) = 1$, then $C(D)$ is the connection polynomial of the shortest LFSR that generates the sequence **s** whose transform is $S(D)$, and

$$L_n(\mathbf{s}) = \max\,[\text{degree of } C(D),\ 1 + \text{degree of } P(D)]. \quad (20)$$

V. Application to Decoding of the BCH Codes

Let $g(X) = g_0 + g_1X + \cdots + g_{r-1}X^{r-1} + X^r$, $g_0 \neq 0$, be a monic polynomial of degree r, $r \geq 1$, with coefficients in some finite field $GF(q)$. Let n be the least integer such that $g(X)$ divides $X^n - 1$. With every n-tuple $\mathbf{f} = [f_0, f_1, \cdots, f_{n-1}]$ of elements of $GF(q)$, associate the polynomial $f(X) = f_0 + f_1X + \cdots + f_{n-1}X^{n-1}$ of degree less than n. Then the *cyclic code* generated by $g(X)$ is the set of n-tuples $\mathbf{f}$ such that $g(X)$ divides $f(X)$. The length is n digits and the code redundancy is r digits.

A Bose–Chaudhuri–Hocquenghem (BCH) code [2] is a cyclic code where $g(X)$ is chosen to be the minimum-degree monic polynomial with coefficients in $GF(q)$ having $\alpha^{m_o}, \alpha^{m_o+1}, \cdots, \alpha^{m_o+d-2}$ as roots where α is a specified nonzero element of $GF(q^m)$, m_o is some positive integer, and d, $d \geq 2$, is any integer such that the $d-1$ specified roots of $g(X)$ are all distinct. We shall call such a code a BCH (α, q, m_o, d) code when we wish to specify the main parameters. It is well known that such a BCH code has minimum distance at least d, and d is sometimes called the *design distance* of the code.

If a codeword $\mathbf{f}$ in a BCH (α, q, m_o, d) code is transmitted, and an n-tuple $\mathbf{r} = [r_0, r_1, \cdots, r_{n-1}]$ of elements from $GF(q)$ is received, then $\mathbf{e} = [e_0, e_1, \cdots, e_{n-1}] = \mathbf{r} - \mathbf{f}$ is called the *error pattern*. Associating polynomials with **e** and **r** as was done with **f**, we have

$$r(X) = f(X) + e(X). \quad (21)$$

With the error polynomial $e(X)$, one associates the weighted power sum symmetric functions $S_1, S_2, \cdots$ defined by

$$S_i = e(\alpha^i), \qquad i = 1, 2, 3, \cdots. \quad (22)$$

Since $g(X)$ divides $f(X)$, all roots of $g(X)$ are also roots of $f(X)$ so that from (21) and (22) it follows that

$$S_i = r(\alpha^i),$$
$$i = m_o, m_o + 1, \cdots, m_o + d - 2, \quad (23)$$

and hence that this set of $d-1$ consecutive S can be formed *at the receiver*. This can be accomplished with simple logical circuitry [2]. The BCH decoding problem simply stated is the following. Given the $d-1$ consecutive S_i defined in (23), find the error pattern $e(X)$.

Let t be the Hamming weight of the error pattern **e**, i.e., the number of nonzero components. If the jth nonzero component in **e** is the digit e_k, then $X_j = \alpha^k$ is called the *locator* of this error and $Y_j = e_k$ is the *error magnitude*. X_j is an element of $GF(q^m)$ and Y_j is an element of $GF(q)$. From (22), it follows that

$$S_i = \sum_{j=1}^{t} Y_jX_j^i, \qquad i = 1, 2, 3, \cdots. \quad (24)$$

For binary codes ($q = 2$), the error locators completely describe the error pattern since $Y_j = 1$, $j = 1, 2, \cdots, t$. For general q, Forney [3] has given a simple procedure for determining the error magnitudes given the error locators. Therefore the essential BCH decoding problem reduces to the following. Given $S_{m_o}, S_{m_o} + 1, \cdots, S_{m_o} + d - 2$, find the error locators $X_1, X_2, \cdots, X_t$.

Following Berlekamp [1], we first observe that

$$\frac{1}{1 - X_jD} = 1 + X_jD + X_j^2D^2 + \cdots. \quad (25)$$

Multiplying by $Y_jX_j^{m_o}$ in (25) and summing, we obtain with the aid of (24)

$$\sum_{j=1}^{t} \frac{Y_jX_j^{m_o}}{1 - DX_j} = S_{m_o} + S_{m_o+1}D + S_{m_o+2}D^2 + \cdots. \quad (26)$$

The left-hand side of (26) is recognized to be the partial fraction expansion of $P(D)/C(D)$, where

$$C(D) = \prod_{j=1}^{t} (1 - X_jD) \quad (27)$$

and

$$P(D) = \sum_{j=1}^{t} Y_jX_j^{m_o} \prod_{\substack{k=1\\k\neq j}}^{t} (1 - X_kD). \quad (28)$$

Therefore, we may write

$$\frac{P(D)}{C(D)} = S_{m_o} + S_{m_o+1}D + S_{m_o+2}D^2 + \cdots \quad (29)$$

where $C(0) = 1$ and where $P(D)$ and $C(D)$ are relatively prime polynomials. This latter property follows from the fact that if $P(D)$ and $C(D)$ had any common factors of degree at least one, then the partial fraction expansion of their ratio must have fewer nonzero terms than the degree of $C(D)$ contrary to (26). From (27) and (28), we see that the degree of $C(D)$ is exactly t, while the degree of $P(D)$ is less than t. From (29) and the corollary of Theorem 4, Theorem 5 then follows.

Theorem 5

The polynomial $C(D)$ defined by (27) is the connection polynomial of the unique shortest LFSR over $F = GF(q^m)$ that generates the sequence $S_{m_0}, S_{m_0} + 1, S_{m_0} + 2, \cdots$.

From (27), it follows that the t roots of $C(D)$ are the reciprocals of the t error locators. Chien [4] has given a simple means for implementing the task of finding the roots from $C(D)$ so that the essential decoding problem for the BCH codes reduces finally to the following. Given $S_{m_0}, S_{m_0} + 1, \cdots, S_{m_0} + d - 2$, find the polynomial $C(D)$ in (27). From Theorem 5 and the corollary of Theorem 3, it follows that the LFSR synthesis algorithm may be used to solve this decoding problem when the error pattern has weight guaranteed correctable by the design distance of the code. We state this fact as the following corollary.

Corollary

When the weight t of the error pattern $\mathbf{e}$ satisfies $2t \leqq d - 1$, then $C(D)$ defined by (27) is the connection polynomial of the unique shortest LFSR over $GF(q^m)$ that generates the sequence $S_{m_0}, S_{m_0} + 1, \cdots, S_{m_0} + d - 2$ and therefore will be produced when the LFSR synthesis algorithm is applied to this $n = d - 1$ digit sequence.

The determination of $C(D)$ from the sequence given in this corollary is precisely the function of the interative algorithm developed by Berlekamp [1]. In fact, the LFSR synthesis algorithm of Section III is (except for the minor variation noted earlier) precisely the Berlekamp algorithm abstracted from its particular application to the decoding of the BCH codes.

The reader is referred to Berlekamp [1] for 1) a discussion of the simplification that occurs when the algorithm is used with binary BCH codes, namely $d = 0$ automatically in step 2) when N is odd, 2) applicability of the algorithm to errors-and-erasures decoding, and 3) modifications by which the algorithm can be extended to correct some errors of weight t with $2t > d - 1$, essentially by postulating additional S_i, $i > m_o + d - 1$, at the receiver.

VI. Additional Applications

There appears to be a number of interesting applications for the LFSR synthesis algorithm of Section III. The most obvious is that of finding a simple digital device to generate a prescribed binary sequence with useful properties in some application. Less obviously, the algorithm might be used as part of a source coder, or data compressor, for a binary data source whose output contains considerable redundancy. For instance, the source digits might be processed by the algorithm in blocks of 127 digits. Each block could then be represented for transmission as a 7-bit block giving the length L of the shortest LFSR that generates the original sequence, followed by L bits to indicate the values of the tap connections and a further L bits giving the initial contents of the LFSR. Therefore, a total of $2L + 7$ bits would be transmitted in place of the original 127 bits. Such a data compression scheme could be expected to perform efficiently only when the underlying constraints producing the source redundancy were with high probability linear relations among the binary source digits.

VII. Remarks

It should be pointed out that although the $\{c_i\}$ and $\{s_i\}$ considered in Sections II and III were assumed to lie in a field F, the proofs of Theorem 1 and Lemma 1 made no use of the existence of a multiplicative inverse in F. Hence Theorem 1 and Lemma 1 remain valid under the weaker hypothesis that the $\{c_i\}$ and $\{s_i\}$ are elements of a commutative ring.

Two developments that have come to our attention since the initial manuscript of this paper was prepared are deserving of mention. H. H. Harris of the Honeywell Corp., St. Petersburg, Fla. (private communication) has simulated a data compression scheme similar to that described in Section VI and reports an approximate 50-percent data reduction for digitized voice data. Zierler [5] has recently described the BCH decoding problem as a problem in ideals over polynomial rings in terms that are formally equivalent to Theorem 5 above.

Acknowledgment

The debt that the author owes to the work of Dr. E. R. Berlekamp of the Bell Telephone Laboratories is both obvious and gratefully acknowledged. Hopefully, the approach in this paper will yield additional insight into, and be of tutorial value for, Berlekamp's work [1]. It is also a pleasure to acknowledge the stimulating interest and helpful suggestions received from numerous colleagues; specific mention must be made of Profs. R. Gallager and M. Goutman of Massachusetts Institute of Technology, Cambridge, and of Dr. G. D. Forney, Jr., of Codex Corporation, Watertown, Mass.

References

[1] E. R. Berlekamp, "Nonbinary BCH decoding," presented at the 1967 Internat'l Symp. on Information Theory, San Remo, Italy. ——*Algebraic Coding Theory.* New York: McGraw-Hill, 1968, chs. 7 and 10.
[2] W. W. Peterson, *Error-Correcting Codes.* Cambridge, Mass: M. I. T. Press, and New York: Wiley, ch. 9, 1961.
[3] G. D. Forney, Jr., "On decoding BCH codes," *IEEE Trans. Information Theory*, vol. IT-11, pp. 549–557, October 1965.
[4] R. T. Chien, "Cyclic decoding procedures for the Bose-Chaudhuri-Hocquenghem codes," *IEEE Trans. Information Theory*, vol. IT-10, pp. 357–363, October 1964.
[5] N. Zierler, "A complete theory for generalized BCH codes," *Proc. 1968 Symp. on Error Correcting Codes*, H. B. Mann, Ed. New York: Wiley, 1968.

Weight Enumeration

V

Editor's Comments on Papers 21, 22, and 23

21 MacWilliams: *A Theorem on the Distribution of Weights in a Systematic Code*

22 Pless: *Power Moment Identities on Weight Distributions in Error Correcting Codes*

23 MacWilliams, Sloane, and Goethals: *The MacWilliams Identities for Nonlinear Codes*

While the construction of codes with good distance properties was seen as a fundamental problem confronting coding theorists, it was also recognized that the evaluation of the performance of these codes under various decoding regimes was of importance. The complete structure of a code is seldom known. It is one thing to find the minimum distance of a code and quite another to determine its structure. Even when a great deal is known about the code structure, it may not be enough to evaluate the performance of the code using a given decoding algorithm. For example, a linear code has the property that the set of distances from one fixed word to all other words in the dictionary is independent of the fixed word chosen. However, for two linear codes this set of distances can be identical and yet each code has different error-correcting properties [Berlekamp (1968), p. 399].

The weight enumerator of a code, which we define as the polynomial

$$A(z)=\sum_i A_i z^i,$$

where A_i is the number of codewords of Hamming weight i, gives much information on the probability of decoding error for many of the common decoding algorithms. For linear codes it completely specifies the set of distances from a fixed codeword to all others, and it may or may not do this for a given nonlinear code. While it may not give the probability of decoding error for the optimal decoding algorithm (e.g., Slepian's standard array algorithm for the binary symmetric channel), it is, nevertheless, a useful and powerful concept in code evaluation.

The three papers on weight enumeration included here are, without question, the most significant contributions of a general nature that have yet appeared. The paper by MacWilliams [21] relates the weight enumerator of a linear code to that of its dual; that of Pless [22] contains some higher-order identies of a similar nature. Since the appearance of these papers, a great deal of effort has been expended on the weight-enumeration problem. Most results have been of a specialized nature, usually pertaining to a particular code, or class of codes, with perhaps the BCH and Reed–Muller codes receiving the most attention. The 1972 contribution of MacWilliams, Sloane, and Goethals [23] generalizes the earlier work of MacWilliams to the case of nonlinear codes by carefully defining the dual of such a code. With the increasing interest in nonlinear codes this work will doubtless be of prominence in future weight-enumeration work.

Reprinted from *Bell System Tech. J.*, **42,** 79–94 (1963)

21

A Theorem on the Distribution of Weights in a Systematic Code†

By JESSIE MACWILLIAMS

(Manuscript received September 4, 1962)

A systematic code of word length n is a subspace of the vector space of all possible rows of n symbols chosen from a finite field. The weight of a vector is the number of its nonzero coordinates; clearly any given code contains a certain finite number of vectors of each weight from zero to n. This set of integers is called the spectrum of the code, and very little is known about it, although it appears to be important both mathematically and as a practical means of evaluating the error-detecting properties of the code.

In this paper it is shown that the spectrum of a systematic code determines uniquely the spectrum of its dual code (the orthogonal vector space). In fact the two sets of integers are related by a system of linear equations. Consequently there is a set of conditions which must be satisfied by the weights which actually occur in a systematic code. If there is enough other information about the code, it is possible to use this result to calculate its spectrum.

In most systems of error correction by binary or multiple level codes the minimum distance between two code words is an important parameter. (The distance between two code words is the number of coordinate places in which they differ.) Much attention has been given to devising codes which have an assigned minimum distance.

The weight of a code word is its distance from the origin. The distance between two code words is the weight of the vector obtained by subtracting one from the other, coordinate by coordinate. If the code words form a vector space, this vector is itself a member of the code. Such codes are called systematic codes. The set of integers specifying the weight of each code word is then exactly the same collection of numbers as the set of integers specifying the distance between each pair of code words.

† This paper formed part of a thesis presented to the Department of Mathematics, Harvard University, in partial fulfillment of the requirements for the degree of Doctor of Philosophy.

Thus it is customary to talk about weight properties rather than distance properties of systematic codes.

In many cases, practically all that is known explicitly about the distribution of weights in a code is that the weight has a certain minimum value. Recent studies have shown that it would be useful (e.g., in the study of real life channels) to have more information. We would like to be able to answer questions of the following sort:

i. Given a method (implemented or theoretical) for constructing a systematic code, how many elements of each weight will be obtained? (It is a safe assumption that nobody will want to write out the code vectors and count them.)

ii. Given a set, $u_1, u_2, \cdots, u_s$, of positive integers, is it possible to construct a systematic code with elements of these weights only?

In theory there exists a method of answering these questions.[1,2,3] Unfortunately this method is quite difficult to apply. The purpose of this paper is to give a different method which is in some ways more useful.

We show that the spectrum of a systematic code determines uniquely the spectrum of the dual code (the orthogonal vector space). In fact, the two sets of integers are related by a system of linear equations. Our main theorem shows how to obtain this system of equations.

In Section I we give definitions and statements of the main theorem and of some corollaries. Section II contains proofs of these theorems. Section III describes how the results of Section I may be applied.

I. DEFINITIONS, NOTATION AND A STATEMENT OF THE MAIN THEOREM

Let F be a finite field of q elements; q is a prime power. Let F^n denote the direct sum of n copies of F. F^n is the set of all possible row vectors of length n, in which each coordinate is an element of F. Addition of two vectors is defined coordinate by coordinate, under the rules prevailing in F.

F^n is a vector space of dimension n over F. Choose a basis consisting of the n vectors

$$\begin{aligned} \epsilon_1 &= (1, 0, 0, \cdots, 0) \\ \epsilon_2 &= (0, 1, 0, \cdots, 0) \\ &\cdots\cdots\cdots\cdots\cdots \\ \epsilon_n &= (0, 0, 0, \cdots, 1). \end{aligned}$$

An element u of F^n is then expressed uniquely as

$$u = \sum_{i=1}^{n} u_i \epsilon_i, \qquad u_i \in F.$$

We write $u = (u_1, u_2, \cdots, u_n)$.

The weight of u is defined to be the number of u_i which actually appear in this sum — i.e., the number of nonzero coordinates in the vector u.

An alphabet is any subspace of F^n; a vector belonging to the alphabet $\mathcal{A}$ is called a letter of $\mathcal{A}$. It may happen that every letter of $\mathcal{A}$ has zero as the jth coordinate — this case is not excluded.

The scalar product of two vectors,

$$u = \sum_{i=1}^{n} u_i \epsilon_i, \quad v = \sum_{i=1}^{n} v_i \epsilon_i, \qquad u_i, v_i \in F,$$

is $u * v = \sum_{i=1}^{n} u_i v_i$, where the multiplication and addition are carried out in F. If F is the field of two elements 0, 1, for example, the scalar product of (1, 1, 0) with itself is $1 \cdot 1 + 1 \cdot 1 = 0$.

Two vectors u, v are orthogonal if their scalar product is zero. In the example above, (1, 1, 0) is orthogonal to itself.

The orthogonal complement of an alphabet $\mathcal{A}$ is the set of all vectors of F^n which are orthogonal to every vector of $\mathcal{A}$. It is clear that these vectors also form an alphabet, say $\mathcal{B}$, which is called the dual alphabet of $\mathcal{A}$. If k is the dimension of $\mathcal{A}$, the dimension of $\mathcal{B}$ is $m = n - k$.

The main result of this paper is as follows (the proof is given in Section II):

Let $\mathcal{A}$ be an alphabet of dimension k, and $\mathcal{B}$ the dual alphabet of dimension m. Let A_i, B_i denote the number of letters of weight i in $\mathcal{A}$, $\mathcal{B}$ respectively. Of course, $A_0 = B_0 = 1$. Set $\gamma = q - 1$. Let z be an indeterminate.

Theorem 1: The quantities defined above are related by the equation

$$\sum_{i=0}^{n} A_i (1 + \gamma z)^{n-i} (1 - z)^i = q^k \sum_{i=0}^{n} B_i z^i.$$

Remarks:

i. The formula above is symmetric in the sense that, setting $(1 - z)/(1 + \gamma z) = \hat{z}$, we obtain by straightforward algebra

$$\sum_{i=0}^{n} B_i (1 + \gamma \hat{z})^{n-i} (1 - \hat{z})^i = q^m \sum_{i=0}^{n} A_i \hat{z}^i.$$

ii. Theorem 1 is a statement about equivalent classes;[3,4] it is still true if $\mathcal{A}$, $\mathcal{B}$ are replaced by equivalent alphabets.

An alphabet $\mathcal{A}$ is said to be decomposable,[3,4] with respect to the basis

$\epsilon_1, \epsilon_2, \cdots, \epsilon_n$ of F^n, if it is the direct sum of two alphabets $\mathcal{A}_1, \mathcal{A}_2$, where $\mathcal{A}_1$ contains n columns of zeros and $\mathcal{A}_2$ occupies these columns only. For example, the alphabet

$$\begin{matrix} 0 & 0 & 0 & 0 \\ 1 & 1 & 0 & 0 \\ 0 & 0 & 1 & 1 \\ 1 & 1 & 1 & 1 \end{matrix}$$

is decomposable, with

$$\mathcal{A}_1 = \begin{matrix} 0 & 0 & 0 & 0 \\ 1 & 1 & 0 & 0 \end{matrix} \qquad \mathcal{A}_2 = \begin{matrix} 0 & 0 & 0 & 0 \\ 0 & 0 & 1 & 1. \end{matrix}$$

In general, $\mathcal{A}_1$ is a k_1-dimensional alphabet in F^{n_1}, and $\mathcal{A}_2$ a k_2-dimensional alphabet in F^{n_2}, with $n_1 + n_2 = n$, $k_1 + k_2 = k$, $k_1 \leqq n_1$, $k_2 \leqq n_2$.

The dual alphabet of a decomposable alphabet is also decomposable; in fact $\mathcal{B} = \mathcal{B}_1 + \mathcal{B}_2$, where $\mathcal{B}_i$ is the dual alphabet of $\mathcal{A}_i$ in F^{n_i}, $i = 1, 2$. (The example above is self-dual.)

Corollary 1.1: If $\mathcal{A}$ is decomposable, say $\mathcal{A} = \mathcal{A}_1 + \mathcal{A}_2$, the equation

$$\sum_{i=0}^{n} A_i(1 + \gamma z)^{n-i}(1 - z)^i = q^k \sum B_i z^i$$

is reducible in the obvious sense; the factors are the equations pertaining to $\mathcal{A}_i, \mathcal{B}_i$ in F^{n_i}, $i = 1, 2$.

For the example above we have

$$[(1 + z)^4 + (1 - z)^2]^2 = 2^2(1 + z^2)^2.$$

Corollary 1.2: A necessary condition for the existence of an alphabet containing letters of weights w_i, $i = 1, 2, \cdots, s$, and no other, is that there exists a set of integers α_i, $i = 1, 2, \cdots, s$, such that the expression

$$(1 + \gamma z)^n + \gamma \sum_{j=1}^{s} \alpha_i(1 + \gamma z)^{n-w_i}(1 - z)^{w_i},$$

when expanded in powers of z, takes the form

$$q^k + \gamma q^k \sum_{i=1}^{n} \beta_i z^i,$$

where the β_i are positive integers.

Unfortunately, this condition is not sufficient. For example,

$$(1 + z)^8 + 7(1 + z)^6(1 - z)^2 + 7(1 + z)^2(1 - z)^6 + (1 - z)^8$$
$$= 2^4(1 + 7z^2 + 7z^6 + z^8),$$

but it is not possible to construct a binary alphabet containing 7 letters of weight 2 and no letters of weight 4.

If $A_1 = A_2 = \cdots = A_{2j} = 0$, every vector of weight $\leqq j$ in F^n appears as a coset leader for $\mathcal{A}$, and conversely. Another way of saying this is that, for all pairs of distinct letters a, a', of $\mathcal{A}$ and any $i \leqq j$, the set of vectors at distance i from a is disjoint from the set of vectors at distance i from a'. In this case we can enumerate these vectors by weights as follows:

Let $f_{s,i}$ denote the number of vectors of weight s in F^n which are at distance i from some letter of $\mathcal{A}$. Write

$$(1 + \gamma z)^{n-i}(1 - z)^i = \sum_{i=0}^{n} \Psi(i, j)z^j$$

Corollary 1.3: If $\mathcal{A}$ contains no letter of weight $< 2j + 1$, then

$$\sum_{s=1}^{n} f_{s,j}x^s = \sum_{i=0}^{n} B_i\Psi(i, j)(1 + \gamma x)^{n-i}(1 - x)^i.$$

The proofs of corollaries 1.1 to 1.3 are given in Section II.

II. PROOFS

If $\mathcal{A}$ is an alphabet of F^n and $\mathcal{B}$ the orthogonal complement of $\mathcal{A}$, the weights of the letters of $\mathcal{B}$ are, of course, uniquely determined by the letters of A. However, a much stronger statement can be made: the set of integers specifying the number of letters of each weight in $\mathcal{B}$ is related by a system of linear equations to the set of integers similarly defined for $\mathcal{A}$. This section will consist of proofs of this statement and of some of its consequences.

Two proofs are given. The first is short and easy; the second is longer and more sophisticated. However, it incidentally produces a more general result and gives some insight into what is going on.

We make the following conventions for notation: $\mathcal{A}$ shall be a k-dimensional alphabet in F^n; $\mathcal{B}$ shall be the orthogonal complement of $\mathcal{A}$ of dimension $m = n - k$; γ shall denote the quantity $q - 1$. A_i, B_i denote the number of letters of weight i in $\mathcal{A}$, $\mathcal{B}$ respectively. The binomial coefficient $\binom{r}{s}$ is understood to be zero if $s > r$.

Let $\epsilon_1, \epsilon_2, \cdots, \epsilon_n$ be the usual basis of F^n. Let $s = (s_1, s_2, \cdots, s_\nu)$ be a set of ν different indices, $1 \leqq s_i \leqq n$, and let $t = (t_1, t_2, \cdots, t_{n-\nu})$ be the complementary set of indices. Denote by F_s^ν, $F_t^{n-\nu}$ the spaces generated by $\epsilon_{s_1}, \cdots, \epsilon_{s_\nu}$ and $\epsilon_{t_1}, \cdots, \epsilon_{t_{n-\nu}}$. Clearly, F_s^ν, $F_t^{n-\nu}$ are orthogonal complements in F^n. Let $| H |$ stand for the number of vectors in a space H.

Lemma 2.0:

$$|\mathcal{A} \cap F_t^{n-\nu}| = q^{k-\nu} |\mathcal{B} \cap F_s^{\nu}|.$$

Proof: The orthogonal complement of $\mathcal{A} \cap F_t^{n-\nu}$ is the smallest space containing $\mathcal{B}$ and F_s^{ν}. This is the lattice theoretic union of $\mathcal{B}$ and F_s^{ν}, which we write $\mathcal{B} \cup F_s^{\nu}$. Then

$$|\mathcal{B} \cup F_s^{\nu}| \cdot |\mathcal{A} \cap F_t^{n-\nu}| = q^n = q^{m+k}.$$

The number of vectors in $\mathcal{B} \cup F_s^{\nu}$ is $q^m q^{\nu}/|\mathcal{B} \cap F_s^{\nu}|$.
Hence

$$q^{m+\nu} |\mathcal{A} \cap F_t^{n-\nu}| = q^{m+k} |\mathcal{B} \cap F_s^{\nu}|$$

or

$$|\mathcal{A} \cap F_t^{n-\nu}| = q^{k-\nu} |\mathcal{B} \cap F_s^{\nu}|.$$

Denote by $\{(\epsilon_{s_1}, \cdots, \epsilon_{s_\nu}), a\}$ a pair consisting of ν basis vectors of F^n and a vector a of $\mathcal{A}$ which is orthogonal to each of $\epsilon_{s_1}, \cdots, \epsilon_{s_\nu}$.

Lemma 2.1:

i. For a fixed set of indices $s_1, \cdots, s_\nu$ the number of pairs

$$\{(\epsilon_{s_1}, \cdots, \epsilon_{s_\nu}), a\} \quad \textit{is} \quad |\mathcal{A} \cap F_t^{n-\nu}|.$$

ii. The total number of such pairs for all choices of ν distinct basis vectors is $\sum_{i=0}^{n} A_i \binom{n-i}{\nu}$.

Proof:

i. $F_t^{n-\nu}$ consists of exactly those vectors of F^n which are orthogonal to $\epsilon_{s_1}, \cdots, \epsilon_{s_\nu}$; hence $\mathcal{A} \cap F_t^{n-\nu}$ consists of exactly those vectors of $\mathcal{A}$ which are orthogonal to $\epsilon_{s_1}, \cdots, \epsilon_{s_\nu}$.

ii. If $a \in \mathcal{A}$ is of weight i, then a is orthogonal to $n - i$ of the vectors $\epsilon_1, \cdots, \epsilon_n$. A set of ν vectors may be chosen from these $n - i$ in $\binom{n-i}{\nu}$ ways. Hence the total number of pairs

$$\{(\epsilon_{s_1}, \cdots, \epsilon_{s_\nu}), \quad a\} \text{ is } \sum_{i=0}^{n} A_i \binom{n-i}{\nu}.$$

Let $\sum_s$ indicate summation over all possible choices of ν indices $s_1, \cdots, s_\nu$; similarly, $\sum_t$ denotes summation over all the complementary sets $t_1, \cdots, t_{n-\nu}$. Lemma 2.1 is equivalent to

$$\sum_t |\mathcal{A} \cap F_t^{n-\nu}| = \sum_{i=0}^{n} A_i \binom{n-i}{\nu}.$$

Replace $\mathfrak{A}$ by $\mathfrak{B}$, ν by $n - \nu$ and s by t. The same argument then gives

$$\sum_s |\mathfrak{B} \cap F_s^{\nu}| = \sum_{i=0}^{n} B_i \binom{n-i}{n-\nu}.$$

Lemma 2.2

$$\sum_{i=0}^{n} A_i \binom{n-i}{\nu} = q^{k-\nu} \sum_{i=0}^{n} B_i \binom{n-i}{n-\nu}.$$

Proof: For a fixed set s (which determines, of course, a fixed set t) we have by 2.0

$$|\mathfrak{A} \cap F_t^{n-\nu}| = q^{k-\nu} |\mathfrak{B} \cap F_s^{\nu}|.$$

Thus

$$\sum_t |\mathfrak{A} \cap F_t^{n-\nu}| = q^{k-\nu} \sum_s |\mathfrak{B} \cap F_s^{\nu}|,$$

which, by 2.1 is the same thing as

$$\sum_{i=0}^{n} A_i \binom{n-i}{\nu} = q^{k-\nu} \sum_{i=0}^{n} B_i \binom{n-i}{n-\nu}.$$

The equation of 2.2 holds for $\nu = 0, 1, \cdots, n-1$. This is, in fact, one form of the promised set of linear equations between the quantities A_i, B_i.

We now give the second proof.

Let $\mathcal{G}$ be a finite Abelian group. A character χ of $\mathcal{G}$ is a homomorphism of $\mathcal{G}$ into the multiplicative group of complex numbers of absolute value 1. The characters of $\mathcal{G}$ form a group $\mathcal{G}^*$ which is isomorphic to $\mathcal{G}$, there being in general no canonical isomorphism.†

If $\mathfrak{A}$ is a subgroup of $\mathcal{G}$, the characters such that $\chi(a) = 1$ for all a of $\mathfrak{A}$ form a subgroup $\mathfrak{B}^*$ of $\mathcal{G}^*$. $\mathfrak{B}^*$ is precisely the character group of $\mathcal{G}$ *mod* $\mathfrak{A}$.

Suppose now that $\mathcal{G}$ is the additive group of a finite field. $\mathcal{G}^*$ is just a multiplicative copy of $\mathcal{G}$, and the characters can be labeled by the elements of $\mathcal{G}$ in a symmetric way; that is, if $r, s, \cdots$ are elements of $\mathcal{G}$ we have

$$\chi_r(s) = \chi_s(r) = \chi(r, s).$$

If $\mathcal{G}$ is the additive group of a prime field of order q, we take r, s etc.

† For prime fields, the proof can be given without mentioning the word character. The presentation here is an uneasy compromise with conscience — we wish to indicate possible extensions to nonprime fields without doing too much work.

to be the integers *mod* q, and set $\chi(r, s) = \zeta^{rs}$ where ζ is a primitive qth root of unity.

We have from the general theory of characters

$$\chi(r, 0) = \chi(0, s) = 1,$$

$$\sum_{r=1}^{\gamma} \chi(r, s) = -1 \qquad \text{if } s \neq 0.$$

Let $\epsilon_1, \epsilon_2, \cdots, \epsilon_n$ be the fixed basis of F^n, and $u = (u_1, u_2, \cdots, u_n)$ the coordinates of a vector of F^n with respect to this basis. The character group of F^n is, of course, a multiplicative copy of F^n. We label the characters by elements of F^n as follows:

$$\psi_u(v) = \prod_{i=1}^{n} \chi(u_i, v_i) = \psi_v(u) = \psi(u, v).$$

Let $\mathcal{A}$ be a subspace of F^n. The characters such that $\psi_b(a) = 1$ for all a of $\mathcal{A}$ form a subgroup $\mathcal{B}^*$ of the character group. $\mathcal{B}^*$ is exactly the character group of F^n *mod* $\mathcal{A}$. The elements b which label these characters form a subspace $\mathcal{B}$ of F^n, isomorphic to F^n *mod* $\mathcal{A}$. In our notation, the equation $\psi(a, b) = 1$ holds for all a of $\mathcal{A}$ and all b of $\mathcal{B}$, and given either $\mathcal{A}$ or $\mathcal{B}$, the other is uniquely† determined by this condition.

Lemma 2.3: Let $\mathcal{A}$, $\mathcal{B}$ be related as above. Then

i. $$\sum_{a \in \mathcal{A}} \psi(v, a) = q^k \text{ if } v \in \mathcal{B}.$$

ii. $$\sum_{a \in \mathcal{A}} \psi(v, a) = 0 \text{ if } v \notin \mathcal{B}.$$

Proof: Part *i* is obvious, since by definition $\psi(v, a) = 1$ if $v \in B$. For *ii* we observe that for $a \in \mathcal{A}$, $\psi(v, a) = \psi_a(v)$ is a character of F^n *mod* $\mathcal{B}$. If $\bar{v}$ denotes a coset of F^n *mod* $\mathcal{B}$, $\sum_{a \in \mathcal{A}} \psi_a(\bar{v}) = 0$ for $\bar{v} \neq \mathcal{B}$. Now $\psi_a(v) = \psi_a(\bar{v})$ for any v in $\bar{v}$; hence

$$\sum_{a \in \mathcal{A}} \psi(v, a) = \sum_{a \in \mathcal{A}} \psi_a(\bar{v}) = 0 \quad \text{if} \quad v \notin \mathcal{B}.$$

Lemma 2.4: If F is a prime field, then $\mathcal{A}$, $\mathcal{B}$ are related as in 2.3 *if and only if they are orthogonal complements.*

Proof: $\psi(a, b) = \zeta^{\Sigma a_i b_i}$ where ζ is a primitive qth root of unity. Hence $\psi(a, b) = 1$ implies that a is orthogonal to b. Since $\mathcal{A}$ is isomorphic to F^n *mod* $\mathcal{B}$ the dimensions of $\mathcal{A}$, $\mathcal{B}$ add up to n. Thus $\mathcal{B}$ is the orthogonal complement of $\mathcal{A}$.

---

† That is, if F is a prime field. Otherwise we must fix the basis of F over its prime field before we claim uniqueness.

Let $f(i, s)$ denote a function of the integers i, s with values in a ring R. The values of $f(i, s)$ may be added and multiplied, and these operations obey the two distributive laws. $f(i, v_i)$ denotes the same function of i and the ith coordinate of v.

Lemma 2.5:

$$\sum_{v \in F^n} \prod_{i=1}^{n} f(i, v_i) = \prod_{i=1}^{n} \sum_{r=0}^{\gamma} f(i, r).$$

Proof: If $n = 1$ the statement is

$$\sum_{s=0}^{\gamma} f(1, s) = \sum_{r=0}^{\gamma} f(1, r),$$

which is obvious. Assume the truth of the lemma for F^{n-1}.

Let F_r^n, $0 \leqq r \leqq \gamma$, denote the set of vectors of F^n which have last coordinate r. Clearly the F_r^n are a partition of F^n. Then

$$\begin{aligned}\sum_{v \in F^n} \prod_{i=1}^{n} f(i, v_i) &= \sum_{r=0}^{\gamma} \sum_{v \in F_r^n} \left[\prod_{i=1}^{n-1} f(i, v_i) f(n, r)\right] \\ &= \sum_{r=0}^{\gamma} f(n, r) \sum_{v \in F^{n-1}} \prod_{i=1}^{n-1} f(i, v_i) \\ &= \sum_{r=0}^{\gamma} f(n, r) \prod_{i=1}^{n-1} \sum_{r=0}^{\gamma} f(i, r) \quad \text{(by induction)} \\ &= \prod_{i=1}^{n} \sum_{r=0}^{\gamma} f(i, r).\end{aligned}$$

Let $z^{(r)}$ be a set of (commuting) indeterminates, $r = 0, 1, \cdots, \gamma$. To each vector $v = (v_1, v_2, \cdots, v_n)$ of F^n associate a monomial $\prod_{i=1}^{n} z^{(v_i)}$. The monomial associated with v describes how many times each field element appears as a component of v. Let R be the ring of polynomials in $z^{(0)}, z^{(1)}, \cdots, z^{(\gamma)}$ over the complex numbers. Let $u = (u_1, u_2, \cdots, u_n)$ be a fixed vector of F^n.

Lemma 2.6:

$$\sum_{v \in F^n} \psi(u, v) z^{(v_1)} z^{(v_2)} \cdots z^{(v_n)} = \prod_{j=1}^{n} \sum_{r=0}^{\gamma} \chi(u_j, r) z^{(r)}.$$

Proof: Set $f(j, v_j) = \chi(u_j, v_m) z^{(v_j)}$, which is in R. Then

$$\psi(u, v) z^{(v_1)} z^{(v_2)} \cdots z^{(v_n)} = \prod_{j=1}^{n} f(j, v_j).$$

By 2.5,

$$\sum_{v \in F^n} \psi(u, v) z^{(v_1)} z^{(v_2)} \cdots z^{(v_n)} = \prod_{j=1}^{n} \sum_{r=0}^{\gamma} f(j, r)$$

$$= \prod_{j=1}^{n} \sum_{r=0}^{\gamma} \chi(u_j, r) z^{(r)}.$$

Lemma 2.7: Let $\mathcal{A}$, $\mathcal{B}$ be orthogonal complements in F^n, as usual. Then

$$\sum_{u \in \mathcal{A}} \prod_{j=1}^{n} \sum_{r=0}^{\gamma} \chi(u_j, r) z^{(r)} = q^k \sum_{v \in \mathcal{B}} z^{(v_1)} z^{(v_2)} \cdots z^{(v_n)}$$

Proof: We evaluate the quantity

$$F(u, v) = \sum_{u \in \mathcal{A}} \sum_{v \in F^n} \chi(u, v) z^{(v_1)} z^{(v_2)} \cdots z^{(v_n)}$$

in two ways, which give the two sides of the equation.
By 2.6

$$F(u, v) = \sum_{u \in \mathcal{A}} \prod_{j=1}^{n} \sum_{r=0}^{\gamma} \chi(u_j, r) z^{(r)}.$$

Also

$$F(u, v) = \sum_{v \in F^n} z^{(v_1)} z^{(v_2)} \cdots z^{(v_n)} \sum_{u \in \mathcal{A}} \psi(u, v).$$

By 2.4 and 2.3

$$\sum_{u \in \mathcal{A}} \psi(u, v) = \begin{cases} q^k \text{ if } v \in \mathcal{B} \\ 0 \text{ if } v \notin \mathcal{B}. \end{cases}$$

Hence

$$F(u, v) = q^k \sum_{v \in \mathcal{B}} z^{(v_1)} z^{(v_2)} \cdots z^{(v_n)}.$$

Theorem 2.8: Let $\mathcal{A}$ be a k-dimensional alphabet of F^n, and $\mathcal{B}$ the orthogonal complement of dimension $m = n - k$. Let A_i, B_i denote the number of letters of weight i in $\mathcal{A}$, $\mathcal{B}$. Then

$$\sum_{i=0}^{n} A_i (1 + \gamma z)^{n-i} (1 - z)^i = q^k \sum_{i=0}^{n} B_i z^i.$$

Proof: In 2.7 set $z^{(r)} = \begin{cases} z & \text{if } r \neq 0 \\ 1 & \text{if } r = 0. \end{cases}$

If $u_j = 0$, $\chi(u_j, r)$ is 1 and $\sum_{r=0}^{\gamma} \chi(u_j, r) z^{(r)}$ becomes $(1 + \gamma z)$.

If $u_j \neq 0$ $\sum_{r=1}^{\gamma} \chi(u_j, r)$ is -1, and $\sum_{r=0}^{\gamma} \chi(u_j, r) z^{(r)}$ becomes $(1 - z)$.

Let $|u|$ denote the number of nonzero u_j.

Then $\prod_{j=1}^{n} \sum_{r=0}^{\gamma} \chi(u_j, r) z^{(r)}$ goes into $(1+\gamma z)^{n-|u|}(1-z)^{|u|}$;

$|u|$ is of course the weight of $u = (u_1, u_2, \cdots, u_n)$, so that the left-hand side of 2.7 becomes

$$\sum_{i=0}^{n} A_i(1+\gamma z)^{n-i}(1-z)^i.$$

The right-hand side of 2.7 is clearly

$$q^k \sum_{i=0}^{n} B_i z^i,$$

which proves the theorem.

Innumerable sets of linear equations between the quantities A_i, B_i may be obtained from theorem 2.8. The following two are sometimes useful.

Lemma 2.9: For $\nu = 0, 1, \cdots, n$,

$$i. \qquad \sum_{i=0}^{n-\nu} A_i \binom{n-i}{\nu} = q^{k-\nu} \sum_{i=0}^{n} B_i \binom{n-i}{n-\nu}.$$

(*These are the equations of* 2.2.)

$$ii. \qquad \sum_{i=\nu}^{n} A_i \binom{i}{\nu} = q^{k-\nu} \sum_{i=0}^{\nu} (-1)^i B_i \gamma^{\nu-i} \binom{n-i}{n-\nu}.$$

i. is obtained by setting $(1+\gamma z)/(1-z) = 1+y$,

ii. by setting $(1-z)/(1+\gamma z) = 1+y$. The algebraic details are easy to verify.

This process is reversible, i.e., (*i*) or (*ii*) imply 2.8. Before exploring the consequences of theorem 2.8, we give a different specialization of 2.7.

Theorem 2.10: Let $B_s^{(1)}$ *be the number of letters of* $\mathcal{B}$ *which contain s coordinates equal to* 1. *Let* A_{0s} *be the number of letters u in* $\mathcal{A}$ *of weight s for which* $\sum_{i=1}^{n} u_i = 0$. *Let* A_{1s} *be the number of letters u in* $\mathcal{A}$ *of weight s for which* $\sum_{i=1}^{n} u_i \neq 0$. (*Clearly* $A_s = A_{0s} + A_{1s}$).
Then

$$\sum_{s=0}^{n} B_s^{(1)} z^s = (A_{0s} - A_{1s}/\gamma)(z-1)^s(z+\gamma)^{n-s}.$$

Proof: In 2.7 set

$$z^{(r)} = \begin{cases} z & \text{if } r = 1 \\ 1 & \text{if } r \neq 1. \end{cases}$$

Then

$$q^k \sum_{\nu \in \mathcal{B}} z^{(\nu_1)} z^{(\nu_2)} \cdots z^{(\nu_n)} \text{ becomes } q^k \sum_{s=0}^{n} B_s^{(1)} z^s,$$

$$\sum_{r=0}^{\gamma} \chi(u_i, r) z^{(r)} \text{ becomes } \chi(u_i, 1)z + \sum_{r=0}^{\gamma} \chi(u_i, r) - \chi(u_i, 1)$$

$$= \chi(u_i, 1)(z - 1) + \sum_{r=0}^{\gamma} \chi(u_i, r)$$

$$= \begin{cases} \chi(u_i, 1)(z-1) & \text{if } u_i \neq 0 \\ z + \gamma & \text{if } u_i = 0 \end{cases}$$

$$\prod_{i=1}^{n} \sum_{r=0}^{\gamma} \chi(u_i, r) z^{(r)} \quad \text{becomes} \quad (z+\gamma)^{n-|u|}(z-1)^{|u|} \prod_{u_i \neq 0} \chi(u_i, 1).$$

Now if u is a letter of A, so are also the letters $2u, \cdots, \gamma u$, and these have the same weight as u. We sum first over these letters

$$\sum_{s=1}^{\gamma} \prod \chi(su_i, 1) = \sum_{s=1}^{\gamma} \chi(s\Sigma u_i, 1)$$

$$= \begin{cases} -1 & \text{if } \Sigma u_i \neq 0 \\ \gamma & \text{if } \Sigma u_i = 0 \end{cases}$$

The sum of

$$(z+\gamma)^{n-|u|}(z-1)^{|u|} \prod_{u_i \neq 0} \chi(u_i, 1)$$

over all letters in $\mathcal{A}$ of the same weight as u is thus

$$(A_{0|u|} - (A_{1|u|}/\gamma))(z+\gamma)^{n-|u|}(z-1)^{|u|}.$$

Hence the left-hand side of 2.7 becomes

$$\sum_{s=0}^{n} (A_{0s} - (A_{1s}/\gamma))(z+\gamma)^{n-s}(z-1)^s.$$

We return now to the consequences of theorem 2.8. As remarked in the proof of 2.10, if $\mathcal{A}$ contains a letter u, it contains also the letters $2u, \cdots, \gamma u$; that is, the number of letters of weight i in $\mathcal{A}$ is divisible by γ for $i > 0$. We have then

Lemma 2.11: A necessary condition for the existence of an alphabet containing letters of weights w_i, $i = 1, 2, \cdots, s$, and no other, is the existence of a set of integers α_i, $i = 1, 2, \cdots, s$, such that the expression

$$(1 + \gamma z)^n + \gamma \sum_{j=1}^{s} \alpha_i (1 + \gamma z)^{n-w_i} (1 - z)^{w_i},$$

when expanded in powers of z, takes the form

$$q^k + \gamma q^k \sum_{i=1}^{n} \beta_i z^i,$$

where the β_i *are positive integers.*

It has been pointed out before that this condition is not sufficient.

Suppose now that $\mathcal{A} = \mathcal{A}_1 + \mathcal{A}_2$ is a decomposable alphabet. $\mathcal{A}_j$ is a k_j-dimensional alphabet in F^{n_j}, $j = 1, 2$, with orthogonal alphabet B_j. $k_1 + k_2 = k$, and $n_1 + n_2 = n$. Let $A_i^{(1)}$, $A_i^{(2)}$ and $B_i^{(1)}$, $B_i^{(2)}$ be the number of letters of weight i in $\mathcal{A}_1$, $\mathcal{A}_2$, $\mathcal{B}_1$, $\mathcal{B}_2$.

Lemma 2.12:

$$\sum_{i=0}^{n} A_i (1 + \gamma z)^{n-i} (1 - z)^i$$

$$= \left[\sum_{i=0}^{n_1} A_i^{(1)} (1 + \gamma z)^{n_1 - i} (1 - z)^i\right]\left[\sum_{i=0}^{n_2} A_i^{(2)} (1 + \gamma z)^{n_2 - i} (1 - z)^i\right]$$

$$= \sum_{i=0}^{n} \left[q^{k_1} \sum_{i=0}^{n_1} B_i^{(1)} z^i\right]\left[q^{k_2} \sum_{i=0}^{n_2} B_i^{(2)} z^i\right].$$

Proof: The number of letters of weight s in $\mathcal{B}_1 + \mathcal{B}_2$ is

$$\sum_{\sigma+\rho=s} B_\sigma^{(1)} B\rho^{(2)},$$

which is the coefficient of z^s in $\sum_{i=0}^{n_1} B_i^{(1)} z^i \sum_{i=0}^{n_2} B_i^{(2)} z^i$. Similarly,

$$\sum_{\sigma+\rho=s} A_\sigma^{(1)} A_\rho^{(2)}$$

is the coefficient of $(1 + \gamma z)^{n-s} (1 - z)^s$ in

$$\left[\sum_{i=0}^{n_1} A_i^{(1)} (1 + \gamma z)^{n_1 - i} (1 - z)^i\right]\left[\sum_{i=0}^{n_2} A_i^{(2)} (1 + \gamma z)^{n_2 - i} (1 - z)^i\right].$$

We define the coset leader of a coset of $\mathcal{A}$ in F^n to be an element of least weight in the coset. The weight of a coset is defined to be the weight of its coset leader.

If $A_i = 0$ for $i = 1, 2, \cdots, 2e$ every vector of weight $\leqq e$ in F^n appears as a coset leader for $\mathcal{A}$ and conversely. Another way of saying this is: for all pairs of distinct letters a, a' of $\mathcal{A}$, the set of vectors at distance $i \leqq e$ from a is disjoint from the set of vectors at distance i from a'.

Let $c_1^{(i)}, c_2^{(i)}, \cdots, c_\nu^{(i)}$ be the cosets of A of weight i; we assume that

$\nu = \gamma^i \binom{n}{i}$, i.e., that all vectors of weight i appear as coset leaders. Let $f_{s,i}$ be the number of vectors of weight s contained in the set-theoretic union $\bigcup_{j=1}^{\nu} c_j^{(i)}$. The polynomial $\sum_{s=0}^{n} f_{s,i} x^s$ is called the enumerator (by weight) of this set of vectors. We propose to show that theorem 2.8 gives a convenient expression for this enumerator. We need the following preliminary lemma.

Lemma 2.13: Let u be a fixed vector of weight i. Let d_{st} be the number of vectors of weight s which are at distance t from u. Then

$$\sum_{s=0}^{n} \sum_{t=0}^{n} d_{st} x^s y^t = (1 + \gamma xy)^{n-i}[x + y + (\gamma - 1)xy]^i.$$

Proof: Suppose first that $u = (u_1, u_2, \cdots, u_i)$ is a vector of weight i in F^i. We show that under these circumstances

$$\sum_{s=0}^{i} \sum_{t=0}^{i} d_{st} = [x + y + (\gamma - 1)xy]^i.$$

This is obvious for $i = 1$; we suppose it true for $i - 1$. Let $v = (v_1, v_2, \cdots, v_{i-1})$ be a vector of weight s distant t from $(u_1, u_2, \cdots, u_{i-1})$ in F^{i-1}. From v we obtain:

i. One vector $(v_1, v_2, \cdots, v_{i-1}, 0)$, weight s, distant $t + 1$ from u.

ii. One vector $(v_1, v_2, \cdots, v_{i-1}, u_i)$ weight $s + 1$ distant t from u.

iii. $\gamma - 1$ vectors $(v_1, v_2, \cdots, v_{i-1}, v_i)$ $v_i \neq 0$, $v_i \neq u_i$ which have weight $s + 1$, and are distant $t + 1$ from u.

Hence the enumerator for i is obtained by multiplying that for $i - 1$ by $[x + y + (\gamma - 1)xy]$, and the lemma is proved for $n = i$.

We now apply induction to $n - i$. Let u be a vector of weight i in F^n, and u' a vector of weight i in F^{n-1} obtained from u by omitting one zero coordinate. Let v' be a vector of F^{n-1} which has weight s and is distant t from u'. From v' we obtain in F^n

i. One vector of weight s, distant t from u, by adding a zero coordinate to v'.

ii. γ vectors of weight $s + 1$, distant $t + 1$ from u, by adding a non-zero coordinate to v'.

This corresponds to multiplication by $(1 + \gamma xy)$. Hence the lemma is proved.

Lemma 2.14: Suppose that $A_i = 0$, $i = 1, 2, \cdots, 2e$, and take $t \leqq e$. Then the enumerator, $\sum_{s=0}^{n} f_{s,t} x^s$, of vectors in cosets of weight t of F^n mod $\mathfrak{A}$ is the coefficient of y^t in

$$\sum_{i=0}^{n} A_i(1 + \gamma xy)^{n-i}(x + y + (\gamma - 1)xy)^i.$$

Proof: The cosets of weight t in F^n *mod* $\mathcal{A}$ are disjoint, and contain all vectors of F^n which are at distance t from some letter of $\mathcal{A}$.

Set $(1 + \gamma z)^{n-i}(1 - z)^i = \sum_{r=0}^{n} \Psi(i, n, r)z^r$. Let $\mathcal{B}$, B_i have their usual meaning. Assume the conditions of 2.14.

Lemma 2.15:†

$$q^m \sum_{s=0}^{n} f_{s,t}x^s = \sum_{i=0}^{n} B_i \Psi(i, n, t)(1 + \gamma x)^{n-i}(1 - x)^i$$

Proof: Set

$$z = \frac{x + y + (\gamma - 1)xy}{1 + \gamma xy},$$

then

$$1 + \gamma z = \frac{(1 + \gamma x)(1 + \gamma y)}{(1 + \gamma xy)}, \qquad 1 - z = \frac{(1 - x)(1 - y)}{1 + \gamma xy}.$$

Make this substitution in the equation

$$\sum_{i=0}^{n} B_i(1 + \gamma z)^{n-i}(1 - z) = q^m \sum_{i=0}^{n} A_i z^i$$

we obtain

$$\sum_{i=0}^{n} B_i(1 + \gamma x)^{n-i}(1 - x)^i(1 + \gamma y)^{n-i}(1 - y)^i$$
$$= q^m \sum_{i=0}^{n} A_i(1 + \gamma xy)^{n-i}(x + y + (\gamma - 1)xy)^i.$$

Equating coefficients of y^t gives us

$$\sum_{i=0}^{n} B_i\Psi(i, n, t)(1 + \gamma x)^{n-i}(1 - x)^i = q^m \sum_{s=0}^{n} f_{s,t}x^s$$

III. APPLICATIONS

The easiest application of theorem 1 is to a generalized Hamming alphabet, that is, a close-packed 1-error correcting alphabet over a field of q elements. Such an alphabet exists for $n = (q^m - 1)/\gamma$, all $m > 1$.[3]

The dual alphabet is of dimension m, and contains $(q^m - 1)$ letters of weight q^{m-1}. The spectrum of a generalized Hamming alphabet is thus given by the expansion of

$$(1 + \gamma z)^n + (q^m - 1)(1 + \gamma z)^{n-u}(1 - z)^u$$

where $n = (q^m - 1)/\gamma$, $u = q^{m-1}$

† A similar formula ($q = 2$) is found by Lloyd[5] for close-packed codes which are not assumed to be group codes.

TABLE I — DISTRIBUTION OF WEIGHTS IN THE TWO GOLAY CODES

The first table is for the 3-error-correcting (23, 12) alphabet over Z_2, the second for the 2-error-correcting (11, 6) alphabet over Z_3. In both cases, i stands for weight, B_i for the number of letters of weight i in the dual alphabet, A_i for the number of letters of this weight in the Golay alphabet.

i	B_i	A_i
0	1	1
7	0	23 × 11
8	23 × 22	23 × 22
11	0	23 × 56
12	23 × 56	23 × 56
15	0	23 × 22
16	23 × 11	23 × 11
23	0	1

i	B_i	A_i
0	1	1
5	0	2 × 66
6	2 × 66	2 × 66
8	0	2 × 165
9	2 × 55	2 × 55
11	0	2 × 12

Theorem 1 may, in fact, be used to calculate the number of letters of each weight in any close-packed code. The results for the two Golay[6] codes are given in Table I.

Anything which is known about the structure of an alphabet or its dual may be used with theorem 1 to limit the number of possible weight distributions. Such items of information are a very diversified character, and no general method has been developed. However, the results obtained by hand calculation indicated that it is probably worthwhile to make a systematic computer study of the known classes of alphabets.

ACKNOWLEDGMENTS

The author would like to thank her friends and associates for their consistently helpful suggestions; and is especially indebted to H. E. Elliott for the elegant proof of lemma 2.5 and to J. B. Kruskal for theorem 2.10. The criticisms and suggestions of Professor A. M. Gleason of Harvard University produced new and better proofs of practically every other theorem in this paper.

REFERENCES

1. Slepian, D., A Class of Binary Signaling Alphabets, B.S.T.J., **35**, January, 1956, pp. 203–234.
2. Bose, R. C., and Kuebler, R. R., Jr., A Geometry of Binary Sequences Associated with Group Alphabets in Information Theory, Ann. Math. Stat. **31**, 1960, p. 113.
3. MacWilliams, F. J., Error-correcting Codes for Multiple-Level Transmission, B.S.T.J., **40**, January, 1961, pp. 281–308.
4. Slepian, D., Some Further Theory of Group Codes, B.S.T.J., **39**, September, 1960, pp. 1219–1252.
5. Lloyd, S. P., Binary Block Coding, B.S.T.J., **36**, March, 1957, pp. 517–535.
6. Golay, N. J. E., Notes on Digital Coding, Proc. I.R.E., **37**, 1949, p. 657.
7. Peterson, W. W., *Error Correcting Codes*, M.I.T. Press and John Wiley and Sons, 1961.

Reprinted from *Inform. Contr.*, **6**, 147–152 (1963)

22

Power Moment Identities on Weight Distributions in Error Correcting Codes

Vera Pless

Air Force Cambridge Research Laboratories, Bedford, Massachusetts

A series of identities relating the weight distribution in any code space to the weight distribution in the orthogonal code space has been given by Mrs. J. MacWilliams (1962). Here we derive a series of power moment identities from the MacWilliams identities. An earlier result of Assmus and Mattson is shown to be equivalent to the third power moment identity.

A unique solution to the power moment identities is given under certain conditions. Applications are given.

I. INTRODUCTION

A. Summary

A series of identitites relating the weight distribution in any code space to the weight distribution in the orthogonal code space has been given by Mrs. J. MacWilliams (1962). Here we derive a series of power moment identities from the MacWilliams identities. An earlier result of Assmus and Mattson (1961) is shown to be equivalent to one of these identities. For $q = 2$, their result was also proven by Zierler (1962).

We study states of partial information on weight distributions such that the power moment identities determine uniquely the remaining values. Applications are made in determining the weight distributions of certain cyclic codes.

B. Definitions

These are the definitions and notations relevant primarily to Section II.

Let V be a vector space of dimension n over $GF(q)$, the Galois field of q elements, q a prime power. Let A be a subspace of V, of dimension k. Two vectors $v = (v_0, v_1, \cdots, v_{n-1})$ and $w = (w_0, w_1, \cdots w_{n-1})$ in V are said to be orthogonal if $\sum_{i=0}^{n-1} v_i w_i$ is 0 in $GF(q)$. Let B be the set of all vectors in V orthogonal to each vector in A. B is called the orthogonal of A.

If $v = (v_0, \cdots, v_{n-1})$ is a vector in V, the weight of v is defined to be the number of nonzero v_i. A_i will denote the number of vectors of weight i in A and B_i will denote the number of vectors in weight i in B.

The following are the facts relevant to Section III.

We recall that a code space is called cyclic if it is invariant under the coordinate permutation corresponding to the mapping $i \to i + 1$ (modulo n) of the coordinate indices. For n prime, a cyclic code is called a quadratic residue code if, for $0 < b < n$, the coordinate permutation corresponding to the mapping $i \to bi$ (modulo n) leaves the code invariant if and only if b is a quadratic residue of n. For the rest of this paragraph and in section III we assume that $q = 2$ and n is a prime, which we will refer to by p, such that $p \equiv -1$ modulo 8. These codes were originally defined in a different context by Gleason (1961) and then in this fashion by Prange. Both Gleason (1961) and Prange from their different points of view demonstrated the following facts about quadratic residue codes A of dimension $(p - 1)/2$.

(a) There are exactly two quadratic residue codes of dimension $(p - 1)/2$. These are equivalent and hence have the same weight distributions.

(b) $A_i = 0$ unless i is a multiple of 4.

(c) If B is the orthogonal of A, B is a quadratic residue code of dimension $(p + 1)/2$, and B is obtained from A by adjoining the all one vector. This implies

(d) $B_i = 0$ unless $i = 0$ or 3 modulo 4.

(e) Prange showed that if the minimum weight in B is m, m must be odd, and $m + 1$ is then the minimum weight in A. From this it follows that the maximum weight in A is $n - m$.

II. THE POWER MOMENTS

Let $\gamma = q - 1$. MacWilliams (1962) binomial moments are

$$\sum_{j=0}^{n} \binom{j}{\nu} A_j = q^{k-\nu} \sum_{j=0}^{n} (-1)^j B_j \gamma^{\nu-j} \binom{n-j}{n-\nu}, \qquad B_1$$

$$\sum_{j=0}^{n} \binom{n-j}{\nu} A_j = q^{k-\nu} \sum_{j=0}^{n} \binom{n-j}{n-\nu} B_j \qquad B_2$$

where $\binom{a}{b} = 0$ if $b > a$; ν can take any integral values from 0 to n so that (B_1) (or (B_2)) is a set of $n + 1$ equations.

The power moment identities which can be derived from these are the following:

$$\sum_{j=0}^{n} (j^r)A_j = \sum_{j=0}^{n} (-1)^j B_j \left(\sum_{\nu=0}^{r} \nu!\, S(r, \nu) q^{k-\nu} \gamma^{\nu-j} \binom{n-j}{n-\nu}\right), \qquad P_1$$

$$\sum_{j=0}^{n} (n-j)^r A_j = \sum_{j=0}^{n} B_j \left(\sum_{\nu=0}^{r} \nu!\, S(r, \nu) q^{k-\nu} \binom{n-j}{n-\nu}\right). \qquad P_2$$

Here $S(r, \nu)$ is a Stirling number of the second kind,

$$S(r, \nu) = \left[\frac{\Delta^j x^r}{\nu!}\right]_{x=0} = \frac{1}{\nu!} \sum_{i=1}^{\nu} (-1)^{\nu-i} \binom{\nu}{i} i^r.$$

We get an infinite system of equations for either P_1 or P_2 as r can take any positive values or zero. P_1 and P_2 are valid for $r > n$ provided we understand that $\binom{a}{b} = 0$ if $b < 0$. If $r < n$, then $\sum_{j=0}^{n}$ on the right sides of P_1 and P_2 can be replaced by $\sum_{j=0}^{r}$.

We will give the derivation of P_2 from B_2. P_1 follows from B_1 similarly. From Jordan (1950) we know that

$$(n-j)^r = \sum_{\nu=0}^{r} \nu! \binom{n-j}{\nu} S(r, \nu)$$

so that

$$\begin{aligned}
\sum_{j=0}^{n} (n-j)^r A_j &= \sum_{j=0}^{n} \left(\sum_{\nu=0}^{r} \nu! \binom{n-j}{\nu} S(r, \nu)\right) A_j \\
&= \sum_{\nu=0}^{r} \nu!\, S(r, \nu) \left(\sum_{j=0}^{n} \binom{n-j}{\nu} A_j\right) \\
&= \sum_{\nu=0}^{r} \nu!\, S(r, \nu) \left(q^{k-\nu} \sum_{j=0}^{n} \binom{n-j}{n-\nu} B_j\right) \qquad \{\text{by } B_2\} \\
&= \sum_{j=0}^{n} B_j \left(\sum_{\nu=0}^{r} \nu!\, S(r, \nu) q^{k-\nu} \binom{n-j}{n-\nu}\right).
\end{aligned}$$

It can also be shown that the binomial identities can be derived from the power moment identities.

Since we are interested in solutions for codes, from now on we assume $A_0 = B_0 = 1$.

The following is a list of the first few equations in P_1 for $q = 2$.

$$r = 0 \qquad \sum A_j = 2^k, \qquad (1)$$

$$r = 1 \qquad \sum jA_j = 2^{k-1}n - 2^{k-1}B_1, \tag{2}$$

$$r = 2 \qquad \sum j^2A_j = 2^{k-2}n\,(n + 1) - 2!\,2^{k-2}nB_1 + 2!\,2^{k-2}B_2, \tag{3}$$

$$\begin{aligned} r = 3 \qquad \sum j^3A_j &= 2^{k-3}\,(n^3 + 3n^2) \\ &\quad - 2^{k-3}\,(3n^2 + 3n - 2)\,B_1 + 3!\,2^{k-3}nB_2 \\ &\quad - 3!\,2^{k-3}B_3, \end{aligned} \tag{4}$$

$$\begin{aligned} r = 4 \qquad \sum j^4A_j &= 2^{k-4}\,(n^4 + 6n^3 + 3n^2 - 2n) \\ &\quad - 2^{k-2}\,(n^3 + 3n^2 - 9n + 7)B_1 \\ &\quad + 2^{k-2}\,(3n^2 + 3n - 4)B_2 - 4!\,2^{k-4}nB_3 \\ &\quad + 4!\,2^{k-4}B_4. \end{aligned} \tag{5}$$

The Assmus-Mattson (1961) result on the sum of the squares of the weights in a code can be stated as follows. Consider the matrix M whose rows are all the elements in A over a general $GF(q)$, q a prime. Assmus and Mattson assume no column of M is 0 (this is equivalent to $B_1 = 0$). Let $c_1, \cdots, c_s$ be a largest set of columns of M such that no two are multiples of each other. Let j_i be the number of columns of M which are nonzero multiples of c_i. Then

$$B_2 = \gamma \sum_{i=1}^{s} \binom{j_i}{2}, \qquad n = \sum_{i=1}^{s} j_i.$$

The equation in P_1 for $r = 2$ is

$$\begin{aligned} \sum i^2A_i &= q^{k-1}\gamma n + q^{k-2}\gamma^2 n(n - 1) + B_2(q^{k-2}2) \\ &= q^{k-1}\gamma n + q^{k-2}\gamma^2 n(n - 1) + \gamma q^{k-2}2 \sum_{i=1}^{s} \binom{j_i}{2} \\ &= q^{k-1}\gamma n + q^{k-2}\gamma^2 n(n - 1) + \gamma q^{k-2} \left(\sum_{i=1}^{s} j_i^2 - \sum_{i=1}^{s} j_i \right) \\ &= q^{k-1}\gamma n - q^{k-2}\gamma^2 n - \gamma q^{k-2} \sum_{i=1}^{s} j_i + q^{k-2}\gamma^2 n^2 + q^{k-2}\gamma \sum_{i=1}^{s} j_i^2 \\ &= q^{k-2}\gamma n(q - \gamma - 1) + q^{k-2}\gamma^2 n^2 + q^{k-2}\gamma \sum_{i=1}^{s} j_i^2 \\ &= \gamma q^{k-2} \left(n^2\gamma + \sum_{i=1}^{s} j_i^2 \right) \end{aligned}$$

which is the Assmus-Mattson identity.

III. APPLICATIONS

THEOREM. *A unique solution to P_1 (hence also to B_1, B_2, P_2) exists under the following conditions. Only s A_i's are unknown and B_1, B_2, $\cdots$, B_{s-1} are known.*

PROOF: The first s equations in P_1 are s equations in s unknowns, whose coefficient matrix is van der Monde. Hence there is a unique solution for the unknown A_i's in the first s equations and each additional equation brings in exactly one additional B_i which can be uniquely solved for.

As an application of the preceding theorem consider the quadratic residue codes, for $q = 2$, of dimension $(p - 1)/2$, $p \equiv -1$ modulo 8. Let A denote such a code and B denote its orthogonal. Mattson-Solomon (1961) give the following bound on the minimum odd weight m in B, $m^2 - m + 1 \geqq p$. By remark (e) we know that the minimum weight in A is $m + 1$. For $p = 7, 23, 31$, or 47 this makes enough B_i's zero so that the conditions of the theorem are fulfilled and the unique solution is precisely the known weight distribution of these codes. Since Muir and Metzler (1930) give simple expressions for van der Monde determinants and also for determinants which are close to van der Monde, the computation of the weights is not arduous.

TABLE 1
WEIGHT DISTRIBUTIONS FOR SOME QUADRATIC RESIDUE CODES

	7	23	31	47	71
A_0	1	1	1	1	1
A_4	7				
A_8		22 × 23	15 × 31		
A_{12}		56 × 23	280 × 31	276 × 47	35 × 71
A_{16}		11 × 23	589 × 31	7590 × 47	2,345 × 71
A_{20}			168 × 31	49588 × 47	186,186 × 71
A_{24}			5 × 31	81720 × 47	4,340,910 × 71
A_{28}				35420 × 47	37,861,505 × 71
A_{32}				3795 × 47	129,893,225 × 71
A_{36}				92 × 47	181,404,764 × 71
A_{40}					103,914,580 × 71
A_{44}					24,093,685 × 71
A_{48}					2,170,455 × 71
A_{52}					71,610 × 71
A_{56}					670 × 71
A_{60}					7 × 71

As an example consider the $p = 23$ case. $m = 7$ so that the only nonzero A_i possible are A_8, A_{12}, and A_{16}. The first three equations in P_1 are

$$A_8 + A_{12} + A_{16} = 23 \times 89,$$

$$8A_8 + 12A_{12} + 16A_{16} = 2^{10} \times 23,$$

$$8^2A_8 + 12^2A_{12} + 16^2A_{16} = 2^9 \times 23 \times 24.$$

Hence $A_8 = 22 \times 23$, $A_{12} = 56 \times 23$, $A_{16} = 11 \times 23$.

When $p = 71$, the general known properties of quadratic residue codes together with the power moment identities do not yield a unique weight distribution for A, but do eliminate all except four possibilities. The correct weight distribution was then determined by examining sets of vectors in the code. All computations were done by hand. This new result is listed along with the cases $p = 7, 23, 31, 47$. These cases had been known previously, the calculation for $p = 47$ being due to Gleason (1961).

Acknowledgments

I wish to thank John Pierce for aid in solving P_1 and Eugene Prange for helpful suggestions and illuminating discussions.

Received: September 28, 1962; revised January 16, 1963

References

Assmus, E. F., and Mattson, H. F. (1961), Error correcting codes: An axiomatic approach. ARM No. 269, Applied Research Laboratory, Sylvania Electronic Systems, Waltham, Mass.

Gleason, A. M. (1961–1962), private communications.

Jordan, C. (1950), "Calculus of Finite Differences," 2nd ed. Chelsea, New York.

MacWilliams, F. J. (1962), A theorem on the distribution of weights in a systematic code. *Bell System Tech., J.* **42,** 79–94.

Mattson, H. F., and Solomon, G. (1961), A new treatment of Bose-Chaudhuri codes. *J. Soc. Ind. Appl. Math.* **9,** 654–669.

Muir, T., and Metzler, W. (1930), "Theory of Determinants." Privately published, Albany, N. Y.

Prange, E. (1961–1962), private communications.

Zeiler, N. (1962), A note on the mean square weight for group codes. *Inform. and Control* **5,** 87–89.

Reprinted from *Bell System Tech. J.*, **51**, 803–819 (1972)

23

The MacWilliams Identities for Nonlinear Codes

By Mrs. F. J. MacWILLIAMS, N. J. A. SLOANE, and J.-M. GOETHALS

(Manuscript received December 13, 1971)

In recent years a number of nonlinear codes have been discovered which have better error-correcting capabilities than any known linear codes. However, very little is known about the properties of such codes. In this paper we study the most basic property, the weight enumerator. The weight of a codeword is the number of its nonzero components; the weight enumerator gives the number of codewords of each weight, and is fundamental for obtaining the error probability when the code is used for error-correction on a noisy channel. In 1963 one of us showed that the weight enumerator of a linear code is related in a simple way to that of the dual code (Jessie MacWilliams, "A Theorem on the Distribution of Weights in a Systematic Code," Bell System Technical Journal, 42, No. 1 (January 1963), pp. 79–94). In the present paper, which is a sequel, we show that the same relationship holds for the weight enumerator of a nonlinear code. Furthermore, a definition is given for the dual $\mathcal{A}^{\perp}$ *of a nonlinear binary code* $\mathcal{A}$ *which satisfies* $(\mathcal{A}^{\perp})^{\perp} = \mathcal{A}$ *provided* $\mathcal{A}$ *contains the zero codeword.*

I. INTRODUCTION

In recent years a number of nonlinear codes have been discovered which have better error-correcting capabilities than any known linear codes (e.g., Refs. 1 and 2). However, very little is known about the properties of such codes. In this paper we study the most basic property, the Hamming weight enumerator (defined in Section II), which gives fundamental information about the error probability when the code is used in various error-correction schemes (Ref. 3, Ch. 16). In 1963 one of us showed that the Hamming and the complete weight enumerators of a linear code are related in a simple way to those of the dual code (Ref. 4; Theorems 1 and 3 below). The requirement that the code be linear is unsatisfactory for two reasons: (*i*) Several pairs of nonlinear codes $\mathcal{A}$, $\mathcal{B}$ are known whose weight enumerators satisfy Theorem 3. One example of such a pair is given by the Preparata[2] and Kerdock[1] codes, another by the code shown in Fig. 1. (*ii*) The important theorem of S. P. Lloyd (giving a necessary condition for the existence of a prefect code) may be deduced for linear codes as a corollary to Theorem 3 (Ref. 4, Lemma 2.15), but may be proved directly without assuming linearity (Ref. 5; Ref. 6, p. 111).

It is the purpose of the present paper, therefore, to define the "weight enumerators of the dual code" so as to make Theorems 1 and 3 (and the corresponding theorem for the Lee weight enumerator, Theorem 2) valid even for nonlinear codes.

Furthermore, if $\mathcal{A}$ is a nonlinear binary code which contains the zero codeword, we define the formal dual $\mathcal{A}^{\perp}$ so as to satisfy:

(*i*) $(\mathcal{A}^{\perp})^{\perp} = \mathcal{A}$,
(*ii*) if $\mathcal{A}$ is linear the two definitions of $\mathcal{A}^{\perp}$ agree.

The paper is arranged as follows. Section II states the three MacWilliams identities (Theorems 1, 2, 3). Section III treats the binary case, when the three theorems coincide. The formal dual of a nonlinear binary code is defined in Section 3.5. Section IV treats the general case, first proving Theorem 1 and then deducing Theorems 2, 3 from it. In Section V we discuss properties of the "weights of the dual code" $B(i)$. However, the problem of finding conditions for the $B(i)$ to be positive integers remains unsolved.

0	0	0	0	0	0	0	0	1	1	1	1	1	1	1	1
1	1	0	0	0	0	0	0	0	0	1	1	1	1	1	1
1	0	1	0	0	0	0	0	0	1	0	1	1	1	1	1
1	0	0	1	0	0	0	0	0	1	1	0	1	1	1	1
1	0	0	0	1	0	0	0	0	1	1	1	0	1	1	1
1	0	0	0	0	1	0	0	0	1	1	1	1	0	1	1
1	0	0	0	0	0	1	0	0	1	1	1	1	1	0	1
1	0	0	0	0	0	0	1	0	1	1	1	1	1	1	0

Fig. 1—The sixteen rows form a nonlinear code $\mathcal{A}$.

II. WEIGHT ENUMERATORS

Let F be a finite field $GF(q)$, where q is a prime power; and let F^n be a vector space of dimension n over F. A *linear code* $\mathcal{A}$ of length n over $GF(q)$ is a subspace of F^n, and $\mathcal{A}^\perp$ denotes the orthogonal subspace or *dual* code of $\mathcal{A}$. A code is *self-dual* if $\mathcal{A} = \mathcal{A}^\perp$. A *nonlinear* code is any subset of F^n. In this paper a code is linear unless stated otherwise.

We propose to describe the code vectors of a code $\mathcal{A}$ in three ways, giving progressively less information (but becoming progressively easier to handle).

2.1 *The Complete Weight Enumerator*

Let the elements of F be $\omega_0 = 0, \omega_1, \omega_2, \cdots, \omega_{q-1}$, in some fixed order. The *composition* of a vector $\mathbf{v} \in F^n$ is defined to be

$$\text{comp}(\mathbf{v}) = \mathbf{s} = (s_0, s_1, \cdots, s_{q-1}), \tag{1}$$

where $s_i = s_i(\mathbf{v})$ is the number of coordinates of $\mathbf{v}$ equal to ω_i. Clearly $\sum_{i=0}^{q-1} s_i = n$.

Let $A(\mathbf{t})$ be the number of vectors $\mathbf{v}$ in $\mathcal{A}$ with comp $(\mathbf{v}) = \mathbf{t}$. The set of integers $\{A(\mathbf{t})\}$ is the *complete weight enumerator* of $\mathcal{A}$.

The first MacWilliams identity relates the complete weight enumerators of $\mathcal{A}$ and $\mathcal{A}^\perp$. (Ref. 4, Lemma 2.7. See also Refs. 7 and 8.)

Theorem 1: If $\mathcal{A}$ is a linear code with complete weight enumerator $\{A(\mathbf{t})\}$, and its dual code $\mathcal{A}^\perp$ has complete weight enumerator $\{B(\mathbf{t})\}$, then

$$\sum_{\mathbf{s}} B(\mathbf{s}) z_0^{s_0} \cdots z_{q-1}^{s_{q-1}} = \frac{1}{|\mathcal{A}|} \sum_{\mathbf{t}} A(\mathbf{t}) \prod_{l=0}^{q-1} \left(\sum_{j=0}^{q-1} \mathcal{X}(\omega_j \omega_l) z_j \right)^{t_l} \tag{2}$$

where the z_i are indeterminates and $\mathcal{X}$ is a character on $GF(q)$ (defined in Section 4.2).

2.2 *The Lee Weight Enumerator*

For $q = 2$ this description coincides with the preceding, and for $q = 2^s$, $s > 1$ it is not defined; so in this section q is assumed to be an odd prime power.

For q prime, we wish to classify the coordinates of the code vectors by magnitude. For example, codewords over $GF(5) = \{0, 1, -1, 2, -2\}$ would be classified according to the number of components which are 0, the number which are ± 1, and the number which are ± 2 (but without regard to the actual number which are $1, -1, 2$, or -2).

In general, for q a prime power, let the elements of F be $\omega_0 = 0$, $\omega_1, \cdots, \omega_\delta, \omega_{-\delta}, \omega_{-\delta+1}, \cdots, \omega_{-1}$, where $\omega_{-i} = -\omega_i$ and $\delta = \frac{1}{2}(q-1)$.

Then the *Lee weight* of a vector $\mathbf{v} \in F^n$ is defined to be

$$\text{Lee}(\mathbf{v}) = (l_0, l_1, \cdots, l_\delta),$$

where $l_i = l_i(\mathbf{v})$ is the number of coordinates of $\mathbf{v}$ equal to either ω_i or $-\omega_i$. In the notation of eq. (1),

$$\begin{aligned} l_0(\mathbf{v}) &= s_0(\mathbf{v}) \\ l_i(\mathbf{v}) &= s_i(\mathbf{v}) + s_{-i}(\mathbf{v}) \quad \text{for} \quad i = 1, \cdots, \delta. \end{aligned} \tag{3}$$

Let $A^L(\mathbf{t})$ be the number of vectors $\mathbf{v}$ in $\mathcal{A}$ with Lee $(\mathbf{v}) = \mathbf{t}$; so that $\{A^L(\mathbf{t})\}$ is the *Lee weight enumerator* of $\mathcal{A}$.

The second MacWilliams identity relates the Lee weight enumerators of $\mathcal{A}$ and $\mathcal{A}^\perp$:

Theorem 2:

$$\sum_{\mathbf{s}} B^L(\mathbf{s}) z_0^{s_0} \cdots z_\delta^{s_\delta}$$

$$= \frac{1}{|\mathcal{A}|} \sum_{\mathbf{t}} A^L(\mathbf{t}) \prod_{l=0}^{\delta} \left(z_0 + \sum_{j=1}^{\delta} (\mathcal{X}(\omega_j \omega_l) + \mathcal{X}(-\omega_j \omega_l)) z_j \right)^{t_l}, \tag{4}$$

where $\{B^L(\mathbf{s})\}$ *is the Lee weight enumerator for* $\mathcal{A}^\perp$.

(Theorem 2 is believed to be new.) The Lee enumerator is important both because it is an appropriate measure for codes to be used in phase-modulation communication schemes (see Lee, Ref. 9; Berlekamp, Ref. 3, p. 205) and as a compromise in giving much more information than the Hamming enumerator, yet requiring only half as many variables as the complete enumerator.

2.3 *The (Hamming) Weight Enumerator*

For the rest of the paper let q be any prime power.

The (Hamming) *weight* of a vector $\mathbf{v}$, $wt(\mathbf{v})$, is the number of its nonzero coordinates, so that

$$wt(\mathbf{v}) = \sum_{i=1}^{q-1} s_i(\mathbf{v}). \tag{5}$$

Let $\mathcal{A}$ be a linear code of length n over $GF(q)$, and let $A(i)$ be the number of vectors $\mathbf{v}$ in $\mathcal{A}$ with $wt(\mathbf{v}) = i$. Then $\{A(i)\}$ is the (Hamming or ordinary) *weight enumerator* of $\mathcal{A}$. Similarly $\{B(i)\}$ denotes the weight enumerator of the dual code $\mathcal{A}^\perp$. The third MacWilliams identity (Ref. 4, Theorem 1) relates $\{A(i)\}$ and $\{B(i)\}$:

Theorem 3:

$$\sum_{i=0}^{n} B(i)z^i = \frac{1}{|\mathcal{A}|} \sum_{i=0}^{n} A(i)(1 + (q - 1)z)^{n-i}(1 - z)^i. \quad (6)$$

2.4 *An Example*

Let $\mathcal{A}$ be the self-dual code of length 2 over $GF(5)$ consisting of the code vectors 0 0, 1 2, 2 −1, −2 1, −1 −2.

The complete, Lee, and Hamming weight enumerators are, respectively,

$$A(20000) = A(01100) = A(01001) = A(01010) = A(00011) = 1,$$

$$A^L(200) = 1, \qquad A^L(011) = 4,$$

and

$$A(0) = 1, \qquad A(2) = 4.$$

In this case, $\mathfrak{X}(\omega_j\omega_l) = \alpha^{jl}$ where $\alpha = e^{(2\pi i)/5} = \cos 72° + i \sin 72°$. Theorems 1, 2, 3 assert (correctly) that

$$\begin{aligned} z_0^2 + z_1z_2 + z_2z_{-1} + z_1z_{-2} + z_{-2}z_{-1} = \tfrac{1}{5}[&(z_0 + z_1 + z_2 + z_{-2} + z_{-1})^2 \\ &+ (z_0 + \alpha z_1 + \alpha^2 z_2 + \alpha^3 z_{-2} + \alpha^4 z_{-1})(z_0 + \alpha^2 z_1 + \alpha^4 z_2 + \alpha z_{-2} + \alpha^3 z_{-1}) \\ &+ (z_0 + \alpha^2 z_1 + \alpha^4 z_2 + \alpha z_{-2} + \alpha^3 z_{-1})(z_0 + \alpha^4 z_1 + \alpha^3 z_2 + \alpha^2 z_{-2} + \alpha z_{-1}) \\ &+ (z_0 + \alpha z_1 + \alpha^2 z_2 + \alpha^3 z_{-2} + \alpha^4 z_{-1})(z_0 + \alpha^3 z_1 + \alpha z_2 + \alpha^4 z_{-2} + \alpha^2 z_{-1}) \\ &+ (z_0 + \alpha^3 z_1 + \alpha z_2 + \alpha^4 z_{-2} + \alpha^2 z_{-1})(z_0 + \alpha^4 z_1 + \alpha^3 z_2 + \alpha^2 z_{-2} + \alpha z_{-1})], \end{aligned}$$

that

$$\begin{aligned} z_0^2 + 4z_1z_2 = \tfrac{1}{5}[&(z_0 + 2z_1 + 2z_2)^2 \\ &+ 4(z_0 + (\alpha + \alpha^4)z_1 + (\alpha^2 + \alpha^3)z_2)(z_0 + (\alpha^2 + \alpha^3)z_1 + (\alpha + \alpha^4)z_2)], \end{aligned}$$

and that

$$1 + 4z^2 = \tfrac{1}{5}[(1 + 4z)^2 + 4(1 - z)^2].$$

III. THE BINARY CASE

All the codes in this section are binary, so that Theorems 1 and 2 coincide with Theorem 3.

3.1 *Preliminaries*

Let $F = GF(2)$; let F^n be a vector space of dimension n over F. For purposes of notation we define a group G which is a multiplicative copy of F^n, as follows. Let $x_1, \cdots, x_n$ be indeterminates satisfying $x_i^2 = 1$ and $x_i x_j = x_j x_i$ for $i, j = 1, \cdots, n$. Then G is the multiplicative group consisting of all products $x_1^{v_1} x_2^{v_2} \cdots x_n^{v_n}$ where v_i is 0 or 1. To each vector

$$\mathbf{v} = (v_1, v_2, \cdots, v_n)$$

in F^n we associate the element

$$x^{\mathbf{v}} = x_1^{v_1} x_2^{v_2} \cdots x_n^{v_n}$$

of G. Thus F^n and G are isomorphic, and addition of vectors in F^n corresponds to multiplication in G.

3.2 *Characters*

Let $\mathfrak{X}_{\mathbf{u}}$, $\mathbf{u} \in F^n$, be a character of G given by

$$\mathfrak{X}_{\mathbf{u}}(x^{\mathbf{v}}) = (-1)^a,$$

where $a = \mathbf{u}\mathbf{v}^T$ is the scalar product of $\mathbf{u}$, $\mathbf{v}$ in $GF(2)$.

Let σ_i be the set of vectors of F^n of weight i. Clearly,

$$|\sigma_i| = \binom{n}{i}.$$

Let

$$X_i = \sum_{\mathbf{v} \in \sigma_i} x^{\mathbf{v}}.$$

(For example, $X_1 = x_1 + x_2 + \cdots + x_n$.) X_i is an element of the group algebra QG of G over the field of rational numbers Q.

$\mathfrak{X}_{\mathbf{u}}$ is extended linearly to elements of QG, for example,

$$\mathfrak{X}_{\mathbf{u}}(X_i) = \sum_{\mathbf{v} \in \sigma_i} \mathfrak{X}_{\mathbf{u}}(x^{\mathbf{v}}).$$

Note that $\mathfrak{X}_{\mathbf{u}}(X_i)$ is a rational integer, not an element of $GF(2)$.

Let S_n be the group of all permutations of n symbols, i.e., the group of all $n \times n$ permutation matrices. $\mathbf{v}\pi$ is the vector obtained from $\mathbf{v}$ by multiplying by the permutation matrix π.

Lemma 3.1:

$$\mathfrak{X}_{\mathbf{u}\pi}(x^{\mathbf{v}}) = \mathfrak{X}_{\mathbf{u}}(x^{\mathbf{v}\pi^T}) \quad \textit{for any } \pi \textit{ in } S_n.$$

Proof:

$$\mathfrak{X}_{\mathbf{u}\pi}(x^{\mathbf{v}}) = (-1)^a,$$

$$a = \mathbf{u}\pi\mathbf{v}^T = \mathbf{u}(\mathbf{v}\pi^T)^T. \qquad \text{Q.E.D.}$$

3.3 *Krawtchouk Polynomials*

The *Krawtchouk polynomial* $P_s(i)$ (a polynomial in s) is defined by

$$(1 + z)^{n-s}(1 - z)^s = \sum_{i=0}^{n} P_s(i)z^i, \tag{7}$$

so that

$$P_s(i) = \sum_{r=0}^{\min(i,s)} (-1)^r \binom{s}{r}\binom{n-s}{i-r} \qquad i = 0, \cdots, n. \tag{8}$$

It follows from the definition that

$$\sum_{i=0}^{n} P_s(i) = 2^n\, \delta_{s,0}\,. \tag{9}$$

Other properties may be found in Refs. 10 and 11.

Let J_s be the vector with $v_1 = v_2 = \cdots = v_s = 1$ and $v_{s+1} = \cdots = v_n = 0$.

Lemma 3.2: If $\mathbf{u}$ *has weight* s,

$$\mathfrak{X}_{\mathbf{u}}(X_i) = P_s(i).$$

Proof: Since X_i is clearly invariant under any permutation in S_n we may suppose, by (3.1), that $\mathbf{u} = J_s$.

Consider the formal sum

$$\sum_{i=0}^{n} \mathfrak{X}_{J_s}(X_i)z^i = \mathfrak{X}_{J_s}\left(\sum_{i=0}^{n} X_i z^i\right).$$

Now

$$\sum_{i=0}^{n} X_i z^i = \prod_{j=1}^{n} (1 + x_j z), \tag{10}$$

and

$$\mathfrak{X}_{J_s}(1 + x_j z) = \begin{cases} 1 - z & \text{if } j = 1, \cdots, s, \\ 1 + z & \text{if } j = s + 1, \cdots, n. \end{cases}$$

Thus

$$\sum_{i=0}^{n} \mathfrak{X}_{J_s}(X_i)z^i = (1 + z)^{n-s}(1 - z)^s. \qquad \text{Q.E.D.}$$

Lemma 3.3:

$$\binom{n}{i} P_i(s) = \binom{n}{s} P_s(i).$$

Proof: By rearranging the binomial coefficients in eq. (8).

3.4 *Definition of B(i) and Proof of Theorem 3*

Let $\mathcal{C}$ be an arbitrary (linear or nonlinear) code, i.e., any subset of F^n; let $A(i)$ be the number of vectors in $\mathcal{C}$ of weight i. Define

$$\boldsymbol{\mathcal{C}} = \sum_{\mathbf{v} \varepsilon \mathcal{C}} x^{\mathbf{v}};$$

$\boldsymbol{\mathcal{C}}$ is an element of QG. Corresponding to $\mathcal{C}$ we define numbers $B(i)$, $i = 0, 1, \cdots, n$, by

$$B(i) = \frac{1}{|\mathcal{C}|} \sum_{\mathbf{u} \varepsilon \sigma_i} \mathfrak{X}_{\mathbf{u}}(\boldsymbol{\mathcal{C}}). \tag{11}$$

Note that $B(i)$ is a rational number, perhaps negative.

With this definition of $B(i)$ we can now prove the binary version of Theorem 3, as follows. Define

$$\boldsymbol{\mathcal{C}}^{\pi} = \sum_{\mathbf{v} \varepsilon \mathcal{C}} x^{\mathbf{v}\pi}.$$

We average $\boldsymbol{\mathcal{C}}$ over all equivalent codes $\boldsymbol{\mathcal{C}}^{\pi}$:

Lemma 3.4:

$$\sum_{\pi \varepsilon S_n} \boldsymbol{\mathcal{C}}^{\pi} = \sum_{i=0}^{n} A(i) i!\, (n - i)!\, X_i .$$

Proof: Let $\mathbf{v}$ be a vector of weight i in $\mathcal{C}$. The $i!$ permutations of the nonzero symbols of $\mathbf{v}$ leave $\mathbf{v}$ unchanged, as do the $(n - i)!$ permutations of the places in which $\mathbf{v}$ contains zero. Thus

$$\sum_{\pi \varepsilon S_n} x^{\mathbf{v}\pi} = i!\, (n - i)!\, X_i . \qquad \text{Q.E.D.}$$

Lemma 3.5:

$$B(j) = \frac{1}{|\mathcal{C}|} \frac{1}{j!\, (n - j)!} \sum_{\pi \varepsilon S_n} \mathfrak{X}_{J_j \pi}(\boldsymbol{\mathcal{C}}).$$

Proof: As π runs through S_n, $J_j\pi$ runs through $j!(n - j)!$ copies of σ_j.

Q.E.D.

Proof of Theorem 3: By (3.5), (3.1):

$$B(j) = \frac{1}{|\mathcal{A}|} \frac{1}{j!\,(n-j)!} \mathfrak{X}_{J_j}\Big(\sum_{\pi \varepsilon S_n} \mathcal{A}^{\pi}\Big)$$

$$= \frac{1}{|\mathcal{A}|} \frac{1}{j!\,(n-j)!} \mathfrak{X}_{J_j}\Big(\sum_{i=0}^{n} A(i) i!\,(n-i)!\, X_i\Big) \quad \text{by (3.4)},$$

$$= \frac{1}{|\mathcal{A}|} \sum_{i=0}^{n} A(i) \frac{i!\,(n-i)!}{j!\,(n-j)!} P_j(i) \quad \text{by (3.2)},$$

$$= \frac{1}{|\mathcal{A}|} \sum_{i=0}^{n} A(i) P_i(j) \quad \text{by (3.3)}.$$

Multiply both sides by z^j and sum on j:

$$\sum_{j=0}^{n} B(j) z^j = \frac{1}{|\mathcal{A}|} \sum_{i=0}^{n} A(i) \sum_{j=0}^{n} P_i(j) z^j$$

$$= \frac{1}{|\mathcal{A}|} \sum_{i=0}^{n} A(i)(1+z)^{n-i}(1-z)^{i}. \qquad \text{Q.E.D.}$$

In the next section we show that in the case $\mathcal{A}$ is linear, $B(i)$ is the usual weight distribution of the dual code.

3.5 *The Dual Code*

If $\boldsymbol{\mathcal{A}} = \sum_{\mathbf{v} \varepsilon F^n} \alpha_{\mathbf{v}} x^{\mathbf{v}}$, $\alpha_{\mathbf{v}} \varepsilon Q$, is any element of QG for which $A(0) = 1$, we define its formal weight distribution to be $\{A(i)\}$, where

$$A(i) = \sum_{\mathbf{v} \varepsilon \sigma_i} \alpha_{\mathbf{v}}, \tag{12}$$

$$|\boldsymbol{\mathcal{A}}| = \sum_{i=0}^{n} A(i), \tag{13}$$

and its formal dual to be

$$\boldsymbol{\mathcal{A}}^{\perp} = \frac{1}{|\boldsymbol{\mathcal{A}}|} \sum_{\mathbf{u} \varepsilon F^n} \mathfrak{X}_{\mathbf{u}}(\boldsymbol{\mathcal{A}}) x^{\mathbf{u}}. \tag{14}$$

It follows from (12) that the formal weight distribution of $\boldsymbol{\mathcal{A}}^{\perp}$ is $\{B(i)\}$, where

$$B(i) = \frac{1}{|\boldsymbol{\mathcal{A}}|} \sum_{\mathbf{u} \varepsilon \sigma_i} \mathfrak{X}_{\mathbf{u}}(\boldsymbol{\mathcal{A}}). \tag{11'}$$

If $\boldsymbol{\mathcal{A}}$ is a linear or nonlinear code, then clearly (12), (13) give the usual weight distribution and total number of codewords, and eq. (11′) for $B(i)$ coincides with eq. (11) of Section 3.4.

Theorem 4: If $\mathfrak{A}$ is a linear code, then the expressions (14), (11′) for its dual code and weight distribution of dual code, coincide with the usual definitions.

Proof: If $\mathbf{u}$ is in the dual subspace to $\mathcal{A}$, then $\mathcal{X}_{\mathbf{u}}(x^{\mathbf{v}}) = 1$ for all $\mathbf{v} \in \mathcal{A}$, so $\mathcal{X}_{\mathbf{u}}(\mathfrak{A}) = |\mathcal{A}|$. If $\mathbf{u} \notin \mathcal{A}^{\perp}$, then $\mathbf{u}\mathbf{v}^T \equiv 1$ (modulo 2) for exactly half the vectors $\mathbf{v} \in \mathcal{A}$, so

$$\mathcal{X}_{\mathbf{u}}(\mathfrak{A}) = 0 \quad \text{for} \quad \mathbf{u} \notin \mathcal{A}^{\perp}.$$

Therefore from (14),

$$\mathfrak{A}^{\perp} = \frac{1}{|\mathcal{A}|} \sum_{\mathbf{u} \in \mathcal{A}^{\perp}} x^{\mathbf{u}}. \qquad \text{Q.E.D.}$$

Combining Theorem 4 with the results of the last section, we have completed the proof of Theorem 3 for binary linear codes.

Theorem 5: Let $\mathfrak{A} = \sum_{\mathbf{v} \in F^n} \alpha_{\mathbf{v}} x^{\mathbf{v}}$, $\alpha_{\mathbf{v}} \in Q$, be any element of QG for which $A(0) = 1$, with formal dual $\mathfrak{A}^{\perp}$ given by eq. (14). Then

(*i*) $|\mathfrak{A}|\,|\mathfrak{A}^{\perp}| = 2^n$,
(*ii*) $(\mathfrak{A}^{\perp})^{\perp} = \mathfrak{A}$.

(Note that by the earlier remarks this theorem includes linear and nonlinear binary codes as a special case.)

Proof: (*i*) Set $z = 1$ in Theorem 3.
(*ii*) From (14), $(\mathfrak{A}^{\perp})^{\perp} = \sum_{\mathbf{u} \in F^n} \beta_{\mathbf{u}} x^{\mathbf{u}}$, where

$$\begin{aligned}
\beta_{\mathbf{u}} &= \frac{1}{|\mathfrak{A}^{\perp}|} \mathcal{X}_{\mathbf{u}}(\mathfrak{A}^{\perp}), \\
&= \frac{1}{2^n} \sum_{\mathbf{v} \in F^n} \mathcal{X}_{\mathbf{v}}(\mathfrak{A}) \mathcal{X}_{\mathbf{u}}(x^{\mathbf{v}}) \quad \text{by } (i), (14), \\
&= \frac{1}{2^n} \sum_{\mathbf{v} \in F^n} \mathcal{X}_{\mathbf{v}}\Big(\sum_{\mathbf{w} \in F^n} \alpha_{\mathbf{w}} x^{\mathbf{w}} \Big) \mathcal{X}_{\mathbf{u}}(x^{\mathbf{v}}), \\
&= \frac{1}{2^n} \sum_{\mathbf{w} \in F^n} \alpha_{\mathbf{w}} \sum_{\mathbf{v} \in F^n} (-1)^{\mathbf{v}(\mathbf{u}+\mathbf{w})^T}, \\
&= \frac{1}{2^n} (2^n \alpha_{\mathbf{u}}),
\end{aligned}$$

since the innermost sum is zero unless $\mathbf{u} = \mathbf{w}$. Q.E.D.

Remarks: In spite of Theorem 5, eq. (14) is not always a satisfactory definition of the dual of a nonlinear code, even in the binary case.

For example, Fig. 1 shows a nonlinear code with weight distribution $A(0) = A(8) = 1, A(2) = A(6) = 7$, and

$$\mathcal{A} = 1 + x_1(x_2 + x_3 + \cdots + x_8) + x_1 \cdots x_8\left(1 + \frac{1}{x_1}\left(\frac{1}{x_2} + \cdots + \frac{1}{x_8}\right)\right).$$

When the weight distribution is substituted in the right-hand side of the MacWilliams identity (6), $B(i)$ is found to be the same as $A(i)$ (Ref. 4, bottom of p. 82) so that this code is in some sense self-dual. However, although eq. (11) correctly gives the weight distribution $B(0) = B(8) = 1, B(2) = B(6) = 7$, eq. (14) gives

$$\mathcal{A}^{\perp} = 1 - \tfrac{1}{2}x_1(x_2 + x_3 + \cdots + x_8) + \tfrac{1}{2} \sum_{2 \leq i < j \leq 8} x_i x_j + \cdots$$

which seems unsatisfactory. A better definition of the dual of a nonlinear code has recently been given by P. Delsarte and J. -M. Goethals (private communication).

IV. THE GENERAL CASE

4.1 *Preliminaries*

Let $q = p^f$, $f \geq 1$, where p is prime; and let $F = GF(q) = \{\omega_0 = 0, \omega_1, \cdots, \omega_{q-1}\}$. Let $x_i^{(\omega_i)}$ be commuting indeterminates satisfying

$$x_i^{(\omega_j)} x_i^{(\omega_k)} = x_i^{(\omega_j + \omega_k)};$$

and let G be the multiplicative group consisting of all products $x_1^{(v_1)} x_2^{(v_2)} \cdots x_n^{(v_n)}$, $v_i \in F$. To each vector $\mathbf{v} = (v_1, \cdots, v_n)$ in F^n we associate the element $x^{(\mathbf{v})} = x_1^{(v_1)} \cdots x_n^{(v_n)}$ of G; as in Section 3.1, G is a multiplicative copy of F^n. Let $\mathcal{C}G$ be the group algebra of G over the complex numbers.

4.2 *Characters*

Let $p(x)$ be a primitive irreducible polynomial of degree f over $GF(p)$, and let α be a root of $p(x)$. Then any element $\lambda \in GF(q)$ has the canonical representation

$$\lambda = \lambda_0 + \lambda_1\alpha + \lambda_2\alpha^2 + \cdots + \lambda_{f-1}\alpha^{f-1}, \qquad \lambda_i \in GF(p).$$

If $GF(q)$ is considered as an additive group, it forms an abelian group, denoted by $(GF(q), +)$, which is isomorphic to the direct product of f copies of $GF(p)$; the isomorphism being given for example by

$$\lambda \leftrightarrow (\lambda_0, \lambda_1, \cdots, \lambda_{f-1}).$$

A character $\mathfrak{X}$ on $GF(q)$ is a homomorphism from $(GF(q), +)$ to the multiplicative group of the complex numbers. Define a fixed character on $GF(q)$ by

$$\mathfrak{X}(\lambda) = \xi^{\lambda_0}$$

where $\xi = e^{(2\pi i)/p}$, and

$$\mathfrak{X}(\lambda + \mu) = \xi^{\lambda_0 + \mu_0}.$$

All characters on $GF(q)$ are now given by

$$\mathfrak{X}_\nu(\lambda) = \mathfrak{X}(\lambda\nu), \qquad \text{all } \nu \in GF(q).$$

All of the following depends on the choices of $p(x)$, α, and $\mathfrak{X}$; this dependence on coordinatization seems inevitable in studying codes over $GF(q)$.

Define a character $\mathfrak{X}_{\mathbf{u}}$ on G by

$$\mathfrak{X}_{\mathbf{u}}(x^{\mathbf{v}}) = \mathfrak{X}(\mathbf{u}\mathbf{v}^T) = \mathfrak{X}\left(\sum_{i=1}^{n} u_i v_i\right) \tag{15}$$

where $\sum_{i=1}^{n} u_i v_i \in GF(q)$. These characters form a group isomorphic to G (and to F^n): $\mathfrak{X}_{\mathbf{u}} \leftrightarrow x^{\mathbf{u}}$. We extend $\mathfrak{X}_{\mathbf{u}}$ to $\mathcal{C}G$ by linearity.

Lemma 4.1:

$$\mathfrak{X}_{\mathbf{u}\pi}(x^{(\mathbf{v})}) = \mathfrak{X}_{\mathbf{u}}(x^{(\mathbf{v}\pi^T)}) \quad \text{for any } \pi \in S_n .$$

The proof is straightforward and is omitted.

4.3 *Generalized Krawtchouk Polynomials.*

Let $\mathbf{s} = (s_0, s_1, \cdots, s_{q-1})$, $\mathbf{t} = (t_0, t_1, \cdots, t_{q-1})$ be compositions as defined in Section 2.1. The *generalized Krawtchouk polynomial* $P_{\mathbf{s}}(\mathbf{t})$ is defined by

$$\prod_{l=0}^{q-1} \left(\sum_{i=0}^{q-1} \mathfrak{X}(\omega_i \omega_l) z_i\right)^{s_l} = \sum_{\mathbf{t}} P_{\mathbf{s}}(\mathbf{t}) z_0^{t_0} z_1^{t_1} \cdots z_{q-1}^{t_{q-1}}. \tag{16}$$

$$\text{Let } X_{\mathbf{t}} = \sum_{\substack{\mathbf{v} \in F^n \\ \text{comp}(\mathbf{v}) = \mathbf{t}}} x^{\mathbf{v}}.$$

Lemma 4.2:

$$\prod_{k=1}^{n} \sum_{i=0}^{q-1} x_k^{(\omega_i)} z_i = \sum_{\mathbf{t}} X_{\mathbf{t}} z_0^{t_0} z_1^{t_1} \cdots z_{q-1}^{t_{q-1}}.$$

This is a straightforward generalization of eq. (10). For example, expand the product ($n = 3$, $q = 4$)

$$(x_1^{(\omega_0)} z_0 + x_1^{(\omega_1)} z_1 + x_1^{(\omega_2)} z_2 + x_1^{(\omega_3)} z_3)$$
$$\cdot(x_2^{(\omega_0)} z_0 + x_2^{(\omega_1)} z_1 + x_2^{(\omega_2)} z_2 + x_2^{(\omega_3)} z_3)$$
$$\cdot(x_3^{(\omega_0)} z_0 + x_3^{(\omega_1)} z_1 + x_3^{(\omega_2)} z_2 + x_3^{(\omega_3)} z_3).$$

Lemma 4.3: For any composition $\mathbf{s}$ *let*

$$\mathbf{u} = (\underbrace{\omega_0\omega_0 \cdots \omega_0}_{s_0}\underbrace{\omega_1\omega_1 \cdots \omega_1}_{s_1} \cdots \underbrace{\omega_{q-1}\omega_{q-1} \cdots \omega_{q-1}}_{s_{q-1}})$$

so that comp $(\mathbf{u}) = \mathbf{s}$. *Then*

$$\mathfrak{X}_{\mathbf{u}}(X_{\mathbf{t}}) = P_{\mathbf{s}}(\mathbf{t}).$$

Proof: Consider the formal sum

$$\sum_{\mathbf{t}} \mathfrak{X}_{\mathbf{u}}(X_{\mathbf{t}}) z_0^{t_0} \cdots z_{q-1}^{t_{q-1}} = \mathfrak{X}_{\mathbf{u}}\left(\prod_{k=0}^{n} \sum_{i=0}^{q-1} x_k^{(\omega_i)} z_i\right) \quad \text{by (4.2)},$$
$$= \prod_{k=1}^{n} \sum_{i=0}^{q-1} \mathfrak{X}_{\mathbf{u}}(x_k^{(\omega_i)}) z_i$$
$$= \prod_{k=1}^{n} \sum_{i=0}^{q-1} \mathfrak{X}(u_k \omega_i) z_i \quad \text{by eq. (15)},$$
$$= \prod_{l=0}^{q-1} \left(\sum_{i=0}^{q-1} \mathfrak{X}(\omega_i \omega_l) z_i\right)^{s_l} \quad \text{by the form of } \mathbf{u},$$
$$= \sum_{\mathbf{t}} P_{\mathbf{s}}(\mathbf{t}) z_0^{t_0} \cdots z_{q-1}^{t_{q-1}}$$

by eq. (16). Q.E.D.

For a composition $\mathbf{s}$, let $\binom{n}{\mathbf{s}}$ denote the multinomial coefficient $n!/(s_0!s_1! \cdots s_{q-1}!)$.

Lemma 4.4:

$$\binom{n}{\mathbf{s}} P_{\mathbf{s}}(\mathbf{t}) = \binom{n}{\mathbf{t}} P_{\mathbf{t}}(\mathbf{s}).$$

Proof: Set $\alpha_l = \sum_{i=0}^{q-1} \mathfrak{X}(\omega_i \omega_l) z_i$, so (16) becomes

$$\prod_{l=0}^{q-1} \alpha_l^{s_l} = \sum_{\mathbf{t}} P_{\mathbf{s}}(\mathbf{t}) \prod_i z_i^{t_i}.$$

Multiply by $\prod_{l=0}^{q-1} \binom{n}{\mathbf{s}} y_l^{s_l}$ and sum on $\mathbf{s}$:

$$\sum_{\mathbf{s}} \binom{n}{\mathbf{s}} \prod_{l=0}^{q-1} (\alpha_l y_l)^{s_l} = \sum_{\mathbf{s},\mathbf{t}} \binom{n}{\mathbf{s}} P_{\mathbf{s}}(\mathbf{t}) \prod_i z_i^{t_i} \prod_l y_l^{s_l}. \tag{17}$$

The left-hand side is

$$(\alpha_0 y_0 + \alpha_1 y_1 + \cdots + \alpha_{q-1} y_{q-1})^n$$

which rearranged becomes

$$(\beta_0 z_0 + \beta_1 z_1 + \cdots + \beta_{q-1} y_{q-1})^n, \tag{18}$$

where

$$\beta_i = \sum_{l=0}^{q-1} \mathfrak{X}(\omega_i \omega_l) y_l .$$

Expanding (18) we get

$$\sum_{\mathbf{t}} \binom{n}{\mathbf{t}} \prod_{i=0}^{q-1} \left(\sum_{l=0}^{q-1} \mathfrak{X}(\omega_i \omega_l) y_l \right)^{t_i} z_i^{t_i} = \sum_{\mathbf{s},\mathbf{t}} \binom{n}{\mathbf{t}} \sum_{\mathbf{s}} P_{\mathbf{t}}(\mathbf{s}) \prod_l y_l^{s_l} \prod_i z_i^{t_i}. \tag{19}$$

Equating coefficients in (17), (19) gives the result. Q.E.D.

4.4 *Definition of* $B(\mathbf{s})$ *and Proof of Theorem 1*

As in Section 2.1, let $\mathcal{A}$ be any code in F^n, with complete weight enumerator $\{A(\mathbf{t})\}$; and let

$$\boldsymbol{\mathcal{A}} = \sum_{\mathbf{v} \varepsilon \mathcal{A}} x^{\mathbf{v}}$$

be the corresponding element of $\mathcal{C}G$. For each composition $\mathbf{s}$ define

$$B(\mathbf{s}) = \frac{1}{|\mathcal{A}|} \sum_{\substack{\mathbf{u} \varepsilon F^n \\ \text{comp}(\mathbf{u}) = \mathbf{s}}} \mathfrak{X}_{\mathbf{u}}(\boldsymbol{\mathcal{A}}). \tag{20}$$

In general $B(\mathbf{s})$ is a complex number. With this definition of $B(\mathbf{s})$ we can now prove Theorem 1.

Remark: If $\mathcal{A}$ is a linear code it follows immediately (as in the proof of Theorem 4) that $\{B(\mathbf{s})\}$ is the composition of the dual code to $\mathcal{A}$.

We first average $\boldsymbol{\mathcal{A}}$ over all equivalent codes. For a vector $\mathbf{u}$ of composition t,

$$\sum_{\pi \varepsilon S_n} x^{\mathbf{u}\pi} = \prod_{i=0}^{q-1} (t_i!) \sum_{\substack{\mathbf{v} \varepsilon F^n \\ \text{comp}(\mathbf{v}) = \mathbf{t}}} x^{\mathbf{v}}.$$

Set $d(\mathbf{t}) = \prod_{i=1}^{q-1} (t_i!)$. Then

$$\sum_{\pi \varepsilon S_n} \boldsymbol{\mathcal{A}}^{\pi} = \sum_{\mathbf{t}} d(\mathbf{t}) A(\mathbf{t}) X_{\mathbf{t}} . \tag{21}$$

Proof of Theorem 1:

From eq. (20),

$$|\mathcal{Q}|\, B(\mathbf{s}) = \sum_{\substack{\mathbf{u}\in F^n \\ \mathrm{comp}\,(\mathbf{u})=\mathbf{s}}} \mathfrak{X}_{\mathbf{u}}(\mathbf{Q})$$

$$= \frac{1}{d(\mathbf{s})} \sum_{\pi \in S_n} \mathfrak{X}_{\mathbf{u}\pi}(\mathbf{Q})$$

$$= \frac{1}{d(\mathbf{s})} \mathfrak{X}_{\mathbf{u}}\Big(\sum_{\pi \in S_n} \mathbf{Q}^{\pi}\Big) \quad \text{by (4.1)},$$

[**u** is now the vector defined in Lemma (4.3)],

$$= \frac{1}{d(\mathbf{s})} \sum_{\mathbf{t}} d(\mathbf{t}) A(\mathbf{t}) \mathfrak{X}_{\mathbf{u}}(X_{t}) \qquad \text{by (21)},$$

$$= \sum_{t} \frac{d(\mathbf{t})}{d(\mathbf{s})} A(\mathbf{t}) P_{\mathbf{s}}(\mathbf{t}) \qquad \text{by (4.3)},$$

$$= \sum_{\mathbf{t}} P_{\mathbf{t}}(\mathbf{s}) A(\mathbf{t}) \qquad \text{by (4.4)}.$$

Multiply both sides by $z_0^{s_0} \cdots z_{q-1}^{s_{q-1}}$ and sum over all compositions **s**.

Q.E.D.

4.5 *Proofs of Theorems 2 and 3.*

We use the notation of Sections 2.2 and 2.3.

Proof of Theorem 2:

In eq. (2) replace z_i by z_i for $1 \leqq i \leqq \delta$. Then using eq. (3), we see that eq. (2) collapses into eq. (4). Q.E.D.

Proof of Theorem 3:

In eq. (2) set $z_0 = 1$, $z_i = z$ for $i \neq 0$, and use eq. (5) to obtain (6).

Q.E.D.

V. DISCUSSION

We return to the binary case, which is easier to visualize.

The Hamming distance between vectors **u**, **v** is the weight of **u** + **v** (the weight of **u** − **v** if not binary). Coding theorists are interested in the distance structure of a code, not just in its weight structure. For linear codes, these are the same; they may also be the same for nonlinear codes, as in the example in Fig. 1. The following lemma is obvious.

Lemma 5.1: The distance and weight structure of a code $\mathcal{A}$ *are the same if and only if the weight structure of* $\mathcal{A} + \mathbf{v}$ *is the same as that of* $\mathcal{A}$ *for all* $\mathbf{v} \in \mathcal{A}$.

A code of this type will be said to have property 5.1. From now on we restrict ourselves to such codes.

A code with property 5.1 clearly contains the vector **0**. The element of QG corresponding to $\mathcal{A} + \mathbf{v}$ is $\boldsymbol{\mathcal{A}} x^{\mathbf{v}}$.

Property 5.1 implies that

$$|\mathcal{A}|\, B(s) = \sum_{\mathbf{u} \in \sigma_s} \mathfrak{X}_{\mathbf{u}}(\boldsymbol{\mathcal{A}}) = \sum_{\mathbf{u} \in \sigma_s} \mathfrak{X}_{u}(\boldsymbol{\mathcal{A}} x^{\mathbf{v}}) \quad \text{for} \quad \mathbf{v} \in \mathcal{A}.$$

Lemma 5.2: Property 5.1 implies that $B(s) \geqq 0$.

Proof: Take the sum over all $\mathbf{v} \in \mathcal{A}$ of the equation

$$|\mathcal{A}|\, B(s) = \sum_{\mathbf{u} \in \sigma_s} \mathfrak{X}_{\mathbf{u}}(\boldsymbol{\mathcal{A}} x^{\mathbf{v}}).$$

$$|\mathcal{A}|^2\, B(s) = \sum_{\mathbf{u} \in \sigma_s} \mathfrak{X}_{\mathbf{u}} \sum_{\mathbf{v} \in \mathcal{A}} (\boldsymbol{\mathcal{A}} x^{\mathbf{v}})$$

$$= \sum_{\mathbf{u} \in \sigma_s} \mathfrak{X}_{\mathbf{u}}(\mathcal{A}) \sum_{\mathbf{v} \in \mathcal{A}} \mathfrak{X}_{\mathbf{u}}(x^{\mathbf{v}})$$

$$= \sum_{\mathbf{u} \in \sigma_s} (\mathfrak{X}_{\mathbf{u}}(\mathcal{A}))^2. \qquad \text{Q.E.D.}$$

Corollary 5.3: If $B(s) = 0$ *then* $\mathfrak{X}_{\mathbf{u}}(\mathcal{A}) = 0$ *for each* $\mathbf{u} \in \sigma_s$.

Property 5.1 does not imply that $B(s)$ is an integer. Since by Theorem 5, $\sum_s B(s) = 2^n/|\mathcal{A}|$, $B(s)$ cannot all be integers unless $|\mathcal{A}| = 2^k$. For example, the code $\begin{pmatrix} 000 \\ 110 \\ 011 \end{pmatrix}$ has property 5.1, but the $B(s)$ are not all integers.

At present we have a satisfactory interpretation for $A(s)$, $B(s)$ if $\sum_{\pi \in S_n} \mathcal{A}^{\pi}$ can be generated by a linear code. ($\mathcal{A}$ need not be linear; any collection of vectors with the same weights as the vectors of a linear code will give the same average.) It would be very desirable to find an explanation for the cases in which $A(s)$, $B(s)$ can be thought of as the weight distribution of nonlinear codes.

VI. ACKNOWLEDGMENT

The authors would like to state that they found the basic idea of this paper, and much more, in J. H. van Lint's book *Coding Theory.*[6]

Added to galley proof:

Since this paper was written, it has come to our attention that Neal Zierler (unpublished) discovered the nonlinear MacWilliams identity for Hamming weight enumerators in 1966.

REFERENCES

1. Kerdock, A. M., "A Class of Low-Rate Nonlinear Codes," to appear in Info. and Control.
2. Preparata, F. P., "A Class of Optimum Nonlinear Double-Error Correcting Codes," Info. and Control, *13*, No. 4 (October 1968), pp. 378–400.
3. Berlekamp, E. R., *Algebraic Coding Theory*, New York: McGraw-Hill, 1968.
4. MacWilliams, F. J., "A Theorem on the Distribution of Weights in a Systematic Code," B.S.T.J., *42*, No. 1 (Janaury 1963), pp. 79–94.
5. Lloyd, S. P., "Binary Block Coding," B.S.T.J., *36*, No. 2 (March 1956), pp. 517–535.
6. van Lint, J. H., *Coding Theory*, New York: Springer-Verlag, 1971.
7. Assmus, E. F., Jr., "Research to Develop the Algrbraic Theory of Codes," Sylvania Electronic Systems, Waltham, Mass., Report AFCRL-67-0365, June 1967; especially Part V.
8. Gleason, A. M., "Weight Polynomials of Self-Dual Codes and the MacWilliams Identities," Actes, Congrès intern. Math., 1970, Vol. 3, pp. 211–215; Paris: Gauthier-Villars, 1971.
9. Lee, C. Y., "Some Properties of Non-binary Error-Correcting Codes," IEEE Trans. Info. Theory, *IT-4*, No. 2 (June 1958), pp. 77–82.
10. Krawtchouk, M., "Sur une généralisation des polynomes d'Hermite," Comptes Rendus, *189*, 1929, pp. 620–622.
11. Szegö, G., *Orthogonal Polynomials*, Colloquium Publications, Vol. 23, New York: American Mathematical Society, revised edition, 1959, pp. 35–37.

Coding and Combinatorics

VI

Editor's Comments on Papers 24, 25, and 26

24 **Rudolph:** *A Class of Majority Logic Decodable Codes*

25 **Lin:** *On A Class of Cyclic Codes*

26 **Assmus and Mattson:** *On Tactical Configurations and Error-Correcting Codes*

The relationships between coding and combinatorics are both deep and numerous. In many instances the techniques of one have produced results applicable to the other. This section contains three papers which make effective use of combinatorial structures in constructing and analyzing codes.

The algorithm given by Reed [5] for decoding Muller's codes [4] used the idea of a finite geometry to produce a majority-logic-decoding algorithm. Massey (1963) later extended this concept to L-step majority logic decoding. The simplicity of majority logic decoding was apparent, but relatively few codes could be used effectively with this decoding algorithm. This condition led to a search for other majority-logic-decodable codes.

Rudolph [24], using a slightly modified form of a majority-logic-decoding algorithm, observed that the incidence matrix of a balanced incomplete block design, if used as a parity-check matrix, will, by its very definition, result in an efficient majority-logic-decodable code. Essentially the technique was to build the code around the parity-check equations. He then observed that, since the hyperplanes of a projective geometry yielded a cyclic incidence matrix, the resultant majority-logic-decodable code was cyclic as well. This idea, although simple, was highly effective in producing an interesting and new class of codes.

Weldon (1967) took a different point of view by defining his Euclidean geometry code as an extension of a cyclic code whose roots are specified. It was then shown that the parity-check matrix may be interpreted as containing all flats of a given dimension in a Euclidean geometry, and this led to a simple majority-logic-decoding algorithm. Since this work was also considered in a later paper (Weldon [28]) we chose not to include it here.

The paper by Lin [25] is the most comprehensive treatment of the finite geometry codes of which I am aware. Although it treats these codes as subclasses of polynomial codes (and thus, in a sense, properly belongs in the next section), it does discuss at some length their geometric properties, as opposed to their algebraic properties, which other authors have tended to stress. It is also a rather critical paper in that it reconciles the approach of Weldon (1967) and Kasami, Lin, and Peterson [29] in the definition of Euclidean geometry codes (see Theorem 5, Lin [25]). The paper by Goethals and Delsarte (1968) also gives an excellent geometric and algebraic treatment of projective geometry codes from quite a different point of view.

The application of combinatorial methods are by no means limited to the above situation. Although there are many excellent papers which examine the relationship between coding and combinatorics, the one by Assmus and Mattson [26] is certainly one of the more interesting ones. In a sense it is representative of much of the work in the area and contains an excellent bibliography.

Some recent work in the Soviet Union by Semakov et al. (1968, 1969) also utilizes combinatorial structures to construct a class of equidistant codes. Goethals has been able to use some of their concepts to construct new t-designs, in work that is yet to be published. Thus the relationships between coding and combinatorics continue to be mutually beneficial.

Reprinted from *I.E.E.E. Trans. Inform. Theory,* **IT-13,** 305–307 (1967)

24

A CLASS OF MAJORITY LOGIC DECODABLE CODES

L. D. RUDOLPH

Correspondence

A Class of Majority Logic Decodable Codes

Finding cyclic block codes that can be decoded efficiently by majority logic is important because the decoders are very easy to implement. This correspondence describes the results of a search for such codes.

Majority logic decoding of block codes was introduced in 1954 by Reed[1] who devised a decoding scheme for the class of codes discovered by Muller.[2] In 1958, Yale[3] and Zierler[4] applied majority logic decoding to the maximal length sequence codes. In 1961 Mitchell[5] applied a similar decoding scheme to the cyclic Hamming codes, the "augmented" maximal length sequence codes, and the (15, 7), (21, 11) and (73, 45) Bose–Chaudhuri–Hocqueghem (BCH) codes. In 1962, Massey[6] devised majority algorithms for block codes during his study of threshold decoding. The majority algorithm given below differs from algorithms described previously in that the estimators (parity checks) are not necessarily orthogonal.

Let $H = (h_{ij})$, $i = 0, 1, \cdots, m-1$; $j = 0, 1, \cdots, n-1$, denote the parity check matrix of a cyclic code over $GF(p)$. Suppose the leftmost column of H contains r nonzero elements. Let $H_0 = (h_{i_k j})$, $k = 1, 2, \cdots, r$; $j = 0, 1, \cdots, n-1$, denote the submatrix consisting of the r rows of H in which $h_{i0} \neq 0$. A received sequence $B = (b_0, b_1, \cdots, b_{n-1})$ is the vector sum of a transmitted code word $C = (c_0, c_1, \cdots, c_{n-1})$ and an error vector $E = (e_0, e_1, \cdots, e_{n-1})$.

To decode received digit b_0, first multiply the received sequence B by H_0 and set the product H_0B equal to zero. The resulting equations are

$$\sum_{j=0}^{n-1} h_{i_k j} b_j = 0, \qquad k = 1, 2, \cdots, r. \tag{1}$$

Treating b_0 as an unknown and solving,

$$b_0 = -h_{i_k 0}^{-1} \sum_{j=1}^{n-1} h_{i_k j} b_j, \qquad k = 1, 2, \cdots, r. \tag{2}$$

Denote the r "estimates" of the first received digit by $b_0^{(k)}$, $k = 1, 2, \cdots, r$. One additional estimator is the identity $b_0^{(0)} = b_0$. Now set the decoded symbol $\hat{C}_0$ equal to that value of $GF(p)$ assumed by the largest fraction of the $r+1$ estimates $\{b_0^{(k)}\}$. In case of ambiguity, preference is given to that value of maximum occurrence assumed by the estimate with minimum superscript.

A scheme for decoding the first digit of a cyclic code also decodes the other $n-1$ digits,[7] so that, in general, the estimators are

$$b_i^{(0)} = b_i$$

$$b_i^{(k)} = -h_{i_k 0}^{-1} \sum_{j=1}^{n-1} h_{i_k j} b_{i+j} \qquad i = 0, 1, \cdots, n-1 \tag{3}$$

$$k = 1, 2, \cdots, r.$$

Subscript addition is performed modulo n. Decoded symbols $\hat{C}_i$ are calculated from the $\{b_i^{(k)}\}$ in the same way $\hat{C}_0$ is calculated from the $\{b_0^{(k)}\}$.

Manuscript received July 5, 1963; revised June 17, 1966.

For example, suppose the parity-check matrix H of a binary code is given as

$$H = \begin{bmatrix} 1&1&1&0&1&0&0 \\ 1&1&0&1&0&0&1 \\ 1&0&1&0&0&1&1 \\ 0&1&0&0&1&1&1 \\ 1&0&0&1&1&1&0 \\ 0&0&1&1&1&0&1 \\ 0&1&1&1&0&1&0 \end{bmatrix}. \tag{4}$$

The rank of H over $GF(2)$ is 3, so the code has block length $n = 7$ with three check digits per block. Represent a received sequence (code word with errors) by the vector

$$B = \begin{bmatrix} b_0 \\ b_1 \\ b_2 \\ b_3 \\ b_4 \\ b_5 \\ b_6 \end{bmatrix}. \tag{5}$$

First, set the product H_0B equal to zero.

$$H_0B = \begin{bmatrix} 1&1&1&0&1&0&0 \\ 1&1&0&1&0&0&1 \\ 1&0&1&0&0&1&1 \\ 1&0&0&1&1&1&0 \end{bmatrix} \begin{bmatrix} b_0 \\ b_1 \\ b_2 \\ b_3 \\ b_4 \\ b_5 \\ b_6 \end{bmatrix} = 0. \tag{6}$$

Treating b_0 as an unknown and solving,

$$\begin{bmatrix} 1 \\ 1 \\ 1 \\ 1 \end{bmatrix} [b_0] = \begin{bmatrix} 1&1&0&1&0&0 \\ 1&0&1&0&0&1 \\ 0&1&0&0&1&1 \\ 0&0&1&1&1&0 \end{bmatrix} \begin{bmatrix} b_1 \\ b_2 \\ b_3 \\ b_4 \\ b_5 \\ b_6 \end{bmatrix}. \tag{7}$$

In equation form with the identity $b_0^{(0)} = b_0$ included, this is

$$\begin{aligned} b_0^{(0)} &= b_0 \\ b_0^{(1)} &= b_1 \oplus b_2 \oplus b_4 \\ b_0^{(2)} &= b_1 \oplus b_3 \oplus b_6 \\ b_0^{(3)} &= b_2 \oplus b_5 \oplus b_6 \\ b_0^{(4)} &= b_3 \oplus b_4 \oplus b_5. \end{aligned} \tag{8}$$

The symbol "$\oplus$" denotes modulo 2 addition. H is cyclic, so that in general

$$\begin{aligned} b_i^{(0)} &= b_i \\ b_i^{(1)} &= b_{i+1} \oplus b_{i+2} \oplus b_{i+4} \\ b_i^{(2)} &= b_{i+1} \oplus b_{i+3} \oplus b_{i+6} \\ b_i^{(3)} &= b_{i+2} \oplus b_{i+5} \oplus b_{i+6} \\ b_i^{(4)} &= b_{i+3} \oplus b_{i+4} \oplus b_{i+5}. \end{aligned} \tag{9}$$

Subscript addition is performed modulo 7. For binary codes with $r + 1$ odd, $\hat{C}_i$ is the ordinary majority function of the estimates:

$$\hat{C}_i = \text{maj } \{b_i^{(0)}, b_i^{(1)}, b_i^{(2)}, b_i^{(3)}, b_i^{(4)}\}. \tag{10}$$

The matrix H chosen for this example is the parity-check matrix of a cyclic single-error-correcting Hamming code. Equation (9) shows that all single errors can be corrected. Since a received digit b_j occurs in no more than two of the five equations, a single error can cause at most two of the estimates $b_i^{(0)}, \cdots, b_i^{(4)}$ to be in error. But then $\hat{C}_i = \text{maj } \{b_i^{(0)}, \cdots, b_i^{(4)}\} = C_i$ for $i = 0, 1, \cdots, n-1$.

The guaranteed error correction of any code (using this algorithm) can be found in like manner by inspection of its parity-check matrix. The characteristics desired in a parity-check matrix from this viewpoint lead one quite naturally to a consideration of combinatorial configurations.

A (b, v, r, k', λ)-configuration[8] is a system of b sets and v elements whose b by v incidence matrix of 0's and 1's has the following properties:

1) Every row has exactly k' 1's.
2) Every column has exactly r 1's.
3) The inner product of any two columns is equal to λ.

For example, the parity-check matrix (4) represents a (b, v, r, k', λ)-configuration with $b = v = 7$, $r = k' = 4$, $\lambda = 2$. If the parity-check matrix H is the incidence matrix of a (b, v, r, k', λ)-configuration, then the r by v submatrix H_0 will have r 1's in its leftmost column and λ 1's in all other columns. This leads to a set of $r + 1$ estimators (including the identity $b_i^{(0)} = b_i$) with each b_j appearing in no more than λ equations. Then the decoding algorithm is capable of correcting any combination of e or fewer errors where

$$e = \left[\frac{r}{2\lambda}\right]. \tag{11}$$

The brackets denote "integer part of."

A (b, v, r, k', λ)-configuration will be called "cyclic" if its incidence matrix is cyclic, i.e., if every cyclic permutation of a row of H is also a row of H. Some cyclic (b,v,r,k',λ)-configuration can be derived from difference sets[8],[11] in combinatorics. One well-known class of cyclic (b, v, r, k', λ)-configurations is associated with projective geometries.

Denote by $PG(s, p^t)$ the projective geometry[9] of dimension s over $GF(p^t)$. With each geometry $PG(s, p^t)$ one can associate $s - 1$ cyclic incidence matrices relating points to lines, points to planes, etc. The (b, v, r, k', λ)-configuration corresponding to the incidence matrix relating points and l-spaces of $PG(s, p^t)$ has the following parameters:

$$\begin{aligned} b &= \frac{(1 + p^t + \cdots + p^{st}) \cdots (p^{lt} + \cdots + p^{st})}{(1 + p^t + \cdots + p^{lt}) \cdots (p^{(l-1)t} + p^{lt})p^{lt}} \\ v &= 1 + p^t + \cdots + p^{st} \\ r &= \frac{(p^t + \cdots + p^{st}) \cdots (p^{lt} + \cdots + p^{st})}{(p^t + \cdots + p^{lt}) \cdots (p^{(l-1)t} + p^{lt})p^{lt}} \\ k' &= 1 + p^t + \cdots + p^{lt} \\ \lambda &= \begin{cases} 1 & \text{if } l = 1 \\ \dfrac{(p^{2t} + \cdots + p^{st}) \cdots (p^{lt} + \cdots + p^{st})}{(p^{2t} + \cdots + p^{lt}) \cdots (p^{(l-1)t} + p^{lt})p^{lt}} & \text{if } l > 1. \end{cases} \end{aligned} \tag{12}$$

This in turn corresponds to a code of block length $n = v$ with error correcting capability $e = [r/2\lambda]$ using majority decoding. The remaining code parameter of interest is the number of information digits per block, k, which was determined by computer analysis for a number of codes.

The number of check digits per code block is equal to the rank of the parity-check matrix H over $GF(p)$. Given a (0, 1)-matrix, the choice of field is arbitrary, but some choices yield efficient codes while others do not. For the incidence matrices associated with a projective geometry $PG(s, p^t)$, the field of p elements appears to be a good choice. The computer analysis was restricted to finding the rank only over $GF(p)$.

Conceptually, the search procedure was as follows:

1) Find the incidence matrix of a given (b, v, r, k', λ)-configuration. (Some configurations were found in Singer[10] and Berman;[11] others were generated on the computer.)
2) Calculate the rank of H over $GF(p)$.

It is possible to take advantage of the cyclic property in finding the rank of a cyclic matrix over a finite field by applying Euclid's algorithm to $x^n - 1$ and the rows (written as polynomials) and observing the degree of the result. This was the method actually implemented on the computer.

A partial list of the codes found is given in Table I. Examination of the leftmost column of the list suggests the following conjecture:

TABLE I
CYCLIC (n, k, e) CODES ASSOCIATED WITH $PG(s, p^t)$

	$l = s - 1$	$l = s - 2$	$l = s - 3$	$l = s - 4$
Binary Codes				
$PG(2, 2)$	(7, 3, 1)			
$PG(3, 2)$	(15, 10, 1)	(15, 4, 3)		
$PG(4, 2)$	(31, 25, 1)	(31, 15, 2)	(31, 5, 7)	
$PG(5, 2)$	(63, 56, 1)	(63, 41, 2)	(63, 21, 5)	(63, 6, 15)
$PG(2, 4)$	(21, 11, 2)			
$PG(3, 4)$	(85, 68, 2)	(85, 24, 10)		
$PG(4, 4)$	(341, 315, 2)	(341, 195, 8)	(341, 45, 42)	
$PG(5, 4)$	(1365, 1328, 2)	(1365, 1063, 8)	(1365, 483, 34)	(1365, 78, 170)
$PG(2, 8)$	(73, 45, 4)			
$PG(3, 8)$	(585, 520, 4)	(585, 184, 36)		
$PG(4, 8)$	(4681, 4555, 4)	(4681, 3105, 32)	(4681, 590, 292)	
$PG(2, 16)$	(273, 191, 8)			
$PG(3, 16)$	(4369, 4112, 8)	(4369, 1568, 136)		
$PG(2, 32)$	(1057, 813, 16)			
$PG(2, 64)$	(4161, 3431, 32)			
Base 3 Codes				
$PG(2, 3)$	(13, 6, 2)			
$PG(3, 3)$	(40, 29, 1)	(40, 10, 6)		
$PG(4, 3)$	(121, 105, 1)	(121, 60, 5)	(121, 15, 20)	
$PG(2, 9)$	(91, 54, 5)			
$PG(2, 27)$	(757, 540, 14)			
Base 5 Codes				
$PG(2, 5)$	(31, 15, 3)			
$PG(3, 5)$	(156, 120, 2)	(156, 38, 15)		
$PG(2, 25)$	(651, 425, 13)			
Base 7 Codes				
$PG(2, 7)$	(57, 28, 4)			

The incidence matrix relating points and $(s-1)$-spaces of $PG(s, p^t)$ has rank $\binom{p+s-1}{s}^t + 1$ over $GF(p)$.

The codes found are either BCH codes or somewhat inferior to the corresponding BCH codes. Even the inferior codes are of interest from a practical standpoint, however, because of the simplicity of decoding.

While this correspondence was in review, E. J. Weldon, Jr., in his study of difference set cyclic codes [12] observed that the codes associated with $PG(2, 2^t)$ had dimension $n - (3^t + 1)$ for $1 \leq t \leq 5$. Subsequently, R. L. Graham and Mrs. F. J. MacWilliams[13] proved that in general the codes associated with $PG(2, p^t)$ have dimension $n - (\binom{p+1}{2}^t + 1)$. This verifies the above conjecture for the important case $s = 2$.

Acknowledgment

I wish to acknowledge the many helpful discussions with M. E. Mitchell and J. P. Lipp of the General Electric Company and to thank C. H. Burton of the SURC Computer Center for his support of this investigation.

L. D. Rudolph
Syracuse University Research Corp.
Syracuse, N. Y.

References

[1] I. S. Reed, "A class of multiple-error-correcting codes and the decoding scheme," *IRE Trans. on Information Theory*, vol. IT-4, pp. 38–49, 1954.

[2] D. E. Muller, "Application of Boolean algebra to switching circuit design and to error detection," *IRE Trans. on Electronic Computers*, vol. EC-3, pp. 6–12, September 1954.

[3] R. B. Yale, "Error correcting codes and linear recurring sequences," M.I.T. Lincoln Lab., Lexington, Mass., Lincoln Laboratory Group Rept. 34-77, 1958.

[4] N. Zierler, "On a variation of the first order Reed-Muller codes," M.I.T. Lincoln Lab., Lexington, Mass., Lincoln Laboratory Group Rept. 34-80, 1958.

[5] M. E. Mitchell et al., "Coding and decoding operations research," for U.S.A.F. Cambridge Research Center, Bedford, Mass. General Electric Co. Final Rept., Contract AF 19(604)-6183, 1961.

[6] J. L. Massey, *Threshold Decoding*. Cambridge, Mass.: M.I.T. Press, 1963.

[7] W. W. Peterson, *Error-Correcting Codes*. Cambridge, Mass.: M.I.T. Press, 1961, pp. 201–204.

[8] H. J. Ryser, *Combinatorial Mathematics*. New York: Wiley, 1963.

[9] H. B. Mann, *Analysis and Design of Experiments*. New York: Dover, 1949.

[10] J. Singer, "A theorem in finite projective geometry and some applications to number theory," *Trans. Amer. Math. Soc.*, vol. 43, 1938.

[11] G. Berman, "Finite projective geometrics," *Canad. J. Math.*, vol. 4, 1952.

[12] E. J. Weldon, Jr., "Difference set cyclic codes," *Bell Sys. Tech. J.*, vol. 7, 1966.

[13] R. L. Graham and F. J. MacWilliams, "On the number of information symbols in difference set cyclic codes," *Bell Sys. Tech. J.*, vol. 7, 1966.

Reprinted from *Error Correcting Codes,* edited by H. Mann,
John Wiley and Sons, Inc., New York, 131–148 (1968)

25

S. LIN

On a Class of Cyclic Codes

ABSTRACT

Two subclasses of polynomial codes have been studied. One subclass of polynomial codes has been proved to contain the Euclidean geometry codes as a proper subclass. Another subclass of polynomial codes has been shown to be closely related to the projective geometry codes. A BCH lower bound on the minimum distance of Euclidean geometry codes has been derived.

1. Introduction

The polynomial approach of Kasami, et al, to a class of cyclic codes (polynomial codes) [1, 2] gives a unified formulation for several important classes of codes, and puts the latter codes into a large framework. Two subclasses of polynomial codes are closely related to the geometry codes which have been studied extensively for the past few years [3, 4, 5, 6, 7, 8, 9, 10, 11, 12, 13, 14, 15, 16, 17]. It is the purpose of this paper to establish the relationship between polynomial codes and geometry codes. Firstly, we shall summarize the important results of polynomial codes. Secondly, we shall prove that one subclass of polynomial codes contains the Euclidean geometry (E. G.) codes [7] and the generalized Reed-Muller codes [6, 10, 12] as subclasses. Thirdly, a relation between one subclass of polynomial codes and finite projective geometry codes [8, 9] will be established. A BCH lower bound on the minimum distance of Euclidean geometry codes is derived. From this BCH bound, we are able to show that E. G. codes are in general more powerful than Weldon's decoding scheme has been able to demonstrate.

This work is based primarily on the results in the first and second references. Where possible the notation and conventions employed therein will be followed here.

2. Summarized Results on Polynomial Codes [1, 2]

Let q be a power of prime, say $q = p^c$, and α be a primitive root of $GF(q^{ms})$ Any non-zero element α^j in $GF(q^{ms})$ can be expressed as

$$\alpha^j = \sum_{i=1}^{m} a_{ij}\alpha^{i-1} \quad \text{for} \quad 0 \leq j < q^{ms} - 1 , \tag{1}$$

where $a_{ij} \varepsilon GF(q^s)$. The correspondence between α^j and the m-tuple $(a_{1j}, a_{2j}, \ldots, a_{mj})$ is one-to-one.

Let b be a factor of q^s-1, and

$$z = (q^s - 1)/b , \qquad n = (q^{ms} - 1)/b . \tag{2}$$

Let $X_1, X_2, \ldots, X_m$ be m variables over $GF(q^s)$ and $\bar{X} = (X_1, X_2, \ldots, X_m)$ Define

$$P(m, s, \mu, b) \tag{3}$$

as the set of polynomials $f(\bar{X}) = f(X_1, X_2, \ldots, X_m)$ in $X_1, X_2, \ldots, X_m$ with coefficients in $GF(q^s)$ such that the sum of the exponents of each term of $f(\bar{X})$ is a multiple of b and the degree of $f(\bar{X})$ is μb or less. Therefore, each polynomial in $P(m, s, \mu, b)$ is of the following form

$$f(\bar{X}) = \sum C X_1^{\nu_1} X_2^{\nu_2} \ldots X_m^{\nu_m} \tag{4}$$

where $C \varepsilon GF(q^s)$, $0 \leq \nu_i < q^s$ and $\sum_{i=1}^{m} \nu_i = jb$ with $0 \leq j \leq \mu$. The parameter μ is at most mz .

Define a vector

$$\bar{v}(f) = (v_0, v_1, v_2, \ldots, v_{n-1}) \tag{5}$$

whose components are in $GF(q^s)$ as follows

$$v_j = f(a_{1j}, a_{2j}, \ldots, a_{mj}) \tag{6}$$

for $0 \leq j < n$, where $f(\bar{X})$ is in $P(m, s, \mu, b)$, $(a_{1j}, a_{2j}, \ldots, a_{mj}) \Longleftrightarrow \alpha^j$, and $n = (q^{ms} - 1)/b$.

Now, define

$$Q(m, s, \mu, b, q) \tag{7}$$

to be the set of all polynomials $f(\bar{X})$ in $P(m, s, \mu, b)$ such that

$$f(a_{1j}, a_{2j}, \ldots, a_{mj}) \in GF(q) \quad (8)$$

for $0 \leq j < n$.

Definition [Kasami, Lin and Peterson]: A q-ary (n, m, s, μ, q)-polynomial code is defined as the set of vectors $\bar{v}(f)$

$$\{\bar{v}(f) \mid f(\bar{X}) \in Q(m, s, \mu, b, q)\} \ . \quad (9)$$

Let h be a non-negative integer less than q^{ms} . Express h in radix-q^s form as follows

$$h = \delta_0 + \delta_1 q^s + \delta_2 q^{2s} + \ldots + \delta_{m-1} q^{(m-1)s} \quad (10)$$

where $0 \leq \delta_i < q^s$ for $0 \leq i < m$. The q^s-weight of h is defined as

$$W_{q^s}(h) = \sum_{i=0}^{m-1} \delta_i \ . \quad (11)$$

Let

$$h' \equiv hq^{\ell} \ (\mathrm{mod}\ q^{ms} - 1) \ . \quad (12)$$

Then the q^s-weight of h' is defined as

$$W_{q^s}(h') = W_{q^s}(hq^{\ell}) \ . \quad (13)$$

Kasami, et al, proved the following two theorems:

Theorem 1: A q-ary (n, m, s, μ, q)-polynomial code is a cyclic code which has the following parameters:

a) Code length $n = (q^{ms} - 1)/b$

b) Number of information digits

$$k = \left\{ \begin{array}{l} \text{the number of non-negative integers } h \text{ less} \\ \text{than } q^{ms}-1 \text{ which are divisible by } b \text{ and such that} \\ \max\limits_{0 \leq \ell < s} W_{q^s}(hq^{\ell}) = jb \text{ with } 0 \leq j \leq \mu \ . \end{array} \right\}$$

c) Minimum distance d_{min}

$$d_{min} \geq [(R+1)q^{Qs}-1]/b$$

where Q and R are quotient and remainder resulting from dividing $m(q^s-1)-\mu b$ by q^s-1 . The generator polynomial $g(X)$ has α^j as a root if and only if there exists an h such that h is divisible by b, and

$$\min_{0 \leq \ell < s} W_{q^s}(hq^{\ell}) = jb$$

with $0 < j < mz - \mu$.

Theorem 2. The q-ary dual code of a (n, m, s, μ, q)-polynomial code has the following parameters:

a) $n = (q^{ms}-1)/b$.

b) Number of information digits

$$k^0 = \left\{ \begin{array}{l} \text{the number of non-negative integers } h \text{ less} \\ \text{than } q^{ms}-1 \text{ which are divisible by } b \text{ and such that} \\ \min_{0 \leq \ell < s} (hq^{\ell}) = jb \text{ with } 0 < j < mz - \mu . \end{array} \right\}$$

The generator polynomial $g^0(X)$ has α^h as a root if and only if h is divisible by b and

$$\max_{0 \leq \ell < s} W_{q^s}(hq^{\ell}) = jb$$

with $0 \leq j \leq \mu$.

A general lower bound on the minimum distance of the dual code of a (n, m, s, μ, q)-polynomial code has not been obtained. But, for some special cases, we are able to derive a BCH lower bound.

3. A Subclass of Polynomial Codes Which is Related to Euclidean Geometry Codes

The class of cyclic codes based on the Euclidean geometry was first introduced by Rudolph [3], and then has been studied extensively by Weldon and others [7,11,14,16,17]. Codes of this class are called Euclidean geometry codes. In general, a code in this class is less efficient than the corresponding BCH code of the same designed minimum distance. But it can be decoded with a relatively modest amount of equipment. In this section, we shall prove that dual codes of the class of polynomial codes with b = 1 contain the E. G. codes as a subclass.

For $b = 1$, a (n, m, s, μ, q)-polynomial code has the following parameters:

$$n = q^{ms} - 1,$$

$$k = \left\{ \begin{array}{l} \text{the number of non-negative integers } h \text{ less} \\ \text{than } q^{ms}-1 \text{ such that } \max\limits_{0 \leq \ell < s} W_{q^s}(hq^{\ell}) = \mu \end{array} \right\},$$

$$d_{min} \geq (q^s - N) q^{(m-D-1)s} - 1$$

where D and N are quotient and remainder resulting from dividing μ by $q^s - 1$, i.e.

$$\mu = D(q^s - 1) + N \tag{14}$$

with $0 \leq N < q^s - 1$. The generator polynomial $g(X)$ has α^h as a root if and only if

$$0 < \min_{0 \leq \ell < s} W_{q^s}(hq^{\ell}) < m(q^s - 1) - \mu . \tag{15}$$

For $s = 1$, this subclass of polynomial codes reduces to generalized Reed-Muller codes [10]. Thus, we may consider this subclass as a new generalization of Reed-Muller codes.

Let L be a t-dimensional subspace of all m-tuples over $GF(q^s)$ and $\{A_1, A_2, \ldots, A_t\}$ be a basis of L. Let Γ be the null space of L. Then the dimension of Γ is $m - t$. The following polynomial

$$f_L(\bar{X}) = \prod_{i=1}^{t} \{1 - (A_t \cdot \bar{X}^T) q^s - 1\} \tag{16}$$

has degree $t(q^s - 1)$, where $A_t \cdot \bar{X}^T$ is the inner product of A_t and $\bar{X}$, i.e.

$$A_t \cdot \bar{X}^T = (a_{1t}, a_{2t}, \ldots, a_{mt})(X_1, X_2, \ldots, X_m)^T$$

$$= \sum_{i=1}^{m} a_{it} X_i \quad (\text{over } GF(q^s)) .$$

Express μ as in Equation (14). For $0 \leq t \leq D$, $f_L(\bar{X})$ is a polynomial in $Q(m, s, \mu, 1, q)$ and vector $\bar{v}(f)$ defined in accordance with Equation (5) and Equation (6) is a code vector in the $(q^{ms}-1, m, s, \mu, q)$-polynomial code. Since

$$f_L(\bar{X}_j) = 1 \text{ for } X_j \in \Gamma$$

and

$$f_L(\bar{X}_j) = 0 \text{ for } \bar{X}_j \notin \Gamma$$

the components of $\bar{v}(f_L)$ are "1" at the locations corresponding to the non-zero m-tuples of Γ and "0" at other locations. The Hamming weight of $\bar{v}(f_L)$ is $q^{(m-t)s}-1$. Let Γ' be a coset with respect to the subspace Γ. Then, the polynomial

$$f'_L(\bar{X}) = \prod_{i=1}^{t} \{1 - [A_i \cdot (\bar{X} - B)^T]^{q^s-1}\} \tag{17}$$

is also in $Q(m, s, \mu, 1, q)$ for $0 \leq t \leq D$, where B is any vector in Γ'. Therefore, the components of $\bar{v}(f'_L)$ are "1" at the locations corresponding to the elements of Γ' and are "0" at other locations. The weight of $\bar{v}(f'_L)$ is $q^{(m-t)s}$.

All the m-tuples over $GF(q^s)$ form a Euclidean geometry over $GF(q^s)$, i.e. $E.G.(m, q^s)$. Every m-tuple is a point in $E.G.(m, q^s)$. The all zero m-tuple is regarded as the point at infinity. Therefore, every bit position of a code vector in a $(q^{ms}-1, m, s, \mu, q)$-polynomial code can be uniquely associated with a point in $E.G.(m, q^s)$ except the point at infinity. Γ is a $(m-t)$-flat through the point at infinity, and Γ' is a $(m-t)$-flat through the point B. For convenience, we call the code vector $\bar{v}(f_L)$ or $\bar{v}(f'_L)$ a $(m-t)$-flat in $E.G.(m, q^s)$. By Equation (16) and Equation (17), we obtain:

Theorem 3. For $0 \leq t \leq D$, every $(m-t)$-flat of $E.G.(m, q^s)$ is a code vector in a $(q^{ms}-1, m, s, \mu, q)$-polynomial code, where $\mu = D(q^s-1) + N$ with $0 \leq N < q^s-1$.

Thus, a $(q^{ms}-1, m, s, \mu, q)$-polynomial code contains all the $(m-D)$-flats, $(m-D+1)$-flats, ..., and m-flats of $E.G(m, q^s)$. The smallest flats which are contained in the code are $(m-D)$-flats. The number of $(m-D)$-flats which intersect on a given $(m-D-1)$-flat is

$$J = 1 + q^s + q^{2s} + \ldots + q^{Ds} = \frac{q^{(D+1)s} - 1}{q^s - 1}. \tag{18}$$

It follows from Weldon's argument [7], that the dual code of a $(q^{ms}-1, m, s, \mu, q)$-polynomial code is $(m-D)$-step orthogonalizable [18] and has minimum distance at least

$$J + 1 = \frac{q^{(D+1)s} - 1}{q^s - 1} + 1 \tag{19}$$

where $\mu = D(q^s - 1) + N$ with $0 \le N < q^s - 1$. Equation (19) gives a lower bound for the minimum distance of the dual code of a $(q^{ms}-1, m, s, \mu, q)$-polynomial code. For large D and S, this bound is very loose. A BCH lower bound can be derived by counting the consecutive roots in the generator polynomial of the dual code.

From Theorem 2, the generator polynomial $g^0(X)$ of the dual of a $(q^{ms}-1, m, s, \mu, q)$-polynomial code has α^h as a root if and only if

$$\max_{0 \le \ell < s} W_{q^s}(hq^{\ell}) \le \mu .$$

Let $\mu = D(q^s-1) + N$ with $0 \le N < q^s - 1$. Consider the following integer

$$\begin{aligned} h_0 = (q^s - 1) + (q^s - 1)q^s + \ldots + (q^s - 1)q^{(D-2)s} \\ + (q-1)q^{(D-1)s} + (\lambda + 1)q^{Ds} \end{aligned} \tag{20}$$

where λ is the quotient resulting from dividing N by q^{s-1}, i.e.

$$N = \lambda q^{s-1} + \sigma \tag{21}$$

with $0 \le \sigma < q^{s-1}$. Since $N < q^s - 1$, therefore, $0 \le \lambda \le q - 1$. For $0 \le \ell < s$, the radix-q^s expansion of $h_0 q^{\ell}$ is

$$\begin{aligned} h_0 q^{\ell} = (q^s - q^{\ell}) + (q^s - 1)q^s + \ldots + (q^s - 1)q^{(D-2)s} \\ + (q^{\ell+1} - 1)q^{(D-1)s} + (\lambda + 1)q^{\ell} q^{Ds} . \end{aligned} \tag{22}$$

The q^s-weight of $h_0 q^{\ell}$ is

$$W_{q^s}(h_0 q^{\ell}) = (D-1)(q^s - 1) + (\lambda + q)q^{\ell} . \tag{23}$$

It is obvious that

$$\max_{0 \le \ell < s} W_{q^s}(h_0 q^{\ell}) = W_{q^s}(h_0 q^{s-1})$$
$$= D(q^s - 1) + \lambda q^{s-1} + 1 \ . \tag{24}$$

It is easy to show that any integer h less than h_0 satisfies

$$\max_{0 \le \ell < s} W_{q^s}(hq^{\ell}) \le D(q^s - 1) + \lambda q^{s-1} + 1 \ . \tag{25}$$

If N is divisible by q^{s-1} $(\sigma = 0)$, then h_0 is the smallest integer such that

$$\max_{0 \le \ell < s} W_{q^s}(h_0 q^{\ell}) = D(q^s - 1) + \lambda q^{s-1} + 1$$
$$= \mu + 1 \ . \tag{26}$$

If N is not divisible by q^{s-1} $(\sigma \ne 0)$, then

$$\max_{0 \le \ell < s} W_{q^s}(h_0 q^{\ell}) \le \mu \ . \tag{27}$$

By Equation (25), Equation (26) and Equation (27), the dual code of a $(q^{ms}-1, m, s, \mu, q)$-polynomial code has at least the following consecutive roots

$$\alpha^0, \alpha^1, \alpha^2, \ldots, \alpha^{h_0 - 1} \tag{28}$$

where α is a primitive root of $GF(q^{ms})$. By Bose's argument [19], the dual code has minimum distance at least

$$h_0 + 1 = (\lambda+1) q^{Ds} + q \cdot q^{(D-1)s} \ .$$

<u>Theorem 4</u>. The dual code of a $(q^{ms}-1, m, s, \mu, q)$-polynomial code has minimum distance at least

$$(\lambda+1) q^{Ds} + q \cdot q^{(D-1)s} \tag{29}$$

where D and λ satisfy the following equations

$$\mu = D(q^s - 1) + N \qquad 0 \le N < q^s - 1 \ ,$$
$$N = \lambda q^{s-1} + \sigma \qquad 0 \le \sigma < q^{s-1} \ .$$

For $\mu = D(q^s - 1) + N$ with $0 \le N < q^s - 1$, Theorem 3 tells us that the smallest flats of E.G. (m, q^s) are $(m-D)$-flats. By Weldon's argument, the dual code of a $(q^{ms}-1, m, s, \mu, q)$-polynomial code is $(m-D)$-step orthogonalizable. The polynomial code of the smallest dimension which contains $(m-D)$-flats is the code with $\mu = D(q^s-1)$, i.e. a $(q^{ms}-1, m, s, D(q^s-1), q)$-polynomial code. Therefore, the dual code of a $(q^{ms}-1, m, s, D(q^s-1), q)$-polynomial code is the code of the largest dimension which contains $(m-D)$-flats of E.G. (m, q^s) in its null space. If Weldon's decoding scheme is used, the dual code of a $(q^{ms}-1, m, s, \mu, q)$-P-code with $\mu = D(q^s-1) + N$ and the dual code of a $(q^{ms}-1, m, s, D(q^s-1), q)$-P-code will correct the same number of errors $\left\lfloor \frac{J}{2} \right\rfloor$*, where

$$J = \frac{q^{(D+1)s}-1}{q^s-1} . \tag{30}$$

In the following, we shall prove that, for $q = 2$, the dual code of a $(2^{ms}-1, m, s, D(2^s-1), 2)$-polynomial code is an E.G. code defined by Weldon [7]. Let h be a positive integer less than $2^{ms}-1$. In radix-2^s expansion

$$h = \sum_{i=0}^{m-1} \delta_i 2^{is} \tag{31}$$

where $0 \leq \delta_i < 2^s$ for $0 \leq i \leq m-1$.
In radix-2 expansion,

$$h = \sum_{i=0}^{m-1} \sum_{t=0}^{s-1} \delta_{it} 2^{t+is} \tag{32}$$

where $\delta_{ij} = 0$ or 1 for $0 \leq i \leq m-1$ and $0 \leq t \leq s-1$. The binary ms-tuple associated with h may contain some disjoint binary representations of multiples of 2^s-1. Now define the s-weight of h, denoted by $W_s(h)$, as the maximum number of such disjoint multiples [7].

Definition (Weldon) [7]. The $(\nu, s)^{th}$ order cyclic E.G. code is a code whose parity check polynomial $h(X)$ contains among its roots all α^h such that $W_s(h) \leq \nu$.

Theorem 5. For a positive integer h less than $2^{ms}-1$, $W_s(h) \leq \nu$ if and only if $\min_{0 \leq \ell < s} W_{2^s}(h2^\ell) < (\nu+1)(2^s-1)$.

Proof: Step 1. We show that $\min_{0 \leq \ell < s} W_{2^s}(h2^\ell) < (\nu+1)(2^s-1) \Longrightarrow W_s(h) \leq \nu$.

---

* The symbol $\lfloor y \rfloor$ denotes the greatest integer contained in y.

Let

$$h = h_0 + h_1 + h_2 + \dots + h_r \tag{33}$$

where

a) $h_0 \geq 0$ and $h_i > 0$ for $1 \leq i \leq r$,

b) h_i is a multiple of 2^s-1, for $1 \leq i \leq r$,

and c) the radix-2 expansions of $h_1, h_2, \dots, h_r$ are mutually disjoint.

Then,

$$W_{2^s}(h_j 2^\ell) = W_{2^s}(h_0 2^\ell) + W_{2^s}(h_1 2^\ell) + \dots + W_{2^s}(h_r 2^\ell) \ . \tag{34}$$

Since $h_i 2^\ell$ is divisible by 2^s-1, $W_{2^s}(h_i 2^\ell)$ is a multiple of 2^s-1 . Let $W_{2^s}(h_i 2^\ell) = k_i(2^s-1)$, for $1 \leq i \leq r$, where $k_i > 0$. Then

$$\begin{aligned} W_{2^s}(h2^\ell) &= W_{2^s}(h_0 2^\ell) + (2^s-1)\sum_{i=1}^{r} k_i \\ &\geq (2^s-1)\sum_{i=1}^{r} k_i \\ &\geq r(2^s-1) \ . \end{aligned} \tag{35}$$

Therefore, we have

$$\nu + 1 > r \ . \tag{36}$$

Since the inequality is true for any possible r, thus

$$\min_{0 \leq \ell < s} W_{2^s}(hq^\ell) < (\nu+1)(2^s-1) \implies W_s(h) \leq \nu \ .$$

Step 2 . We show that $W_s(h) \leq \nu \implies \min_{0 \leq \ell < s} W_{2^s}(h2^\ell) < (\nu+1)(2^s-1)$.

Let $W_s(h) = j$ where $j \leq \nu$. Then h can be expressed as a sum follows

$$h = h_0 + h_1 + \dots + h_j \tag{37}$$

where a) $h_0 \geq 0$ and $h_i > 0$ for $1 \leq i \leq j$,

b) h_i is a multiple of 2^s-1, for $1 \leq i \leq j$,

c) The radix-2 expansions of $h_1, \ldots, h_j$ are mutually disjoint.

By the definition of s-weight, the 2^s-weight of h_i must be equal to 2^s-1 exactly,

$$W_{2^s}(h_i) = (2^s-1)$$

for $1 \leq i \leq j$. Also $W_{2^s}(h_0) < 2^s-1$. Otherwise, h_0 can be partitioned into at least two disjoint parts such that one part is divisible by 2^s-1 . This is a contradiction to the assumption that j is the maximum number of disjoint multiples of (2^s-1) . Therefore, we obtain

$$W_{2^s}(h) < (j+1)(2^s-1) . \tag{38}$$

Equation (38) implies that $W_{2^s}(h) < (\nu+1)(2^s-1)$ if $W_s(h) \leq \nu$. The theorem is then proved. Q.E.D.

Theorem 5 gives a relation between E.G. codes and polynomial codes. By Theorem 1, the definition of Weldon's E.G. codes and Theorem 5, we have

Theorem 6. The dual code of a binary $(2^{ms}-1, m, s, D(2^s-1), 2)$-polynomial code with an overall parity check ($\alpha^0=1$ as a root in its generator polynomial) is a $[m-D-1, s]^{th}$ order E.G. code.

For $\mu = D(q^s-1)$, we may consider the dual of a $(q^{ms}-1, m, s, D(q^s-1), q)$-P-code with an overall parity check as a generalized $(m-D-1, s)^{th}$ order Euclidean geometry code.

Weldon's decoding scheme only demonstrates that a $(m-D-1, s)^{th}$ order E.G. code can correct $\left\lfloor \frac{J}{2} \right\rfloor$ errors, where

$$J = \frac{q^{(D+1)s} - 1}{q^s - 1} . \tag{39}$$

Theorem 4 tells us that a q-ary $(m-D-1, s)^{th}$ order E.G. code has error correcting capability at least $\left\lfloor \frac{h_0-1}{2} \right\rfloor$ where

$$h_0 = q^{Ds} + q \cdot q^{(D-1)s} - 1 \tag{40}$$

The difference is

$$(q-2)q^{(D-1)s} + (q^s-2)q^{(D-2)} + \ldots + (q^s-2) . \tag{41}$$

Therefore, a $(\nu, s)^{th}$ order E.G. code is, in general, more powerful than Weldon's decoding scheme has been able to demonstrate.

Consider an integer less than $q^{ms}-1$

$$h = \delta_0 + \delta_1 q^s + \dots + \delta_{m-1} q^{(m-1)s} .$$

where $0 \leq \delta_i < q^s$ for $0 \leq i \leq m-1$. We define $h' = \delta'_0 + \delta'_1 q^s + \dots + \delta'_{m-1} q^{(m-1)s}$ as a descendant of h if and only if

$$0 \leq \delta'_i \leq \delta_i \quad \text{for} \quad 0 \leq i \leq m-1 .$$

The necessary and sufficient condition for an argumented primitive cyclic code of length q^{ms} to be invariant under the affine group of permutations is that, for every α^h which is a root of the generator polynomial, $\alpha^{h'}$ is also a root, where α is a primitive element of $GF(q^{ms})$ [6]. By Theorem 1, the argumented $(q^{ms}-1, m, s, \mu, q)$-P-code is doubly-transitive invariant under the affine group of permutations.

4. A Subclass of Polynomial Codes Which is Related to the Projective Geometry Codes

Projective geometry codes were first introduced by L. D. Rudolph [3] in 1964. Since then, this class of codes have been studied estensively by many authors, especially by Weldon [4, 7, 17] and Goethals, et al, [9]. In this section, we shall prove the relationship between projective geometry codes and a certain subclass of polynomial codes.

For $b = q^s-1$, a q-ary $(\frac{q^{ms}-1}{q^s-1}, m, s, \mu, q)$-polynomial code has the following parameters

$$n = \frac{q^{ms}-1}{q^s-1} ,$$

$$k = \left\{ \begin{array}{l} \text{The number of non-negative integers } h \text{ less} \\ \text{than } q^{ms}-1 \text{ which is divisible by } q^s-1 \text{ and} \\ \max\limits_{0 \leq \ell < s} W_{q^s}(hq^\ell) = j(q^s-1) \text{ with } 0 \leq j \leq \mu . \end{array} \right\} ,$$

$$d = [q^{(m-\mu)s}-1]/q^s-1 .$$

The generator polynomial has α^h as a root if and only if

$$\min_{0 \le \ell < s} W_{q^s}(hq^\ell) = j(q^s - 1)$$

with $0 < j < m-\mu$.

In the following, we shall prove the relationship between a $(\frac{q^{ms}-1}{q^s-1}, m, s, \mu, q)$-polynomial code and projective geometry code.

Let h be a non-negative integer which is less than $q^{ms}-1$. Express h in both radix-q^s form and radix-q form as follows:

(1) Radix-q^s form

$$h = \sum_{i=0}^{m-1} \delta_i q^{is} \tag{42}$$

where $0 \le \delta_i < q^s$ for $0 \le i \le m-1$,

(2) Radix-q form

$$h = \sum_{i=0}^{m-1} \sum_{t=0}^{s-1} \delta_{it} q^{t+is} \tag{43}$$

where $0 \le \delta_{it} < q$ for $0 \le i \le m-1$ and $0 \le t \le s-1$.

The q^s-weight of h is $W_{q^s}(h) = \sum_{i=0}^{m-1} \delta_i$ and the q-weight of h is $W_q(h) = \sum_{i=0}^{m-1} \sum_{t=0}^{s-1} \delta_{it}$.

Theorem 7:

$$(q^s-1) W_q(h) = (q-1) \sum_{\ell=0}^{s-1} W_{q^s}(hq^\ell) . \tag{44}$$

Proof. Let

$$\sigma_t = \sum_{i=0}^{m-1} \delta_{it} . \tag{45}$$

Then,

$$W_{q^s}(h) = \sum_{t=0}^{s-1} \sigma_t q^t . \tag{46}$$

For any $0 \leq \ell < s$, we have

$$W_{q^s}(hq^\ell) = q^\ell \sum_{t=0}^{s-\ell-1} \sigma_t q^t + q^{-(s-\ell)} \sum_{t=s-\ell}^{s-1} \sigma_t q^t . \tag{47}$$

Then

$$\sum_{\ell=0}^{s-1} W_{q^s}(hq^\ell) = \sum_{\ell=0}^{s-1} q^\ell \sum_{t=0}^{s-\ell-1} \sigma_t q^t + \sum_{\ell=0}^{s-1} q^{-(s-\ell)} \sum_{t=s-\ell}^{s-1} \sigma_t q^t$$

$$= \frac{q^s-1}{q-1} \sum_{t=0}^{s-1} \sigma_t$$

$$= \frac{q^s-1}{q-1} W_q(h) . \tag{48}$$

Rearranging Equation (48), we obtain

$$(q^s-1) W_q(h) = (q-1) \sum_{\ell=0}^{s-1} W_{q^s}(hq^\ell) . \tag{49}$$

Q.E.D.

Corollary 8

$$(q^s-1) W_q(h) \geq s(q-1) \min_{0 \leq \ell < s} W_{q^s}(hq^\ell) \tag{50}$$

and

$$(q^s-1) W_q(h) \leq s(q-1) \max_{0 \leq \ell < s} W_{q^s}(hq^\ell) . \tag{51}$$

Definition (Weldon) [8]. The binary μs^{th} order projective geometry code (or non-primitive Reed-Muller code as it was called) is a cyclic code of length $\frac{2^{ms}-1}{2^s-1}$ whose generator polynomial contains among its roots all α^h such that h is divisible by 2^s-1 and

$$W_2(h) \leq \mu s . \tag{52}$$

From Equation (50) and Equation (51), it is obvious that

$$W_2(h) \leq \mu s \Longrightarrow \mu(2^s-1) \geq \min_{0 \leq \ell < s} W_{2^s}(h2^\ell) \tag{53}$$

where he is divisible by 2^s-1.

Equation (53) implies the following theorem

Theorem 9: A binary $(\frac{2^{ms}-1}{2^s-1}, m, s, m-\mu-1, 2)$-polynomial code with α^0 in the generator polynomial is a subcode of a μs order binary P.G. code.

By Theorem 2, the generator polynomial of the dual of a $(\frac{2^{ms}-1}{2^s-1}, m, s, \mu, 2)$-polynomial code has α^h as a root if and only if h is divisible by 2^s-1 and

$$\max_{0 \leq \ell < s} W_{2^s}(h2^{\ell}) = j(2^s-1) \tag{54}$$

with $0 \leq j \leq \mu$.

From Equation (51) and Equation (54) we obtain

Theorem 10: The μs^{th} order binary P.G. code is a subcode of the duel of a binary $(\frac{2^{ms}-1}{2^s-1}, m, s, \mu, 2)$-polynomial code.

In reference [2], it has been shown that all the $(m-\mu-1)$-flats, ..., $(m-1)$-flats of P.G. $(m-1, 2^s)$ are contained in the $(\frac{2^{ms}-1}{2^s-1}, m, s, \mu, 2)$-polynomial code. Therefore, the dual code of a $(\frac{2^{ms}-1}{2^s-1}, m, s, \mu, 2)$-p-code is $(m-\mu-1)$-step orthogonalizable, and can correct $\lfloor \frac{J}{2} \rfloor$ errors, where

$$J = \frac{2^{(\mu+1)s}-1}{2^s-1} . \tag{55}$$

$\lfloor \frac{J}{2} \rfloor$ is also the error correcting ability of a μs^{th} order P.G. code of Weldon. Therefore, for the same designed error-correcting ability, the dual code of a polynomial code with $b = 2^s-1$ is more efficient than the corresponding P.G. code.

Projective geometry codes over GF(p) have been studied by Goethals and Delsarte [9] where p is a prime. Let $C(m-\mu-1, m-1, p^s)$ be the code generated by the set of all $(m-\mu-1)$-flats of P.G. $(m-1, p^s)$. They defined a projective geometry code as follows:

Definition (Goethals and Delsarte) [9]: A μth order projective geometry code over GF(p) of length $\frac{p^{ms}-1}{p^s-1}$ is the dual code of $C(m-\mu-1, m-1, p^s)$. Its generator polynomial is proved to have $\alpha^{t(p^s-1)}$ as a root if and only if

$$W_{p^s}[t(p^s-1)p^{\ell}] \leq \mu(p^s-1)$$

for any ℓ, $0 \leq \ell < s$, where $0 < t \leq \frac{p^{ms}-1}{p^s-1}$.

From the above definition, Theorem 1 and Theorem 2, it is obvious that the $C(m-\mu-1, m-1, p^s)$ code is identical with the $(\frac{p^{ms}-1}{p^s-1}, m, s, \mu, p)$-polynomial code, or the μth order projective geometry code of Goethals and Delsarte is the dual code of a $(\frac{p^{ms}-1}{p^s-1}, m, s, \mu, p)$-polynomial code.

Now, we define a μth order q-ary P. G. code as the dual of a $(\frac{q^{ms}-1}{q^s-1}, m, s, \mu, q)$-polynomial code. A μth order q-ary P. G. code contains all the $(m-\mu-1)$-flats of P. G. $(m-1, q^s)$ in its null space. Thus, this code is $(m-\mu-1)$-step orthogonalizable [18].

5. Conclusion

In this paper, we have studied two subclasses of polynomial codes. The relationships between these two subclasses of polynomial codes and geometry codes have been established. A BCH lower bound on the minimum distance of Euclidean geometry codes has been obtained. This bound indicates that E.G. codes are in general more powerful than Weldon's majority-logic decoding scheme has been able to demonstrate. More efficient use of E.G. codes depends on a more effective decoding scheme. A general formula for the number of parity check digits of a polynomial code studied in this paper has not yet been obtained.

Acknowledgments

Part of the actual work described herein was done by Drs. T. Kasami, W. W. Peterson and the author jointly. The author wishes to thank Dr. E. J. Weldon, Jr. of the University of Hawaii, for several stimulating discussions on geometry codes.

REFERENCES

1. Lin, S., Peterson, W. W., and Weldon, E. J., Jr., "Problems in Information Processing," Final Report, Air Force Cambridge Research Labs., Bedford, Massachusetts, (1967).

2. Kasami, T., Lin, S., and Peterson, W. W., "Polynomial Codes," Accepted for publication in IEEE Trans. on Information Theory (1968).

3. Rudolph, L. D., "Geometric Configuration and Majority Logic Decodable Codes," MEE Thesis, University of Oklahoma (1964).

4. Weldon, E. J., Jr., "Difference-Set Cyclic Codes," BSTJ, 45, 1045-1055 (1966).

5. Graham, R. L., and J. MacWilliams, "On the Number of Parity Checks in Difference-Set Cyclic Codes," BSTJ, 45, 1046-1070 (1966).

6. Kasami, T., Lin, S., and Peterson, W. W., "Some Results on Cyclic Codes Which are Invariant Under the Affine Group," Scientific Report AFCRL-66-622, Air Force Cambridge Research Labs., Bedford, Massachusetts, (1966).

7. Weldon, E. J., Jr., "Euclidean Geometry Cyclic Codes, "Proceedings of Symposium of Combinatorial Mathematics at the University of North Carolina, Chapel Hill, North Carolina, (April 1967).

8. Weldon, E. J., Jr., "Non-primitive Reed-Muller Codes," to be published IEEE Trans., IT-13, (March 1968).

9. Goethals, J. M., and Delsarte, P., "On a Class of Majority-Logic Decodable Cyclic Codes," IEEE Trans, IT-13, (March 1968).

10. Kasami, T., Lin, S., and Peterson, W. W., "New Generalizations of the Reed-Muller Codes - Part I: Primitive Codes," to be published, IEEE Trans., Vol. IT-13, (March 1968).

11. MacWilliams, F. J., and Mann, H. B., "On the p-Rank of the Design Matrix of a Difference Set," MRC Technical Summary Report No. 803, Mathematics Research Center, United States Army, University of Wisconsin, (October 1967).

12. Kasami, T., Lin, S., and Peterson, W.W., "Further Results on Generalized Reed-Muller Codes" Accepted to publish in the Journal of the Institute of Communications Engineers of Japan, (1968).

13. Berlekamp, E. R., Algebraic Coding Theory, McGraw Hill, (1968).

14. Delsarte, P., "A Geometric Approach to a Class of Cyclic Codes," Report R68, M.B.L.E., Laboratoire de Recherches, Bruxelles, Belgium, October, 1967.

15. Smith, K. J. C., "Majority Decodable Codes Derived from Finite Geometries," Institute of Statistics Mimeo Series No. 561, University of North Carolina, Chapel Hill, North Carolina, December 1967.

16. Peterson, W. W., and Weldon, E. J., Jr., Error-Correcting Codes, Second Edition, Wiley (1969).

17. Weldon, E. J., Jr., "Some Results on Majority Logic Decoding" To appear in the Proceedings of Symposium on Error Correcting Codes, Mathematics Research Center, U. S. Army, University of Wisconsin, May 1968.

18. Massey, J. L., Threshold Decoding, MIT Press, (1963).

19. Bose, R. C., and Ray, D. K. - Chaudhuri, "On a Class of Error Correcting Binary Group Codes," Information and Control, Vol. 3, pp. 68-79, (1960).

This work was supported in part by NASA Grant NGR-12-001-046.

Reprinted from *J. Combinatorial Theory*, **2**, 243–257 (1967)

26

On Tactical Configurations and Error-Correcting Codes*

E. F. Assmus, Jr.[†] and H. F. Mattson, Jr.

Applied Research Laboratory, Sylvania Electronic Systems, Waltham, Massachussetts

Communicated by Marshall Hall

1. Introduction

The study of error-correcting codes began in 1948 when Shannon's Fundamental Theorem pointed out their importance. In 1956 Paige [37] found a close connection between an error-correcting code and a classical combinatorial design. One purpose of this paper is to prove several results casting further light on the connection between error-correcting codes and combinatorial designs.

It was Steiner [1] in 1852 who gave a succinct statement to the combinatorial problems with which we are concerned. These problems have never been fully solved,[1] and our other purpose is to point out previously unrecognized difficulties of the problems. Briefly, in stating the problem, Steiner drew out some questionable necessary conditions. The problem as he stated it seems to be much more difficult than apparently he supposed, as we explain in Parts 5 and 6.

Theorems 1 and 2, proved here, were stated without proof in [39], where the interested reader will find a brief history of these problems. Propositions 1 and 2 appeared without proof in [38]; a full discussion of the Golay (11,6) code and its connection with the Mathieu groups M_{11} and M_{12} can be found there. This code, together with the (23, 12) Golay

* The work reported herein was partially supported by Air Force Contract No. AF19(604)-8516.

† Lehigh University, Bethlehem, Pennsylvania.

[1] The solution in [9] is unjustifiably claimed to be complete.

code, is an example of a so-called quadratic-residue code. Quadratic-residue codes form a doubly infinite class of codes with interesting design properties; the authors will discuss these in a forthcoming paper.

Theorem 1 has several interesting corollaries which are included in Part 4 as remarks and examples.

The bibliography, though incomplete, does contain several not often cited earlier papers on combinatorial problems and a large part of the literature on perfect codes. We have commented briefly on some of the entries and cited the review in *Mathematical Reviews* where known.

2. Tactical Configurations and Codes

A *tactical configuration* on a set S of cardinality n is a collection $\mathscr{D}$ of subsets of S such that each member of $\mathscr{D}$ has cardinality d, and such that every subset of S of cardinality t is contained in precisely λ distinct members of $\mathscr{D}$. The positive integers λ, t, d, and n are the parameters of $\mathscr{D}$, and to avoid trivial configurations we usually take $0 < t < d < n$. We shall speak of such a tactical configuration as a $\lambda; t - d - n$ configuration or as a tactical configuration of type $\lambda; t - d - n$. This combinatorial notion encompasses (a) finite affine planes (take $\lambda = 1$, $t = 2$, $d = m$, $n = m^2$), (b) finite projective planes (take $\lambda = 1$, $t = 2$, $d = m + 1$, $n = m^2 + m + 1$), (c) balanced incomplete block designs (take $t = 2$), etc. For details we refer the reader to Ryser [8].

A *Steiner system* on S is a $1; t - d - n$ configuration; that is, every subset of t elements of S is contained is precisely one of the chosen subsets in $\mathscr{D}$ (of d elements each).

A *code* is most generally defined as a pair (A, S) where A is a (non-empty) set (to be thought of as " words") and S is a (non-empty) finite set of functions defined on A with values in some "alphabet" set F containing at least two elements. The functions in S are subject to the restriction that if $a, b \in A$, then $(a)f = (b)f$ for all $f \in S$ implies $a = b$. That is, the functions must distinguish the points of A. More concretely, one may order the functions of S as $f_0, \ldots, f_{n-1}$ and then consider the *concrete realization* of (A, S), namely, the set of all ordered n-tuples over F of the form

$$((a)f_0, (a)f_1, \ldots, (a)f_{n-1}), \qquad a \in A \tag{1}$$

The f in S are the *coordinate functions* of the code.

A most important special case arises when F is a field, A is a vector space of finite dimension k over F, and S is a set of linear functionals on A. We then call (A, S) an (n, k) *code over* F. In the present paper we shall also find the non-linear case useful.

If (A, S) is a code and $a, b \in A$, the *distance* between a and b is defined as the number of functions f in S such that $(a)f \neq (b)f$. In the concrete realization, the distance between a and b is the number of coordinate-places where they differ. The *weight* of a code word is its distance from (0, 0, ..., 0) when F is a field; in other words, the weight is then simply the number of non-0 coordinates. If the minimum distance between any two distinct code words of A is d, then the code is capable of correcting any e or fewer errors, where $e = [(d-1)/2]$. This means that, if during transmission of the symbols $(a)f_0, \ldots, (a)f_{n-1}$, e or fewer of them are changed, then we can in principle correct them. The reason is that the spheres of radius e about the set of all code-points $((b)f_0, \ldots, (b)f_{n-1})$, $b \in A$, are disjoint. Thus on changing e or fewer coordinates of the center point of such a sphere, we obtain a point still in that sphere.

A code is called *perfect* if this set of (disjoint) spheres of radius e about the code-points entirely exhausts the containing space $F^n = Fx \ldots xF(n$ times). In this case d is necessarily odd.

3. Relations between Codes and Tactical Configurations

Let (A, S) be a linear (n, k) code over the field $F = GF(q)$. Let d be the minimum distance between code-vectors; in this linear case, d is also the minimum non-0 weight in A because the distance function is invariant under translation. Let $\mathscr{D}$ be the collection of all d-sets[2] $D \subset S$ such that there is a minimum-weight vector of A with its d non-0 coordinates just those of D. Then we have:

Theorem 1. *The linear code (A, S) of minimum distance $d = 2e + 1$ is perfect if and only if $\mathscr{D}$ is a tactical configuration on S of type $\lambda = (q-1)^e$; $(e+1)$ - - d - - n.*

Proof. Let the code be perfect. An $(e + 1)$-set can be filled with non-0 coordinate-values in $(q-1)^{e+1}$ ways. Each such choice yields a vector

[2] A j-set is simply a set of cardinality j. We shall use this terminology throughout.

of weight $e + 1$ in F^n, which must be at distance e or less from some code-vector (in the concrete realization). This code-vector must be of weight d, and it must agree with the original vector of weight $e + 1$ on all of the non-0 coordinates of the latter. Therefore the code-vectors arising in this way are all distinct; their sets D of non-0 coordinates, however, are $(q - 1)^e$ in number because scalar multiplication preserves these relationships. (Notice that we also use the linearity to prove that a d-set can "hold" at most one code-vector up to scalar multiplication.)

For the converse, we need only show that any point of F^n is within distance e of some code-point. If not, let x be a point of F^n of smallest weight which is not within distance e of any code-vector. Then x has weight at least $e + 1$; let E be a set of $e + 1$ non-0 coordinates of x. By assumption E is contained in precisely $(q - 1)^e$ sets D in $\mathscr{D}$. On scalar multiplying we obtain, from each such D, $q - 1$ code-vectors of weight d giving a total of $(q - 1)^{e+1}$ distinct code-vectors of weight d with the associated sets D containing E. These code-vectors are necessarily all different from each other on E; otherwise there would be a non-0 code-vector of weight less than d (by linearity). Therefore one of these code-vectors, say a, agrees precisely with x on E, so that on subtraction we reduce the weight of x by at least 1, contradicting our choice of x, since if $x - a$ is in the sphere of radius e about b (in A), then x is in the sphere of radius e about $a + b$. Q.E.D.

REMARK. When $q = 2$, then half of Theorem 1 holds even for non-linear codes (containing 0). That is, *a perfect code over $GF(2)$ of minimum distance $d = 2e + 1$ yields a Steiner system of type $(e + 1)$ - - d - - n in the above way* (provided $(0, 0, ..., 0)$ is in the concrete realization). The proof is even simpler than before and will be omitted.

Notice that without linearity the converse cannot hold in general, because the code consisting of the weight-d vectors and 0 from a perfect linear code would still yield the proper configuration but of course would no longer be perfect in general.

We now prove the following:

PROPOSITION 1. Let (A, S) be a not necessarily linear perfect code over $GF(2)$ containing 0. Set $f_\infty = \Sigma_{f \in S} f$ and $S_\infty = S \cup f_\infty$. Let $\mathscr{D}_\infty$ be the sets of non-zero coordinates of the minimum-weight vectors of (A, S_∞). Then $\mathscr{D}_\infty$ is a Steiner system on S_∞ of type $(e + 2)$- -$(d + 1)$ - -$(n + 1)$.

PROOF. The result is true when $d = n$, if one admits trivial configurations. Take $d < n$ and observe that f_∞ is not already in S, for, if it were, then the minimum weight would be even (since, by a counting argument, for any $f \in S$ there is a minimum-weight code-point which is 0 at f). Now an $(e + 2)$-set E from S_∞ containing f_∞ is obviously contained in exactly one member of $\mathscr{D}_\infty$; if, on the other hand, $E \subset S$, and if E is not contained in any member of $\mathscr{D}$, then the vector in F^n with 1's at E must lie in a sphere of radius e about a code-point of weight $d + 1 = 2e + 2$ from (A, S). Q.E.D.

We also have

PROPOSITION 2. Let (A, S) be a code over $GF(q)$, containing 0 but not necessarily linear. Suppose the minimum distance is $d > 1$ and that the set $\mathscr{D}$ of non-0 coordinates of weight-d code-points forms a λ; t-d-n configuration on S. Set $S' = S - \{f\}$, where f is any coordinate function of (A, S). Then (A, S') is a code for which $\mathscr{D}'$ (defined analogously) forms a λ; $(t - 1)$- -$(d - 1)$- -$(n - 1)$ configuration on S'.

PROOF. To show that (A, S') is a code we need only show that for any two code words a and b of A, the statement that $af' = bf'$ holds for all $f' \in S'$ implies that $a = b$. We know this is true if also $af = bf$, and if not then we would have the distance between a and b in (A, S) equal to 1, contrary to our assumption that $d > 1$.

Obviously the new code has minimum distance $d - 1$; the members of $\mathscr{D}'$ are those of $\mathscr{D}$ which contain f, with f then removed, so to speak; the rest is self-evident. Q.E.D.

4. EXAMPLES AND REMARKS

1. *The H-Golay Codes.*[3] Let F be $GF(q)$ and let A be an m-dimensional vector-space over F, $m \geq 2$. The set of all non-0 F-linear functionals on A is partioned into subsets of $q - 1$ elements by the action of scalar multiplication. Let S be the set of $n = (q^m - 1)/(q - 1)$ functionals

[3] These codes are usually called Hamming codes, but Hamming [22] invented only the one for $n = 7$, $q = 2$. Golay [21] preceded this publication with his announcement that this (7, 4) binary code could be generalized to a perfect, single-error-correcting $(n, n - m)$ code as above for any prime q.

consisting of some one from each of the above subsets. Then (A, S) is an (n, m) code. We choose a concrete realization of (A, S) by ordering S, and we define the subspace B of F^n orthogonal to the concrete realization of (A, S) under the usual inner product as the $(n, n-m)$ H-Golay code over $GF(q)$. That is, B is the space of all linear relations on the functionals in S. Thus the minimum distance d in B is at least 3. Now we can either show easily that $d = 3$ and then count to show that B is perfect, or we can show directly that for every pair $f, g\ \omega \in S$ there are exactly $q-1$ different $h \in S$ such that $\alpha f + \beta g + \gamma h = 0$ for $\alpha, \beta, \gamma \in F^{\times}$. In either case we arrive immediately at the following class of tactical configurations:

$$q-1; 2 \text{ - - } 3 \text{ - - } \frac{q^m-1}{q-1} \qquad q \text{ prime power}; m = 2, 3, 4, \ldots$$

(For $q = 2$, these are Steiner triple systems.) This configuration is of course that arising from the points and lines of projective $(m-1)$-dimensional space over F; that every line consists of $q+1$ points is equivalent to the above result.

2. *Vasil'ev Codes.* Vasil'ev [31] has recently discovered a large class of non-linear perfect codes over $GF(2)$ with $n = 2^m - 1$ and $d = 3$. Let B be any perfect code over $GF(2)$ with $d = 3$ which contains 0; then B has $2^m - 1$ coordinate places and 2^k code-points, where $k = 2^m - 1 - m$. Let π be any function from B to $GF(2)$ such that $\pi(0) = 0$. Let p on F^n be parity (the sum of all coordinates). Then the new code C is the set of all words of the form $(v; v + a; p(v) + \pi(a))$ of $2n + 1 = 2^{m+1} - 1$ coordinates, where $v \in F^n$ and $a \in B$. It is easy to show that that this code has minimum distance d at least 3 and that it consists of $2^n \cdot 2^{n-m}$ words; it is therefore perfect and $d = 3$. C contains 0, but if B is linear and π is non-linear, then C is non-linear.

3. *The Golay and Related Codes.* Let z be a primitive n-th root of unity over $GF(q) = F$ (assume $(n, q) = 1$ of course). Let $K = F(z)$ and let T be the trace from K to F. Define a set S of linear functional on $A = F \times K$ by $S = \{f_0, \ldots, f_{n-1}\}$, where

$$(c_0, c)f_i = c_0 + T(cz^i), \qquad c_0 \in F,\ c \in K; \qquad i = 0, 1, \ldots, n-1.$$

Then (A, S) is a linear (n, k) code, where $k - 1$ is the degree of K over F.

There are two special cases of this construction which are known to yield perfect codes. They are

(a) $q = 2$ and $n = 23$. Then $k = 12$ and $d = 7$ [25]. The count

$$2^{12} \cdot \left(1 + 23 + \binom{23}{2} + \binom{23}{3}\right) = 2^{23}$$

then shows that the code is perfect. This code yields a Steiner system of type 4-7-23 by Theorem 1, as Paige [37] observed, though he presented the code in quite a different form.

Applying Propositions 1 and 2 to this code we obtain Steiner systems of types 5-8-24 and 3-6-22. Applying Proposition 2 again one obtains a Steiner system of type 2-5-21, which is, of course, the projective plane of order 4. Presumably the fact that this plane has such unusual " extensions" explains its peculiarities. See [42].

(b) $q = 3$ and $n = 11$. Then $k = 6$ and $d = 5$ [38]. Since

$$3^{6} \cdot \left(1 + 2.11 + 4 \cdot \binom{11}{2}\right) = 3^{11},$$

the code is perfect. It yields then a tactical configuration of type 4; 3-5-11.

The set $\mathscr{D}$ for this code also yields a Steiner system of type 4-5-11. For a given 4-set E, when assigned coordinates ± 1, is a weight-4 vector which must lie in a sphere of radius 2 about a code-vector of weight 5 or 6. If all sixteen choices of $\pm$ 1's yielded weight-6 vectors there would be two code-vectors of weight 6 with five non-0 coordinate places in common; this could not happen because it would imply $d < 5$. In fact, every 4-set is contained in a unique member of $\mathscr{D}$ and in three distinct 6-sets arising from weight-6 code-vectors.

If to this (11,6) code we append a functional $f_\infty = - \sum_{i=0}^{10} f_i$, then the resulting code gives rise to a Steiner system of type 5-6-12. That the minimum weight in the new code is 6 and that $a f_\infty = 0$ if a has weight 6 ($a \in A$) require some proof (see [38]). Assuming these facts, it suffices to consider a 5-set E of $S = f_0, \ldots, f_{10}$. We have just observed that any 4-subset of E is contained in 5- and 6-subsets of S belonging to code-vectors of the perfect code; the extra coordinates exhaust the seven remaining coordinates. Hence E is contained in exactly one of these, which either is or gives rise to, a weight-6 vector of the new code (cf. [38]).

4. There are no other perfect codes known. See [27, 28, 29, 30, 32, 33].

In particular there are no known perfect codes with $d > 7$. Correspondingly, there are no known tactical configurations with $t \geq 6$.

5. It is well known and easy to show that a necessary condition for the existence of a tactical configuration of type λ; t-d-n is that

$$\frac{\lambda \binom{n-h}{t-h}}{\binom{d-h}{t-h}}$$

should be an integer for $h = 0, 1, \ldots, t - 1$. Thus Theorem 1 yields *necessary conditions for the existence of a perfect code over* $GF(q)$; these conditions for $h = 0$ and $h = e$ in the case $q = 2$ have been found by Shapiro and Slotnick [28] and for all h by Lloyd [26].

6. For $h = 0$, the expression above is simply the number of d-sets in the configuration. Thus, *the number of code-vectors of minimal weight in a linear perfect code is*

$$(q-1)\frac{\lambda \binom{n}{e+1}}{\binom{d}{e+1}} = (q-1)^{e+1}\frac{\binom{n}{e+1}}{\binom{d}{e+1}}$$

7. Still another application of Theorem 1 allows us to show that *the code-vectors of minimal weight in a linear perfect code span the code-space*, for they span a subspace of the code-space with the same minimal weight vectors and hence yield the tactical configuration, whence a perfect code. But, for a perfect linear code, d and n determine the dimension k and hence they span the whole code.

5. Closed Steiner Systems

A problem going back to Steiner [1] is to find collections of subsets of a given set S of n elements with the following properties:

$\mathscr{D}_3$ is a collection of 3-sets of S forming a Steiner triple system on S (i.e., a 1; 2-3-n configuration);

$\mathscr{D}_4$ is a collection of 4-sets of S such that every 3-set of S not a member of $\mathscr{D}_3$ is contained in exactly one member of $\mathscr{D}_4$, and no element of $\mathscr{D}_4$ contains an element of $\mathscr{D}_3$;

$\mathscr{D}_5$ is a collection of 5-sets of S such that every 4-set of S not containing any member of $\mathscr{D}_3 \cup \mathscr{D}_4$ is contained in exactly one member of $\mathscr{D}_5$, and no element of $\mathscr{D}_5$ contains an element of $\mathscr{D}_3 \cup \mathscr{D}_4$; and so on, up to $\mathscr{D}_k$.

Such a system is called *closed* if $\mathscr{D}_k$ is non-empty but every k-set of S contains some member of $\mathscr{D}_3 \cup \mathscr{D}_4 \cup \dots \cup \mathscr{D}_k$.

We may define the above k-th *order Steiner system* more compactly as follows: $\mathscr{D}_i$ is a collection of non-empty i-sets of S, $i = 3, 4, \dots, k$; a *free* t-set of S is one not containing any member of $\mathscr{D}_3 \cup \dots \cup \mathscr{D}_t$; for $t = 2, 3, \dots, k - 1$, every free t-set is contained in precisely one member of $\mathscr{D}_{t+1}$ and every proper subset of every element of $\mathscr{D}_{t+1}$ is free. Then the system is closed if and only if there are no free k-sets.

We have followed the terminology of [9] except that there a free t-set is one not containing any member of $\mathscr{D}_3 \cup \dots \cup \mathscr{D}_{t-1}$; they then speak of free t-sets not members of $\mathscr{D}_t$, which are our free t-sets.

We now present a theorem from [9]; the result and our proof were discovered independently.

THEOREM 2. *There exists a closed k-th order Steiner system on* $n = 2^{k-1} - 1$ *points, for* $k = 3, 4, \dots$.

PROOF. We realize the desired system by means of the H-Golay code B over $GF(2)$ of $n = 2^{k-1} - 1$ coordinates. Recall that B is the linear space of all n-tuples over $GF(2)$ which are relations on the n non-0 functionals (in some fixed order) on $(k - 1)$-dimensional space over $GF(2)$. Let S be this set of functionals. We define $\mathscr{D}_t$ as the collection of all t-sets on S which are linearly dependent but which have all proper subsets linearly independent. This is the same as saying that $\mathscr{D}_t$ is the collection of all sets M of non-0 positions of code-vectors of weight t such that no code-vector of smaller weight has its non-0 positions included in M. Obviously now a free t-set is a linearly independent t-set.

Now that every free t-set of S is contained in exactly one member of $\mathscr{D}_{t+1}$ follows immediately from the definition of the code, for $t = 2, 3, \dots, n - 1$. (That is, a free t-set cannot "hold" a code-vector; thus the sum of all those t functionals is a non-0 functional not one of the t we started with; these must now constitute a member of $\mathscr{D}_{t+1}$.)

$\mathscr{D}_{k+1}$ is empty because every set of k functionals is linearly dependent; thus there are no free k-sets.

We now generalize Theorem 2, by finding the design properties of the general H-Golay code B over $GF(q)$ for any prime power q, where

$$n = (q^{k-1} - 1)/(q - 1).$$

From Theorem 1 we know that the set $\mathscr{D}_3$ of non-0 positions of weight-3 code-vectors yields a λ; 2-3-n configuration on the set S of the n functionals, where $\lambda = q - 1$. A *free* t-set is defined as any set of t functionals of S not holding a code-vector (means in this case: linearly independent!). Now, since the code is perfect with $d = 3$, every assignment of non-0 coordinates to the places of a free t-set E produces a code-vector at distance 1 from the resulting weight-t vector in $F^n(F = GF(q))$; the code-vector can only be of weight $t + 1$. It therefore agrees with the weight-t vector at each point of E; this means that a different $(t + 1)$-st functional must arise from different assignments of coordinates to E (except for scalar multiplication throughout). Therefore every free t-set is contained in exactly $(q - 1)^{t-1} = \lambda^{t-1}$ different members of $\mathscr{D}_{t+1}$. This agrees with Theorem 1; and there are no free k-sets, as before.

We are thus led to define a k-th *order tactical configuration* as a collection of subsets $\mathscr{D}_3, \mathscr{D}_4, \ldots, \mathscr{D}_k$ of a set S of n points such that $\mathscr{D}_i$ consists of i-sets of S and, for $t = 2, 3, \ldots, k - 1$,

(i) every free t-set on S is contained in exactly λ^{t-1} members of $\mathscr{D}_{t+1}$;

(ii) every proper subset of every member of $\mathscr{D}_{t+1}$ is free. A free t-set is one not containing any member of $\mathscr{D}_3 \cup \mathscr{D}_4 \cup \cdots \cup \mathscr{D}_t$. Such a system is called *closed* if $\mathscr{D}_k$ is non-empty and there are no free k-sets. We have proved

THEOREM 3. Let q be a prime power and set $\lambda = q - 1$. If

$$n = (q^{k-1} - 1)/(q - 1),$$

then there exists a closed k-th order tactical configuration on n points.

We now count the *number* N_t *of free t-sets* connected with a k-th order tactical configuration on a set S of n points. The "multiplicity of inclusion" is λ. Obviously we have

$$N_2 = \binom{n}{2} = \frac{1}{2!}\, n(n - 1)$$

We attempt to find a design on the collection of all free sets, since every subset of a free set is free. Thus, how many free 3-sets is a given pair contained in? The pair is in exactly λ different triples of $\mathscr{D}_3$; therefore it is in exactly $n - \lambda - 2$ free 3-sets. Therefore

$$3N_3 = (n - \lambda - 2)N_2\,,$$

$$N_3 = \frac{1}{3!}\, n(n-1)\,(n - \lambda - 2).$$

Now a given free 3-set is contained in exactly λ^2 different members of $\mathscr{D}_4$; however, each of its 3 pairs is contained in λ members of $\mathscr{D}_3$. If all these resulting fourth points are different from each other (as happens in the H-Golay codes, where "free" reduces to "linearly independent"), then the free triple is contained in exactly $n - \lambda^2 - 3\lambda - 3$ free 4-sets. But in general we can only conclude that this is a lower bound. Therefore

$$4N_4 \geq (n - \lambda^2 - 3\lambda - 3)N_3$$

with equality holding in the linear situation of Theorem 3. Note that equality holds here also when $\lambda = 1$, because no two members of $\mathscr{D}_3$ may intersect in two points. But even for $\lambda = 1$ equality need not hold at $t = 5$.

In general, with a free t-set we must exclude perhaps as many as

$$\binom{t}{3}\lambda + \binom{t}{3}\lambda^2 + \cdots + \binom{t}{t}\lambda^{t-1}$$

of the remaining $n - t$ points; that is,

$$(t+1)\,N_{t+1} \geq \left(n - \frac{(\lambda+1)^t - 1}{\lambda}\right) N_t\,,$$

with equality in the linear case previously mentioned. Therefore

PROPOSITION 3. The number N_t of free t-sets associated with a k-th order tactical configuratiin satisfies

$$N_t \geq \frac{1}{t!}\prod_{i=0}^{t-1}\left(n - \frac{(\lambda+1)^i - 1}{\lambda}\right), \quad t = 4, \ldots, k; \tag{2}$$

and equality holds when the system is realized from a generalized

H-Golay code as in Theorem 3. In all cases we have equality in (2) for $t = 2$ and 3, and also for $t = 4$ when $\lambda = 1$.[4]

Remark. The free sets associated with the linear case are the linearly independent sets from projective $(k - 1)$-space over $GF(q)$, as in Theorem 3. We have proved that the number of such t-sets is given by equality in (2). Furthermore, the free t-sets form the following design: For $\lambda = q - 1$, every pair on S is contained in exactly $n - \lambda - 2$ free 3-sets; every free 3-set is contained in exactly $n - \lambda^2 - 3\lambda - 3$ free 4-sets; . . . ; every free t-set is contained in exactly $n - ((\lambda + 1)^t - 1)/\lambda$ free $(t + 1)$-sets, for $t = 2, 3, ..., k - 1$.

Since $n = ((\lambda + 1)^{k-1} - 1)/\lambda$, the process stops at $t = k - 1$.

6. A. Counterexample

In posing the problem of the existence of k-th order Steiner systems, Steiner [1] gave what he claimed are necessary conditions. The conditions are correct for $k = 3$ (so-called Steiner triple systems) and Kirkman [10] proved them sufficient. The conditions are also correct for $k = 4$, and Hanani [5] has proved them sufficient. These conditions, however, may very well be incorrect for $k > 4$. Not Steiner [1], Netto [3, pp. 202–204], or Hanani and Schonheim [9] offer any proof that these conditions, namely, that n is odd and that

$$\frac{1}{t!} \prod_{i=0}^{t-2} (n + 1 - 2^i) \text{ is an integer for } 3 \leq t \leq k, \tag{3}$$

are in fact necessary. Notice that equality in (2) for $\lambda = 1$ would imply (3). Steiner may well have assumed this equality, and it is asserted in [9, p. 140, (2)]. This equality is false in general (as the following counterexample shows), and for complete Steiner systems it is unproved.

Suppose B is any perfect code over $GF(2)$ containing 0 with $n = 15$. Consider $\mathscr{D}_3$, $\mathscr{D}_4$, and $\mathscr{D}_5$, the weight-3, weight-4, and weight-5 code-points. It is easy to see that they furnish a fifth-order Steiner system.

---

[4] In the case $\lambda = 1$, Steiner [1] seems to have asserted equality in (2) for all t. Netto [3] quoted Steiner's assertion verbatim, and in [9] equality is asserted. As our example will show, there exist fifth-order Steiner systems where the strict inequality holds in (2).

If B is linear it is in fact closed since any five of the coordinate functionals of the orthogonal complement are linearly dependent. Now, Vasil'ev's construction (see Part 4.2) furnishes many non-linear perfect codes with $n = 15$. We next show that for some of these there are free 5-sets. Thus, the number of free 5-sets depends on how $\mathscr{D}_4$ and $\mathscr{D}_5$ are introduced.

So, take the perfect code over $GF(2)$ with $n=7$; it consists of (0000000), (1111111), (1101000) together with its cyclic shifts, and (0010111) together with its cyclic shifts. Take π to be 0 except on one weight-4 vector, say (0010111). One checks easily that the following 5-sets are free in the constructed code:

1000000; 0010111; 0

0100000; 0010111; 0

0001000; 0010111; 0

We have shown, therefore, that there are two fifth-order Steiner systems on a set with 15 elements, one having free 5-sets, the other not.

References

On Tactical Configurations

1. J. Steiner, Combinatorische Aufgabe, *J. Reine Angew. Math.* **45** (1853), 181–182.
2. E. H. Moore, Tactical Memoranda, *Amer. J, Math.* **18** (1896), 264–303.
3. E. Netto, *Lehrbuch der Kombinatorik*, 2nd ed., Teubner, Leipzig, 1927; reprinted by Chelsea, New York.
4. A. Emch, Triple and Multiple Systems, Their Geometric Configurations and Groups, *Trans. Amer. Math. Soc.* **31** (1929), 25–42.
5. H. Hanani, On Quadruple Systems, *Canad. J. Math.* **12** (1960), 145–157.
6. H. Hanani, The Existence and Construction of Balanced Incomplete Block Designs, *Ann. Math. Statist.* **32** (1961), 361–368. (*MR* **29** (1965), ♯4161).
7. H. Hanani, On Some Tactical Configurations, *Canad. J. Math.* **15** (1963), 702–722 (*MR* **28** (1964), ♯1136).
8. H. J. Ryser, *Combinatorial Mathematics*, Wiley, New York, 1963. A standard reference; contains a rather large bibliography.
9. H. Hanani and J. Schonheim, On Steiner Systems, *Israel J. Math.* **2** (1964), 139–142 (*MR* **31** (1966), ♯73).

On Steiner Triple Systems

10. REV. THOMAS KIRKMAN, On a Problem in Combinations, *Cambridge and Dublin Math. J.* **2** (1847), 191–204.

11. A. CAYLEY, On the Triadic Arrangements of Seven and Fifteen Things, *Philos. Mag.* **37** (1850), 50–53 (*Collected Mathematical Papers*, I, 481–484).

12. A. CAYLEY, On a Tactical Theorem Relating to the Triads of Fifteen Things, *London, Edinburgh, and Dublin Philos. Mag. and J.* **25** (1863), Ser. 4, 59–61 (*Collected Mathematical Papers*, V. 95–97).

13. J. J. SYLVESTER, Elementary Researches in the Analysis of Combinatorial Aggregation, *Philos. Mag.* **24** (1844), 285–296 (*Collected Mathematical Papers*, I, 91–102).

14. J. J. SYLVESTER, Note on the Historical Origin of the Unsymmetrical Six-Valued Function of Six Letters, *Philos. Mag.* **21** (1861), 369–377 (*Collected Mathematical Papers*, II, 264–271).

15. J. J. SYLVESTER, Remark on the Tactic of Nine Elements, *Philos. Mag.* **22** (1861), 144–147 (*Collected Mathematical Papers*, II, 286–289).

16. J. J. SYLVESTER, Note on a Nine Schoolgirls Problem, *Messenger of Math.* **22** (1892–1893), 159–160, and (Correction) 192 (*Collected Mathematical Papers*, IV, 732–733).

17. M. REISS, Über eine Steinersche Combinatorische Aufgabe Welche im 45sten Bande Dieses Journals, Seite 181, Gestellt Worden ist, *Crelle's J.* **56** (1859), 326–344.

18. W. BURNSIDE, On an Application of the Theory of Groups to Kirkman's Problem, *Messenger Math.* **23** (1893–1894), 137–143.

19. E. H. MOORE, Concerning Triple Systems, *Math. Ann.* **43** (1893), 271–285.

20. M. HALL, JR., AND J. D. SWIFT, Determination of Steiner Triple Systems of Order 15, *Math. Tables Aids Comput.* **9** (1955), 146–152 (*MR* **18** (1957), p. 192).

On Error-Correcting Codes

21. M. J. E. GOLAY, Notes on Digital Coding, *Proc. Inst. Radio Engrs.* **37** (1949), Correspondence, 657.

22. R. W. HAMMING, Error Détecting and Error Correcting Codes, *Bell System Tech. J.* **29** (1950), 147–160 (*MR* **12** (1951), p. 35).

23. M. J. GOLAY, Binary Coding, *IRE Trans.* PGIT–4 (1954), 23–28.

24. W. W. PETERSON, *Error-Correcting Codes*, The M.I.T. Press Cambridge, Mass. 1961. (*MR* **22** (1961), ♯12003). The standard reference; contains a large bibliography.

25. H. F. MATTSON AND G. SOLOMON, A New Treatment of Bose-Chaudhuri Codes, *J. Soc. Indust. Appl. Math.* **9** (1961), 654–669 (*MR* **24** (1962), ♯B1705). Contains the first non-exhaustive proof that the (23, 12) code of Golay is perfect.

On Perfect Codes

26. S. P. LLOYD, Binary Block Coding, *Bell System Tech. J.* **36** (1957), 517–535 (*MR* **19** (1958), p. 465).

27. M. J. E. GOLAY, Notes on the Penny-Weighing Problem, Lossless Symbol Coding with Nonprimes, etc., *IRE Trans. Inform. Theory* **4** (1958), 103–109 (*MR* **22** (1961), # 13354).

28. H. S. SHAPIRO AND D. L. SLOTNICK, On the Mathematical Theory of Error-Correcting Codes, *IBM J. Res. Develop.* **3** (1959), 25–34 (*MR* **20** (1959), # 5092).

29. C. ENGLEMAN, On Close-Packed Double Error-Correcting Codes on P Symbols, *IRE Trans. Inform. Theory* **7** (1961), 51–52 (*MR* **23** (1962), # B3075).

30. S. JOHNSON, On Perfect Error-Correcting Codes, *Memorandum RM*-3403-*PR* The RAND Corporation, Santa Monica, Calif. (1962).

31. JU. L. VASIL'EV, On Nongroup Close-Packed Codes, *Probl. Cybernetics* **8** (1962), 337–339 (*MR* **29** (1965), # 5661).

32. E. L. COHEN, A Note on Perfect Double Error-Correcting Codes on *q* Symbols, *Information and Control* **7** (1964), 381–384 (*MR* **29** (1965), # 5656). Contains a bibliography on the diophantine equations arising out of this problem. See also *Amer. Math. Soc. Notices*, **13** (1966), 245–246.

33. M. H. MCANDREW, An Algorithm for Solving a Polynomic Congruence, and Its Application to Error-Correcting Codes, *Math. Comp.* **19** (1965), 68–72 (*MR* **30** (1965), # 4612).

34. V. K. LEONT'EV, On a Problem of Close-Packed Codes, *Diskret. Analiz.* (Russian), No. 2 (1964), 56–58 (*MR* **29** (1965) # 4621).

Related Works

35. E. WITT, Die 5-fach Transitiven Gruppen von Mathieu, *Abh. Math. Sem. Hansischen Univ.* **12** (1936), 256–264.

36. E. WITT, Über Steinersche Systeme, *Abh. Math. Sem. Hansischen Univ.* **12** (1936), 265–275.

37. L. J. PAIGE, A Note on the Mathieu Groups, *Canad. J. Math.* **9** (1956), 15–18 (*MR* **18** (1957), p. 871).

38. E. F. ASSUMS, JR., AND H. F. MATTSON, Perfect Codes and the Mathieu Groups, *Arch. Math.* **17** (1966), 121–135.

39. E. F. ASSMUS, JR., AND H. F. MATTSON, JR., Steiner Systems and Perfect Codes, *Univ. N. Carolina, Inst. Statistics Mimeo Ser.*

40. R. D. CARMICHAEL, *Introduction to the Theory of Groups of Finite Order*, Ginn. Boston, 1937; reprinted by Dover, New York, 1956.

41. R. H. BRUCK, What is a Loop?, in *Studies in Modern Algebra*, Vol. 2 of M.A.A. Studies in Mathematics, Prentice-Hall, Englewood Cliffs, N.J., 1963, pp. 59–99.

42. W. L. EDGE, Some Implications of the Geometry of the 21-Point Plane, *Math. Z.* **87** (1965), 348–362 (*MR* **30** (1965), # 5209).

Printed in Italy by I.T.E.C. - Milan

Generalized Codes

VII

Editor's Comments on Papers 27, 28, and 29

27 Kasami, Lin, and Peterson: *New Generalizations of the Reed–Muller Codes; Part I: Primitive Codes*

28 Weldon: *New Generalizations of the Reed–Muller Codes; Part II: Nonprimitive Codes*

29 Kasami, Lin, and Peterson: *Polynomial Codes*

In 1968 Kasami, Lin, and Peterson [27] were successful in generalizing the binary Reed–Muller codes to the nonbinary case in an entirely natural way. This work is important not only for the new codes, but also for its construction technique and insight into the structure of Reed–Muller codes. It is, as far as I know, the first work in which the roots of the generating polynomial of the cyclic versions of these codes are specified in terms of the q-weight $w_q(h)$ of an integer h; that is,

$$w_q(h) = \sum_{i=0}^{m-1} \delta_i \quad \text{if } h = \sum_{i=0}^{m-1} \delta_i q^i;\ 0 \leqslant h \leqslant q^m - 1,\ 0 \leqslant \delta_i \leqslant q-1 .$$

Some previous work of Weldon (1967) used a different integer weight to the same end. The nonprimitive generalized Reed–Muller codes were analyzed by Weldon [28], who recognized the class of projective geometry codes as a subclass.

The most general class of cyclic codes known at this time are, undoubtedly, the polynomial codes defined and analyzed by Kasami, Lin, and Peterson [29], which includes, as subclasses, BCH, generalized Reed–Muller, Reed–Solomon, and the duals of the Euclidean and projective geometry codes. This paper is somewhat intricate and requires careful reading. The construction technique is essentially an extension of the polynomial method of Mattson and Solomon [18] to polynomials of n variables, with the code generator again specified in terms of the q-weights of integers. An interesting paper by Delsarte (1970) is very helpful in clarifying the definition of polynomial codes and placing them in a better perspective.

The three papers presented here, together with the paper by Lin in the previous section, represent only a small fraction of the work done on polynomial and finite geometry codes. The reader might also consult the paper by Gore and Cooper (1970), which corrects certain errors in Kasami, Lin, and Peterson [29].

It is my understanding that some recent work by Goppa (1971) of the U.S.S.R. constructs an important new class of linear codes. Unfortunately, this work was not available in translated form at the time of publication of this volume.

27

New Generalizations of the Reed-Muller Codes Part I: Primitive Codes

TADAO KASAMI, MEMBER, IEEE, SHU LIN, MEMBER, IEEE, AND W. WESLEY PETERSON, FELLOW, IEEE

***Abstract*—First it is shown that all binary Reed–Muller codes with one digit dropped can be made cyclic by rearranging the digits. Then a natural generalization to the nonbinary case is presented, which also includes the Reed–Muller codes and Reed–Solomon codes as special cases. The generator polynomial is characterized and the minimum weight is established. Finally, some results on weight distribution are given.**

I. Introduction

IT IS WELL KNOWN that the first-order Reed–Muller codes with one digit dropped can be made cyclic by rearranging the digits.[1] In fact, for the cyclic form, the all 1's vector and the maximal length sequences of length $2^m - 1$ generate the code. We have observed that the entire class of Reed–Muller codes are cyclic. This has led to a new generalization which is better in many cases than previous ones. This newly found mathematical structure for this class of codes has made it possible to find some other new facts about Reed–Muller codes.

First we prove that the binary Reed–Muller codes are all cyclic, in order to show the very simple ideas involved. Then in the next section more general results are given. In the final section, results on the weight distribution of Reed–Muller codes are given.

Let v_I denote a vector of length $2^m - 1$ over $GF(2)$ consisting of all 1's, and let $v_1, v_2, \cdots, v_m$ denote m linearly independent "maximal length sequences" generated by the same linear shift register. It is well known that the code generated by these $m + 1$ vectors is cyclic and is equivalent to the first-order Reed–Muller code with one digit dropped.[1] Multiplication of vectors is defined as follows. If

$$u = (u_1, u_2, \cdots, u_n)$$

$$v = (v_1, v_2, \cdots, v_n),$$

then

$$uv = (u_1v_1, u_2v_2, \cdots, u_nv_n).$$

The νth order Reed–Muller code (with one digit dropped) is generated by $v_I, v_1, v_2, \cdots, v_m$ and products of any ν or fewer of these vectors. An equivalent statement is that a vector v is a code vector in a νth order Reed–Muller code (with one digit dropped) if and only if it can be expressed as a polynomial of degree ν or less in $v_I, v_1, v_2, \cdots, v_m$.

Let T denote the operation of shifting cyclically one place to the right. Then a code is a cyclic code if and only if for each code vector v, Tv is also a code vector.

The key idea in the proof is simply the observation that T commutes not only with the addition of vectors but also with multiplication. That is,

$$T(v_1 + v_2) = Tv_1 + Tv_2, \tag{1}$$

and also

$$T(v_1v_2) = (Tv_1)(Tv_2). \tag{2}$$

Consider any code vector in the νth order Reed–Muller code with one digit dropped. It can be expressed as a polynomial of degree ν or less in $v_I, v_1, v_2, \cdots, v_m$

$$v = \sum C_i v_I^{n_{Ii}} v_1^{n_{1i}} v_2^{n_{2i}} \cdots v_m^{n_{mi}}. \tag{3}$$

Then, because of the commutative property

$$Tv = \sum C_i v_I^{n_{Ii}} (Tv_1)^{n_{1i}} (Tv_2)^{n_{2i}} \cdots (Tv_m)^{n_{mi}}. \tag{4}$$

Since the first-order Reed–Muller code with one digit dropped is cyclic, $Tv_1, Tv_2, \cdots, Tv_m$ are code vectors and hence are linear combinations of $v_I, v_1, v_2, \cdots, v_m$. It follows that Tv is a polynomial of the same degree as ν in $v_I, v_1, v_2, \cdots, v_m$ and hence is a code vector, and the νth order Reed–Muller code with one digit dropped is cyclic.

For any cyclic code over $GF(q)$ of length n relatively prime to q, there is a set of three closely related codes—the original code, another cyclic code, and a code found by adding an overall parity check to one of the cyclic codes. Let us assume that the original code includes the all 1's word as a code vector. The extended code, of length $n + 1$, is found by adding an overall check digit. The other cyclic code is found by taking the subset of code vectors in the original code whose symbols add to zero, i.e., the code words which have 1 as a root. Given any one of the three codes, the others can be found easily. In the binary case, at least, given the weight distribution of one code, the weight distribution of the others can be found in a trivial way. The generalized Reed–Muller codes as defined here are cyclic codes in which the all 1's vector is a code vector, and their dual codes have 1 as a root of every code word. These codes have length $q^m - 1$.

Manuscript received March 6, 1967. This paper was presented at the 1967 International Symposium on Information Theory, San Remo, Italy. This work was supported in part by the Air Force Cambridge Research Laboratories, Office of Aerospace Research, Bedford, Mass., under Contract AF 19(628)-4379.

T. Kasami is with the Department of Control Engineering, Faculty of Engineering Science, Osaka University, Toyonaka, Japan.
S. Lin and W. W. Peterson are with the Department of Electrical Engineering, University of Hawaii, Honolulu, Hawaii.

The codes corresponding most closely to the original Reed–Muller codes are the extended codes, of length q^m, obtained by adding an overall parity check to the code which includes the all 1's vector. Bearing this in mind will help minimize the confusion in reading the following sections.

II. A Generalization of Reed–Muller Codes

Let α be a primitive element of the Galois field $GF(q^m)$. Consider the following matrix

$$G = \begin{bmatrix} 1 & 1 & 1 & \cdots & 1 \\ 1 & \alpha & \alpha^2 & \cdots & \alpha^{n-1} \end{bmatrix}, \tag{5}$$

where $n = q^m - 1$. Each element α^j of $GF(q^m)$ can be expressed as a linear combination of $1, \alpha, \alpha^2, \cdots, \alpha^{m-1}$

$$\alpha^j = \sum_{i=0}^{m-1} a_{ij}\alpha^i \quad \text{for} \quad 0 \leq j \leq q^m - 2, \tag{6}$$

where $a_{ij} \in GF(q)$ for $0 \leq i \leq m - 1$, since $GF(q^m)$ can be considered a vector space over the field $GF(q)$, with $1, \alpha, \alpha^2, \cdots, \alpha^{m-1}$ as basis elements. The matrix G can then be rewritten as

$$G = \begin{bmatrix} v_I \\ v_0 \\ v_1 \\ \vdots \\ v_{m-1} \end{bmatrix} = \begin{bmatrix} 1 & 1 & 1 & \cdots & 1 \\ a_{00} & a_{01} & a_{02} & \cdots & a_{0,n-1} \\ a_{10} & a_{11} & a_{12} & \cdots & a_{1,n-1} \\ \vdots & \vdots & \vdots & & \vdots \\ a_{m-1,0} & a_{m-1,1} & a_{m-1,2} & \cdots & a_{m-1,n-1} \end{bmatrix}. \tag{7}$$

Now the vector product of two vectors is defined as

$$\begin{aligned} u &= (a_0, a_1, a_2, \cdots, a_{n-1}) \\ v &= (b_0, b_1, b_2, \cdots, b_{n-1}) \\ uv &= (a_0b_0, a_1b_1, a_2b_2, \cdots, a_{n-1}b_{n-1}). \end{aligned} \tag{8}$$

Let us construct a matrix from the matrix G such that the first row is v_I and other rows are all products of the powers of the elementary vectors $v_0, v_1, v_2, \cdots, v_{m-1}$

$$v_0^{k_0}v_1^{k_1} \cdots v_{m-1}^{k_{m-1}} \tag{9}$$

with $\sum_{i=0}^{m-1} k_i \leq \nu$, where $0 \leq k_i \leq q - 1$. We denote this matrix G_ν

$$G_\nu = \begin{bmatrix} v_I \\ v_0^{k_0}v_1^{k_1} \cdots v_{m-1}^{k_{m-1}} \end{bmatrix}. \tag{10}$$

Since $(v_i)^q = v_i$, ν is at most equal to $m(q - 1)$. Let the number of rows of G_ν be k. We shall call the code generated by G_ν the νth order code. It will be shown then that its dual (or the null space) is related to the code of order $\mu = m(q - 1) - \nu - 1$ generated by G_μ. For the binary case, $q = 2$, the code generated by G_ν is just the Reed–Muller code of order ν modified by dropping one digit, and its dual consists of the even-weight code vectors of the code generated by G_μ, where $\mu = m - \nu - 1$.[2] Thus we shall call these codes primitive generalized Reed–Muller (GRM) codes. This generalization is different from the generalization of Dwork and Heller,[3] and a comparision of the data in the Appendix with formulas given by Dwork and Heller shows that the codes presented here are better, at least in many cases. The GRM codes are cyclic also, and this fact along with information about the generator polynomial is given by the next theorem.

Let h be an integer such that

$$0 \leq h \leq q^m - 1. \tag{11}$$

We can express h in radix-q form

$$h = \delta_0 + \delta_1 q + \delta_2 q^2 + \cdots + \delta_{m-1}q^{m-1}, \tag{12}$$

where $0 \leq \delta_i \leq q - 1$. Define the weight of h as

$$W(h) = \sum_{i=0}^{m-1} \delta_i. \tag{13}$$

Theorem 1

The code C_d that is the null space of G_ν (i.e., C_d is the dual code of the νth order GRM code) is cyclic and α^h is a root of the generator polynomial $g_d(X)$ if and only if the weight of h is less than or equal to ν.

Proof: The first element of the jth column of G_ν is 1 and the other elements are of the form

$$a_{0j}^{k_0}a_{1j}^{k_1} \cdots a_{(m-1)j}^{k_{m-1}}, \tag{14}$$

where $\sum_{i=0}^{m-1} k_i \leq \nu$. For simplicity, name the elements in the jth column $b_{1j}(=1), b_{2j}, b_{3j}, \cdots, b_{kj}$, where k is the number of rows in G_ν.

Consider those integers h such that

$$W(h) = \sum_{i=0}^{m-1} \delta_i \leq \nu, \tag{15}$$

then

$$(\alpha^j)^h = (\alpha^j)^{\sum_{i=0}^{m-1} \delta_i q^i} \tag{16}$$

Substituting (6) into (16) gives

$$\begin{aligned} (\alpha^j)^h &= \left(\sum_{i=0}^{m-1} a_{ij}\alpha^i\right)^{\sum_{i=0}^{m-1} \delta_i q^i} \\ &= \prod_{l=0}^{m-1} \left\{\left(\sum_{i=0}^{m-1} a_{ij}\alpha^i\right)^{\delta_l q^l}\right\} \\ &= \prod_{l=0}^{m-1} \left\{\left(\sum_{i=0}^{m-1} a_{ij}\alpha^{iq^l}\right)^{\delta_l}\right\} \end{aligned} \tag{17}$$

$$= \prod_{l=0}^{m-1} \left\{\sum a_{i_1 j}a_{i_2 j} \cdots a_{i_{\delta_l} j}\alpha^{(i_1 + i_2 + \cdots + i_{\delta_l})q^l}\right\}$$

$$0 \leq i_1, i_2, \cdots, i_{\delta_l} \leq m - 1. \tag{18}$$

In the t-th factor of (18), each term in the summation has as coefficient $a_{i_1 t} a_{i_2 t} \cdots a_{i_s t}$, which is a product of δ_t of the factors $a_{0t}, a_{1t}, \cdots, a_{(m-1)t}$ with repetitions permitted. The expanded product is of the form

$$(\alpha^j)^h = \sum_l b_{lj} \alpha^{\beta_l(h)}, \qquad (19)$$

where each b_{lj} is equal to a product of powers of $a_{0j}, a_{1j}, \cdots, a_{m-1,j}$ with the sum of the exponents equal to $\delta_0 + \delta_1 + \cdots + \delta_{m-1} \leq \nu$ and $\beta_l(h)$ depends only on h and l, but does not depend on j. The same notation b_{lj} as for a row of G_ν is used here because the elements b_{lj} are the elements of G_ν.

A vector $u = (u_0, u_1, \cdots, u_{n-1})$ is a code vector of the null space of G_ν, where $n = q^m - 1$, if and only if

$$\sum_{j=0}^{n-1} u_j b_{lj} = 0 \qquad \text{for } 1 \leq l \leq k, \qquad (20)$$

where k is the number of rows of G_ν.

Consider α^h with $W(h) \leq \nu$.

$$S_h = \sum_{j=0}^{q^m-2} u_j (\alpha^j)^h = \sum_{j=0}^{q^m-2} u_j \sum_l b_{lj} \alpha^{\beta_l(h)} \qquad (21)$$

by (33), and interchanging the order of summation,

$$S_h = \sum_l \alpha^{\beta_l(h)} \sum_{j=0}^{q^m-2} u_j b_{lj} = 0 \qquad (22)$$

if $u = (u_0, u_1, u_2, \cdots, u_{n-1})$ is a code word. This implies that if $W(h) \leq \nu$, α^h is a root of every code word when considered as a polynomial. Thus this code is contained in the cyclic code that has as roots of its generator polynomial every α^h, where $W(h) \leq \nu$. The number of such roots is the same as the number of rows of G_ν, and therefore the cyclic code has at least as many check digits, and at most as many information digits as C_d. Therefore, the cyclic code and C_d are the same, i.e., C_d is cyclic and has as roots of its generators exactly all α^h for which $W(h) \leq \nu$. Q.E.D.

Corollary 2

The generator polynomial of the νth order GRM code has as roots those α^h such that

$$0 < W(h) \leq \mu, \qquad (23)$$

where $\mu = m(q - 1) - \nu - 1$. The following corollary follows from Theorem 1 and Corollary 2.

Corollary 3

Let $g_d(X)$ be the generator polynomial of the dual of the νth order GRM code. Then the generator polynomial of the GRM code of order $\mu = m(q - 1) - \nu - 1$ is $g_\mu(X) = g_d(X)/(X - 1)$.

There seems to be no simple formula for the number of information symbols in the νth order GRM code, but it is easily computed, for example, by enumerating the number of numbers h satisfying (11) and (13) for each choice of $W(h)$. The Appendix, which lists n, k, and the minimum distance for all codes of length $q^m - 1 \leq 1500$, $m > 1$, was computed in this manner.

A recursion formula for the number of numbers of weight u can be derived as follows. Consider

$$a(t) = \sum a_i t^i = (1 + t + t^2 + \cdots + t^{q-1})^m. \qquad (24)$$

When the right side is expanded, but before terms of like powers in t are collected, there are q^m terms altogether, each of the form

$$t^{\delta_0} t^{\delta_1} t^{\delta_2} \cdots t^{\delta_{m-1}} = t^i.$$

There is one term like this for each possible choice of $\delta_0, \delta_1, \cdots, \delta_{m-1}$, and thus there is a one-to-one correspondence between such terms and integers h, where

$$h = \delta_0 + \delta_1 q + \delta_2 q^2 + \cdots + \delta_{m-1} q^{m-1}.$$

Furthermore i is the weight of h. Then when terms are collected, the coefficient a_i of t^i must be the number of integers less than q^m of weight i. Then

$$a(t) = \left(\frac{1 - t^q}{1 - t}\right)^m \qquad (25)$$

so that

$$(1 - t)^m a(t) = (1 - t^q)^m$$

or

$$a(t) \sum_{j=0}^{m} (-1)^j \binom{m}{j} t^j = \sum_{j=0}^{m} (-1)^j \binom{m}{j} t^{qj}. \qquad (26)$$

When this is expanded, the coefficients of t^u must be the same on both sides, and the resulting equation is

$$a_u + \sum_{i=1}^{u} (-1)^i \binom{m}{i} a_{u-i} = 0 \text{ if } q \text{ does not divide } u$$
$$= (-1)^k \binom{m}{k} \text{ if } u = kq. \qquad (27)$$

We know that $a_0 = 1$. Knowing a_i, $i < u$, we can compute a_u from this formula, and thus we can calculate all a_u recursively.

Corollary 4

The extended GRM codes of length q^m are invariant under the doubly transitive affine group of permutations.

Proof: Kasami *et al.*[4] give as a necessary and sufficient condition for invariance of an extended cyclic code under the affine group the condition that if α^h is a root of the generator polynomial,

$$h = \delta_0 + \delta_1 q + \cdots + \delta_{m-1} q^{m-1}.$$

Then for every h'

$$h' = \delta_0' + \delta_1' q + \cdots + \delta_{m-1}' q^{m-1},$$

for which $\delta_i' \leq \delta_i$ for $0 \leq i < m$, $\alpha^{h'}$ must also be a root of the generator polynomial. By Theorem 1 the GRM codes clearly satisfy this condition, since $W(h') \leq W(h)$.

It has been shown[5] that certain balanced incomplete block (BIB) designs can be derived from binary codes which are invariant under a doubly transitive group of permutations, and thus certain BIB designs can be derived from the Reed–Muller codes of length 2^m.

III. The Minimum Weight in GRM Codes

Theorem 5

The minimum weight in the νth order GRM code is exactly $(R + 1)q^Q - 1$, where R is the remainder and Q the quotient resulting from dividing

$$\mu + 1 = m(q - 1) - \nu$$

by $q - 1$. The νth order GRM code is a subcode of the BCH code whose generator polynomial has $\alpha, \alpha^2, \cdots, \alpha^{(R+1)q^Q-2}$ as all its roots.

Proof: Let h_0 be the smallest integer such that $W(h_0) = \mu + 1$. Then all the integers less than h_0 have weight less than or equal to μ. If we write h_0 in radix-q form, it must be of the following form

$$h_0 = Rq^Q + (q - 1)q^{Q-1} + \cdots + (q - 1)q + (q - 1) \tag{28}$$
$$= (R + 1)q^Q - 1,$$

where Q and R are the quotient and remainder resulting from dividing $\mu + 1$ by $q - 1$, i.e.,

$$\mu + 1 = Q(q - 1) + R \qquad 0 \leq R < q - 1. \tag{29}$$

Therefore, there are

$$(R + 1)q^Q - 1 \tag{30}$$

successive integers $h = 0, 1, 2, \cdots, (R + 1)q^Q - 2$ which have weight less than or equal to μ. By Corollary 2, the generator polynomial $g_\nu(X)$ of the νth order GRM code C_ν must have the following successive roots

$$\alpha, \alpha^2, \cdots, \alpha^{(R+1)q^Q-2}, \tag{31}$$

and therefore C_ν is a subcode of a BCH code with minimum distance

$$d_{\min} \geq (R + 1)q^Q - 1. \tag{32}$$

It remains to show that this minimum actually occurs. By definition, each code vector of a νth order GRM code generated by G, is a linear combination of the elementary vectors $v_I, v_0, v_1, \cdots, v_{m-1}$ and their vector products of powers defined in (9), with the sum of exponents in each product less than or equal to ν. Each code vector can then be expressed as a polynomial Γ of degree ν or less in $v_I, v_0, v_1, \cdots, v_{m-1}$, and every such polynomial is a code vector

$$u = \Gamma\,(v_I, v_0, v_1, \cdots, v_{m-1}). \tag{33}$$

Let $u = (u_0, u_1, \cdots, u_{n-1})$ be a code vector. Then each digit u_j can be expressed as polynomial Γ of degree ν or less in $a_{0j}, a_{1j}, \cdots, a_{m-1,j}$, the elements of the matrix G given by (7)

$$u_j = \Gamma\,(1, a_{0j}, a_{1j}, \cdots, a_{m-1,j}). \tag{34}$$

Each term in (45) is of the form

$$a_{0j}^{k_1} a_{1j}^{k_2} \cdots a_{m-1,j}^{k_{m-1}} \tag{35}$$

with $\sum_{i=0}^{m-1} k_i \leq \nu$.

Let V_m be the m-dimensional vector space of m-tuples over $GF(q)$. Then $(a_{0j}, a_{1j}, \cdots, a_{m-1,j})$ is a vector in V_m. The weight of a code vector u in the νth order GRM code is just the number of the nonzero vectors in V_m such that

$$u_j = \Gamma\,(1, a_{0j}, a_{1j}, \cdots, a_{m-1,j}) \neq 0. \tag{36}$$

Let $\mu = m(q - 1) - \nu - 1$. Let Q and R be the quotient and remainder resulting from dividing $\mu + 1$ by $q - 1$. Thus

$$\nu = (m - Q)(q - 1) - R. \tag{37}$$

Now, consider the vector $u = (u_0, u_1, \cdots, u_{n-1})$ with

$$u_j = (a_{0j}^{q-1} - 1)(a_{1j}^{q-1} - 1) \cdots (a_{m-Q-1,j}^{q-1} - 1)$$
$$\cdot(a_{m-Q,j} - \beta_1)(a_{m-Q,j} - \beta_2) \cdots$$
$$\cdot(a_{m-Q,j} - \beta_{q-1-R}), \tag{38}$$

where $\beta_1, \beta_2, \cdots, \beta_{q-1-R}$ are $q - 1 - R$ different nonzero elements of $GF(q)$. The degree of the polynomial of (38) is

$$(q - 1)(m - Q - 1) + (q - 1 - R)$$
$$= (q - 1)(m - Q) - R$$
$$= \nu. \tag{39}$$

Thus, u is a polynomial of degree ν in $v_I, v_0, \cdots, v_{m-1}$ and is therefore a code vector of the νth order GRM code. Since $(a_{ij})^{q-1} = 1$ for any nonzero element a_{ij} in $GF(q)$ and $(a_{ij})^{q-1} = 0$ for $a_{ij} = 0$, it follows that $\mu_j \neq 0$ if and only if

$$a_{0j} = a_{ij} = \cdots = a_{m-Q-1,j} = 0$$
$$a_{m-Q,j} \notin \{\beta_1, \beta_2, \cdots, \beta_{q-1-R}\}. \tag{40}$$

Nonzero vectors satisfying (36) can be constructed only by choosing $a_{ij} = 0$ for $i < m - Q$, $a_{m-Q,j}$, one of the $R + 1$ values satisfying (40), and the other Q components arbitrarily, except that the all-zero choice must be excluded. Thus there are exactly

$$(R + 1)q^Q - 1$$

nonzero vectors $(a_{0j}, a_{1j}, \cdots, a_{m-1,j})$ in V_m such that $u_j \neq 0$ and (36) is satisfied. Therefore, with this choice, the code vector defined in (38) has weight equal to

$$(R + 1)q^Q - 1.$$

This proves that the bound derived in the first part of this proof is the true minimum weight. Q.E.D.

For the special case $m = 1$ of the above generalization of the Reed–Muller codes, the code length is $n = q - 1$, the code symbols are from the field $GF(q)$. Then, the νth order GRM code is generated by

$$G_\nu = \begin{bmatrix} 1 & 1 & 1 & \cdots & 1 \\ 1 & \alpha & \alpha^2 & \cdots & \alpha^{q-2} \\ 1 & \alpha^2 & \alpha^4 & \cdots & (\alpha^2)^{q-2} \\ \vdots & & & & \\ 1 & \alpha^\nu & \alpha^{2\nu} & \cdots & (\alpha^\nu)^{q-2} \end{bmatrix}. \tag{41}$$

By Corollary 3, the parity check matrix of the $u = (q - 1) - (\nu + 1)$th order GRM code is

$$H = \begin{bmatrix} 1 & \alpha & \alpha^2 & \cdots & \alpha^{q-2} \\ 1 & \alpha^2 & \alpha^4 & \cdots & (\alpha^2)^{q-2} \\ \vdots & & & & \\ 1 & \alpha^\nu & \alpha^{2\nu} & \cdots & (\alpha^\nu)^{q-2} \end{bmatrix}. \tag{42}$$

It is evident that, for $m = 1$, the μth order GRM code is just the Reed–Solomon code with ν parity check digits and minimum distance $d = \nu + 1$.

IV. On the Weight Distribution for GRM Codes

The complete weight distributions of the Reed–Solomon codes are given by the following formula[4],[6],[7]

$$W_j = \binom{n}{j} \sum_{h=0}^{j-r-1} (-1)^h \binom{j}{h} (q^{j-h-r} - 1), \tag{43}$$

where W_j is the number of code vectors of weight j, n is the code length, and r is the number of parity check digits. This formula actually applies to any code whose minimum distance equals $1 + r$.

In the following some results on the number of minimum-weight code words in GRM codes are derived.

Lemma 6

The sum $S_j = \sum y^j$ of the jth powers of all elements of $GF(q)$ is zero unless j is greater than zero and divisible by $q - 1$. In this latter case, $\sum y^j = q - 1$.

Proof: In the case $j = 0$, (considering $0^0 = 1$) $\sum y^j = q = 0$. In the case $j = s(q - 1)$, $0^j = 0$ but $y^j = 1$ if $y \neq 0$, and $\sum y^j = q - 1$. Let α be a primitive element of $GF(q)$, and let n be the order of α^j. Then the sum $\sum y^j$ consists of 0, and $(q - 1)/n$ repetitions of each $\alpha^j, \alpha^{2j}, \cdots, \alpha^{(n-1)j}$. But $\alpha^j, \alpha^{2j}, \cdots, \alpha^{(n-1)j}$ are the roots of $X^n - 1$ in $GF(q)$. The sum of the roots is the negative of the coefficient of X^{n-1}, which is zero. Hence

$$\alpha^j + \alpha^{2j} + \cdots + \alpha^{(n-1)j} = 0,$$

and hence $\sum y^j = 0$. Q.E.D.

Lemma 7

If j is divisible by $q - 1$, $W(j)$ is at least $q - 1$.

Proof: In the base q system

$$j = d_0 + d_1 q + d_2 q^2 + \cdots$$

and

$$W(j) = d_0 + d_1 + d_2 + \cdots,$$

then

$$j - W(j) = d_1(q - 1) + d_2(q^2 - 1) + \cdots,$$

and since each term is divisible by $q - 1$, $j - W(j)$ is divisible by $q - 1$. Since j is divisible by $q - 1$, then $W(j)$ must be divisible by $q - 1$. But if $j \neq 0$, $W(j) > 0$, and hence must be at least $q - 1$. Q.E.D.

Lemma 8

If $C_m^n \neq 0$ modulo p, a prime, then each digit in the radix-q expression for n is at least as great as the corresponding digit in the radix-q system for m, where $q = p^l$.

Proof: Let the expressions for m and n in a radix-p system be

$$n = \sum_i \sum_{j=0}^{l-1} n_{i,j} p^{il+j}$$
$$m = \sum_i \sum_{j=0}^{l-1} m_{i,j} p^{il+j}. \tag{44}$$

Then the expression in a radix-q system is

$$n = \sum_i n_i q^i$$
$$m = \sum_i m_i q^i, \tag{45}$$

where

$$n_i = \sum_{j=0}^{l-1} n_{i,j} p^j$$
$$m_i = \sum_{j=0}^{l-1} m_{i,j} p^j. \tag{46}$$

By a well-known theorem[9]

$$C_m^n = \prod_i \prod_{j=0}^{l-1} C_{m_{i,j}}^{n_{i,j}} \quad \text{(modulo } p\text{)}, \tag{47}$$

and by the same theorem

$$C_{m_i}^{n_i} = \prod_{j=0}^{l-1} C_{m_{i,j}}^{n_{i,j}}. \tag{48}$$

Therefore

$$C_m^n = \prod_i C_{m_i}^{n_i}. \tag{49}$$

If C_m^n is not zero, then each term in the product is not zero, and hence $n_i \geq m_i$. Q.E.D.

It follows from Pele's Theorem 1[8] that if $X_1, X_2, \cdots, X_{q^h}$ are all the elements of an h-dimensional subspace of $GF(q^m)$ over $GF(q)$, then the power sum symmetric functions

$$S_j = \sum_{i=1}^{q^h} X_i^j$$

are zero for $j = 1, 2, \cdots, q^h - 1$. Now let us consider what other S_j are equal to zero. Let $A_1, A_2, \cdots, A_h$ be a basis for the subspace. Then each element of the subspace can be written

$$X_i = a_{1i}A_1 + a_{2i}A_2 + \cdots + a_{hi}A_h, \qquad (50)$$

where $a_{li} \in GF(q)$. Then

$$S_j = \sum (a_{1i}A_1 + a_{2i}A_2 + \cdots + a_{hi}A_h)^j, \qquad (51)$$

where the summation is overall choices of each coefficient a_{li}. But

$$(a_{1i}A_1 + a_{2i}A_2 + \cdots + a_{hi}A_h)^j$$
$$= \sum \frac{j!}{j_1!\, j_2! \cdots j_h!} a_{1i}^{j_1}A_1^{j_1}a_{2i}^{j_2}A_2^{j_2} \cdots a_{hi}^{j_h}A_h^{j_h}, \qquad (52)$$

where this summation is overall possible ways of choosing $j_1, j_2, \cdots, j_h$ so that $j_1 + j_2 + \cdots + j_h = j$.

Changing the order of summation gives

$$S_j = \sum C A_1^{j_1} A_2^{j_2} \cdots A_h^{j_h}, \qquad (53)$$

where the summation is overall possible choices of $j_1, j_2, \cdots, j_h$, and C depends on $j_1, j_2, \cdots, j_h$ and $a_{1i}, a_{2i}, \cdots, a_{hi}$ in the following way:

$$C = \frac{j!}{j_1!\, j_2! \cdots j_h!} \sum a_{1i}^{j_1} a_{2i}^{j_2} \cdots a_{hi}^{j_h} \qquad (54)$$

and this summation is overall values of each a_{li} from $GF(q)$. This can be rewritten

$$C = \frac{j!}{j_1!\, j_2! \cdots j_h!} \sum a_{1i}^{j_1} \sum a_{2i}^{j_2} \cdots \sum a_{hi}^{j_h}, \qquad (55)$$

where each summation is overall elements of $GF(q)$. By Lemma 6, C is zero unless each j_i is nonzero a multiple of $q - 1$.

The first factor in (55) can be written

$$\frac{j!}{j_1!\, j_2! \cdots j_h!} = \binom{j}{j_1}\binom{j - j_1}{j_2}\binom{j - j_1 - j_2}{j_3} \cdots. \qquad (56)$$

Let us assume that $C \neq 0$. Then each factor must not be congruent to zero modulo p, the characteristic of the field. Therefore, by Lemma 8, each digit in the radix-q expression of j is at least as large as the corresponding digit of j_1. It follows that $W(j) = W(j_1) + W(j - j_1)$. Similarly, $W(j - j_1) = W(j_2) + W(j - j_1 - j_2)$. Thus

$$W(j) = W(j_1) + W(j_2) + \cdots + W(j_h). \qquad (57)$$

But if $C \neq 0$, then each j_i is a multiple of $q - 1$ and hence by Lemma 7 has weight at least $q - 1$, thus

$$W(j) \geq h(q - 1). \qquad (58)$$

If this condition is violated, it is impossible for C to be nonzero. Therefore, the following theorem holds.

Theorem 9

The power sum symmetric function S_j of the elements of any h-dimensional subspace of $GF(q^m)$ over $GF(q)$ is zero if $W(j) < h(q - 1)$.

In the following, the weight distribution of the binary Reed–Muller codes will be considered. Consider the νth order modified Reed–Muller code which has minimum distance d

$$d = 2^h - 1,$$

where $h = m - \nu$.

Let $X_1, X_2, \cdots, X_d$ be the location numbers of a minimum-weight code vector. By Corollary 2, we have

$$S_j = \sum_{i=1}^{d} X_i^j = 0 \qquad (59)$$

for those j such that $1 \leqq W(j) < h$. Let us consider the set A of integers d which can be expressed in the following binary form

$$d = 2^{h-1} + 2^{h-2} + \cdots + 2^{h-t}, \qquad (60)$$

where $1 \leqq t \leqq h$. It can be shown that

$$W(2^h - 1 + d) = h. \qquad (61)$$

On the other hand, if an integer i $(1 \leqq i \leqq 2^h - 1)$ is not in A, then

$$W(2^h - 1 + i) < h. \qquad (62)$$

The power sum symmetric functions S_i and the elementary symmetric functions σ_i of $X_1, X_2, \cdots, X_d$ are related by Newton's identities

$$S_j + \sum_{i=1}^{j-1} S_{j-i}\sigma_i + \sigma_j = 0 \quad \text{for odd } j,$$
$$S_j + \sum_{i=1}^{j-1} S_{j-i}\sigma_i = 0 \quad \text{for even } j. \qquad (63)$$

Lemma 10

Let $X_1, X_2, \cdots, X_{2^h-1}$ be the location numbers of a minimum-weight code vector of a νth order modified binary Reed–Muller code, where $h = m - \nu$. Then

$$\sigma_i = 0$$

for each i which is less than 2^h and is not in A.

Proof: First, we know that

$$\sigma_i = 0 \quad \text{for} \quad i > 2^h - 1. \qquad (64)$$

Consider

$$S_{2^h} + \sigma_1 S_{2^h-1} + \sigma_2 S_{2^h-2} + \cdots + \sigma_{2^h-1} S_1 = 0. \qquad (65)$$

Since $W(2^h - i) < h$ for $1 < i \leq 2^h - 1$, then $S_{2^h - i} = 0$ by (59). Thus, we have

$$S_{2^h} + \sigma_1 S_{2^h-1} = 0, \tag{66}$$

where $S_{2^h-1} \neq 0$; otherwise $\alpha^{2^{h-1}}$ will be a root of the generator polynomial of the code. Since $S_{2^h} = 0$, it follows that

$$\sigma_1 = 0. \tag{67}$$

Suppose for an integer $r < 2^h - 1$, it is true that

$$\sigma_i = 0 \qquad 1 \leq i \leq r \tag{68}$$

for each i which is not in A. Now, we want to show that the aforementioned hypothesis is also true for $r' = r + 1$. Consider

$$S_{r'+2^h-1} + \sigma_1 S_{r'-1+2^h-1} + \cdots + \sigma_i S_{r'-i+2^h-1} + \cdots + \sigma_{r'} S_{2^h-1} = 0. \tag{69}$$

Let us examine the term

$$\sigma_i S_{r'-i+2^h-1} \qquad 1 \leq i \leq r'. \tag{70}$$

1) If i is in A, then $r' - i$ can not be in A, because $r' \leq 2^h - 1$ and every element of A is equal to or greater than 2^{h-1}. This implies

$$W(r' - i + 2^h - 1) < h \tag{71}$$

and

$$S_{r'-i+2^h-1} = 0. \tag{72}$$

2) If $r' - i$ is in A, then i can not be in A. By hypothesis, we have

$$\sigma_i = 0 \qquad 1 \leq i \leq r \tag{73}$$

for i which is not in A.

Substituting (72) and (73) into (69) gives

$$S_{r'+2^h-1} + \sigma_{r'} S_{2^h-1} = 0. \tag{74}$$

If r' is not in A, then $W(r' + 2^h - 1) < h$ and

$$S_{r'2^h-1} = 0. \tag{75}$$

This implies

$$\sigma_{r'} = 0 \tag{76}$$

for r' not in A. If r' is in A, then $W(r' + 2^h - 1) = h$ and $S_{r'+2^h-1}$ may not be zero. This implies that $\sigma_{r'}$ may not be zero for $r' \in A$. Thus, we have proved the hypothesis is also true for $r' = r + 1$. Q.E.D.

By Lemma 10, we have

$$\begin{aligned} H(X) &= \prod_{i=1}^{2^h-1} (X - X_i) \\ &= X^{2^h-1} + \sigma_1 X^{2^h-2} + \cdots + \sigma_{2^h-2} X + \sigma_{2^h-1} \\ &= X^{2^h-1} + \sigma_{2^h-s} X^{2^h-s-1} + \sigma_{2^h-s+2^h-s} X^{2^{h-s}-1} \\ &\quad + \cdots + \sigma_{2^h-2} X + \sigma_{2^h-1}, \end{aligned} \tag{77}$$

where

$$\sigma_{2^h-1} = \prod_{i=1}^{2^h-1} X_i \neq 0.$$

By Pele's Theorem 1,[8] $X_1, X_2, \cdots, X_{2^h-1}$ must be the nonzero elements of a h-dimensional subspace of $GF(2^m)$. By Theorem 9, we know if $X_1, X_2, \cdots, X_{2^h-1}$ are the nonzero elements of a h-dimensional subspace of $GF(2^m)$, then

$$S_j = \sum_{i=1}^{2^h-1} X_i^j = 0 \tag{78}$$

for $W(j) < h$. This implies that $X_1, X_2, \cdots, X_{2^h-1}$ are the location numbers of a minimum-weight code vector of a νth order modified binary Reed–Muller code, where $h = m - \nu$. Thus, we have Theorem 11.

Theorem 11

For a νth order modified binary Reed–Muller code, there exists a one-to-one correspondence between the minimum-weight code vectors ($d = 2^h - 1$) and the distinct h-dimensional subspaces of $GF(2^m)$, where $h = m - \nu$.

In $GF(2^m)$, there are

$$\prod_{i=0}^{h-1} \left\{ \frac{(2^m - 2^i)}{(2^h - 2^i)} \right\} \tag{79}$$

distinct h-dimensional subspaces. Thus, the number of minimum-weight code vectors in a νth order modified binary Reed–Muller code is equal to

$$N_{2^h-1} = \prod_{i=0}^{h-1} \frac{(2^m - 2^i)}{(2^h - 2^i)}. \tag{80}$$

By a similar argument, it can be shown that the following theorem holds.

Theorem 12

For a νth order modified binary Reed–Muller code, there exists a one-to-one correspondence between the code vectors of weight 2^h (next to the minimum weight) and the distinct co-sets of all the h-dimensional subspaces of $GF(2^m)$, where $h = m - \nu$. Thus,

$$N_{2^h} = \left\{ \prod_{i=0}^{h-1} \frac{(2^m - 2^i)}{(2^h - 2^i)} \right\} \cdot (2^{m-h} - 1). \tag{81}$$

V. Conclusion

It has been shown that all Reed–Muller codes with one digit dropped can be rearranged into cyclic form. This newly discovered property of these codes has led to a new natural generalization to the nonbinary case and to some new information on weight distribution of Reed–Muller codes.

Appendix

Parameters of Generalized Reed–Muller Codes

Q= 2	M= 3	N= 7		
			K= 4	D= 3.
Q= 2	M= 4	N= 15		
			K= 11	D= 3.
			K= 5	D= 7.
Q= 2	M= 5	N= 31		
			K= 26	D= 3.
			K= 16	D= 7.
			K= 6	D= 15.
Q= 2	M= 6	N= 63		
			K= 57	D= 3.
			K= 42	D= 7.
			K= 22	D= 15.
			K= 7	D= 31.
Q= 2	M= 7	N= 127		
			K= 120	D= 3.
			K= 99	D= 7.
			K= 64	D= 15.
			K= 29	D= 31.
			K= 8	D= 63.
Q= 2	M= 8	N= 255		
			K= 247	D= 3.
			K= 219	D= 7.
			K= 163	D= 15.
			K= 93	D= 31.
			K= 37	D= 63.
			K= 9	D= 127.
Q= 2	M= 9	N= 511		
			K= 502	D= 3.
			K= 466	D= 7.
			K= 382	D= 15.
			K= 256	D= 31.
			K= 130	D= 63.
			K= 46	D= 127.
			K= 10	D= 255.
Q= 2	M= 10	N=1023		
			K= 1013	D= 3.
			K= 968	D= 7.
			K= 848	D= 15.
			K= 638	D= 31.
			K= 386	D= 63.
			K= 176	D= 127.
			K= 56	D= 255.
			K= 11	D= 511.
Q= 3	M= 2	N= 8		
			K= 6	D= 2.
			K= 3	D= 5.
Q= 3	M= 3	N= 26		
			K= 23	D= 2.
			K= 17	D= 5.
			K= 10	D= 8.
			K= 4	D= 17.
Q= 3	M= 4	N= 80		
			K= 76	D= 2.
			K= 66	D= 5.
			K= 50	D= 8.
			K= 31	D= 17.
			K= 15	D= 26.
			K= 5	D= 53.
Q= 3	M= 5	N= 242		
			K= 237	D= 2.
			K= 222	D= 5.
			K= 192	D= 8.
			K= 147	D= 17.
			K= 96	D= 26.
			K= 51	D= 53.
			K= 21	D= 80.
			K= 6	D= 161.
Q= 3	M= 6	N= 728		
			K= 722	D= 2.
			K= 701	D= 5.
			K= 651	D= 8.
			K= 561	D= 17.
			K= 435	D= 26.
			K= 294	D= 53.
			K= 168	D= 80.
			K= 78	D= 161.
			K= 28	D= 242.
			K= 7	D= 485.
Q= 4	M= 2	N= 15		
			K= 13	D= 2.
			K= 10	D= 3.
			K= 6	D= 7.
			K= 3	D= 11.
Q= 4	M= 3	N= 63		
			K= 60	D= 2.
			K= 54	D= 3.
			K= 44	D= 7.
			K= 32	D= 11.
			K= 20	D= 15.
			K= 10	D= 31.
			K= 4	D= 47.
Q= 4	M= 4	N= 255		
			K= 251	D= 2.
			K= 241	D= 3.
			K= 221	D= 7.
			K= 190	D= 11.
			K= 150	D= 15.
			K= 106	D= 31.
			K= 66	D= 47.
			K= 35	D= 63.
			K= 15	D= 127.
			K= 5	D= 191.
Q= 4	M= 5	N=1023		
			K= 1018	D= 2.
			K= 1003	D= 3.
			K= 968	D= 7.
			K= 903	D= 11.
			K= 802	D= 15.
			K= 667	D= 31.
			K= 512	D= 47.
			K= 357	D= 63.
			K= 222	D= 127.
			K= 121	D= 191.
			K= 56	D= 255.
			K= 21	D= 511.
			K= 6	D= 767.
Q= 5	M= 2	N= 24		
			K= 22	D= 2.
			K= 19	D= 3.
			K= 15	D= 4.
			K= 10	D= 9.
			K= 6	D= 14.
			K= 3	D= 19.
Q= 5	M= 3	N= 124		
			K= 121	D= 2.
			K= 115	D= 3.
			K= 105	D= 4.
			K= 90	D= 9.
			K= 72	D= 14.
			K= 53	D= 19.
			K= 35	D= 24.
			K= 20	D= 49.
			K= 10	D= 74.
			K= 4	D= 99.
Q= 5	M= 4	N= 624		
			K= 620	D= 2.
			K= 610	D= 3.
			K= 590	D= 4.
			K= 555	D= 9.
			K= 503	D= 14.
			K= 435	D= 19.
			K= 355	D= 24.
			K= 270	D= 49.
			K= 190	D= 74.
			K= 122	D= 99.
			K= 70	D= 124.
			K= 35	D= 249.
			K= 15	D= 374.
			K= 5	D= 499.
Q= 7	M= 2	N= 48		
			K= 46	D= 2.
			K= 43	D= 3.
			K= 39	D= 4.
			K= 34	D= 5.
			K= 28	D= 6.
			K= 21	D= 13.
			K= 15	D= 20.
			K= 10	D= 27.
			K= 6	D= 34.
			K= 3	D= 41.
Q= 7	M= 3	N= 342		
			K= 339	D= 2.
			K= 333	D= 3.
			K= 323	D= 4.
			K= 308	D= 5.
			K= 287	D= 6.
			K= 259	D= 13.
			K= 226	D= 20.
			K= 190	D= 27.
			K= 153	D= 34.
			K= 117	D= 41.
			K= 84	D= 48.
			K= 56	D= 97.
			K= 35	D= 146.
			K= 20	D= 195.
			K= 10	D= 244.
			K= 4	D= 293.
Q= 8	M= 2	N= 63		
			K= 61	D= 2.
			K= 58	D= 3.
			K= 54	D= 4.
			K= 49	D= 5.
			K= 43	D= 6.
			K= 36	D= 7.
			K= 28	D= 15.
			K= 21	D= 23.
			K= 15	D= 31.
			K= 10	D= 39.
			K= 6	D= 47.
			K= 3	D= 55.
Q= 8	M= 3	N= 511		
			K= 508	D= 2.
			K= 502	D= 3.
			K= 492	D= 4.
			K= 477	D= 5.
			K= 456	D= 6.
			K= 428	D= 7.
			K= 392	D= 15.
			K= 350	D= 23.
			K= 304	D= 31.
			K= 256	D= 39.
			K= 208	D= 47.
			K= 162	D= 55.
			K= 120	D= 63.
			K= 84	D= 127.
			K= 56	D= 191.
			K= 35	D= 255.
			K= 20	D= 319.
			K= 10	D= 383.
			K= 4	D= 447.
Q= 9	M= 2	N= 80		
			K= 78	D= 2.
			K= 75	D= 3.
			K= 71	D= 4.
			K= 66	D= 5.
			K= 60	D= 6.
			K= 53	D= 7.
			K= 45	D= 8.
			K= 36	D= 17.
			K= 28	D= 26.
			K= 21	D= 35.
			K= 15	D= 44.
			K= 10	D= 53.
			K= 6	D= 62.
			K= 3	D= 71.
Q= 9	M= 3	N= 728		
			K= 725	D= 2.
			K= 719	D= 3.
			K= 709	D= 4.
			K= 694	D= 5.
			K= 673	D= 6.
			K= 645	D= 7.
			K= 609	D= 8.
			K= 564	D= 17.
			K= 512	D= 26.
			K= 455	D= 35.
			K= 395	D= 44.
			K= 334	D= 53.
			K= 274	D= 62.
			K= 217	D= 71.
			K= 165	D= 80.
			K= 120	D= 161.
			K= 84	D= 242.

APPENDIX (*Cont'd*)

```
   K=   56 D=  323.
   K=   35 D=  404.
   K=   20 D=  485.
   K=   10 D=  566.
   K=    4 D=  647.
Q=  11 M=    2 N= 120
   K=  118 D=    2.
   K=  115 D=    3.
   K=  111 D=    4.
   K=  106 D=    5.
   K=  100 D=    6.
   K=   93 D=    7.
   K=   85 D=    8.
   K=   76 D=    9.
   K=   66 D=   10.
   K=   55 D=   21.
   K=   45 D=   32.
   K=   36 D=   43.
   K=   28 D=   54.
   K=   21 D=   65.
   K=   15 D=   76.
   K=   10 D=   87.
   K=    6 D=   98.
   K=    3 D=  109.
Q=  11 M=    3 N=1330
   K= 1327 D=    2.
   K= 1321 D=    3.
   K= 1311 D=    4.
   K= 1296 D=    5.
   K= 1275 D=    6.
   K= 1247 D=    7.
   K= 1211 D=    8.
   K= 1166 D=    9.
   K= 1111 D=   10.
   K= 1045 D=   21.
   K=  970 D=   32.
   K=  888 D=   43.
   K=  801 D=   54.
   K=  711 D=   65.
   K=  620 D=   76.
   K=  530 D=   87.
   K=  443 D=   98.
   K=  361 D=  109.
   K=  286 D=  120.
   K=  220 D=  241.
   K=  165 D=  362.
   K=  120 D=  483.
   K=   84 D=  604.
   K=   56 D=  725.
   K=   35 D=  846.
   K=   20 D=  967.
   K=   10 D= 1088.
   K=    4 D= 1209.
Q=  13 M=    2 N= 168
   K=  166 D=    2.
   K=  163 D=    3.
   K=  159 D=    4.
   K=  154 D=    5.
   K=  148 D=    6.
   K=  141 D=    7.
   K=  133 D=    8.
   K=  124 D=    9.
   K=  114 D=   10.
   K=  103 D=   11.
   K=   91 D=   12.
   K=   78 D=   25.
   K=   66 D=   38.
   K=   55 D=   51.
   K=   45 D=   64.
   K=   36 D=   77.
   K=   28 D=   90.
   K=   21 D=  103.
   K=   15 D=  116.
   K=   10 D=  129.
   K=    6 D=  142.
   K=    3 D=  155.
Q=  16 M=    2 N= 255
   K=  253 D=    2.
   K=  250 D=    3.
   K=  246 D=    4.
   K=  241 D=    5.
   K=  235 D=    6.
   K=  228 D=    7.
   K=  220 D=    8.
   K=  211 D=    9.
   K=  201 D=   10.
   K=  190 D=   11.
   K=  178 D=   12.
   K=  165 D=   13.
   K=  151 D=   14.
   K=  136 D=   15.
   K=  120 D=   31.
   K=  105 D=   47.
   K=   91 D=   63.
   K=   78 D=   79.
   K=   66 D=   95.
   K=   55 D=  111.
   K=   45 D=  127.
   K=   36 D=  143.
   K=   28 D=  159.
   K=   21 D=  175.
   K=   15 D=  191.
   K=   10 D=  207.
   K=    6 D=  223.
   K=    3 D=  239.
Q=  17 M=    2 N= 288
   K=  286 D=    2.
   K=  283 D=    3.
   K=  279 D=    4.
   K=  274 D=    5.
   K=  268 D=    6.
   K=  261 D=    7.
   K=  253 D=    8.
   K=  244 D=    9.
   K=  234 D=   10.
   K=  223 D=   11.
   K=  211 D=   12.
   K=  198 D=   13.
   K=  184 D=   14.
   K=  169 D=   15.
   K=  153 D=   16.
   K=  136 D=   33.
   K=  120 D=   50.
   K=  105 D=   67.
   K=   91 D=   84.
   K=   78 D=  101.
   K=   66 D=  118.
   K=   55 D=  135.
   K=   45 D=  152.
   K=   36 D=  169.
   K=   28 D=  186.
   K=   21 D=  203.
   K=   15 D=  220.
   K=   10 D=  237.
   K=    6 D=  254.
   K=    3 D=  271.
Q=  19 M=    2 N= 360
   K=  358 D=    2.
   K=  355 D=    3.
   K=  351 D=    4.
   K=  346 D=    5.
   K=  340 D=    6.
   K=  333 D=    7.
   K=  325 D=    8.
   K=  316 D=    9.
   K=  306 D=   10.
   K=  295 D=   11.
   K=  283 D=   12.
   K=  270 D=   13.
   K=  256 D=   14.
   K=  241 D=   15.
   K=  225 D=   16.
   K=  208 D=   17.
   K=  190 D=   18.
   K=  171 D=   37.
   K=  153 D=   56.
   K=  136 D=   75.
   K=  120 D=   94.
   K=  105 D=  113.
   K=   91 D=  132.
   K=   78 D=  151.
   K=   66 D=  170.
   K=   55 D=  189.
   K=   45 D=  208.
   K=   36 D=  227.
   K=   28 D=  246.
   K=   21 D=  265.
   K=   15 D=  284.
   K=   10 D=  303.
   K=    6 D=  322.
   K=    3 D=  341.
Q=  23 M=    2 N= 528
   K=  526 D=    2.
   K=  523 D=    3.
   K=  519 D=    4.
   K=  514 D=    5.
   K=  508 D=    6.
   K=  501 D=    7.
   K=  493 D=    8.
   K=  484 D=    9.
   K=  474 D=   10.
   K=  463 D=   11.
   K=  451 D=   12.
   K=  438 D=   13.
   K=  424 D=   14.
   K=  409 D=   15.
   K=  393 D=   16.
   K=  376 D=   17.
   K=  358 D=   18.
   K=  339 D=   19.
   K=  319 D=   20.
   K=  298 D=   21.
   K=  276 D=   22.
   K=  253 D=   45.
   K=  231 D=   68.
   K=  210 D=   91.
   K=  190 D=  114.
   K=  171 D=  137.
   K=  153 D=  160.
   K=  136 D=  183.
   K=  120 D=  206.
   K=  105 D=  229.
   K=   91 D=  252.
   K=   78 D=  275.
   K=   66 D=  298.
   K=   55 D=  321.
   K=   45 D=  344.
   K=   36 D=  367.
   K=   28 D=  390.
   K=   21 D=  413.
   K=   15 D=  436.
   K=   10 D=  459.
   K=    6 D=  482.
   K=    3 D=  505.
Q=  25 M=    2 N= 624
   K=  622 D=    2.
   K=  619 D=    3.
   K=  615 D=    4.
   K=  610 D=    5.
   K=  604 D=    6.
   K=  597 D=    7.
   K=  589 D=    8.
   K=  580 D=    9.
   K=  570 D=   10.
   K=  559 D=   11.
   K=  547 D=   12.
   K=  534 D=   13.
   K=  520 D=   14.
   K=  505 D=   15.
   K=  489 D=   16.
   K=  472 D=   17.
   K=  454 D=   18.
   K=  435 D=   19.
   K=  415 D=   20.
   K=  394 D=   21.
   K=  372 D=   22.
   K=  349 D=   23.
   K=  325 D=   24.
   K=  300 D=   49.
   K=  276 D=   74.
   K=  253 D=   99.
   K=  231 D=  124.
   K=  210 D=  149.
   K=  190 D=  174.
```

APPENDIX (*Cont'd*)

```
K=  171 D=  199.
K=  153 D=  224.
K=  136 D=  249.
K=  120 D=  274.
K=  105 D=  299.
K=   91 D=  324.
K=   78 D=  349.
K=   66 D=  374.
K=   55 D=  399.
K=   45 D=  424.
K=   36 D=  449.
K=   28 D=  474.
K=   21 D=  499.
K=   15 D=  524.
K=   10 D=  549.
K=    6 D=  574.
K=    3 D=  599.
Q=  27 M=   2 N= 728
K=  726 D=    2.
K=  723 D=    3.
K=  719 D=    4.
K=  714 D=    5.
K=  708 D=    6.
K=  701 D=    7.
K=  693 D=    8.
K=  684 D=    9.
K=  674 D=   10.
K=  663 D=   11.
K=  651 D=   12.
K=  638 D=   13.
K=  624 D=   14.
K=  609 D=   15.
K=  593 D=   16.
K=  576 D=   17.
K=  558 D=   18.
K=  539 D=   19.
K=  519 D=   20.
K=  498 D=   21.
K=  476 D=   22.
K=  453 D=   23.
K=  429 D=   24.
K=  404 D=   25.
K=  378 D=   26.
K=  351 D=   53.
K=  325 D=   80.
K=  300 D=  107.
K=  276 D=  134.
K=  253 D=  161.
K=  231 D=  188.
K=  210 D=  215.
K=  190 D=  242.
K=  171 D=  269.
K=  153 D=  296.
K=  136 D=  323.
K=  120 D=  350.
K=  105 D=  377.
K=   91 D=  404.
K=   78 D=  431.
K=   66 D=  458.
K=   55 D=  485.
K=   45 D=  512.
K=   36 D=  539.
K=   28 D=  566.
K=   21 D=  593.
K=   15 D=  620.
K=   10 D=  647.
K=    6 D=  674.
K=    3 D=  701.
Q=  29 M=   2 N= 840
K=  838 D=    2.
K=  835 D=    3.
K=  831 D=    4.
K=  826 D=    5.
K=  820 D=    6.
K=  813 D=    7.
K=  805 D=    8.
K=  796 D=    9.
K=  786 D=   10.
K=  775 D=   11.
K=  763 D=   12.
K=  750 D=   13.
K=  736 D=   14.
K=  721 D=   15.
K=  705 D=   16.
K=  688 D=   17.
K=  670 D=   18.
K=  651 D=   19.
K=  631 D=   20.
K=  610 D=   21.
K=  588 D=   22.
K=  565 D=   23.
K=  541 D=   24.
K=  516 D=   25.
K=  490 D=   26.
K=  463 D=   27.
K=  435 D=   28.
K=  406 D=   57.
K=  378 D=   86.
K=  351 D=  115.
K=  325 D=  144.
K=  300 D=  173.
K=  276 D=  202.
K=  253 D=  231.
K=  231 D=  260.
K=  210 D=  289.
K=  190 D=  318.
K=  171 D=  347.
K=  153 D=  376.
K=  136 D=  405.
K=  120 D=  434.
K=  105 D=  463.
K=   91 D=  492.
K=   78 D=  521.
K=   66 D=  550.
K=   55 D=  579.
K=   45 D=  608.
K=   36 D=  637.
K=   28 D=  666.
K=   21 D=  695.
K=   15 D=  724.
K=   10 D=  753.
K=    6 D=  782.
K=    3 D=  811.
Q=  31 M=   2 N= 960
K=  958 D=    2.
K=  955 D=    3.
K=  951 D=    4.
K=  946 D=    5.
K=  940 D=    6.
K=  933 D=    7.
K=  925 D=    8.
K=  916 D=    9.
K=  906 D=   10.
K=  895 D=   11.
K=  883 D=   12.
K=  870 D=   13.
K=  856 D=   14.
K=  841 D=   15.
K=  825 D=   16.
K=  808 D=   17.
K=  790 D=   18.
K=  771 D=   19.
K=  751 D=   20.
K=  730 D=   21.
K=  708 D=   22.
K=  685 D=   23.
K=  661 D=   24.
K=  636 D=   25.
K=  610 D=   26.
K=  583 D=   27.
K=  555 D=   28.
K=  526 D=   29.
K=  496 D=   30.
K=  465 D=   61.
K=  435 D=   92.
K=  406 D=  123.
K=  378 D=  154.
K=  351 D=  185.
K=  325 D=  216.
K=  300 D=  247.
K=  276 D=  278.
K=  253 D=  309.
K=  231 D=  340.
K=  210 D=  371.
K=  190 D=  402.
K=  171 D=  433.
K=  153 D=  464.
K=  136 D=  495.
K=  120 D=  526.
K=  105 D=  557.
K=   91 D=  588.
K=   78 D=  619.
K=   66 D=  650.
K=   55 D=  681.
K=   45 D=  712.
K=   36 D=  743.
K=   28 D=  774.
K=   21 D=  805.
K=   15 D=  836.
K=   10 D=  867.
K=    6 D=  898.
K=    3 D=  929.
Q=  32 M=   2 N=1023
K= 1021 D=    2.
K= 1018 D=    3.
K= 1014 D=    4.
K= 1009 D=    5.
K= 1003 D=    6.
K=  996 D=    7.
K=  988 D=    8.
K=  979 D=    9.
K=  969 D=   10.
K=  958 D=   11.
K=  946 D=   12.
K=  933 D=   13.
K=  919 D=   14.
K=  904 D=   15.
K=  888 D=   16.
K=  871 D=   17.
K=  853 D=   18.
K=  834 D=   19.
K=  814 D=   20.
K=  793 D=   21.
K=  771 D=   22.
K=  748 D=   23.
K=  724 D=   24.
K=  699 D=   25.
K=  673 D=   26.
K=  646 D=   27.
K=  618 D=   28.
K=  589 D=   29.
K=  559 D=   30.
K=  528 D=   31.
K=  496 D=   63.
K=  465 D=   95.
K=  435 D=  127.
K=  406 D=  159.
K=  378 D=  191.
K=  351 D=  223.
K=  325 D=  255.
K=  300 D=  287.
K=  276 D=  319.
K=  253 D=  351.
K=  231 D=  383.
K=  210 D=  415.
K=  190 D=  447.
K=  171 D=  479.
K=  153 D=  511.
K=  136 D=  543.
K=  120 D=  575.
K=  105 D=  607.
K=   91 D=  639.
K=   78 D=  671.
K=   66 D=  703.
K=   55 D=  735.
K=   45 D=  767.
K=   36 D=  799.
K=   28 D=  831.
K=   21 D=  863.
K=   15 D=  895.
K=   10 D=  927.
K=    6 D=  959.
K=    3 D=  991.
```

ACKNOWLEDGMENT

The tables in the Appendix were computed on the IBM 7040 at the University of Hawaii Statistical and Computing Center. The programming was done by S. Y. C. Chen.

REFERENCES

[1] N. Zierler, "On a variation of the first-order Reed–Muller codes," M.I.T. Lincoln Lab., Group Rept. 34–80, Lexington, Mass., October 1958.

[2] W. W. Peterson, *Error Correcting Codes.* New York: Wiley, 1961, p. 74.

[3] B. M. Dwork and R. M. Heller, "Results of a geometric approach to the theory and construction of non-Binary, multiple error and failure correcting codes," *IRE Nat'l Conv. Rec.*, pt. 4, pp. 123–129, 1959.

[4] T. Kasami, S. Lin, and W. W. Peterson, "Some results on cyclic codes which are invariant under the affine group," AF Cambridge Research Labs., Bedford, Mass., Scientific Rept. AFCRL-66-622, 1966; a part of this paper has been published in the *J. Inst. Elect. Commun. Engrs.* (Japan), vol. 50, pp. 1617–1622, September 1967.

[5] T. Kasami and S. Lin, "Some codes which are invariant under a doubly-transitive permutation group and their connection with balanced incomplete block designs," AF Cambridge Research Labs., Bedford, Mass., Scientific Rept. AFCRL-66-142, January 28, 1966.

[6] E. F. Assmus, H. F. Mattson and R. Turyn, "Cyclic codes," AF Cambridge Research Labs., Bedford, Mass., Sci. Rept. AFCRL-65-322, April 28, 1965.

[7] G. D. Forney, Jr., "Concatenated codes," Sc.D. dissertation, Dept. Elec. Engrg., M.I.T., Cambridge, Mass., June 1965.

[8] R. L. Pele, "Some remarks on the vector subspaces of a finite field," AF Cambridge Research Labs., Bedford, Mass., Scientific Rept. AFCRL-66-477; submitted to *Can. J. Math.*

[9] E. Lucas, "Sur les congruences des nombres euleriennes et des coefficients differentiels des fonctions trigonometriques, suivant un module premier," *Bull. Soc. Math.* (France), vol. 6, pp. 49–54, 1878.

28

New Generalizations of the Reed–Muller Codes Part II: Nonprimitive Codes

EDWARD J. WELDON, JR., MEMBER, IEEE

***Abstract*—In this paper a class of nonprimitive cyclic codes quite similar in structure to the original Reed–Muller codes is presented. These codes, referred to herein as nonprimitive Reed–Muller codes, are shown to possess many of the properties of the primitive codes. Specifically, two major results are presented. First the code length, number of information symbols, and minimum distance are shown to be related by means of a parameter known as the order of the code. These relationships show that for given values of code length and rate the codes have relatively large minimum distances. It is also shown that the codes are subcodes of the BCH codes of the same length and guaranteed minimum distance; thus in general the codes are not as powerful as the BCH codes. However, for most interesting values of code length and rate the difference between the two types of codes is slight.**

The second result is the observation that the codes can be decoded with a variation of the original algorithm proposed by Reed for the Reed–Muller codes. In other words, they are L-step orthogonalizable. Because of their large minimum distances and the simplicity of their decoders, nonprimitive Reed–Muller codes seem attractive for use in error-control systems requiring multiple random-error correction.

I. Introduction

The Reed–Muller codes have been shown to be equivalent to primitive cyclic codes (codes of length $2^m - 1$) with an overall parity check added.[7] Recently some interesting and useful properties of these codes have been unearthed.[8]

In this paper a class of nonprimitive cyclic codes similar to the Reed–Muller (primitive) cyclic codes is investigated. This class of codes contains as subclasses the primitive Reed–Muller codes, Rudolph's projective geometry codes,[9] and Weldon's difference-set codes.[10] It is shown that all of these codes can be decoded fairly simply with a modification of the original Reed algorithm[2] for Muller's codes.[3] One simplification of this algorithm is presented; others are suggested by the structure of the codes. Although only the binary case is treated explicitly, the generalization to codes over an arbitrary prime field is not difficult. This point, as well as another generalization of the binary codes, is discussed briefly.

Manuscript received March 6, 1967. This paper was presented at the 1967 International Symposium on Information Theory, San Remo, Italy.
The author is with the Department of Electrical Engineering, University of Hawaii, Honolulu, Hawaii.

II. Primitive Reed–Muller Codes

The original Reed–Muller codes have length 2^m and are noncyclic. Reordering of the code digits enables the code to be put in the form of a cyclic code with an overall check bit added. In this paper both the cyclic code of length $2^m - 1$ and this code with the overall check bit added will be referred to as Reed–Muller codes. Context will make clear which is meant in a given situation.

Let α denote a primitive element of $GF(2^m)$ and consider the generator matrix of the first-order Reed–Muller code

$$G_1 = \begin{bmatrix} 1 & 1 & 1 & 1 & \cdots & 1 \\ 0 & \alpha^0 & \alpha^1 & \alpha^2 & \cdots & \alpha^{2^m-2} \end{bmatrix}. \qquad (1)$$

With 0 and the powers of α expressed as m-place column vectors, i.e.,

$$\alpha^j = \sum_{i=0}^{m-1} a_{ij}\alpha^i; \qquad a_{ij} = 0, 1, \qquad (2)$$

this matrix can be rewritten as

$$G_1 = \begin{bmatrix} 1 & 1 & 1 & \cdots & 1 \\ 0 & a_{00} & a_{01} & \cdots & a_{0\lambda} \\ 0 & a_{10} & a_{11} & \cdots & a_{1\lambda} \\ \vdots & & & & \\ 0 & a_{(m-1)0} & a_{(m-1)1} & \cdots & a_{(m-1)\lambda} \end{bmatrix} = \begin{bmatrix} V_I \\ V_0 \\ V_1 \\ \vdots \\ V_{m-1} \end{bmatrix}, \quad (3)$$

where $\lambda = 2^m - 2$.

Every column position in this matrix can be associated uniquely with a point in an m-dimensional Euclidean geometry over $GF(2)$, i.e., $EG(m, 2)$. It can be verified easily that the set of points associated with nonzero positions in any row of G_1 (except the first) form an $(m - 1)$-dimensional affine space, or $(m - 1)$-flat, in $EG(m, 2.)$ The first row, of course, forms the entire m-flat. Also any linear combination of rows of G_1 yields an $(m - 1)$-flat and conversely every $(m - 1)$-flat in $EG(m, 2)$ is in the row space of G_1. For emphasis this is stated as Lemma 1.

Lemma 1

Every $(m - 1)$-flat in $EG(m, 2)$ is a code word in the first-order Reed–Muller code.

Now consider the matrix whose rows are the vector products of the elementary vectors $V_I, V_0, V_1, \cdots, V_{m-1}$ taken ν or fewer at a time

$$G_\nu = \begin{bmatrix} V_I \\ V_0 \\ \vdots \\ V_{m-1} \\ V_0 V_1 \\ V_0 V_2 \\ \vdots \\ V_{m-2} V_{m-1} \\ V_0 V_1 V_2 \\ \vdots \end{bmatrix}, \quad (4)$$

where vector product is defined in (8) of Part I of this paper. Interpreted geometrically, vector product is simply intersection. Thus, for example, the row space of G_2 contains the intersections of every two $(m - 1)$-flats, i.e., every $(m - 2)$-flat. This is a consequence of the fact that vector multiplication distributes over addition; that is, for any vectors V_1, V_2, and V_3,

$$V_1 \times V_2 + V_1 \times V_3 = V_1 \times (V_2 + V_3). \quad (5)$$

In general the following is true.

Lemma 2

Every flat of $(m - \nu)$ or more dimensions is a code word in the νth order Reed–Muller code. Then since every flat of dimensionality greater than $m - \nu$ can be expressed as a linear combination of $(m - \nu)$-flats, we have Lemma 3.

Lemma 3

The set of $(m - \nu)$-flats spans the code space of the νth order Reed–Muller code.

In an m-dimensional Euclidean geometry, the intersection of an $(m - \nu)$-flat and a $(\nu + 1)$-flat is either empty or a flat of at least one dimension. Therefore the inner product of two such flats must be zero. This gives Lemma 4.

Lemma 4

Every flat of at least $(\nu + 1)$ dimensions is in the null space of the νth order code.

From the preceding two lemmas it follows that the next result holds.

Lemma 5

The $(m - \nu - 1)$th order code is the dual of the νth order code. In Kasami *et al.*,[8] it is shown that the matrix of (4) has the same row space as

$$G_\nu = \begin{bmatrix} \alpha^0 & \alpha^0 & \alpha^0 & \alpha^0 & \cdots & \alpha^0 \\ 0 & \alpha^0 & \alpha^{e_1} & \alpha^{2e_1} & \cdots & \alpha^{\lambda e_1} \\ 0 & \alpha^0 & \alpha^{e_2} & \alpha^{2e_2} & \cdots & \alpha^{\lambda e_2} \\ \vdots & & & & & \\ 0 & \alpha^0 & \alpha^{e_i} & \alpha^{2e_i} & \cdots & \alpha^{\lambda e_i} \end{bmatrix}, \quad (6)$$

where $\lambda = 2^m - 2$ and the e_i are all possible integers less than $2^m - 1$ whose weight is ν or less.

By Lemma 5 the parity check matrix of the $(m - \nu - 1)$th order code, $H_{m-\nu-1}$, has the same row space as the generator matrix of the νth order code. It is with the properties of $H_{m-\nu-1}$ that we shall be concerned in the following presentation of the nonprimitive Reed–Muller codes.

III. Nonprimitive Reed–Muller Codes

A nonprimitive cyclic code is one in which every root of the generator polynomial is a power of a given nonprimitive root β of $GF(2^m)$. The length of such a code n, which is the order of β, must divide $2^m - 1$. Thus let

$$r = \frac{2^m - 1}{n} \quad (7)$$

and consider the rows of G_ν for which r divides e_i. Every such row consists of the first digit and r replications of the second through $(n + 1)$th digits, for letting $t_i = e_i/r$ gives

$$(\alpha^{e_i})^{nv} = (\alpha^{rn})^{t_i v} = 1; \qquad v = 1, 2, \cdots, r - 1, \quad (8)$$

so that

$$\alpha^{i e_j} = \alpha^{e_j(n s + i)}; \quad i = 0, 1, \cdots, n - 1. \quad (9)$$

Let us denote by $H'_{m-\nu-1}$ the $(n - k) \times n$ matrix whose rows are identical to the second through $(n + 1)$th digits of the rows of G_ν for which r divides e_i. That is

$$H'_{m-\nu-1} = \begin{bmatrix} \alpha^0 & \alpha^{rt_1} & \alpha^{2rt_1} & \cdots & \alpha^{(n-1)rt_1} \\ \alpha^0 & \alpha^{rt_2} & \alpha^{2rt_2} & \cdots & \alpha^{(n-1)rt_2} \\ \vdots & \vdots & \vdots & & \vdots \\ \alpha^0 & \alpha^{rt_{n-k}} & \alpha^{2rt_{n-k}} & \cdots & \alpha^{(n-1)rt_{n-k}} \end{bmatrix}, \quad (10)$$

where $w(rt_i) \leq \nu$ for $1 \leq i \leq n - k$. Setting $\beta = \alpha^r$ gives

$$H'_{m-\nu-1} = \begin{bmatrix} \beta^0 & \beta^{t_1} & \beta^{2t_1} & \cdots & \beta^{(n-1)t_1} \\ \beta^0 & \beta^{t_2} & \beta^{2t_2} & \cdots & \beta^{(n-1)t_2} \\ \vdots & \vdots & \vdots & & \vdots \\ \beta^0 & \beta^{t_{n-k}} & \beta^{2t_{n-k}} & \cdots & \beta^{(n-1)t_{n-k}} \end{bmatrix} = \begin{bmatrix} B^{t_1} \\ B^{t_2} \\ \vdots \\ B^{t_{n-k}} \end{bmatrix}. \quad (11)$$

This matrix can be employed as either the generator or the parity check matrix of a cyclic code of length n. For reasons which will become apparent later, the parity check interpretation is the more interesting of the two.

Various facts are well known about the null space of the matrix $H'_{m-\nu-1}$. First of all, if B^t is a row of $H'_{m-\nu-1}$, so are B^{2t}, B^{4t}, $\cdots$. Thus the null space is the space of all polynomials which are multiples of the binary polynomial

$$g(x) = \prod_{i=1}^{n-k} (x + \beta^{t_i}). \quad (12)$$

That is, $g(x)$ is the generator polynomial of the (n, k) cyclic code of which $H'_{m-\nu-1}$ is the parity check matrix. The number of parity checks in this code is given by

$$n - k = \begin{bmatrix} \text{number of integers } < 2^m - 1 \text{ which are} \\ \text{divisible by } r \text{ whose weight is } \nu \text{ or less} \end{bmatrix}. \quad (13)$$

Such a code will be referred to as a nonprimitive Reed–Muller code.

Definition

The cyclic $[(2^m - 1)/r, k]$ code whose generator polynomial contains all roots α^{rt_i} such that $w(rt_i) \leq \nu$ will be referred to as the νth order nonprimitive Reed–Muller code.
When

$$r = 2^s - 1 \quad (14)$$

for some integer s, these codes have several useful properties. In this case

$$n = \frac{2^m - 1}{2^s - 1} \quad (15)$$

so that m/s is an integer; call it $z + 1$. Thus

$$n = 2^{sz} + 2^{s(z-1)} + \cdots + 2^s + 1. \quad (16)$$

It is well known that for such values of n every bit position can be associated uniquely with a point in a z-dimensional projective geometry over $GF(2^s)$, i.e., in $PG(z, 2^s)$. A $(z - 1)$-flat in this geometry consists of the $2^{s(z-1)} + 2^{s(z-2)} + \cdots + 2^s + 1$ elements α^j of $GF(2^m)$ such that

$$b_1 \alpha^{c_1} + b_2 \alpha^{c_2} + \cdots + b_z \alpha^{c_z} = \alpha^j, \quad (17)$$

where b_i is an element of $GF(2^s)$ and the c_i are chosen so that the z powers of α are linearly independent. Two elements α^i and α^j which differ only by a factor of an element of $GF(2^s)$ are taken as the same point. That is, since $GF(2^s)$ is a subfield of $GF(2^m)$, for every element β of $GF(2^s)$

$$\beta^{2^s-1} = 1. \quad (18)$$

It follows that $\beta = \alpha^n$ for some primitive element α in $GF(2^m)$. Consequently the elements α^j, α^{j+n}, α^{j+2n}, $\cdots$, $\alpha^{j+(r-1)n}$ are interpreted as the same point, and there are r replications of each of the points of the $(z - 1)$-flat in $GF(2^m)$. With regard to these replications the following lemma will be useful in the sequel.

Lemma 6

The points in the r replications of an $(z - 1)$-flat of $PG(z, 2^s)$ in $GF(2^m)$ plus the point at infinity form an $(m - s)$-flat in $EG(m, 2)$.

Proof: An $(m - s)$-flat in $EG(m, 2)$ consists of all points α^j such that

$$a_0 + a_1 \alpha^{e_1} + a_2 \alpha^{e_2} + \cdots + a_{m-s} \alpha^{e_{m-s}} = \alpha^j, \quad (19)$$

where $a_i = 0, 1$ and the $(m - s)$ points α^{e_i} are linearly independent. If the flat in question passes through the point at infinity, $a_0 = 0$. But the coefficients b_i in (17) can be regarded as binary s-tuples. Since $m - s = zs$, the r replications of the $(z - 1)$-flat plus the point at infinity can be regarded as consisting of all points which are linear combinations [over $GF(2)$] of $m - s$ linearly independent points. But since the latter are independent, the 2^{m-s} points satisfy the form of (19) and form an $(m - s)$-flat in $EG(m, 2)$.

The converse statement, that every $(m - s)$-flat in $EG(m, 2)$ passing through the point at infinity consists of r replications of an $(z - 1)$-flat in $PG(z, 2^s)$, is not true, however, for there are many more such $(m - s)$-flats in $EG(m, 2)$ than $(z - 1)$-flats in $PG(z, 2^s)$.

In Section IV of this paper a decoding algorithm for these codes is presented. This algorithm is essentially an adaptation of the original Reed algorithm for Muller's codes to the cyclic case and to projective rather than Euclidean geometries. As such it relies heavily on the fact that certain flats are in the null spaces of the codes. The reader may now see the connection between the replications of the flats and the replications in the matrix G_r. By defining the codes as we have, we ensure that the

$$J = \frac{(2^{sz} + 2^{s(z-1)} + \cdots + 2^s + 1) - (2^{s(z-h-1)} + 2^{s(z-h-2)} + \cdots + 2^s + 1)}{(2^{s(z-h)} + 2^{s(z-h-1)} + \cdots + 2^s + 1) - (2^{s(z-h-1)} + 2^{s(z-h-2)} + \cdots + 2^s + 1)}$$
$$= 2^{sh} + 2^{s(h-1)} + \cdots + 2^s + 1 \qquad (20)$$

flats necessary for decoding are in the null space. The result proved in the following two lemmas specifies which flats are in the various null spaces.

Lemma 7

The parity check matrix H_{sh} of the (sh)th order nonprimitive Reed–Muller code has every $(z-1)$-flat of $PG(z, 2^s)$ in its row space when $r = 2^s - 1$. (Note: the prime on H_{sh} is no longer necessary and is omitted hereafter.)

Proof: The primitive sth order code has every $(m-s)$-flat of $EG(m, 2)$ in its row space by Lemma 2. By definition, H_s contains in its row space the first replication of the rows of G_s which are expressible in terms of powers of $\beta = \alpha^{2^s-1}$. In their binary representation the rows of G_s which can be expressed in powers of β, contain all binary rows which consist of r replications of a given pattern and the point at infinity, for the cyclic shifts of the $2^m - 1$ digits of such a binary row (without the point at infinity) form a cyclic subspace. It is well known that this subspace is spanned by rows which are expressible in powers of β. Therefore by Lemma 6 every $(z-1)$-flat of $PG(z, 2^s)$ is in the row space of H_s. It is a straightforward matter to generalize this result to give Lemma 8.

Lemma 8

The parity check matrix H_{sh} of the (sh)th order nonprimitive Reed–Muller code has every $(z-h)$-flat of $PG(z, 2^s)$ in its row space.

Proof: The (sh)th order primitive Reed–Muller code contains all vector products of sh or fewer of the vectors $V_I, V_0, V_1, \cdots, V_{m-1}$. It therefore contains all vector products of h or fewer of the vector products of $V_I, V_0, V_1, \cdots, V_{m-1}$ taken s or fewer at a time. With the aid of Lemma 7 this says that the matrix H_{sh} has in its row space the vector products of h or fewer of the $(z-1)$-flats of $PG(z, 2^s)$. But every $(z-h)$-flat can be regarded as the intersection of h flats of dimension $(z-1)$. This proves the stated result.

Rudolph[9] has considered cyclic codes based on projective geometries in which all flats of various dimensions are in the null spaces. Lemma 8 establishes that Rudolph's codes and the nonprimitive Reed–Muller codes with $r = 2^s - 1$ are the same.

The preceding lemmas characterize the structure and dimensionality of the nonprimitive Reed–Muller codes fairly well. We now use this structure to determine a lower bound on the minimum distance of these codes. By Lemma 8 an (sh)th order nonprimitive Reed–Muller code contains every $(z-h)$-flat of $PG(z, 2^s)$ in its null space. In particular the null space contains the $(z-h)$-flats which intersect only in a given $(z-h-1)$-flat. That it is always possible to construct J such flats follows from the definition of an $(z-h)$-flat; it consists of the points linearly dependent on an $(z-h-1)$-flat and another point not in the $(z-h-1)$-flat. After constructing a number of $(z-h)$-flats intersecting only on a given $(z-h-1)$-flat, either all the points of the geometry are employed, in which case (20) holds, or they are not, in which case another $(z-h)$-flat can be constructed with the $(z-h-1)$-flat and one of the remaining points.

The parity check equations corresponding to the $(z-h)$-flats are said to be orthogonal (in the sense of Massey[12]) on the $(z-h-1)$-flat. Clearly it is possible to construct a set of J orthogonal equations for every $(z-h-1)$-flat in $PG(z, 2^s)$. As was originally realized by Reed[2] for the Muller codes, if $[J/2]$[1] or fewer errors occurred, the parity check sum corresponding to any $(z-h-1)$-flat is correctly determined by the majority of the parity check sums orthogonal on it. For if the $(z-h-1)$-flat sum is a 1, then there are at most $[J/2] - 1$ errors in the orthogonal $(z-h)$-flats which are not in the $(z-h)$-flat. Consequently at least $J + 1 - [J/2]$ (a majority) of the orthogonal sums must be 1. Similarly if the $(z-h-1)$-flat sum is 0, at most $[J/2]$ of the orthogonal sums can be 1. But $J - [J/2]$ constitutes a majority, so that the assertion is proved that every $(z-h-1)$-flat check sum is known when $[J/2]$ or fewer errors occur.

Now, given the $(z-h-1)$-flat check sums it is possible to determine the $(z-h-2)$-flat check sums in the same manner, for the number of $(z-h-1)$-flats orthogonal on an $(z-h-2)$-flat is greater than J, and therefore the argument given earlier holds here as well. In general, the number of orthogonal flats increases as the dimensionality of the flats decreases [see (20)], so that it is finally possible to determine the 0-flats and thus to correct up to and including $[J/2]$ errors. Thus the following is true.

Lemma 9

The minimum distance of a Reed–Muller code of order sh is at least

$$d \geq 2^{sh} + 2^{s(h-1)} + \cdots + 2^s + 2. \qquad (21)$$

Proof: For since $[J/2]$ errors are correctable, d is at least $2[J/2] + 1$ or J. But as every generator polynomial contains the root α^0, d is even and at least $J + 1$.

The preceding results can now be summarized to give Theorem 1.

[1] The symbol $[x]$ denotes the greatest integer contained in x.

Theorem 1

The nonprimitive order-ν Reed–Muller codes for which $r = 2^s - 1$ have the following parameters:

$$n = (2^m - 1)/(2^s - 1)$$

$$k = \begin{bmatrix}\text{number of integers less than } 2^m - 1 \text{ which} \\ \text{are divisible by } 2^s - 1 \text{ and whose weight exceeds } \nu\end{bmatrix}$$

$$d \geqq 2^{sh} + 2^{s(h-1)} + \cdots + 2^s + 2,$$

where $h = [\nu/s]$. The generator polynomial of such a code contains as roots every power of α whose exponent has weight ν or less and is divisible by $2^s - 1$.

Proof: The collection of lemmas establishes the result for $\nu = sh$. For $sh \leqq \nu < s(h+1)$, the code of order ν is a subcode of the code of order sh; clearly its minimum distance cannot be less than that of the code of order sh. This proves the theorem.

A superficial examination of the set of $n - k$ numbers divisible by $2^s - 1$ whose weight is sh or less is rewarding. From Peterson[13] we know that the weight of any multiple of $2^s - 1$ is at least s. For since $w(2^s - 1) = s$, assume $w(i(2^s - 1)) \leqq s$ for all integers i less than j, where

$$j(2^s - 1) = A_0 + A_1 2^s + A_2 2^{2s} + \cdots + A_z 2^{zs};$$

$$0 \leqq A_i \leqq 2^s - 1. \qquad (22)$$

Clearly

$$j(2^s - 1) = \sum_{i=0}^{z} A_i + \sum_{i=0}^{z} A_i(2^{si} - 1), \qquad (23)$$

therefore $\sum_{i=0}^{z} A_i$ is a multiple of $2^s - 1$. But as

$$w(j(2^s - 1)) = \sum_{i=0}^{z} w(A_i) \geqq w\left(\sum_{i=0}^{z} A_i\right) \geqq s \qquad (24)$$

the weight of any multiple of $2^s - 1$ is at least[2] s.

Now for any value of j less than $2^{hs} + 2^{(h-1)s} + \cdots + 2^s + 1$, it is true that

$$w(j(2^s - 1)) \leqq hs, \qquad (25)$$

for

$$2^{(h+1)s} - 1 = j(2^s - 1)$$
$$+ (2^{hs} + 2^{(h-1)s} + \cdots + 2^s + 1 - j)(2^s - 1) \qquad (26)$$

and thus

$$(h+1)s = w(j(2^s - 1))$$
$$+ w[(2^{hs} + 2^{(h-1)s} + \cdots + 2^s + 1 - j)(2^s - 1)]. \qquad (27)$$

But by (24) the second term on the right side of (27) is at least s; (25) follows directly.

This result states that counting 0, $g(x)$ has $2^{hs} + 2^{(h-1)s} + \cdots + 2^s + 1$ consecutive roots. Since

$$w(2^m - 1 - 2^s + 1) = m - s > hs; \qquad k < n$$

and

$$w(2^{hs} + 2^{h(s-1)} + \cdots + 2^s + 1) = (h+1)s > hs, \qquad (28)$$

it has exactly this many consecutive roots. Thus the minimum distance of the nonprimitive Reed–Muller codes with $r = 2^s - 1$ is guaranteed by the BCH bound.[4],[5]

Corollary 1

Every nonprimitive Reed–Muller code of order sh is a subcode (subset) of the nonprimitive BCH code of the same length and guaranteed minimum distance.

The proof given here is, in essence, the same as that given in Kasami *et al.*[8] for the nonprimitive codes. It is included because it introduces some notation necessary in the sequel.

Theorem 1 gives $n - k$ for the codes of order sh as the number of integers less than $2^m - 1$ which are divisible by $2^s - 1$ and whose weight is sh or less. It would be desirable to replace this rather cumbersome definition with a simple combinatorial expression. If one does not quibble over the meaning of "simple," this can be done.

We wish to find the number of integers j in (22) such that $\sum_{i=0}^{z} A_i$ is a multiple of $2^s - 1$ and such that

$$w(j(2^s - 1)) \leqq sh. \qquad (29)$$

Letting $N(w)$ denote the number of integers $j(2^s - 1)$ of weight w gives

$$n - k = \sum_{w=0}^{sh} N(w). \qquad (30)$$

We now proceed to find $N(w)$.

A number $j(2^s - 1)$ of weight w can, by means of (22) and (23), be regarded as a $(z+1)$-by-s binary array containing exactly w ones. Every different array corresponds to a different value of j. Let W_i, $0 \leqq i \leqq s - 1$, denote the number of ones in the ith column of the array. Clearly

$$w = \sum_{i=0}^{s-1} W_i. \qquad (31)$$

Also, perhaps not so clearly, (23) implies that

$$\sum_{i=0}^{s-1} W_i 2^i = 0 \qquad (\text{mod } 2^s - 1). \qquad (32)$$

Thus it follows that

$$N(w) = \sum_{W_i} \binom{z+1}{W_0}\binom{z+1}{W_1} \cdots \binom{z+1}{W_{s-1}}, \qquad (33)$$

where the sum is over all values of W_i such that (31) and (32) hold. For $s = 2$, for example, this gives

$$N(w) = \sum_k \binom{z+1}{3k - w}\binom{z+1}{2w - 3k}, \qquad (34)$$

where the sum is performed on all values of k for which

[2] The stated result is clearly a special case of $w(j(p^s - 1) \geqq p - 1)s$; p a prime.

TABLE I

Binary Nonprimitive Reed-Muller Codes of Length less than 4000 for which $r = 2^s - 1$. Codes for $s = 1$ are listed in Part I of this Paper.

s	h	n	k	d
2	1	21	11	6
	1	85	68	6
	2		24	22
	1	341	315	6
	2		195	22
	3		45	86
	1	1365	1328	6
	2		1063	22
	3		483	86
	4		78	342
3	1	73	45	10
	1	585	520	10
	2		184	74
4	1	273	191	18
5	1	1057	813	34

the arguments are positive integers. Finally from (30) and (33), we have,

$$n - k = \sum_{w=0}^{sh} \sum_{W_i} \prod_{i=0}^{s-1} \binom{z+1}{W_i}, \quad (35)$$

which hardly seems simpler than the verbal expression for $n - k$ given in Theorem 1.

Table 1 contains a list of all binary nonprimitive Reed–Muller codes of length less than 4000 for which $r = 2^s - 1$. Primitive codes ($s = 1$) are listed in Part I of this paper.

IV. Decoding

Since the generator polynomials of the Reed–Muller codes are known, their encoders and syndrome calculators can be constructed readily. (See Peterson,[1] for example.) Then the argument employed in lower bounding the minimum distance suggests how to construct the decoders, at least in theory.

Since every $(z - h)$-flat is in the null space of the order $-(sh)$ code, the corresponding check sum can be formed by adding appropriate digits of the syndrome. Taking the majority vote of the J check sums orthogonal on a given $(z - h - 1)$-flat produces the check sum associated with that flat. Repeating this process a total of $z - h$ times produces the 0-flats (points, noise bits), and error correction is accomplished by adding these to the appropriate information bits.

The above decoding procedure has been referred to as $(z - h)$-step orthogonalization.[12] We employ this terminology in stating the result as Theorem 2.

Theorem 2

The Reed–Muller codes of order (sh) can be $(z - h)$-step orthogonalized when $r = 2^s - 1$. The modifier "nonprimitive" has been omitted here since the result also holds when $r = 1$, (i.e., for primitive Reed–Muller codes).

Fig. 1 shows the general form of the majority-logic decoder. It illustrates the fact that $z - h$ levels of majority logic are employed. Since each node of the resulting "tree" has J branches, a total of

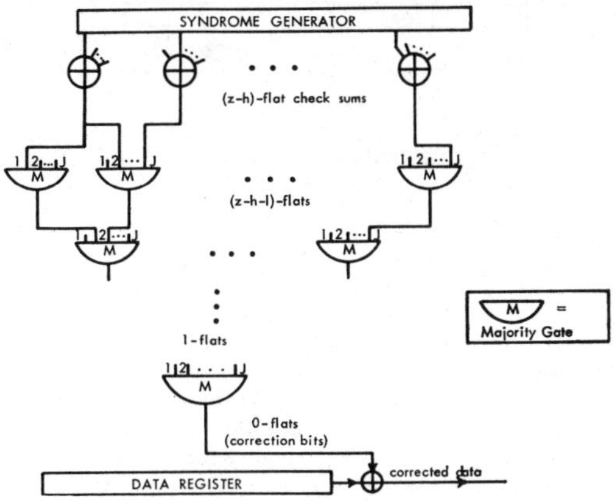

Fig. 1. A $(z - h)$-step majority-logic decoder.

$$\sum_{i=0}^{z-h-1} (2^{sh} + 2^{s(h-1)} + \cdots + 2^s + 1)^i \quad (36)$$

majority gates are required. Clearly if sh or $z - h$ is large, the decoder is likely to be unreasonably complex. Fortunately, for many of the most useful nonprimitive Reed–Muller codes, these parameters are quite small. For example, if $z - h = 1$, only a single majority gate is required. (The difference set codes of Weldon[10] are of this type.) As another example, if $s = 1$, the codes are primitive Reed–Muller codes and approximately $n/2$ majority gates are required.

The majority-gate tree necessary in the Reed algorithm can in all cases be replaced by a single majority gate. However, the inputs to this gate, which are the $(z - h)$-flat check sums, must each be weighted by an appropriate factor. If a particular $(z - h)$-flat is employed i times in the decoder of Fig. 1, then it has a weighting factor of i in the single-gate decoder of Fig. 2. It has been shown that no majority-gate decision in the original decoder will be incorrect, provided $J/2$ or fewer errors occurred. Clearly the majority of inputs to the ensemble of first layer gates will be correct, and a single, weighted input, majority gate suffices.

Corollary 2

The Reed–Muller codes with $r = 2^s - 1$ can be decoded with a single, weighted-input, majority gate.

Unfortunately in many cases this result is less useful than it sounds since the weighting factors tend to become quite large. This difficulty does not seem to be inherent in the procedure, however, and may possibly be eliminated by a more nearly complete understanding of the codes and their geometries.

As mentioned earlier, the nonprimitive Reed–Muller codes with $r = 2^s - 1$ were first discovered by Rudolph.[9]

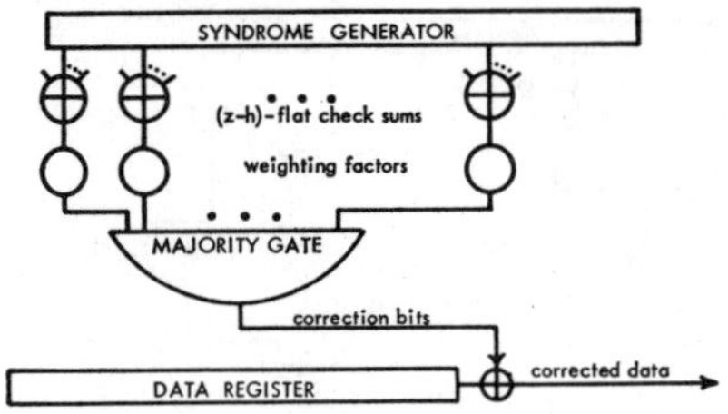

Fig. 2. A single-gate, $(z - h)$-step, majority-logic decoder.

His decoding procedure employs a set of $(z - h)$-flats orthogonal on a 0-flat. Unfortunately, unless $z - h = 1$, it is not possible to realize the full error-correcting ability of the codes with this procedure. For if $z - h > 1$, it is easy to show that $PG(z, 2^s)$ cannot contain $2^{sh} + 2^{s(h-1)} + \cdots + 2^s + 1$ or more $(z - h)$-flats orthogonal on a 0-flat.

Up to this point only binary codes have been considered. We now briefly consider the more general nonbinary case. In Part I of this paper the notion of a cyclic Reed–Muller code is generalized so that the codes can be constructed with symbols chosen from $GF(q)$. Theorems 1 and 2 can be extended to codes with symbols chosen from $GF(p)$, p a prime, without difficulty; the generalization to $GF(q)$ is not an obvious one. In the p-any case each majority gate produced at its output the field element which appears at the majority of its inputs. Provided the number of errors does not exceed half the number of orthogonal equations, correct decoding results. Codes produced in this manner are not generally as efficient as the binary codes (in an information-theoretic sense) and, in fact, are sometimes inferior to codes derived from binary codes in the following manner.

Consider the parity check matrix of a binary Reed–Muller code of length n in binary form (4). This matrix can be regarded as the parity check matrix of a code of length n with symbols chosen from an $GF(2^d)$. Each row of the binary matrix specifies a parity check of the nonbinary code according to the rule: the check digit is the sum of the information digits corresponding to ones in the row of the binary matrix. The key points are that only majority gates are employed in decoding (see Fig. 1) and that the set of orthogonal equations can be used to correct the same number of errors regardless of whether the symbols are binary or otherwise, provided the majority rule is modified as above. It should be noted that although the number of information symbols in a nonbinary code derived from an (n, k) binary code is not less than k, in general it may be greater.

V. Summary

A new class of random-error-correcting codes has been presented. These codes are called nonprimitive Reed–Muller codes because of their close relationship to the primitive Reed–Muller codes.[8] It is shown that the class of nonprimitive Reed–Muller codes contains the projective geometry codes discovered by Rudolph[9] as a subclass. The latter codes are investigated in detail and two results proved. First, the codes are moderately efficient random-error correctors for practical values of code length and rate. Second, they can be decoded with a relatively modest amount of equipment. It appears that these codes may be suitable for use in error-control systems requiring random-error correction.

References

[1] W. W. Peterson, *Error-Correcting Codes.* New York: Wiley, 1961.

[2] I. S. Reed, "A class of multiple-error-correcting codes and the decoding scheme," *IRE Trans. Information Theory,* vol. IT-4, pp. 38–49, September 1954.

[3] D. D. Muller, "Applications of Boolean algebra to switching circuit design and to error detection," *IRE Trans. Electronic Computers,* vol. EC-3, pp. 6–12, September 1954.

[4] R. C. Bose and C. K. Ray-Chaudhuri, "On a class of error correcting binary group codes," *Information and Control,* vol. 3, pp. 68–79, March 1960.

[5] A. Hocquenghem, "Codes correcteurs d'erreurs," *Chiffres,* vol. 2, pp. 147–156, 1959.

[6] W. W. Peterson, "Encoding and error-correction procedures for the Bose-Chaudhuri codes," *IRE Trans. Information Theory,* vol. IT-6, pp. 459–70, 1960.

[7] T. Kasami and S. Lin, "Some codes which are invariant under a doubly-transitive permutation group and their connection with balanced incomplete block designs," USAF Cambridge Research Center, Bedford, Mass., Sci. Rept. AFCRL-66-142, 1961.

[8] T. Kasami, S. Lin, and W. W. Peterson, "Some results on cylic codes which are invariant under the affine group," AF Cambridge Research Lab., Bedford, Mass., Sci. Rept. AFCRL-66-622, 1966.

[9] L. D. Rudolph, "A class of majority logic decodable codes," *IEEE Trans. Information Theory,* vol. IT-13, pp. 305–307, April 1967.

[10] E. J. Weldon, Jr., "Difference-set cylic codes," *Bell Sys. Tech. J.,* vol. 45, pp. 1045–1055, September 1966.

[11] R. L. Graham and J. MacWilliams, "On the number of parity checks in difference-set cyclic codes," *Bell Sys. Tech. J.,* vol. 45, pp. 1056–1070, September 1966.

[12] J. L. Massey, *Threshold Decoding.* Cambridge, Mass.: M.I.T. Press, 1963.

[13] W. W. Peterson, private communication, 1966.

29

Polynomial Codes

TADAO KASAMI, MEMBER, IEEE, SHU LIN, MEMBER, IEEE, AND W. WESLEY PETERSON, FELLOW, IEEE

***Abstract*—A class of cyclic codes is introduced by a polynomial approach that is an extension of the Mattson–Solomon method and of the Muller method. This class of codes contains several important classes of codes as subclasses, namely, BCH codes, Reed–Solomon codes, generalized primitive Reed–Muller codes, and finite geometry codes. Certain fundamental properties of this class of codes are derived. Some subclasses are shown to be majority-logic decodable.**

I. Introduction

THIS PAPER presents a class of cyclic codes and their dual codes that contains many well-known classes of codes as subclasses, such as BCH codes [1], [2], Reed–Solomon codes [2], [3], generalized Reed–Muller codes [4], [5], projective geometry codes [6], [7], [8], and Euclidean geometry codes [6], [9]. This class of codes is introduced by a polynomial approach that is an extension of the Mattson–Solomon method [10] and the Muller method [11]. The generator polynomial of any code in this class is characterized, and the minimum distance is lower bounded. A code in this class is proved to be either a BCH code or a subcode of a BCH code of the same minimum distance. For some subclasses of this class of codes, the exact minimum distance is established. Some useful geometric properties are also proved. Because of the geometric structure, certain subclasses of this class of codes are shown to be majority-logic decodable.

II. Preliminaries

Let $GF(q^{ms})$ be the extension field of $GF(q^s)$, where m and s are positive integers and q is a power of prime. Let α be a primitive element of $GF(q^{ms})$. Then $\{\alpha^0, \alpha^1, \alpha^2, \cdots, \alpha^{m-1}\}$ form a basis of $GF(q^{ms})$ over $GF(q^s)$. Any nonzero element α^j in $GF(q^{ms})$ can be expressed as

$$\alpha^j = \sum_{i=1}^{m} a_{ij}\alpha^{i-1} \quad \text{for} \quad 0 \leq j \leq q^{ms} - 2 \tag{1}$$

where $a_{ij} \varepsilon GF(q^s)$. For convenience, we shall call the m-tuple $(a_{1j}, a_{2j}, \cdots, a_{mj})$ as the *coordinate vector* of α^j.

It is known that $q^{ms} - 1$ is divisible by $q^s - 1$. Suppose that b is a factor of $q^s - 1$. Let

$$z = (q^s - 1)/b, \tag{2}$$

and

$$n = (q^{ms} - 1)/b. \tag{3}$$

Let $X_1, X_2, \cdots, X_m$ be m variables over $GF(q^s)$ and $\bar{X} = (X_1, \cdots, X_m)$. We define $P(m, e, \mu, b)$ as the set of polynomials $f(\bar{X}) = f(X_1, X_2, \cdots, X_m)$ in $X_1, X_2, \cdots, X_m$ with coefficients in $GF(q^s)$ such that the sum of powers in each term of $f(\bar{X})$ is a multiple of b and the degree of $f(\bar{X})$ is μb or less. That is,

$$P(m, e, \mu, b) = \Big\{ f(\bar{X}) \mid \text{for each term } CX_1^{\nu_1}X_2^{\nu_2} \cdots X_m^{\nu_m} \text{ of } f(\bar{X}), \; C \text{ is in } GF(q^s) \text{ and } \sum_{i=1}^{m} \nu_i = jb \text{ with } 0 \leq j \leq \mu \Big\}. \tag{4}$$

Since $X_i^{q^s} = X_i$, we assume that ν_i is less than q^s and μ is at most equal to mz, where $z = (q^s - 1)/b$.

From (3), we have

$$(\alpha^{ln})^{q^s-1} = 1, \quad \text{for} \quad 0 \leq l < b.$$

Therefore, α^{ln} is an element of $GF(q^s)$. It follows from (1) that

$$a_{i\,ln+j} = \alpha^{ln}a_{ij}, \quad \text{for} \quad 1 \leq i \leq m. \tag{5}$$

For $0 \leq l < b$, let θ_l denote the set of vectors

$$\theta_l = \{(a_{1\,j+ln}, a_{2\,j+ln}, \cdots, a_{m\,j+ln}) \mid 0 \leq j < n\}. \tag{6}$$

Manuscript received December 8, 1967; revised March 11, 1968. This work was supported in part by the USAF Cambridge Research Laboratory, Office of Aerospace Research, Contract AF 19(628)-4379.

T. Kasami is with the Department of Control Engineering, Faculty of Engineering Science, Osaka University, Toyonaka, Osaka, Japan.

S. Lin and W. W. Peterson are with the Department of Electrical Engineering, University of Hawaii, Honolulu, Hawaii 96822

From (5), each vector in θ_l is equal to the product of the scalar α^{ln} and some vector in θ_0.

In [4], it has been shown that the sum $S_j = \sum_{X_i \in GF(q^s)} X_i^j$ of jth powers of all elements of $GF(q^s)$ is zero unless j is equal to $q^s - 1$. Therefore, we have the following.

Lemma 1

Consider $X_1^{\nu_1}X_2^{\nu_2} \cdots X_m^{\nu_m}$ with $0 \leq \nu_i \leq q^s - 1$. Then

$$\sum_{\substack{X_i \in GF(q^s) \\ 1 \leq i \leq m}} X_1^{\nu_1}X_2^{\nu_2} \cdots X_m^{\nu_m} = 0 \tag{7}$$

unless $\nu_1 = \nu_2 = \cdots = \nu_m = q^s - 1$.

Theorem 2

Assume that $\sum_{i=1}^{m} \nu_i = jb$ with $0 \leq \nu_i < q^s$ and $0 < j < mz$. Then,

$$\sum_{\bar{X} \in \theta_0} X_1^{\nu_1}X_2^{\nu_2} \cdots X_m^{\nu_m} = 0. \tag{8}$$

Proof: From (5) and (6) we obtain

$$\begin{aligned} \sum_{\substack{X_i \in GF(q^s) \\ 1 \leq i \leq m}} X_1^{\nu_1}X_2^{\nu_2} \cdots X_m^{\nu_m} &= \sum_{l=0}^{b-1} \sum_{\bar{X} \in \theta_l} X_1^{\nu_1}X_2^{\nu_2} \cdots X_m^{\nu_m} \\ &= \sum_{l=0}^{b-1} (\alpha^{ln})^{jb} \sum_{\bar{X} \in \theta_0} X_1^{\nu_1}X_2^{\nu_2} \cdots X_m^{\nu_m} \\ &= b \sum_{\bar{X} \in \theta_0} X_1^{\nu_1}X_2^{\nu_2} \cdots X_m^{\nu_m}. \end{aligned} \tag{9}$$

Since b is relatively prime to the characteristic of $GF(q^s)$, it follows from Lemma 1 and (9) that

$$\sum_{\bar{X} \in \theta_0} X_1^{\nu_1}X_2^{\nu_2} \cdots X_m^{\nu_m} = 0. \qquad \text{Q.E.D.}$$

A direct consequence of Theorem 2 is the following.

Corollary 3

If $f(\bar{X}) \in P(m, e, \mu, b)$ such that $f(0, 0, \cdots, 0) = 0$ and $\mu < mz$, then $\sum_{\bar{X} \in \theta_0} f(\bar{X}) = 0$.

Let e be a multiple of s, $e = rs$. For any polynomial $f(\bar{X}) = f(X_1, X_2, \cdots, X_m)$ in $P(m, e, \mu, b)$, we define a vector

$$\bar{v}(f) = (v_0, v_1, \cdots, v_{n-1})$$

whose components are in $GF(q^e)$ as follows:

$$v_j = f(a_{1j}, a_{2j}, \cdots, a_{mj}) \tag{10}$$

for $0 \leq j < n$, where $(a_{1j}, a_{2j}, \cdots, a_{mj})$ is the coordinate vector of α^j (a vector in θ_0).

Let $f_1(\bar{X})$ and $f_2(\bar{X})$ be two polynomials in $P(m, e, \mu, b)$. Then, the inner product of $\bar{v}(f_1)$ and $\bar{v}(f_2)$ is

$$\bar{v}(f_1)\bar{v}(f_2)^T = \sum_{\bar{X} \in \theta_0} f_1(\bar{X})f_2(\bar{X}). \tag{11}$$

Lemma 4

Let $f_1(\bar{X})$ and $f_2(\bar{X})$ be two polynomials in $P(m, e, \mu, b)$ with $\mu < mz$. Then, $\bar{v}(f_1) = \bar{v}(f_2)$, if and only if $f_1(\bar{X}) = f_2(\bar{X})$.

Proof: The "if part" is obvious. Assume that $\bar{v}(f_1) = \bar{v}(f_2)$. By (5) we have

$$f_1(a_{1j}, \cdots, a_{mj}) = f_1(a_{1\,j+ln}, \cdots, a_{m\,j+ln}),$$
$$f_2(a_{1j}, \cdots, a_{mj}) = f_2(a_{1\,j+ln}, \cdots, a_{m\,j+ln}),$$
$$\text{for } 0 \leq j < n \text{ and } 0 < l < b.$$

Hence, for any $(a_1, \cdots, a_m) \neq (0, \cdots, 0)$ with $a_i \in GF(q^s)$,

$$f_1(a_1, \cdots, a_m) = f_2(a_1, \cdots, a_m).$$

Therefore, $f_1(\bar{X}) - f_2(\bar{X}) = [(f_1(0, \cdots, 0) - f_2(0, \cdots, 0)] \cdot \prod_{i=1}^{m} (1 - X_i^{q^s-1})$. Since the degree of $f_1(\bar{X}) - f_2(\bar{X})$ is at most μb and $\mu b < mzb = m(q^s - 1)$, we have $f_1(0, 0, \cdots, 0) = f_2(0, 0, \cdots, 0)$. Consequently,

$$f_1(\bar{X}) = f_2(\bar{X}). \qquad \text{Q.E.D.}$$

Consider any non-negative integer h that is less than q^{ms}. In radix-q^s form,

$$h = \delta_1 + \delta_2 q^s + \delta_3 q^{2s} + \cdots + \delta_m q^{(m-1)s} \tag{12}$$

where $0 \leq \delta_i < q^s$ for $1 \leq i \leq m$. The weight of h, $W_{q^s}(h)$, is defined by

$$W_{q^s}(h) = \sum_{i=1}^{m} \delta_i. \tag{13}$$

Lemma 5

The weight of a non-negative integer $h (h < q^{ms})$ is divisible by b if and only if h is divisible by b.

Proof: Consider

$$\begin{aligned} h - W_{q^s}(h) &= \sum_{i=1}^{m} \delta_i q^{(i-1)s} - \sum_{i=1}^{m} \delta_i \\ &= \sum_{i=2}^{m} \delta_i (q^{(i-1)s} - 1). \end{aligned} \tag{14}$$

Since the right-hand side of (14) is divisible by b, therefore, $h - W_{q^s}(h)$ must be divisible by b. The lemma is then obvious. Q.E.D.

III. Definition of the Codes

Let $Q(m, s, \mu, b, q)$ be the subset of all polynomials $f(\bar{X})$ in $P(m, s, \mu, b)$ such that

$$f(a_{1j}, a_{2j}, \cdots, a_{mj}) \in GF(q)$$

for $0 \leq j < n$; that is

$$\begin{aligned} Q(m, s, \mu, b, q) = \{f(\bar{X}) \mid\ & f(\bar{X}) \in P(m, s, \mu, b) \\ & \text{and } f(a_{1j}, a_{2j}, \cdots, a_{mj}) \\ & \text{in } GF(q) \text{ for } 0 \leq j < n\}. \end{aligned} \tag{15}$$

Definition

An (n, m, s, μ, q)-polynomial code is defined as a set of vectors $\bar{v}(f)$

$$\{\bar{v}(f) \mid f(\bar{X}) \in Q(m, s, \mu, b, q)\}. \tag{16}$$

It follows from the definition that an (n, m, s, μ, q)-polynomial code is a linear code of length $n = (q^{ms} - 1)/b$ with code symbols from $GF(q)$.

IV. Characterization of Polynomial Codes as a Subclass of Cyclic Codes

The following theorem and corollary are the principal results of this paper.

Theorem 6

An (n, m, s, μ, q)-polynomial code over $GF(q)$ is cyclic, and α^h is a root of the generator polynomial $g(X)$ if and only if b divides h and

$$\min_{0 \le l < s} W_{q^s}(hq^l) = jb \tag{17}$$

with $0 < j < mz - \mu$.

Corollary 7

The dual code [over $GF(q)$] of an (n, m, s, μ, q)-polynomial code has α^h as a root in its generator polynomial $g_d(X)$ if and only if b divides h and

$$\max_{0 \le l < s} W_{q^s}(hq^l) = jb \tag{18}$$

with $0 \le j \le \mu$.

The remainder of this section is devoted to the proof, which is rather lengthy and requires a number of lemmas. By definition, any polynomial in $P(m, s, \mu, b)$ is of the following form:

$$f(\bar{X}) = \sum c_{\nu_1, \nu_2, \cdots, \nu_m} X_1^{\nu_1} X_2^{\nu_2} \cdots X_m^{\nu_m}$$

where $0 \le \nu_i < q^s$ and $\sum_{i=1} \nu_i = jb$ with $0 \le j \le \mu$.

By Lemma 4, it is easy to prove the following theorem.

Theorem 8

A necessary and sufficient condition for $f(\bar{X})$ in $Q(m, s, \mu, b, q)$ is

$$f^q(\bar{X}) = f(\bar{X})$$

or equivalently that the equation

$$c^q_{\nu_1, \nu_2, \cdots, \nu_m} = c_{\nu_1 q, \nu_2 q, \cdots, \nu_m q} \tag{19}$$

holds for every $(\nu_1, \nu_2, \cdots, \nu_m)$.

Let U_b denote the set of m-tuples

$$U_b = \left\{ (\nu_1, \nu_2, \cdots, \nu_m) \;\middle|\; 0 \le \nu_i < q^s \text{ and } \sum_{i=1}^{m} \nu_i = jb \text{ with } 0 \le j < mz \right\}. \tag{20}$$

By letting $(\nu_1, \nu_2, \cdots, \nu_m)$ correspond to $h = \sum_{i=1}^{m} \nu_i q^{(i-1)s}$, it can be seen by Lemma 5 that the number of m-tuples of U_b is the number of the non-negative integers less than $q^{ms} - 1$ that are divisible by b (the all-zero m-tuples should be included). Hence the number is

$$(q^{ms} - 1)/b = n. \tag{21}$$

For any m-tuple $(\nu_1, \nu_2, \cdots, \nu_m)$ in U_b, we define an equivalence relation as follows:

$$(\nu_1', \nu_2', \cdots, \nu_m') \sim (\nu_1, \nu_2, \cdots, \nu_m)$$

if and only if there is an integer l with $0 \le l < s$ such that

$$\nu_i' \equiv \nu_i q^l \pmod{q^s - 1} \qquad 1 \le i \le m \tag{22a}$$

for $0 \le \nu_i < q^s - 1$ and

$$\nu_i' = \nu_i \tag{22b}$$

for $\nu_i = q^s - 1$. If $(\nu_1, \nu_2, \cdots, \nu_m) \in U_b$ and $(\nu_1', \nu_2', \cdots, \nu_m') \sim (\nu_1, \nu_2, \cdots, \nu_m)$, then $(\nu_1', \nu_2', \cdots, \nu_m')$ is also in U_b.

Let us define the following terms.

1) $C(\nu_1, \nu_2, \cdots, \nu_m)$ is the set of all m-tuples in U_b that are equivalent to $(\nu_1, \nu_2, \cdots, \nu_m)$;
2) $\lambda(\nu_1, \nu_2, \cdots, \nu_m)$ is the number of elements in $C(\nu_1, \nu_2, \cdots, \nu_m)$;
3) $\bar{D}(\nu_1, \nu_2, \cdots, \nu_m)$ is the maximum value of $\sum_{i=1}^{m} \nu_i'$ for $(\nu_1', \nu_2', \cdots, \nu_m')$ in $C(\nu_1, \nu_2, \cdots, \nu_m)$;
4) $\underline{D}(\nu_1, \nu_2, \cdots, \nu_m)$ is the minimum value of $\sum_{i=1}^{m} \nu_i'$ for $(\nu_1', \nu_2', \cdots, \nu_m')$ in $C(\nu_1, \nu_2, \cdots, \nu_m)$.

Assume that $f(\bar{X}) \in Q(m, s, \mu, b, q)$. Then it follows from (19) that $c^{q^{\lambda(\nu_1, \cdots, \nu_m)}}_{\nu_1, \cdots, \nu_m} = c_{\nu_1, \cdots, \nu_m}$. This implies that

$$c_{\nu_1, \cdots, \nu_m} \in GF(q^{\lambda(\nu_1, \cdots, \nu_m)}). \tag{23}$$

If $\bar{D}(\nu_1, \cdots, \nu_m) > \mu b$, then coefficient $c_{\nu_1, \nu_2, \cdots, \nu_m}$ must be zero. Otherwise, $c_{\nu_1, \cdots, \nu_m}$ can take any value of $GF(q^{\lambda(\nu_1, \cdots, \nu_m)})$. Therefore, the dimension of an (n, m, s, μ, q)-P-code, $k(\mu, b)$, is equal to the number of m-tuples $(\nu_1, \cdots, \nu_m)$ in the U_b with $0 \le \bar{D}(\nu_1, \cdots, \nu_m) \le \mu b$. Let $r(\mu, b)$ denote the number of check digits of an (n, m, s, μ, q)-P-code. That is,

$$r(\mu, b) = n - k(\mu, b).$$

Then, by (21), $r(\mu, b)$ is equal to the number of m-tuples $(\nu_1, \cdots, \nu_m)$ in U_b such that

$$\mu b < \bar{D}(\nu_1, \cdots, \nu_m) < mzb = m(q^s - 1). \tag{24}$$

For any $(\nu_1, \nu_2, \cdots, \nu_m) \ne (0, 0, \cdots, 0)$ in U_b, it is obvious that

$$(q^s - 1 - \nu_1, \cdots, q^s - 1 - \nu_m)$$

is also in U_b. If $\mu b < \bar{D}(\nu_1, \nu_2, \cdots, \nu_m) < m(q^s - 1)$, then

$$0 < \underline{D}(q^s - 1 - \nu_1, q^s - 1 - \nu_2, \cdots, q^s - 1 - \nu_m) < m(q^s - 1) - \mu b$$

By letting $(q^s - 1 - \nu_1, \cdots, q^s - 1 - \nu_m)$ correspond to $(\nu_1, \cdots, \nu_m)$, we have the following lemma.

Lemma 9

The number of parity check digits of an (n, m, s, μ, q)-polynomial code, $r(\mu, b)$, is equal to the number of m-tuples $(\nu_1, \nu_2, \cdots, \nu_m)$'s in U_b such that

$$0 < \underline{D}(\nu_1, \nu_2, \cdots, \nu_m) < (mz - \mu)b. \tag{25}$$

Consider an integer $h (0 \le h < q^{ms} - 1)$ whose weight is a multiple of b:

$$W_{q^s}(h) = jb$$

with $0 < j < mz$. Write h in radix-q^s form

$$h = \sum_{i=1}^{m} \delta_i q^{(i-1)s} \tag{26}$$

where $0 \le \delta_i < q^s$. For a non-negative integer l, let

$$hq^l \equiv \sum_{i=1}^{m} \delta_i^{(l)} q^{(i-1)s} \pmod{q^{ms}-1}$$

where $0 \le \delta_i^{(l)} \le q^s - 1$. Thus the weight of hq^l is

$$W_{q^s}(hq^l) = \sum_{i=1}^{m} \delta_i^{(l)}. \tag{27}$$

Let

$$\bar{\delta}_i^{(l)} \equiv \delta_i q^l \pmod{q^s - 1}$$

for $0 \le \delta_i < q^s - 1$, and

$$\bar{\delta}_i^{(l)} = \delta_i$$

for $\delta_i = q^s - 1$. Then, it can be verified easily that

$$\sum_{i=1}^{m} \delta_i^{(l)} = \sum_{i=1}^{m} \bar{\delta}_i^{(l)}. \tag{28}$$

Thus, it follows from the definition of $\underline{D}(\delta_1, \delta_2, \cdots, \delta_m)$ and (28) that the smallest of $\sum_{i=1}^{m} \delta_i^{(l)}$ for $0 \le l < s$ is equal to $\underline{D}(\delta_1, \delta_2, \cdots, \delta_m)$.

Let $E(h)$ denote the smallest of $\sum_{i=1}^{m} \delta_i^{(l)}$ for $0 \le l < s$, i.e.,

$$E(h) = \min_{0 \le l < s} W_{q^s}(hq^l). \tag{29}$$

It follows from the definition of $\underline{D}(\delta_1, \delta_2, \cdots, \delta_m)$ and (28) that $E(h)$ is equal to $\underline{D}(\delta_1, \delta_2, \cdots, \delta_m)$. Then the following lemma holds.

Lemma 10

The number of integer h's such that the $E(h) = jb$ with $0 < j < mz - \mu$ is equal to the number of check digits of an (n, m, s, μ, q)-P-code, $r(\mu, b)$.

Consider the polynomial

$$f_h(\bar{X}) = \left(\sum_{i=1}^{m} \alpha^{i-1} X_i\right)^h = \left(\sum_{i=1}^{m} \alpha^{i-1} X_i\right)^{\sum_{t=1}^{m} \delta_t q^{(t-1)s}}$$

$$= \prod_{t=1}^{m} \left(\sum_{i=1}^{m} \alpha^{(i-1)q^{(t-1)s}} X_i\right)^{\delta_t}. \tag{30}$$

The degree of $f_h(\bar{X})$ is equal to $W_{q^s}(h) = \sum_{t=1}^{m} \delta_t$. By (1), the vector $\bar{v}(f_h)$ defined in accordance with (10) is

$$\bar{v}(f_h) = (1, \alpha^h, \alpha^{2h}, \cdots, \alpha^{(n-1)h}). \tag{31}$$

Let $f(\bar{X})$ be a polynomial in $Q(m, s, \mu, b, q)$ and let $\bar{v}(f) = (v_0, v_1, \cdots, v_{n-1})$. Then it follows from (11) and (31) that

$$\bar{v}(f)\bar{v}(f_h)^T = \sum_{i=1}^{n} v_{i-1}\alpha^{(i-1)h}$$

$$= \sum_{\bar{X} \neq \theta_0} f(\bar{X}) f_h(\bar{X}). \tag{32}$$

Now, suppose that $E(h) = jb$ with $0 < j < mz - \mu$. Then, there is a positive integer h' such that

$$W_{q^s}(h') = jb, \quad 0 < j < mz - \mu \tag{33}$$

and, for some integer l,

$$h' \equiv hq^l \pmod{q^{ms} - 1}. \tag{34}$$

By (33), we have that $f(\bar{X})f_{h'}(\bar{X}) \in P(m, ms, mz - 1, b)$. Since $f_{h'}(0, \cdots, 0) = 0$, it follows from Corollary 3 and (32) that

$$\sum_{i=1}^{n} v_{i-1}\alpha^{(i-1)h'} = 0.$$

By using (34) and the fact that $v_i \in GF(q)$, we have

$$\sum_{i=1}^{n} v_{i-1}\alpha^{(i-1)h} = 0. \tag{35}$$

Thus, α^h is a root of the polynomial $\sum_{i=0}^{n-1} v_i X^i$ that corresponds to the code vector $\bar{v}(f) = (v_0, v_1, \cdots, v_{n-1})$. By Lemma 10 and (35), Theorem 6 is proved. For convenience, we restate Theorem 6.

Theorem 6

An (n, m, s, μ, q)-P-code over $GF(q)$ is cyclic, and α^h is a root of the generator polynomial $g(X)$ if and only if b divides h and

$$E(h) = \min_{0 \le l < s} W_{q^s}(hq^l) = jb \quad \text{with} \quad 0 < j < mz - \mu.$$

Corollary 7 is a direct consequence of Theorem 6.

Let h_0 be the smallest integer divisible by b with $E(h_0) = (mz - \mu)b$. Then h_0 must be

$$h_0 = (q^s - 1) + (q^s - 1)q^s$$
$$+ \cdots + (q^s - 1)q^{(Q-1)s} + Rq^{Qs}, \tag{36}$$

where Q and R are the quotient and remainder resulting from dividing $m(q^s - 1) - \mu b$ by $q^s - 1$, i.e.,

$$m(q^s - 1) - \mu b = Q(q^s - 1) + R \tag{37}$$

with $0 \le R < q^s - 1$. R is obviously divisible by b. It can be easily shown that all positive integer h's that are less than h_0 and divisible by b satisfy $E(h) < m(q^s - 1) - \mu b$. By Theorem 6, the generator polynomial of an (n, m, s, μ, q)-P-code must have the following roots:

$$\beta, \beta^2, \cdots, \beta^{(h_0 \backslash b) - 1}$$

where $\beta = \alpha^b$. We thus have Theorem 11.

Theorem 11

An (n, m, s, μ, q)-polynomial code has minimum distance at least

$$[(R + 1)q^{Qs} - 1]/b.$$

Theorem 12

If one of the following conditions holds: 1) $R = 0$; 2) an R/b-BCH code of length z exists and has minimum weight R/b; or 3) $s = 1$, a (n, m, s, μ, q)-P-code has minimum distance exactly $[(R + 1)q^{Qs} - 1]/b$, and is a subcode of a BCH code with the same minimum distance.

Proof: 1) For the case $R = 0$, let $f_0(X_1) = X_1^z - 1$. 2) If condition 2) holds, there exists a polynomial $f_0(X_1)$ of degree $z - R/b$ such that $f_0(X_1^b)$ is in $Q(1, s, q^s - 1 - R, b, q)$ and has $q^s - 1 - R$ different nonzero roots in $GF(q^s)$. This follows from Mattson–Solomon formulation. 3) For the case $s = 1$, let

$$f_0(X_1) = \prod_{j=1}^{(q-1-R)/b} (X_1 - \gamma^{jb})$$

where γ is a primitive element of $GF(q)$. Then $f_0(X_1^b)$ has $q - 1 - R$ nonzero different roots in $GF(q)$.

Now, consider the following polynomial:

$$f_{\min}(\bar{X}) = f_0(X_1^b)(X_2^{q^s-1} - 1) \cdots (X_{m-Q}^{q^s-1} - 1). \quad (38)$$

Obviously, $f_{\min}(\bar{X}) \in Q(m, s, \mu, b, q)$. It can be shown easily that the weight of vector $\bar{v}(f_{\min})$ is equal to $[(R+1)q^{Qs} - 1]/b$. Q.E.D.

Conversely, it can be shown that if the minimum weight is equal to $[(R + 1)q^{Qs} - 1]/b$, then condition 2) must hold. Theorem 12 implies that for some classes of BCH codes the BCH bound gives the minimum weight exactly. Some examples are given in Table I.

TABLE I

q	m	s	μ	b	Polynomial Code	BCH Code	Exact Minimum Distance
2	3	2	1	3	(21, 10)	(21, 12)	5
2	3	3	1	7	(73, 28)	(73, 49)	9
2	4	2	2	3	(85, 61)	(85, 69)	5
2	2	4	8	1	(255, 16)	(255, 29)	95

V. Important Subclasses of Polynomial Codes

In this section, we shall study some subclasses of polynomial codes and show that several classes of well-known codes are subclasses of polynomial codes.

Subclass 1—BCH Codes [1], [2]

For $m = 1$ and $\mu = n - d$, an $(n, 1, s, n - d, q)$-polynomial code has

$$\beta(=\alpha^b), \beta^2, \cdots, \beta^{d-1}$$

and their conjugates as all roots of its generator polynomial (Theorem 6), where α is a primitive element of $GF(q^s)$. Thus, an $(n, 1, s, n - d, q)$-polynomial code is a d-BCH[1] code $(m_0 = 1)$ of length $n = (q^s - 1)/b$ [2].

Subclass 2—Reed–Solomon Codes [3]

For $m = 1$ and $s = 1$, an $(n, 1, 1, n - d, q)$-polynomial code is a Reed–Solomon code of length $n = (q - 1)/b$ and distance d.

Subclass 3—Generalized Reed–Muller Codes [4], [5], [15]

For $s = 1$ and $b = 1$, a $(n, m, 1, \mu, q)$-polynomial code is a primitive μth order GRM code of length $n = q^m - 1$ and distance $d = (R + 1)q^Q - 1$ where Q and R are quotient and remainder resulting from dividing $m(q - 1) - \mu$ by $q - 1$.

Subclass 4—Generalized Projective Geometry Codes [6]–[8]

For $b = q^s - 1$, an (n, m, s, μ, q)-polynomial code has the following parameters:

$$n = (q^{ms} - 1)/(q^s - 1) = 1 + q^s + q^{2s} + \cdots + q^{(m-1)s}$$

$r(\mu, q^s - 1)$

$$= \left[\begin{array}{l}\text{the number of non-negative integers } h \text{ less than} \\ q^{ms} - 1 \text{ that are divisible by } q^s - 1 \text{ and } \min_{0\le l<s} \\ W_{q^s}(hq^l) = j(q^s - 1) \text{ with } 0 < j < m - \mu\end{array}\right]$$

$$d = (q^{(m-\mu)s} - 1)/(q^s - 1).$$

[1] A d-BCH code is a BCH code with designed distance d.

By Theorem 6, the generator polynomial $g(X)$ has α^h as a root if and only if

$$\min_{0\le l<s} W_{q^s}(hq^l) = j(q^s - 1)$$

with $0 < j < m - \mu$. For $q = p$ (prime), an (n, m, s, μ, q)-polynomial code of this subclass is identical to a $C(m - 1 - \mu, m - 1, p^s)$ code defined by Goethals and Delsarte [i.e., the code generated by PG $(m - 1 - \mu, p^s)$] [8]. Therefore, the *dual code* [over $GF(p)$] of an (n, m, s, μ, p)-polynomial code over $GF(p)$ is a projective geometry code. In the following, we shall derive some geometric properties of this fourth subclass of polynomial codes over $GF(q)$.

The m-tuples of θ_0 defined by (6) form all the distinct points in an $(m - 1)$-dimensional projective geometry over $GF(q^s)$, i.e., PG $(m - 1, q^s)$. Thus θ_0 is a PG $(m - 1, q^s)$. From the definition of a $[(q^{ms} - 1)/(q^s - 1), m, s, \mu, q]$-polynomial code, every bit position of a code vector can be associated uniquely with a point in PG $(m - 1, q^s)$. Let L be a t-dimensional subspace [or t-flat] in PG $(m - 1, q^s)$ and $\{A_0, A_1, \cdots, A_t\}$ be the set of $t + 1$ independent points that generate L. Let Γ be the set of points $\bar{X}_i$ in θ_0 such that, for any $\bar{X}_i$ in Γ, the inner product

$$A_i \cdot \bar{X}_i^T = 0 \quad \text{for} \quad 0 \le i \le t. \quad (39)$$

Then, by (5), Γ is an $(m - t - 2)$-dimensional subspace in PG $(m - 1, q^s)$.

Consider the following polynomial:

$$f_L(\bar{X}) = \prod_{i=0}^{t} \{1 - (A_i\bar{X}^T)^{q^s-1}\}. \quad (40)$$

The degree of $f_L(\bar{X})$ is $(t + 1)(q^s - 1)$, and

$$f_L(\bar{X}_i) = 1 \quad \text{if} \quad \bar{X}_i \in \Gamma,$$
$$f_L(\bar{X}_i) = 0 \quad \text{if} \quad \bar{X}_i \notin \Gamma. \quad (41)$$

If $0 \le t < \mu$, $f_L(\bar{X})$ is a polynomial in $Q(m, s, \mu, q^s - 1, q)$ and $v(f_L)$ is a code vector in a $[(q^{ms}-1)/(q^s-1), m, s, \mu, q]$-polynomial code. From (41), the components of vector $\bar{v}(f_L)$ are 1 at the locations corresponding to the points of Γ and are 0 at other locations. Thus, we have the following theorem.

Theorem 13

For every $(m - t - 2)$-dimensional space Γ in PG $(m - 1, q^s)$ with $0 \le t < \mu$, there is a code vector $\bar{v}(f_L)$ of weight $[q^{(m-t-1)s} - 1]/q^s - 1$ in the $[(q^{ms} - 1)/(q^s - 1), m, s, \mu, q]$-$P$-code such that the components of $\bar{v}(f_L)$ are

1 at the locations corresponding to the points of Γ and are 0 at other locations. Thus, $\bar{v}(f_L)$ corresponds to an $(m - t - 2)$-flat in PG $(m - 1, q^s)$. For simplicity, we say that $\bar{v}(f_L)$ is an $(m - t - 2)$-flat. Q.E.D.

From the above theorem, we know that all the $(m - \mu - 1)$-flats of PG $(m-1, q^s)$ are in the $[(q^{ms}-1)/(q^s-1), m, s, \mu, q]$-polynomial code. The number of $(m - \mu - 1)$-flats in PG $(m - 1, q^s)$ that intersect only on a given $(m - \mu - 2)$-flat is

$$J = \frac{q^{(m-\mu-1)s} + q^{(m-\mu)s} + \cdots + q^{(m-1)s}}{q^{(m-\mu-1)s}}$$

$$= 1 + q^s + q^{2s} + \cdots + q^{\mu s} = \frac{q^{(\mu+1)s} - 1}{q^s - 1}. \quad (42)$$

It follows from the arguments in [7] and [8] that the dual code [over $GF(q)$] of a $[(q^{ms} - 1)/(q^s - 1), m, s, \mu, q]$-polynomial code can be $(m - \mu - 1)$-step orthogonalized [12]. Thus, we obtain Theorem 14.

Theorem 14

For $b = q^s - 1$, the dual code over $GF(q)$ of a $[(q^{ms} - 1)/(q^s - 1), m, s, \mu, q]$-polynomial code is a q-ary projective geometry code (PG code) that has minimum distance at least

$$J + 1 = \frac{q^{(\mu+1)s} - 1}{q^s - 1} + 1 \quad (43)$$

and is $(m - \mu - 1)$-step orthogonalizable.

Let $g(X)$ be the generator polynomial of a $[(q^{ms} - 1)/(q^s - 1), m, s, \mu, q]$-polynomial code, and $Q_0(m, s, \mu, q^s - 1, q)$ be the set of polynomials in $Q(m, s, \mu, q^s - 1, q)$ such that for any $f(\bar{X})$ in $Q_0(m, s, \mu, q^s - 1, q)$

$$f(0, 0, \cdots, 0) = 0.$$

Thus, by (31), (32), and (35), the code

$$C_0 = \{\bar{v}(f) \mid f(\bar{X}) \in Q_0(m, s, \mu, q^s - 1, q)\} \quad (44)$$

is a subcode of a $[(q^{ms} - 1)/(q^s - 1), m, s, \mu, q)$-polynomial code and its generator polynomial is

$$g_0(X) = (X - 1)g(X). \quad (45)$$

It is easy to prove the following theorem.

Theorem 15

All the μ-flats of a PG $(m - 1, q^s)$ are in the null space of code C_0. This implies that C_0 is μ-step orthogonalizable.

Subclass 5—Difference-Set Codes [13], [14]

For $b = q^s - 1$, $m = 3$, and $\mu = 1$, the dual code of an $(n, 3, s, 1, q)$-polynomial code is a q-ary difference-set code that is one-step majority-logic decodable.

Subclass 6—Generalized Euclidean Geometry Codes [6], [9]

For $b = 1$ and $\mu = D(q^s - 1)$, an $(n, m, s, D(q^s - 1), q)$-polynomial code has the following parameters:

$$n = q^{ms} - 1$$

$r(\mu, 1)$

$$= \begin{bmatrix} \text{the number of non-negative integers } h \text{ less than} \\ q^{ms} - 1 \text{ such that } 0 < \min_{0 \le l < s} W_{q^s}(hq^l) < (m - \\ D)(q^s - 1) \end{bmatrix}$$

$$d = q^{(m-D)s} - 1.$$

For $b = 1$, θ_0 consists of all the nonzero m-tuples over $GF(q^s)$. Let L be a t-dimensional subspace of all m-tuples over $GF(q^s)$ and $\{A_1, A_2, \cdots, A_t\}$ be a basis of L. Let Γ be the null space of L. The dimension of Γ is $m - t$. Consider the following polynomial:

$$f_L(\bar{X}) = \prod_{i=1}^{t} \{1 - (A_i \cdot \bar{X}^T)^{q^s-1}\}. \quad (46)$$

The degree of $f_L(\bar{X})$ is $t(q^s - 1)$. For $0 \le t \le D$, $f_L(\bar{X})$ is a polynomial in $Q(m, s, D(q^s - 1), 1, q)$ and $\bar{v}(f_L)$ is a code vector in a $(q^{ms} - 1, m, s, D(q^s - 1), q)$-polynomial code. Since

$$f_L(\bar{X}) = 1 \quad \text{for} \quad \bar{X} \in \Gamma,$$

and

$$f_L(\bar{X}) = 0 \quad \text{for} \quad \bar{X} \notin \Gamma,$$

the components of $\bar{v}(f_L)$ are 1 at the locations corresponding to the nonzero m-tuples of Γ and are 0 at other locations. Let Γ' be a coset with respect to the subspace Γ. Then the polynomial

$$f_L'(\bar{X}) = \prod_{i=1}^{t} \{1 - [A \cdot (\bar{X} - B)^T]^{q^s-1}\} \quad (47)$$

is also in $Q(m, s, D(q^s - 1), 1)$, where B is any element of Γ'. Therefore, the components of the vector $\bar{v}(f_L')$ are 1 at the locations corresponding to the elements of Γ' and are 0 at other locations. The weight of $\bar{v}(f_L')$ is $q^{(m-D)s}$.

Geometrically, each m-tuple in θ_0 can be associated uniquely to a point of an m-dimensional Euclidean geometry over $GF(q^s)$, i.e., EG (m, q^s). Thus, θ_0 consists of all the points of EG (m, q^s) except the point at infinity (all zero m-tuple). Therefore, Γ is an $(m - t)$-flat through the point at infinity and Γ' is an $(m - t)$-flat. From (46) and (47), we obtain Theorem 16.

Theorem 16

For every $(m - t)$-flat $\Gamma(\Gamma')$ in EG (m, q^s) with $0 \le t \le D$, there exists a code vector in a $(q^{ms} - 1, m, s, D(q^s - 1), q)$-polynomial code whose components are 1 at the locations corresponding to the points Γ (or Γ') and are 0 at other locations.

From the above theorem, a $(q^{ms} - 1, m, s, D(q^s - 1), q)$-polynomial code contains all the $(m - D)$-flats of EG (m, q^s). The number of $(m - D)$-flats that intersect on a given $(m - D - 1)$-flat is

$$J = \frac{q^{(m-D)s} + \cdots + q^{ms}}{q^{(m-D)s}} = 1 + q^s + \cdots + q^{Ds}. \quad (48)$$

It follows from Weldon's argument [9] that the dual code of a $(q^{ms} - 1, m, s, D(q^s - 1), q)$-polynomial code is

$(m - D)$-step orthogonalizable and has minimum distance at least

$$J + 1 = \frac{q^{(D+1)s} - 1}{q^s - 1} + 1. \quad (49)$$

From Theorem 6, a $(q^{ms} - 1, m, s, D(q^s - 1), q)$-polynomial code as α^h as a root of its generator polynomial $g(X)$ if and only if

$$0 < \min_{0 \le l < s} W_{q^s}(hq^l) < (m - D)(q^s - 1).$$

For $q = 2$, $(X - 1)g(X)$ is just the generator polynomial of the dual code of a (D, s)th-order Euclidean geometry code defined by Weldon [9]. Therefore, we call the dual code of a $(q^{ms} - 1, m, s, D(q^s - 1), q)$-polynomial code as a (D, s)th-order generalized Euclidean geometry code.

Equation (49) gives a lower bound on the minimum distance of a (D, s)th-order generalized Euclidean geometry code. This bound becomes very loose as D or s (or both) becomes large. From Corollary 7, we know that α^h is a root of the generator polynomial of a (D, s)th-order generalized Euclidean geometry code if and only if

$$0 \le \max_{0 \le l < s} W_{q^s}(hq^l) \le D(q^s - 1) \quad (50)$$

where α is a primitive element of $GF(q^{ms})$. Let h_0 be the smallest integer such that

$$\max_{0 \le l < s} W_{q^s}(h_0 q^l) = D(q^s - 1) + 1.$$

Then h_0 must be

$$\begin{aligned} h_0 &= (q^s - 1) + (q^s - 1)q^s + \cdots \\ &\quad + (q^s - 1)q^{(D-2)s} + (q - 1)q^{(D-1)s} + q^{Ds} \quad (51) \\ &= q^{Ds} + q \cdot q^{(D-1)s} - 1. \end{aligned}$$

It can be shown easily that any positive integer h that is less than h_0 satisfies the condition of (50). Therefore, the generator polynomial of a (D, s)th-order EG code must have the following roots:

$$\alpha^0, \alpha^1, \alpha^2, \cdots, \alpha^{h_0 - 1}. \quad (52)$$

Thus, we have Theorem 17.

Theorem 17

The BCH bound on the minimum distance of a (D, s)th order EG code is

$$d_{BCH} = q^{Ds} + q \cdot q^{(D-1)s}. \quad (53)$$

The difference between BCH bound and the bound of (49) is

$$(q^s - 2) + (q^s - 2)q^s + \cdots + (q^s - 2)q^{(D-2)s}. \quad (54)$$

This indicates that EG codes are actually more powerful than the decoding scheme proposed by Weldon [9] has been able to demonstrate. In other words, Weldon's decoding scheme for EG codes is very inefficient for large D and s.

TABLE II

q	m	s	D	μ	EG Code*	$J+1$	d_{BCH}
2	3	2	1	3	(63, 47)	6	6
2	3	2	2	3	(63, 12)	22	24
2	2	3	1	7	(63, 36)	10	10
2	4	2	2	6	(255, 129)	22	22
2	4	2	3	9	(255, 19)	86	96

* The generator polynomial has $\alpha^0 = 1$ as a root, or the null space contains all the $(m - D)$-flats of an EG(m, q^s) [including the $(m - D)$-flats through the point at infinity].

TABLE III

q	m	s	D	μ	Polynomial Code	d_{BCH}
2	3	2	1	3	(63, 16)	15
2	3	2	2	3	(63, 51)	3
2	2	3	1	7	(63, 27)	7
2	4	2	2	6	(255, 126)	15
2	4	2	3	9	(255, 236)	3

Tables II and III give some examples of EG codes and polynomial codes.

It can be shown that the code C_0 generated by $(X - 1) \cdot g(X)$ has all the D-flats in its null space, and therefore, the code C_0 is D-step orthogonalizable.

VI. Conclusion

In this paper, a class of codes is introduced by a polynomial approach. This class of codes is proved to be a subclass of cyclic codes. The generator polynomial of any code in this class is characterized and the minimum distance is lower bounded. For some subclasses, the exact minimum distance is established. Some classes of well-known codes are shown to be subclasses of codes. Certain geometric properties are derived.

References

[1] R. C. Bose and D. K. Chaudhuri Ray, "On a class of error correcting binary group codes," *Information and Control*, vol. 3, pp. 68–79, 1960.

[2] W. W. Peterson, *Error-Correcting Codes.* New York: Wiley, 1961.

[3] I. S. Reed and G. Solomon, "Polynomial codes over certain finite fields," *J. SIAM*, vol. 8, pp. 300–304, 1960.

[4] T. Kasami, S. Lin, and W. W. Peterson, "New generalizations of the Reed-Muller codes—Part I: Primitive codes," *IEEE Trans. Information Theory*, vol. IT-14, pp. 189–199, March 1968.

[5] ——, "Generalized Reed-Muller codes," *J. Inst. Commun. Engineers Japan.*, vol. 51-c, pp. 98–105, March 1968.

[6] L. D. Rudolph, "Geometric configuration and majority logic decodable codes," M.E.E. thesis, University of Okalahoma, Norman, 1964.

[7] E. J. Weldon, Jr., "New generalizations of the Reed-Muller codes—Part II: Nonprimitive codes," *IEEE Trans. Information Theory*, vol. IT-14, pp. 199–205, March 1968.

[8] J. M. Goethals and P. Delsarte, "On a class of majority-logic decodable cyclic codes," *IEEE Trans. Information Theory*, vol. IT-14, pp. 182–188, March 1968.

[9] E. J. Weldon, Jr., "Euclidean geometry cyclic codes," *1967 Proc. Symp. on Combinatorial Mathematics.*

[10] H. F. Mattson and G. Solomon, "A new treatment of Bose–Chadhuuri codes," *J. SIAM*, vol. 9, pp. 654–669, December 1961.
[11] D. E. Muller, "Applications of Boolean algebra to switching circuit design and to error detection," *IRE Trans. Electronic Computers*, vol. EC-3, pp. 6–12, September 1954.
[12] J. L. Massey, *Threshold Decoding*. Cambridge, Mass.: M.I.T. Press, 1963.
[13] E. J. Weldon, Jr., "Difference-set cyclic codes," *Bell Sys. Tech. J.*, vol. 45, pp. 1045–1055, 1966.
[14] R. L. Graham and J. MacWilliams, "On the number of parity checks in difference-set cyclic codes," *Bell Sys. Tech. J.*, vol. 45, pp. 1046–1070, 1966.
[15] S. Lin, W. W. Peterson, and E. J. Weldon, Jr., "Problems in information processing," USAF Cambridge Research Labs., Bedford, Mass., Final Rept., 1967.

Nonlinear Codes

VIII

Editor's Comments on Papers 30 Through 33

30 Vasil'yév: *On Nongroup Close-Packed Codes*

31 Nordstrom and Robinson: *An Optimum Nonlinear Code*

32 Schönheim: *On Linear and Nonlinear Single-Error-Correcting* q-nary *Perfect Codes*

33 Preparata: *A Class of Optimum Nonlinear Double-Error-Correcting Codes*

Because of the interesting results that were obtained by restricting attention to linear and cyclic codes, the class of nonlinear codes was largly neglected in the sixties. The four papers collected here represent, almost *in toto*, the progress that has been made to date on these codes.

The class of codes obtained by Vasil'yév [30] is particularly interesting. It is a class of perfect single-error-correcting codes which contains both linear and nonlinear codes in its definition and Hamming codes as a subclass. In 1967 Nordstrom and Robinson [31] discovered a binary nonlinear systematic code of length 15, distance 5, with 256 codewords, which is optimal in the sense that no other code of the same length and distance can have more codewords. It contains twice as many codewords as the best linear code of the same length and distance. Furthermore, two previously described nonlinear codes [Green (1966) and Nadler (1962)] were shown to be shortened versions of this interesting code. Finally, Preparata [33] was successful in showing that the Nordstrom–Robinson code is actually a member of a quite general class of binary nonlinear systematic double-error-correcting codes. Codes in this class have length $2^n - 1$ and $2^{2^n - 2n}$ codewords, which makes them optimal. Schönheim [32] was successful in extending Vasil'yév's results to the construction of perfect single-error-correcting codes, linear and nonlinear, over GF(q). As mentioned earlier, these codes, together with the Hamming, Golay, binary repetition, and translates of perfect linear codes, are the only perfect codes.

Hopefully these rather isolated results will be forerunners of more general classes of nonlinear codes and decoding algorithms. The four papers here are made even more interesting by the lack of any of the related material which usually accompanied such developments.

Reprinted from *Prob. Cybernet.*, **8**, 337–339 (1962)

30

VIII. BRIEF REPORTS

NONGROUP CLOSE-PACKED CODES

Pages 337-339

Yu. L. Vasil'yev
(Novosibirsk

1. The study of group codes and close packed codes (c.p. codes) [2] occupies an important place in the theory of correcting codes*. These classes of codes permit different approaches to the solution of the fundamental problem in the theory of correcting codes, namely, the construction, for given n and d, of (n, d)-codes of maxi-

---

*Let us recall the basic concepts of the theory of correcting codes. Given binary sequences $(\sigma_1, \ldots, \sigma_n)$ and $(\tau_1, \ldots, \tau_n)$, the number of places in which these sequences do not coincide is called the <u>distance</u> ρ between the sequences.

The set of all sequences whose distance from a given sequence is not greater than ℓ is called a <u>sphere of radius</u> ℓ.

A set $\{(\tau_1, \ldots, \tau_n)\}$ of binary sequences is called an (n, d)-<u>code</u> if the distance between any two sequences in this set is not less than d. If this set is a group with respect to the operation

$$(\tau_1, \ldots, \tau_n) + (\nu_1, \ldots, \nu_n) = (\tau_1 + \nu_1 \pmod 2), \ldots, \tau_n + \nu_n \pmod 2)),$$

the code is called a <u>group</u> code. If an (n, $2\ell + 1$)-code generates a partition of an n-dinensional unit cube E^n into spheres of radius ℓ which are pairwise disjoint and whose union exhausts E^n, then it is called a <u>close packed</u> (c. p.) code.

mal power. Group codes possess a certain symmetry which makes them more transparent and makes it possible, within the class of these codes, to make substantial progress in the solution of the fundamental problem [3]. In close packed codes, maximality is provided for by the definition itself.

Let us note two circumstances to which the results of the present note are relevant.

a) All known methods for the synthesis of c.p. codes [1, 2, 3, 4] yield group codes. It is trivially simple to construct nongroup c.p. codes which are shifts* of group c.p. codes. However, such

---

* The code $C = \{(\tau_1, \ldots, \tau_n)\}$ is called a shift of the code $C = \{(\gamma_1, \ldots, \gamma_n)\}$ if there exists a sequence $\xi = (\xi_1, \ldots, \xi_n)$ such that $C = \{(\gamma_1 \oplus \xi_1, \ldots, \gamma_n \oplus \xi_n)\}$ where $\oplus$ denotes addition mod 2.

nongroup codes are not of independent interest because their "internal structure" is exactly the same as that of group codes. A code which is neither a group code or a shift of a group code will be called a strong nongroup code. This raises the question of whether strong nongroup codes exist.

b) In [4], during an investigation of the symmetry of c.p. codes, the following theorem was proved: in a c.p. $(n, 2\ell + 1)$-code, the number of code points located at a distance r, $2\ell + 1 \leq r \leq n$, from a given code point z depends on neither the choice of the point z nor on the choice of the code. In connection with this theorem, the suggestion was put forward in [4] that for given n and d there exists, essentially, only one c.p. (n, d)-code; i. e. given two c.p. (n, d)-codes σ_1 and σ_2, there exists a symmetry of the cube E^n which maps σ_1 into σ_2. In other words, it was suggested in [4] that for given n and d, all c.p. codes are of one and the same type [5].

In the present note, we propose a method for the construction of c.p. (n, 3)-codes. In contrast to other methods [1,2,3,4] which permit one to construct c.p. (n, 3)-codes, our method yields not only group codes, but strong nongroup codes as well. This means that the answer to the question raised above is positive, and the supposition regarding c.p. codes is, in general, untrue.

Moreover, in this note we give a lower bound for ratio of the number of types of strong nongroup c.p. (n, 3)-codes to the number of all group (n, 3)-codes. This estimate is expressed by the number $2^{2^{n((1/2) - \delta)}}$, where $n = 2^q - 1$, $q = 4, 5, 6, \ldots$, and

$\delta \to 0$ as $n \to \infty$.

2. Let $p = 2^q - 1$, $q = 1, 2, \ldots$. As is well known [1], c.p. $(p, 3)$-codes exist only for specified values of p (in a c.p. $(p, 3)$-code, there must be $2^p/(p+1)$ sequences).

Let $C^p = \{(\tau_1, \ldots, \tau_p)\}$ be a c.p. $(p, 3)$-code containing the null sequence, let E^p be the set of all sequences $(\alpha_1, \ldots, \alpha_p)$ of length p, and let $\lambda(\tau)$ be an arbitrary function taking the values 0 and 1 on the sequences $\tau = (\tau_1, \ldots, \tau_p)$ in C^p, where $\lambda(0, \ldots, 0) = 0$. For the sequence $\alpha = (\alpha_1, \ldots, \alpha_p)$ we set $|\alpha| = \alpha_1 \oplus \ldots \oplus \alpha_p$.

With the pair of sequences $\alpha = (\alpha_1, \ldots, \alpha_p)$ and $\tau = (\tau_1, \ldots, \tau_p)$, $\alpha \in E^p$, $\tau \in C^p$, we will associate the sequence $(\alpha_1, \ldots, \alpha_p, \alpha_1 \oplus \tau_1, \ldots, \alpha_p \oplus \tau_p, |\alpha| \oplus \lambda(\tau))$, which we will denote by $\alpha \times \tau$. We will let $E^p \times C^p$ denote the set of all sequences $\alpha \times \tau$, where $\alpha \in E^p$ and $\tau \in C^p$.

Theorem. *For any function* $\lambda(\tau)$, *the set* $E^p \times C^p$, $p = 2^q - 1$, *is a* c.p. $(2^{q+1} - 1, 3)$-*code containing the null sequence*.

Proof. It is evident that:

(a) the number of places in the sequence $\alpha \times \tau$ is equal to $2p + 1 = 2^{q+1} - 1$;

(b) the number of all sequences $\alpha \times \tau$ in the set $E^p \times C^p$ is equal to $2^{2p}/(p+1)$ — the number of sequences in a c.p. $(2^{q+1} - 1, 3)$-code;

(c) the null sequence $(0, \ldots, 0)$ of length $2p + 1$ is in $E^p \times C^p$.

Therefore, we must now prove that that the distance ρ between any two sequences in the set $E^p \times C^p$ is not less than three. Let $\delta \times \tau$ and $\beta \times \nu$ be arbitrary noncoinciding sequences in $E^p \times C^p$, $\alpha = (\alpha_1, \ldots, \alpha_p)$, $\beta = (\beta_1, \ldots, \beta_p)$, $\tau = (\tau_1, \ldots, \tau_p)$, $\nu = (\nu_1, \ldots, \nu_p)$, $\alpha, \beta \in E^p$, $\tau, \nu \in C^p$. By definition,

$$\alpha \times \tau = (\alpha_1, \ldots, \alpha_p, \alpha_1 + \tau_1, \ldots, \alpha_p + \tau_p, |\alpha| + \lambda(\tau)),$$
$$\beta \times \nu = (\beta_1, \ldots, \beta_p, \beta_1 + \nu_1, \ldots, \beta_p + \nu_p, |\beta| + \lambda(\nu)).$$

If $\tau \neq \nu$, then, by hypothesis, $\rho(\tau, \nu) \geqslant 3$. Then for $\rho(\alpha, \beta) = 0, 1, 2, 3$ we have

$$\rho[\alpha_1 \times \tau_1, \ldots, \alpha_p + \tau_p), (\beta_1 + \nu_1, \ldots, \beta_p + \nu_p)] \geqslant 3, 2, 1, 0.$$

respectively, and consequently, $\rho(\alpha \times \tau, \beta \times \nu) \geqslant 3$.

If $\tau = \nu$, then $\alpha \neq \beta$ and $\lambda(\tau) = \lambda(\nu)$. Two cases are possible: $|\alpha| \neq |\beta|$ and $|\alpha| = |\beta|$. In the first case

$$\rho(\alpha, \beta) \geqslant 1, \rho[(\alpha_1 + \tau_1, \ldots), (\beta_1 + \nu_1, \ldots)] \geqslant 1, \quad |\alpha| + \lambda(\tau) \neq |\beta| + \lambda(\nu).$$

whence $\rho(\alpha \times \tau, \beta \times \nu) \geqslant 3$. In the second case, $\rho(\alpha, \beta) \geqslant 2$ (since $|\alpha| = |\beta|$, but $\alpha \neq \beta$) and $\rho((\alpha_1 \oplus \tau_1, \ldots), (\beta_1 \oplus \nu_1, \ldots)) \geqslant 2$, whence $\rho(\alpha \times \tau, \beta \times \nu) \geqslant 4$. Thus, the theorem is proved.

3. Let $G^p = \{(\tau_1, \ldots, \tau_p)\}$ be a c.p. group (p, 3)-code.

Let us consider the function $\lambda(\tau)$, $\tau \in G^p$, that is involved in the construction of the set $E^p \times G^p$. If we set $\lambda(\tau) \equiv 0$ for all $\tau \in G^p$, then the set $E^p \times G^p$ will be a group c.p. code. If we fix any two non-null sequences

$$\mu = (\mu_1, \ldots, \mu_p) \text{ and } \nu = (\nu_1, \ldots, \nu_p), \quad \mu \neq \nu, \quad \mu, \nu \in G^p.$$

and set $\lambda(\mu \oplus \nu) \neq \lambda(\mu) \oplus \lambda(\nu)$, then the set $E^p \times G^p$ will be a strong nongroup c.p. code.

As is well known, the (3, 3)-code and the (7, 3)-code are group codes. Moreover, for any $p = 2^q - 1$, $q = 4, 5, 6, \ldots$, one can indicate group as well as nongroup c.p. (p, 3)-codes.

Let us consider some estimates. In order to construct a strong nongroup c.p. code from the group c.p. code $G = \{(\tau_1, \ldots, \tau_p)\}$, we must prescribe the function $\lambda(\tau)$, $\tau \in C^p$, on four sequences — on the null sequence ($\lambda(0, \ldots, 0) = 0$) and on the abovementioned sequences μ, ν, and $\mu \oplus \nu$.

The function $\lambda(\tau)$ can be arbitrary on the remaining sequences in G^p. The number of these sequences is $(2^p/(p+1)) - 4$; the number of different sets $E^p \times G^p$ which are strong nongroup c.p. (2p+1, 3)-codes will be not less than $2^{(2p/(p+1)) - 4}$; the number of types of strong nongroup c.p. (2p+1, 3)-codes will be not less than

$$\frac{\dfrac{2p,}{2^{p-1}} - 4}{(2p+1)!2^{2p+1}} .$$

Since each group (2p+1, 3)-code is completely determined by its generators, whose number does not exceed 2p, the number of group (2p+1, 3)-codes does not exceed $C^{2p}_{2^{2p+1}} \leqslant 2^{5p^2}$. Comparison of the last two estimates yields the result indicated at the very end of Sec. 1.

In conclusion, the author expresses his deep gratitude to V. Glagolev for a number of valuable suggestions.

BIBLIOGRAPHY

1. Hamming R. W., Error detection and error correction codes, BSTJ 29, 1950, 147-160. (Russian translation: Kody s obnaruzheniyem i ispravleniyem oshibok, Moscow, IL, 1956, 7-22).
2. Lloyd S. P., Binary block coding, BSTJ 36, 2, 1957, 517-535. (Russian translation: Cybernetic collection 1, Moscow, IL, 1960, 206-226).
3. Bose R. C. and Ray-Chaudhuri, On a class of error correcting binary group codes, Inf. and Control 3, 1, 1960, 68-79. (Russian translation: Cybernetic collection 2, Moscow, IL, 1960, 83-94).
4. Shapiro H. S. and Slotnick D. L., On the theory of error correcting codes, IBM J. Res. and Development, 3, 1, 1959, 25-34 (Russian translation: Cybernetic collection 5, Moscow, IL, 1962).
5. G. Polya, J. Symb. Logic 5, 3, 1940, 98.

Received 17 November 1961

Reprinted from *Inform. Contr.*, **11,** 613–616 (1967)

31

An Optimum Nonlinear Code*

Alan W. Nordstrom

United Township High School, East Moline, Illinois 61244

AND

John P. Robinson

Department of Electrical Engineering, University of Iowa, Iowa City, Iowa 52240

A systematic nonlinear code having length 15, minimum distance 5, and 256 code words is given in Boolean form. This is the maximum possible number of words for length 15 and distance 5. The distance spectrum of all pairs of code words is an exact multiple of the weight spectrum of the code words.

LIST OF SYMBOLS

$A(n, d)$ = the maximum number of binary n-tuples such that any two n-tuples differ in at least d places.

$B(n, d)$ = the maximum number of code words possible in a linear code of block length n such that the Hamming distance between any two code words is at least d.

$A_0A_1 \cdots B_0B_1 \cdots Z_0Z_1 \cdots$ = Boolean variables.

$\oplus$ = connective symbolizing modulo 2 addition.

I. INTRODUCTION

The optimum code resulted from a study of Nadler's (1962) nonlinear 32-word code and Green's (1966) 64-word nonlinear code. These codes have a minimum Hamming distance of 5, with the former having length 12 and the latter length 13. Both have twice as many words as the best linear code with the same length and minimum distance (Wagner, 1965).

* The research reported in this paper was supported by the National Science Foundation under the Research Participation Program for Exceptional Secondary Students and Grant GK-816.

TABLE I

Weight Spectra of the Optimum Code and its Shortened Versions

Length	0	5	6	7	8	9	10	15
12	1	11	13	2	1	3	1	0
13	1	18	24	4	3	10	4	0
14	1	28	42	8	7	28	14	0
15	1	42	70	15	15	70	42	1

II. THE CODE

The code is systematic with 8 information bits denoted by X_0, $X_1, \cdots, X_7$ and 7 redundant bits denoted by $Y_0, Y_1, \cdots, Y_6$. Each Y, as a Boolean function of the X's, is in the same equivalence class. Y_0 is as follows:

$$Y_0 = X_7 \oplus X_6 \oplus X_0 \oplus X_1 \oplus X_3 \oplus (X_0 \oplus X_4)(X_1 \oplus X_2 \oplus X_3 \oplus X_5) \oplus (X_1 \oplus X_2)(X_3 \oplus X_5), \quad (1)$$

where $\oplus$ denotes modulo 2 addition.

The remaining Y's are found by cyclically shifting X_0 through X_6; i.e., for Y_j substitute $X_{i+j(\text{mod } 7)}$ for X_i in (1) where $i = 0, 1, \cdots, 6$ for each $j, j = 0, 1, \cdots, 6$.

Examining (1) we see that all products X_iX_j, $i \neq j$ in $X_0, \cdots, X_5$ appear, except X_0X_4, X_1X_2, X_3X_5. In these 3 missing pairs all 6 variables appear with one linear term in (1) in each pair.

In the general classification method of Roos (1965) this code would be termed a quadratic code.

Shortening this code by dropping one information bit yields a 128-word code having length 13 with minimum distance 5. Deleting two information bits results in Green's (1966) code, three bits yields Nadler's (1962) code. Next we compare these codes from (1) with Johnson's (1962) bound and Wagner's (1965) determination of $B(n, d)$. $B(n, d)$ is $A(n, d)$ restricted to the class of linear codes. We conclude that $A(15, 5) = 256$, using Johnson's upper bound and (1).

$$A(15, 5) = 256 = 2B(15, 5)$$

$$131 \geqq A(14, 5) \geqq 128 = 2B(14, 5)$$

$$70 \geqq A(13, 5) \geqq 64 = 2B(13, 5)$$

$$39 \geqq A(12, 5) \geqq 32 = 2B(12, 5)$$

Surprisingly, these codes have the weight–distance relationship of a linear code. Any given code word can be normalized to the all-zero word by complementing all words in the appropriate positions. The resulting code is the same as the original code with possibly a permutation of the positions. Thus the weight spectrum of the code gives the distance properties. In Table I we list the weight spectra of these codes obtained from (1).

III. METHODOLOGY

In order to determine the distance properties of Nadler's (1962) code and Green's (1965) code, a computer program was written to calculate the distance for all possible combinations of pairs of words. The distance spectra of these two codes were found to be very similar and it was hypothesized that the codes were closely related.

The next step was to put the codes in the same form. Green's code was in a normalized form in that it had one all-zero word. Nadler's code was normalized to this form by changing all the ones in a particular word to zeroes and then changing the corresponding positions in all of the other words.

As the two codes appeared very similar, it was suggested that Nadler's code was contained in Green's code. This was checked by dropping one of the columns of Green's so that it had the same length and form as Nadler's. The shortened 64-word code was then divided into two 32-word codes. By switching certain columns of the first, a 32-word code that was exactly the same as Nadler's was found. By taking the second and normalizing it, and then switching certain columns, Nadler's code was again found in this half of the 64-word code.

Since the column that was dropped contained 32 ones and 32 zeroes, the idea for building a 128-word nonlinear code came to mind. By taking two of Green's codes and adding an extra column of ones to one of them, an extra column of zeroes to the other and then changing all the numbers of certain columns of the first one, a 128-word code could be built.

A suitable program was written which tried all possible combinations of complementing columns. It was successful and a new 128-word nonlinear code was discovered. Extending the technique another step resulted in the optimum code (1) with 256 words of length 15 and distance 5.

IV. CONCLUDING REMARKS

This note presents an optimum nonlinear code with considerable structure and regularity. It is felt that the structure should allow reasonable decoding algorithms. A quasi-algebraic decoding scheme has been proposed (Robinson, 1968). The functional form suggests that a general class of good nonlinear codes may exist.

RECEIVED: August 21, 1967. REVISED: November 24, 1967.

REFERENCES

GREEN, M. W. (1966), Two heuristic techniques for block-code construction. *IEEE Trans. Inform. Theory* **IT-12,** 273.

JOHNSON, S. M. (1962), A new upper bound for error-correcting codes. *IRE Trans. Inform. Theory* **IT-8,** 203–207.

NADLER, M. (1962), "Topics in Engineering Logic." Macmillan, New York.

ROBINSON, J. P. (1968), Analysis of Nordstrom's optimum quadratic code. Proceedings of the 1968 Hawaii International Conference on System Sciences, Honolulu, Hawaii.

ROOS, JAN-ERIK (1965), An algebraic study of group and nongroup error-correcting codes. *Inform. Control* **8,** 195–214.

WAGNER, T. J. (1965), "Some New Values of $B(n, d)$." University of Texas Report, Austin, Texas.

Reprinted from *Inform. Contr.*, **12**, 23–26 (1968)

32

On Linear and Nonlinear Single-Error-Correcting q-nary Perfect Codes

J. Schönheim

Tel-Aviv University, Israel and the University of Calgary, Canada

Communicated by E. R. Berlekamp

Vasiliev's method (Vasiliev, 1963) for the construction of linear and nonlinear, single error-correcting binary perfect codes is generalized to every prime power q.

1. INTRODUCTION

The mathematical theory of coding has become popular in the last 15 years. Most of the results up to 1961 are available in Peterson (1961). More recent developments are cited in Assmus and Mattson (1967) and Roos (1965).

Let us review only the results which are most relevant for our purpose. For the corresponding definitions, see the next section.

Linear single-error-correcting perfect binary codes have been introduced by Hamming (1950); the possibility of generalizing them was mentioned by Golay (1949). Shapiro and Slotnick (1959) constructed the q-nary analogs to Hamming codes; they even assumed that no other (i.e. nonlinear) perfect codes exist. This assumption was disproved by Vasiliev (1963). His alternative construction of binary perfect codes also leads to nonlinear ones.

We will extend Vasiliev's results to the q-nary case, q being an arbitrary prime power. The construction is simply by induction.

2. DEFINITIONS, NOTATION AND KNOWN RESULTS

Let q be a power of a prime and let x_i $(i = 1, 2, \cdots, n)$ be elements of the Galois field $GF(q)$. Denote by V_n the vector space of n-vectors over $GF(q)$. A code is a subset C_n of V_n. If C_n is a subspace on V_n the code C_n is called linear.

The distance between two vectors is the number of the nonzero components of their difference. The distance between a vector v and the vector zero is called the weight of ν and is denoted by $w(v)$.

Let $S_n = \{s_i\}_{i=0}^{n}$ be the set of vectors in V_n having at most a single nonzero component; $s_0 = 0$ the vector zero. If for every $c, c' \in C_n$ and $s, s' \in S_n$ $c + s = c' + s' \Rightarrow c = c'$ and $s = s'$, the code C_n is said to be single-error-correcting.

A necessary and sufficient condition for a code to have this property is that the distance between two different vectors of C_n is $\geqq 3$.

If, moreover, for every $v \in V_n$ there exist $c \in C_n$ and $s \in S_n$ so that $v = c + s$, C_n is called perfect.

Denote by $|A|$ the number of elements of the set A. Clearly $|S_n| = (q-1)n + 1$. If C_n is perfect, $|S_n|$ must divide q^n and therefore for some m

$$n = \frac{q^m - 1}{q - 1} \quad \text{and} \quad |C_n| = \frac{q^n}{n(q-1)+1} = q^{n-m}.$$

Conversely if

$$n = \frac{q^m - 1}{q - 1}, \qquad |C_n| = q^{n-m}$$

and C_n is single error correcting, it is perfect.

DEFINITION. Let $v_i = (x_{i1}, x_{i2}, \cdots, x_{in})$, $i = 1, 2, \cdots, q-1$ be arbitrary n-vectors over $GF(q)$ and let α_i $(i = 1, 2, \cdots, q-1)$ be the nonzero elements of $GF(q)$ in a fixed but arbitrary order. The generalized parity function $p\{v_i\}$ of the set $\{v_i\}$ is

$$p\{v_i\} = \sum_{i=1}^{q-1} \alpha_i \sum_{\nu=1}^{n} x_{i\nu}.$$

THEOREM. *If q is a power of a prime and m an integer >2, then for every*

$$n = \frac{q^m - 1}{q - 1}$$

there exist linear and nonlinear, single-error-correcting, q-nary perfect codes.

Proof. (Our proof will be constructive and by induction on m.) For $m = 2$, we assume the existence of a linear, single-error-correcting, perfect q-nary code.[1]

[1] The restriction $m > 2$ is used in the proof only in the nonlinear case. Thus, the assumed code for $m = 2$ may also be constructed by the given method. Starting with $m = 1$, $C_1 = \{0\}$ is a (trivial) linear, single error-correcting, perfect, q-nary code, and C_N with $N = q + 1$ is the required code.

We will show that if we have already constructed for $m = M$ a linear code with the needed properties, we are able to construct such linear and nonlinear codes for $m = M + 1$.

Let C_n be a linear, single-error-correcting perfect code with elements $0 = c_0, c_1, \cdots, c_{s-1}$,

$$n = \frac{q^M - 1}{q - 1}; \quad |C_n| = s = q^{n-M}.$$

Let v_i ($i = 1, 2, \cdots, q - 1$) and $p\{v_i\}$ as in the above definition and let $f(c)$ be a vector function defined on C_n with values in $GF(q)$, $f(0) = 0$.

Now define C_N as

$$C_N = \left\{\left(v_1, v_2, \cdots, v_{q-1}, c + \sum_{i=1}^{q-1} v_i, p\{v_i\} + f(c)\right)\right\}$$

where each $v_i \in V_n$ and $c \in C_n$.

N, the number of components of vectors in C_N is $N = nq + 1 = (q^{M+1} - 1)/(q - 1)$. The number of vectors in C_N is $|C_N| = q^{n(q-1)} \times q^{n-M} = q^{N-M-1}$. Therefore if C_N is single-error-correcting, it must be perfect.

We will prove that C_N is single-error-correcting by showing that the distance between two different vectors of C_N is $\geqq 3$.

Let u and u' be two different vectors of C_N; thus

$$u = \left(v_1, v_2, \cdots, v_{q-1}, c + \sum_{i=1}^{q-1} v_i, p\{v_i\} + f(c)\right)$$

$$u' = \left(v_1', v_2', \cdots, v_{q-1}', c' + \sum_{i=1}^{q-1} v_i', p\{v_i'\} + f(c')\right)$$

We have to show that $w(u - u') \geqq 3$.

Denote $w(v_1 - v_1', v_2 - v_2', \cdots, v_{q-1} - v_{q-1}')$ by W_1,

$$w\left(c - c' + \sum_{i=1}^{q-1} (v_i - v_i')\right) \quad \text{by} \quad W_2$$

and $w(p\{v_i\} - p\{v_i'\} + f(c) - f(c'))$ by W_3. Then $w(u - u') = W_1 + W_2 + W_3$.

If $W_1 \geqq 3$, there is nothing to prove. If $W_1 = 0$, $W_2 = w(c - c') \geqq 3$.

Suppose, first, $W_1 = 1$. If $c \neq c'$, then $w(c - c') \geqq 3$ and therefore $W_2 \geqq 2$; hence $W_1 + W_2 \geqq 3$ and the assertion is proved. If $c = c'$, then $W_2 = 1$ and $W_3 = w(p\{v_i\} - p\{v_i'\}) = 1$; hence $W_1 + W_2 + W_3 = 3$ and again $w(u - u') \geqq 3$.

Suppose, finally, $W_1 = 2$. This situation can occur only if or for some k, $1 \leqq k \leqq q - 1$, $w(v_k - v_k') = 2$ or for some pair $k \neq j$

$$w(v_k - v_k') = w(v_j - v_j') = 1. \tag{1}$$

In the first case $W_2 \geqq 1$, hence $W_1 + W_2 \geqq 3$. In the second case, if $c \neq c'$, again $W_2 \geqq 1$. If $c = c'$, we prove that

$$W_2 = W_3 = 0 \tag{2}$$

is impossible. Indeed (1) means that for some μ and some ν, both between 1 and n,

$$x_{k\mu} - x_{k\mu}' \neq 0 \quad \text{and} \quad x_{j\nu} - x_{j\nu}' \neq 0 \tag{3}$$

but (2) implies that

$$\begin{cases} x_{k\mu} - x_{k\mu}' + x_{j\nu} - x_{j\nu}' = 0 \\ \alpha_k(x_{k\mu} - x_{k\mu}') + \alpha_j(x_{j\nu} - x_{j\nu}') = 0 \end{cases}$$

which implies

$$(\alpha_k - \alpha_j)(x_{k\mu} - x_{k\mu}') = 0$$

and this contradicts (3) because $\alpha_k \neq \alpha_j$ unless $k = j$.

Since C_n is linear, it contains the zero vector, so C_N also contains the zero vector. Therefore, C_N will be linear or nonlinear according as $f(c)$ is linear or nonlinear. Q. E. D.

ACKNOWLEDGMENT

Thanks are due to E. R. Berlekamp for several suggestions.

RECEIVED: January 28, 1967; REVISED: January 15, 1968.

REFERENCES

VASILIEV, JR. L. (1963), On nongroup close-packed codes. *Probl. Cybernet.* (*USSR*) **8,** 337–339.

PETERSON, W. W. (1961), Error-correcting codes. MIT Press, Cambridge, Massachusetts and Wiley, New York.

ASSMUS, E. F. JR., AND MATTSON, H. F. JR., (1967), On tactical configurations and error-correcting codes. *J. Combinatorial Theory,* **2,** 243–257.

ROOS, J. E. (1965), An algebraic study of group and nongroup error-correcting codes, *Information and Control* **8,** 195–214.

HAMMING, R. W. (1950), Error detecting and error correcting codes, *Bell System Tech. J.* **29,** 147–160.

GOLAY, M. J. E. (1949), Notes on digital coding, *Proc. IRE* **37,** Corresp. 657.

SHAPIRO, H. S. AND SLOTNICK, D. L. (1959), On the mathematical theory of error-correcting codes. *IBM J. Res. Develop.* **3,** 25–37.

Reprinted from *Inform. Contr.*, **13**, 378–400 (1968)

33

A Class of Optimum Nonlinear Double-Error-Correcting Codes*

FRANCO P. PREPARATA

Coordinated Science Laboratory and Department of Electrical Engineering, University of Illinois at Urbana, Illinois 61801

Nonlinear double-error-correcting block codes of length $(2^n - 1)$ (n even) are presented in this paper. They have the largest possible number of code-words for their length and minimum distance and are formed by adjoining to a certain linear code (referred to as the "kernel") a specific subset of its cosets. The kernel results from the juxtaposition and superposition of Bose–Chaudhuri–Hocquenghem codes of length $(2^{n-1} - 1)$. The presented codes are systematic and are comparable to the corresponding linear codes with regard to the complexity of the encoding and decoding operations.

1. INTRODUCTION

Some examples of nonlinear binary codes have been reported in the literature over the past years (Vasil'ev, 1962; Nadler, 1962; Green, 1966). Particularly interesting for its structure and generality was the class discovered by Vasil'ev (1962), i.e., a class of perfect single-error-correcting group and nongroup codes containing the Hamming codes.

Recently some interest in nonlinear codes has been revived by the discovery made by Nordstrom and Robinson (1967) of a (15, 8) nonlinear double-error-correcting code, of which previously reported (12, 5) (Nadler, 1962) and (13, 6) (Green, 1966) nongroup codes were shortened versions. The (15, 8) code had the interesting features of being systematic and of meeting the Johnson's upper bound (1962) on the number of code words in a code of length 15 and distance 5. Subsequently the (15, 8) code has been described in terms of polynomial (i.e., linear) codes over GF(2) (Preparata, 1968a): This description proved to be a useful framework, since it led to the formal demonstration (Preparata,

* This work was supported in part by the Joint Services Electronics Program under contract DDAB-07-67-C-0199 and in part by NSF Grant GK-2339.

1968b) of the distance properties of the code, previously heuristically assessed.

A question which was first asked by Nordstrom and Robinson (1967) was whether the (15, 8) code was a member of a class of codes. The purpose of this paper is to answer this question in the affirmative. Nongroup double-error-correcting $(2^n - 1, 2^n - 2n)$ codes exist for each even $n \geqq 4$, and contain the (15, 8) code as a special case. Here again the polynomial description has been the essential device in the construction of these codes.

The interesting features of these codes can be summarized as follows: 1) They contain twice as many code words as the double-error-correcting BCH codes of the same length, which is the largest number of code words possible for given length and distance, i.e., they are optimal; 2) their decoding can be based on the calculation of syndrome-like quantites and its complexity is comparable to the corresponding BCH codes; 3) the codes are systematic and encoding can be accomplished very simply by shift-registers in as many time units as are required by the serial transmission of the information digits.

The following sections are devoted to the description of the codes and to the demonstration of the properties stated above.

2. DESCRIPTION OF THE CODES[1]

In the sequel all polynomials considered belong to the algebra A_{n-1} of polynomials over GF(2) modulo $(x^{2^{n-1}-1} + 1)(n \geqq 4)$. Given $a(x) \in A_{n-1}$, $W[a(x)]$ denotes the number of nonzero coefficients of $a(x)$; given $b(x) \in A_{n-1}$, $d[a(x), b(x)] = W[a(x) + b(x)]$ is the Hamming distance between $a(x)$ and $b(x)$. By the symbol $a(x)$ we shall also denote the row vector $[a_{2^{n-1}-2}, a_{2^{n-1}-3}, \cdots, a_0]$ where $a(x) = \sum a_j x^j$.

Let $\{m(x)\}$ be a single-error-correcting BCH code of length $2^{n-1} - 1$, generated by $g_1(x)$, a primitive polynomial of degree $(n - 1)$; that is, if by α we denote a primitive element of $GF(2^{n-1})$, $g_1(\alpha) = 0$. Consider now the code $\{s(x)\}$, whose generator polynomial has roots α, α^3 and 1: clearly $\{s(x)\}$ is a BCH code of minimum weight 6 (see Peterson (1961), p. 167) and $\{s(x)\} \subset \{m(x)\}$. Clearly $\{s(x)\}$ exists only for $2^{n-1} - 1 \geqq 2(n - 1) + 1$, i.e., for $n \geqq 4$; when $n = 4$, $s(x)$ is identically 0. Finally by $u(x)$ we denote the polynomial $(x^{2^{n-1}-1}+1)/(x + 1)$.

[1] There is some overlap between this section and (Preparata, 1968b) since this work is a conceptual and chronological generalization of the latter.

Given two polynomials $a(x)$ and $b(x)$, $(a(x) \in A_{n-1}, b(x) \in A_{n-1})$ and a binary parameter i, we construct $(2^n - 1)$-component vectors over GF(2) of the form

$$[a(x), i, b(x)].$$

Given $m(x) \in \{m(x)\}$, $s(x) \in \{s(x)\}$ and arbitrary i, we now set $a(x) = m(x)$ and $b(x) = m(x) + (m(1) + i)u(x) + s(x)$. We obtain

$$\mathbf{v} = [m(x), i, m(x) + (m(1) + i)u(x) + s(x)] \tag{1}$$

We claim that

LEMMA 1. *The vectors* $\mathbf{v}$ *given by* (1) *form a linear code* $\mathcal{C}_n$.

Proof. The statement follows immediately from the verification that $\mathcal{C}_n$ is a group with respect to addition over GF(2). In fact: i) $\mathcal{C}_n$ contains the additive unity [0, 0, 0], obtained by setting in (1) $m(x) = 0$, $s(x) = 0$, $i = 0$; ii) $\mathcal{C}_n$ is closed with respect to addition, since both $\{m(x)\}$ and $\{s(x)\}$ are group codes.

Q.E.D.

LEMMA 2. *The minimum distance between any two code words of* $\mathcal{C}_n$ *is at least* 6.

Proof. Since $\mathcal{C}_n$ is a linear code its minimum distance coincides with the minimum weight W of its nonzero code words, which we now determine. Assume first that $m(x) = 0$. If also $i = 0$, then $W = W[s(x)] \geqq 6$. If $i = 1$, then $W = 1 + W[u(x) + s(x)] \geqq 1 + W[u(x)] - \max W[s(x)]$ We know that $W[u(x)] = 2^{n-1} - 1$ and that $\max W[s(x)]$ is $2^{n-1} - 6$ for $n > 4$ or is 0 for $n = 4$ (since $\max W[s(x)]$ is the maximum *even* weight of code words of the double-error-correcting BCH code); hence $W \geqq 1 + 2^{n-1} - 1 - 2^{n-1} + 6 = 6$.

Assume now that $m(x) \neq 0$. If $m(x) \notin \{s(x)\}$, then $m^*(x) = m^*(x) + (m(1) + i)u(x) + s(x) \neq 0$ and $m^*(x) \in \{m(x)\}$. It follows that $W \geqq W[m(x)] + W[m^*(x)] \geqq 3 + 3 = 6$, since both $m(x)$ and $m^*(x)$ are nonzero and $\{m(x)\}$ has minimum weight 3. If, alternatively, $m(x) \in \{s(x)\}$, then $W[m(x)] \geqq 6$ and $W \geqq W[m(x)] \geqq 6$. Q.E.D.

The number of information bits of $\mathcal{C}_n$ is readily obtained when one considers that the independently selectable $m(x)$, $s(x)$ and i contribute $(2^{n-1} - n)$, $(2^{n-1} - 2n)$ and 1 information bits, respectively. Therefore $\mathcal{C}_n$ is a $(2^n - 1, 2^n - 3n + 1)$ linear code of minimum distance 6.

Consider now the polynomial $\varphi(x) = (x^{2^{n-1}-1} + 1)/g_1(x)$, i.e., a minimum degree maximum length sequence of length $(2^{n-1} - 1)$. We first show that

LEMMA 3. *There exists an* $s(0 \leqq s \leqq 2^{n-1} - 2)$ *such that* $(x^s\varphi(x))^2 = x^s\varphi(x)$.

Proof. We compute the product $\varphi(x)\varphi(x)$. Since $\varphi(x)$ is not divided by $g_1(x)$, $\varphi^2(x)$ is not zero; moreover, $\varphi^2(x)$ belongs to the code generated by $\varphi(x)$, i.e.

$$\varphi^2(x) = x^r\varphi(x) \tag{2}$$

for some r, $0 \leqq r \leqq 2^{n-1} - 2$. If we multiply (2) by x^{2s} we have $x^{2s}\varphi^2(x) = x^{r+2s}\varphi(x)$, i.e. $(x^s\varphi(x))^2 = x^s\varphi(x)\cdot x^{r+s}$. The lemma follows if $x^{r+s} = 1$, i.e., if $r + s = 0 \pmod{2^{n-1} - 1}$, or, equivalently, $s = 2^{n-1} - 1 - r \mod (2^{n-1} - 1)$. Q.E.D.

We define $f(x) \triangleq x^s\varphi(x)$.

A polynomial $q(x) = ax^j(a = 0, 1; j = 0, 1, \cdots, 2^{n-1} - 2)$ is clearly a minimum weight coset leader of $\{m(x)\}$ for $a = 1$. We now construct vectors $\mathbf{u}$ of the form

$$\mathbf{u} = [q(x), 0, q(x)f(x)]. \tag{3}$$

We have the following lemmas:

LEMMA 4. *The polynomial* $q(x) + q(x)f(x)$ *belongs to* $\{m(x)\}$.

Proof. The assertion follows immediately from Lemma 3, since

$$f(x)\{q(x) + q(x)f(x)\} = f(x)q(x) + f^2(x)q(x) = 0$$

i.e., $q(x) + q(x)f(x)$, being orthogonal to $f(x)$, is divided by $g_1(x)$. Q.E.D.

LEMMA 5. *The sum of two vectors* $\mathbf{u}_1$ *and* $\mathbf{u}_2$ *of the form* (3) *admits of the representation* $(n \geqq 4)$

$$\mathbf{u}_1 + \mathbf{u}_2 = \mathbf{v} + \mathbf{q} + \mathbf{p} \tag{4}$$

with

$$\mathbf{v} = [m'(x), 0, m'(x) + m'(1)u(x)], m'(x) \in \{m(x)\} \quad \text{i.e. } \mathbf{v} \in \mathcal{C}_n,$$

$$\mathbf{q} = [q(x), 0, q(x)], \tag{5}$$

$$\mathbf{p} = [0 \quad , 0, m''(x) \;], m''(x) \in \{m(x)\}. \tag{6}$$

If $q(x) = 0$, *then* $m'(x) = 0$; *if* $q(x) \neq 0$ *then either* $m'(x) = 0$ *or* $m'(x)$ *is a trinomial.*

Proof. Let $\mathbf{u}_1 = [q_1(x), 0, q_1(x)f(x)]$ and $\mathbf{u}_2 = [q_2(x), 0, q_2(x)f(x)]$. We have

$$\mathbf{u}_1 + \mathbf{u}_2 = [q_1(x) + q_2(x), 0, (q_1(x) + q_2(x))f(x)]. \quad (7)$$

Let $q(x)f(x) = (q_1(x) + q_2(x))f(x)$ and $m'(x) \triangleq q_1(x) + q_2(x) + q(x)$. Clearly, since $(q(x) + q_1(x) + q_2(x))f(x) = 0$, $m'(x) \in \{m(x)\}$. If $q_1(x) = q_2(x)$, it follows that $q(x) = 0$ and $m'(x) = 0$. If $q_1(x) \neq q_2(x)$, either $q(x) = q_i(x) (i = 1, 2)$ or $q_1(x)$, $q_2(x)$, $q(x)$ are nonzero and distinct: in the former case $m'(x) = 0$, in the latter $m'(x)$ is a trinomial.

This given we can write

$$q_1(x) + q_2(x) = m'(x) + q(x)$$

and rewrite (7) as

$$\mathbf{u}_1 + \mathbf{u}_2 = [m'(x), 0, m'(x) + m'(1)u(x)] + [q(x), 0, q(x)] + [0, 0, q(x) + q(x)f(x) + m'(x) + m'(1)u(x)].$$

It is now evident that $m''(x) \triangleq (q(x) + q(x)f(x)) + m'(x) + m'(1)u(x) \in \{m(x)\}$ since it is the sum of polynomials belonging to $\{m(x)\}$. Q.E.D.

LEMMA 6. *For any trinomial* $m(x) \in \{m(x)\}$

$$m(\alpha^3) = \alpha^{3s}(\alpha^h + \alpha^{2h})$$

where s and h are integers modulo $2^{n-1} - 1$, $h \neq 0$.

Proof. Let $m(x) = x^s + x^i + x^j$, with distinct s, i, j. Then $m(\alpha^3) = \alpha^{3s} + \alpha^{3i} + \alpha^{3j}$. Recalling that $m(\alpha) = \alpha^s + \alpha^i + \alpha^j = 0$ we have

$$\alpha^{3s} = (\alpha^i + \alpha^j)^3 = \alpha^{3i} + \alpha^{3j} + \alpha^i\alpha^j(\alpha^i + \alpha^j)$$

or equivalently $m(\alpha^3) = \alpha^i\alpha^j\alpha^s = \alpha^{3s}\alpha^{i-s}\alpha^{j-s}$. But $\alpha^{j-s} = 1 + \alpha^{i-s}$; hence, letting $i - s = h \neq 0$, the assertion is proved. Q.E.D.

Consider now the matrices:

$$H_1 = [\alpha^{2^{n-1}-2}, \cdots, \alpha, 1],$$

$$H_3 = [(\alpha^3)^{2^{n-1}-2}, \cdots, \alpha^3, 1],$$

$$U = [1, \cdots, 1, 1].$$

The matrix $H = [H_1^T, H_3^T, U^T]^T$ is the parity check matrix of $\{s(x)\}$ (the superscript T denotes "transpose"). Given a polynomial $h(x)$, we define $h(x)H^T \triangleq [\beta_1, \beta_3, c]$ as the *characteristics* of $h(x)$. For arbitrary $s(x) \in \{s(x)\}$

$$W[h(x) + s(x)] \geqq W[k(x)]$$

where $k(x)$ is a minimum weight member of the coset of $\{s(x)\}$ to which $h(x)$ belongs. We now calculate the characteristics of some polynomials which we shall frequently use in the sequel:

$$\begin{cases} q(x) = \alpha^s & q(x)H^T = [\alpha^s, \alpha^{3s}, 1] \\ m(x) \in \{m(x)\} & m(x)H^T = [0, m(\alpha^3), m(1)] \\ m''(x) \text{ (see (6))} & m''(x)H^T = [0, m'(\alpha^3) + \alpha^{3s}, 1]. \end{cases} \tag{8}$$

The first relation is straightforward. The second follows from $m(\alpha) = 0$, since $m(x) \in \{m(x)\}$. To prove the last relation, recall that $m''(x) = q(x) + q(x)f(x) + m'(1)u(x) + m'(x)$, and that: $q(x)f(x)H^T = [\alpha^s, 0, 0]$, since $f(x)$ is divided by the minimum function $g_3(x)$ of α^3 and by $(x + 1)$; $m'(1)u(x)H^T = [0, 0, m'(1)]$, since $u(x)$ is divided by $g_3(x)$ and the minimum function $g_1(x)$ of α.

LEMMA 7. *For* $m''(x) = q(x) + q(x)f(x) + m'(1)u(x) + m'(x)$, *and arbitrary* $s(x) \in \{s(x)\}$,

$$W[m''(x) + q(x) + s(x)] \geqq \begin{cases} 4 \text{ for even } n \\ 2 \text{ for odd } n. \end{cases}$$

Proof. The characteristics of $(m''(x) + q(x))$ is $[\alpha^3, m'(\alpha^3), 0]$ (see (8)). $W[m''(x) + q(x) + s(x)]$ is the minimum number of columns of H which add to $[\alpha^3, m'(\alpha^3), 0]^T$. Since $c = 0$, this number is even. Let us assume that there are two elements x_1 and x_2 of $GF(2^{n-1})$ which satisfy the equations

$$\begin{cases} x_1 + x_2 = \alpha^s \\ x_1^3 + x_2^3 = m'(\alpha^3). \end{cases}$$

Since $\alpha^s \neq 0$ we make the substitution $y_1 = (x_1/\alpha^s)$, $y_2 = (x_2/\alpha^s)$. After easy manipulations we recognize that y_1 and y_2 are the solutions of the single equation

$$y^2 + y + 1 + m'(\alpha^3)/\alpha^{3s} = 0.$$

Since either $m'(x) = 0$ or $m'(x)$ is a trinomial (Lemma 5), Lemma 6 yields $m'(\alpha^3)/\alpha^{3s} = \alpha^h + \alpha^{2h}$ $(0 \leqq h \leqq 2^{n-1}-2)$ and the previous equation becomes

$$(y + \alpha^h)^2 + (y + \alpha^h) + 1 = 0$$

or, equivalently, letting $y + \alpha^h = z$

$$z^2 + z + 1 = 0. \tag{9}$$

But solutions of (9) are primitive cube roots of unity, whence (9) has solutions in $GF(2^{n-1})$ only for odd n. Q.E.D.

We now construct $(2^n - 1)$-components vectors of the form

$$\mathbf{w} = [m(x) + q(x), i, m(x) + q(x)f(x) + (m(1) + i)u(x) + s(x)] \tag{10}$$

where $m(x)$, $q(x)$, i, $s(x)$ are independently chosen and contribute $(2^{n-1} - n)$, $(n - 1)$, 1, $(2^{n-1} - 2n)$ information bits respectively, for a total of $(2^n - 2n)$ information bits. The vectors $\mathbf{w}$ form a $(2^n - 1, 2^n - 2n)$ code $\mathcal{K}_n$: the generic vector $\mathbf{w}$ can be decomposed as

$$\mathbf{w} = \mathbf{v} + \mathbf{u} \tag{11}$$

where $\mathbf{v}$ and $\mathbf{u}$ are defined by relations (1) and (3), respectively. Let $\mathbf{w}_1 = \mathbf{v}_1 + \mathbf{u}_1$ and $\mathbf{w}_2 = \mathbf{v}_2 + \mathbf{u}_2$ be two distinct code words of $\mathcal{K}_n$. Using relations (4) (Lemma 5) we have

$$\mathbf{w}_1 + \mathbf{w}_2 = (\mathbf{v}_1 + \mathbf{v}_2) + (\mathbf{u}_1 + \mathbf{u}_2) = \mathbf{v}_1 + \mathbf{v}_2 + \mathbf{v} + \mathbf{q} + \mathbf{p}$$

or

$$\mathbf{w}_1 + \mathbf{w}_2 = \mathbf{v}' + \mathbf{q} + \mathbf{p} \tag{12}$$

where $\mathbf{v}' \triangleq \mathbf{v}_1 + \mathbf{v}_2 + \mathbf{v}$. Clearly $\mathbf{v}'$ is an arbitrary member of $\mathcal{C}_n$, but $\mathbf{q} + \mathbf{p}$ can be decomposed as

$$\begin{aligned}\mathbf{q} + \mathbf{p} &= [q(x), 0, q(x)f(x) + m'(x) + m'(1)u(x)] \\ &= [q(x), 0, q(x)f(x)] + [0, 0, m'(x) + m'(1)u(x)] \\ &= \mathbf{u}' + [0, 0, m'(x) + m'(1)u(x)].\end{aligned}$$

When $m'(x) \neq 0$, we recall that $m'(x) \notin \{s(x)\}$ (Lemma 5), that is $[0, 0, m'(x) + m'(1)u(x)] \notin \mathcal{C}_n$: hence $\mathcal{K}_n$ is a nonlinear code. Furthermore, in (11) each nonzero $\mathbf{u}$ identifies a coset of $\mathcal{C}_n$, since $q(x) \neq 0$ identifies a coset of $\{m(x)\}$. Hence $\mathcal{K}_n$ can be seen as the set union of $\mathcal{C}_n$ and of a subset of its cosets, whose cardinality is $2^{n-1} - 1$.

Let W denote the weight of $(\mathbf{w}_1 + \mathbf{w}_2)$. We can now prove the central result of this paper.

THEOREM 1. *For even* $n \geq 4$, $\mathcal{K}_n$ *is a nonlinear* $(2^n - 1, 2^n - 2n)$ *code of minimum distance* 5.

Proof. If $q(x) = 0$, $\mathbf{w}_1 + \mathbf{w}_2 \in \mathcal{C}_n$ and, by Lemma 2, $W \geq 6$. Assume now that $q(x) = x^s$ $(0 \leq s \leq 2^{n-1} - 2)$. In general, W is given by

$$W = i + W[m(x) + q(x)] + W[m(x) + (m(1) + i)u(x) + s(x) + m''(x) + q(x)]. \tag{13}$$

Depending upon the values of $m(1)$ and i we distinguish three cases:

1) $m(1) = 0$, $i = 1$. Relation (13) becomes

$$\begin{aligned} W &\geq 1 + W[m(x) + q(x)] + W[u(x) + s(x) + m''(x)] \\ &\quad - W[m(x) + q(x)] \\ &= 1 + W[u(x) + s(x) + m''(x)]. \end{aligned}$$

From relations (8) and $u(x)H^T = [0, 0, 1]$, we obtain $(u(x) + m''(x))H^T = [\beta_1, \beta_3, c] = [0, m'(\alpha^3) + \alpha^{3s}, 0]$. But by Lemma 5 and Lemma 6 $m'(\alpha^3) = \alpha^{3s}(\alpha^k + \alpha^{2k})$, $0 \leq k \leq 2^{n-1} - 2$. Hence, $m'(\alpha^3) + \alpha^{3s} = \alpha^{3s}(1 + \alpha^k + \alpha^{2k}) \neq 0$ in GF(2^{n-1}), n even: it follows that $W[u(x) + s(x) + m''(x)] \neq 0$. Furthermore, $W[u(x) + s(x) + m''(x)]$ is even and >3, since $c = 0$ and $\beta_1 = 0$ (H_1 is the parity check matrix of a single-error-correcting code). We conclude that $W \geq 1 + 4 = 5$.

2) $m(1) = 0$, $i = 0$. If $m(x) \neq 0$, relation (13) yields

$$W \geq W[m(x)] + W[m(x) + m''(x) + s(x)] - 2W[q(x)].$$

Relations (8) give $(m(x) + m''(x))H^T = [\beta_1, \beta_3, c] = [0, m(\alpha^3) + m'(\alpha^3) + \alpha^{3s}, 1]$, that is, $W[m(x) + m''(x) + s(x)]$ is odd $(c = 1)$ and ≥ 3 $(\beta_1 = 0)$. Furthermore, $m(x) \neq 0$ and $m(1) = 0$ imply $W[m(x)] \geq 4$, whence $W \geq 4 + 3 - 2 = 5$. If $m(x) = 0$ relation (13) becomes

$$W \geq W[q(x)] + W[m''(x) + q(x) + s(x)].$$

From Lemma 7, we have that $W[m''(x) + q(x) + s(x)] \geq 4$ for n even, whence $W \geq 1 + 4 = 5$.

3) $m(1) = 1$. In this case $W[m(x) + q(x)]$ is even and ≥ 2. Assume at first that $W[m(x) + q(x)] = 2$: this implies that $m(x) = x^s + x^i + x^j$ (s, i, j distinct), whence by Lemma 6 $m(\alpha^3) = \alpha^{3s}(\alpha^h + \alpha^{2h})$ for some h, $1 \leq h \leq 2^{n-1} - 2$. Relation (13) yields

$$W = i + 2 + W[m(x) + (1 + i)\, u(x) + m''(x) + q(x) + s(x)].$$

For simplicity we let $k(x) \triangleq m(x) + (1 + i)u(x) + m''(x) + q(x) + s(x)$. With the help of relations (8) the characteristics of $k(x)$ is readily obtained as

$$k(x)H^T = [\beta_1, \beta_3, c] = [\alpha^s, m(\alpha^3) + m'(\alpha^3), i].$$

Since $m'(\alpha^3) = \alpha^{3s}(\alpha^k + \alpha^{2k})(0 \leqq k \leqq 2^{n-1} - 2)$, it follows that $m(\alpha^3) + m'(\alpha^3) = \alpha^{3s}(\alpha^r + \alpha^{2r})$ with $r = h + k$. Therefore, from Lemma 7, $W[k(x)] > 2$ for even n, that is, $W[k(x)] \geqq 4 - i$ (since i is the parity of $W[k(x)]$). We conclude that

$$W \geqq i + 2 + 4 - i = 6.$$

Finally assume that $W[m(x) + q(x)] \geqq 4$. Since $\beta_1 = \alpha^s$, we obtain $W[k(x)] \geqq 1$, whence $W \geqq i + 4 + 1 = i + 5$.

Q.E.D.

Note. It is interesting to consider the problem of extending the method employed for the construction of $\mathcal{K}_n$ to other values of the number of correctable errors, namely to $t = 1$ or to $t > 2$.

Two distinct schemes appear to be candidates for successful generalizations. Consider again relation (10) which describes the double-error-correcting $\mathcal{K}_n$, i.e.,

$$\mathbf{w} = [m(x) + q(x), i, m(x) + (m(1) + i)u(x) + s(x) + q(x)f(x)].$$

Here $\mathcal{K}_n$ is constructed in terms of two codes, i.e., $\{m(x)\}$ and $\{s(x)\}$, with $\{s(x)\} \subset \{m(x)\}$. Specifically, if α is primitive in $GF(2^{n-1})$, then $\{m(x)\}$ is characterized by the root α, and $\{s(x)\}$ by the roots $1, \alpha, \alpha^3$. Therefore two potential generalizations for t-error-correction are:

A. $\{m(x)\}$ has root α, and $\{s(x)\}$ has roots $1, \alpha, \alpha^3, \cdots, \alpha^{2t-1}$.

B. $\{m(x)\}$ has roots $\alpha, \alpha^3, \cdots, \alpha^{2t-3}$, and $\{s(x)\}$ has roots $1, \alpha, \alpha^3, \cdots, \alpha^{2t-1}$.

For $t = 1$, both schemes are successful and generate the same codes, as can be easily shown. Specifically with scheme B, $m(x)$ is the generic member of A_{n-1}, and $\{s(x)\}$ has $(x + 1)g_1(x)$ as its generator, which gives the code $\mathcal{K}_n^{(1)}$

$$\mathbf{w}_B = [m(x), i, m(x) + (m(1) + i)u(x) + s(x)]. \tag{14}$$

Surprisingly, $\mathcal{K}_n^{(1)}$ is a group code, as is apparent from (14). Moreover, it can be shown that it coincides with a Vasil'ev code (1962). In fact (14) can be expressed as

$$\mathbf{w}_B = [m(x), i, m(x) + p(x)]$$

where

$$p(x) = (m(1) + i)u(x) + s(x).$$

If we now impose the condition that $p(x)$ belong to the code generated by $g_1(x)$, this relation becomes an equation in the unknowns $s(x)$ and i, which can always be solved if

$$i = \text{parity } W[m(x)] + \text{parity } W[p(x)]$$

thereby yielding a linear Vasil'ev code (equivalent to a Hamming code).

For $t > 2$, the question whether either of the two outlined schemes produces a viable generalization remains entirely open.

3. THE FORM OF THE REDUNDANCY FUNCTIONS

Consider the expression (10) of the generic vector of $\mathcal{K}_n$, that is

$$\mathbf{w} = [m(x) + q(x), i, m(x) + (m(1) + i)u(x) + s(x) + q(x)f(x)].$$

It is easily seen that $\mathcal{K}_n$ can be encoded as a systematic code, i.e., $(2^n - 2n)$ binary information digits can be arbitrarily assigned in fixed positions and the remaining $(2n - 1)$ redundant digits can be computed as functions of the information digits. In this section we investigate the nature of these functions. For convenience, we now represent $\mathbf{w}$ as

$$\mathbf{w} = [i^{(0)}_{2^{n-1}-2}, \cdots, i^{(0)}_1, i^{(0)}_0, i, i^{(1)}_{2^{n-1}-2}, \cdots, i^{(1)}_{2n-1}, p_{2n-2}, \cdots, p_1, p_0]$$

where i's and p's denote information and redundancy digits, respectively.

Assume for a moment that $s(x) = 0$. Then the leftmost 2^{n-1} digits $[i^{(0)}_{2^{n-1}-2}, \cdots, i]$ completely determine the $2^{n-1} - 1$ rightmost ones; we denote the latter ones by $[\varphi_{2^{n-1}-2}, \cdots, \varphi_0]$, and analyze their dependence upon the former set. Let

$$i(x) \triangleq \sum i^{(0)}_j x^j, \qquad q(x)f(x) \triangleq c(x) = \sum c_j x^j, \qquad f(x) = \sum f_j x^j,$$
$$m(x) = \sum m_j x^j$$

where all summations run for $j = 0, 1, \cdots, 2^{n-1} - 2$. If $q(x) = 0$, then $c_j = 0$ for every j. If $q(x) = x^s$, then due to the unique property of the maximum length sequence (see Peterson (1961), p. 148), $c_{s+b}c_{s+b-1} \cdots c_{s+b-n+2} = 1$ and $c_{j+b} \cdots c_{j+b-n+2} = 0$ for $j \neq s$, where b is such that $f_b f_{b-1} \cdots f_{b-n+2} = 1$. We readily have

$$q(x) = \sum c_{j+b} \cdots c_{j+b-n+2} x^j, \qquad m_j = i^{(0)}_j + c_{j+b} \cdots c_{j+b-n+2}$$

and

$$\varphi_j = i^{(0)}_j + c_{j+b} \cdots c_{j+b-n+2} + i + \sum_k (i^{(0)}_k + c_{k+b} \cdots c_{k+b-n+2}) + c_j$$

or, after regrouping the terms

$$\varphi_j = \{\sum_{k\neq j} i_k^{(0)} + i + c_j\} + \{\sum_{h\neq j+b} c_h \cdots c_{h-n+2}\}. \quad (15)$$

We now recall that, since $c(x) = q(x)f(x) = i(x)f(x)$, $c_j = \sum f_{j-k} i_k^{(0)}$ is a linear function of the variables $i_0^{(0)}, i_1^{(0)}, \cdots, i_{2^n-1-2}^{(0)}$. Specifically, since $f(x)$ is a maximum length sequence, for distinct r and s there is a t such that $c_r + c_s = c_t$. If $g_1(x)$ is a trinomial, for $s = (r + n - 1)$, t satisfies the relations $r < t < r + n - 1$. Hence

$$c_h c_{h-1} \cdots c_{h-n+2} + c_{h-1} c_h \cdots c_{h-n+1}$$
$$= c_{h-1} \cdots c_{h-n+2}(c_h + c_{h-n+1}) = c_{h-1} \cdots c_{h-n+2}.$$

It follows that in the last term of (15), which is the sum of $2^{n-1} - 2$ products, each pair of consecutive products of $(n - 1)$ factors is contracted into a single product of $(n - 2)$ factors, for a total of $2^{n-2} - 1$ products. In conclusion we obtain

$$\varphi_j = \left\{ i + i_j^{(0)} + \sum_{k=0}^{2^{n-1}-2} (1 + f_{j-k}) i_k^{(0)} \right\} + \sum_{h=0}^{2^{n-2}-2} \prod_{s=0}^{n-3} c_{j+b+1+2h-s} \quad (16)$$

which shows that $\varphi_j(i_{2^{n-1}-2}^{(0)}, \cdots, i_0^{(0)}, i)$ is the sum of a strictly linear function and of a nonlinear function of degree at most $(n - 2)$. As a check, for the (15, 8) code, $n = 4$, the latter function is quadratic.

This given, let h_{ij} be the generic entry of the parity check matrix H^* of $\{s(x)\}$ in systematic form, i.e., the $(2n - 1)$ rightmost columns of H^* form the unity matrix and the index j runs from right to left. Then the relations

$$p_i = \varphi_i + \sum_{j=2n-1}^{2^{n-1}-2} h_{ij}(i_j^{(1)} + \varphi_j) \qquad (i = 0, 1, \cdots, 2n - 2) \quad (17)$$

give the sought redundancy functions.

Expressions (15) and (17) are suggestive of a very simple implementation of encoding. In fact, φ_j is a cyclic function of its arguments. Hence it can be realized by a recirculating nonlinear convolutional encoder consisting of a cyclic shift register and of a combinational circuit realizing $\varphi = \varphi_{2^{n-1}-2}$ (see Figure 1). The complete encoder consists of three shift-registers SR1, SR2, SR3 with 1, $(2^{n-1} - 1)$ and $(2n - 1)$ stages, respectively. The operation is organized in four phases G_1, G_2, G_3, G_4, whose durations are 1, $(2^{n-1} - 1)$, $(2^{n-1} - 2n)$ and $(2n - 1)$ time

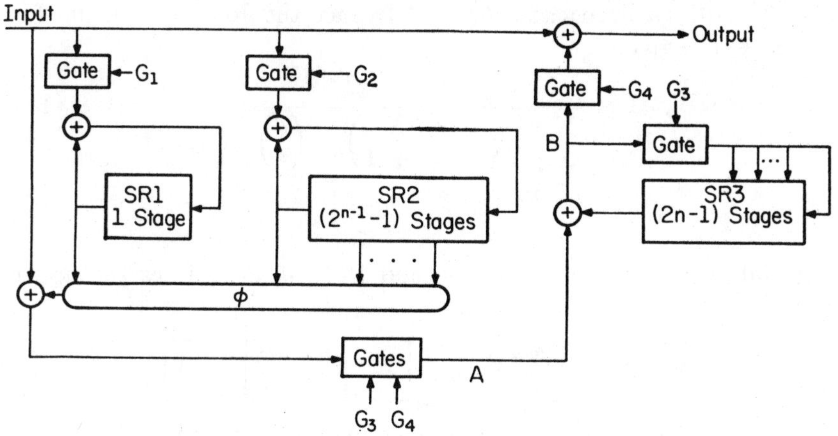

FIG. 1. Encoder for the $\mathcal{K}_n$ code.

units, respectively. The indicated gates are permissive when the applied signals are active. All registers are initially set to 0. The information digits are fed in the sequence $i, i^{(0)}_{2^{n-1}-2}, \cdots, i^{(0)}_0, i^{(1)}_{2^{n-1}-2}, \cdots, i^{(1)}_{2n-1}$, one per time unit. Then during G_1 the digit i is fed to SR1 and during G_2 $i^{(0)}_{2^{n-1}-2}, \cdots, i^{(0)}_0$ are fed to SR2 (while they are concurrently sent to the output): both SR1 and SR2 are recirculating, as shown. During phase G_3, $\varphi_j + i^{(1)}_j$ appears at point A to be fed to SR3, which is a feedback shift register performing the division of a polynomial by $(x + 1)g_1(x) \cdot g_3(x)$ (see Peterson (1961), p. 149); then at the end of G_3 SR3 contains the parity checks $\sum_j h_{ij}(i^{(1)}_j + \varphi_j)$. During G4 the input is 0 and at point B the functions p_i are formed and fed to the output. Therefore the calculation of the redundant digits takes no longer than the serial transmission of the information digits.

4. OPTIMALITY OF THE $\mathcal{K}_n$ CODES

A code $\mathcal{K}_n$ is a $(2^n - 1, 2^n - 2n)$ double-error-correcting code. It contains one information digit more than the corresponding linear code, i.e., the BCH double-error-correcting code of the same length (which is a $(2^n - 1, 2^n - 1 - 2n)$ code).

In this section we prove a stronger statement, namely, that a $\mathcal{K}_n$ code has the largest number of code words for its length and minimum distance, since it meets the Johnson's bound $A(N, d)$ (Johnson, 1962) for

$N = 2^n - 1$ (n even) and $d = 5$.[2] In fact the Johnson's bound for $d = 2t + 1$ is given by

$$A(N, d) \leqq \frac{2^N}{\sum_{j=0}^{t} \binom{N}{j} + \dfrac{\binom{N}{t+1} - \binom{d}{t} R(N, d, t)}{\left[\dfrac{N}{t+1}\right]}} \tag{18}$$

where $[a]$ is the integral part of a, and $R(N, d, t)$ satisfies the upper bound

$$R(N, d, t) \leqq \left[\frac{N}{d}\left[\frac{N-1}{d-1}\left[\cdots\left[\frac{N-t}{d-t}\right]\cdots\right]\right]\right]. \tag{19}$$

When $N = 2^n - 1$ and $t = 2$, relations (18) and (19) specialize as

$$A(2^n - 1, 5) \leqq \frac{2^{2^n-1}}{1 + \binom{2^n-1}{1} + \binom{2^n-1}{2} + \dfrac{\binom{2^n-1}{3} - \binom{5}{2} R(2^n - 1, 5, 2)}{\left[\dfrac{2^n-1}{3}\right]}} \tag{20}$$

$$R(2^n - 1, 5, 2) \leqq \left[\frac{2^n-1}{5}\left[\frac{2^n-2}{4}\left[\frac{2^n-3}{3}\right]\right]\right]. \tag{21}$$

Consider relation (21). For even n, $(2^n - 4)$ is divisible by 3, hence $[(2^n - 3)/3] = (2^n - 4)/3$. Moreover $(2^n - 4)$ is divisible by 4. We must now show that $(2^n - 1)(2^n - 2)(2^{n-2} - 1)$ is divisible by 5. This follows immediately from the observation that the residues modulo 5 of 2^n (n even) alternate as 1 and 4, i.e., the residue of $(2^n - 1)$ alternate as 0 and 3: since $(2^n - 1)(2^n - 2)(2^{n-2} - 1)$ contains two consecutive even powers of 2, we have

$$R(2^n - 1, 5, 2) \leqq \frac{(2^n - 1)(2^n - 2)(2^n - 4)}{60}$$

[2] The observation that $A(2^n - 1, 5)$ (n even) is a power of 2 is originally due to J. P. Robinson. Prior to this, the author formulated a conjecture, based on rather fuzzy geometric arguments, that nonlinear codes of length $(2^n - 1)$ and distance 5, analogous to the (15, 8) code, existed only for even n (private communications, Jan. and March 1968).

from which we readily obtain for even n

$$\frac{\binom{2^n-1}{3} - \binom{5}{2} R(2^n-1, 5, 2)}{\left[\frac{2^n-1}{3}\right]} \geqq 3\,\frac{\frac{(2^n-1)(2^n-2)}{6}}{(2^n-1)} = 2^{n-1} - 1. \tag{22}$$

We then conclude that

$$A(2^n-1, 5) \leqq \frac{2^{2^n-1}}{2^{2n-1} - 2^{n-1} + 1 + (2^{n-1}-1)} = 2^{2^n-2n} \qquad (n \text{ even})$$

which is exactly the number of code words of $\mathcal{K}_n$. Clearly for odd n, ratio (22) is strictly larger than $2^{n-1} - 1$, since $[(2^n-3)/3] = (2^n-5)/3$: which also shows, from a different angle, the unrealizability of $\mathcal{K}_n$ codes for odd n.

5. DECODING OF A $\mathcal{K}_n$ CODE

In this section we show that decoding of a $\mathcal{K}_n$ code can be easily accomplished through the calculation and examination of syndrome-like quantities.

With the vector

$$\mathbf{e} = [e_0(x), e, e_1(x)]$$

we represent an error pattern, where $e_j(x) \in A_{n-1}$ and e is a binary parameter. The distance properties of $\mathcal{K}_n$ give the following condition on $\mathbf{e}$ for correctability

$$W[e_0(x)] + W[e_1(x)] + e \leqq 2. \tag{23}$$

In general the received vector is $\mathbf{r} = [r_0(x), r, r_1(x)] = \mathbf{w} + \mathbf{e}$, with $\mathbf{w} \in \mathcal{K}_n$. We now compute the following functions:

$$\begin{cases} \sigma_0 \triangleq r_0(x)H_1^T, \\ \sigma_1 \triangleq r_1(x)H_1^T, \\ \sigma \triangleq (r_0(x) + r_1(x))H_3^T, \\ d \triangleq r + r_1(x)U^T. \end{cases}$$

Since $r_0(x) = m(x) + q(x) + e_0(x)$, and $m(x)H_1^T = 0$, letting $q(x) = bx^s$, we have $\sigma_0 = b\alpha^s + e_0(\alpha)$. Similarly, from $r_1(x) = m(x) +$

$(m(1) + i)u(x) + q(x)f(x) + s(x) + e_1(x)$ and $u(x)H_1^T = 0$, $s(x)H_1^T = 0$, $q(x)f(x)H_1^T = b\alpha^s$ we obtain $\sigma_1 = b\alpha^s + e_1(\alpha)$. From $r_0(x) + r_1(x) = q(x) + q(x)f(x) + s(x) + e_0(x) + e_1(x) + (m(1) + i)u(x)$, recalling that $s(x)H_3^T = 0$, $f(x)H_3^T = 0$, $u(x)H_3^T = 0$, we obtain $\sigma = b\alpha^{3s} + e_1(\alpha^3) + e_0(\alpha^3)$. Finally since $W[q(x)f(x)]$, $W[s(x)]$, $W[m(x) + m(1)u(x)]$ are even, $d = r + i + e_1(1) = e + e_1(1)$. This is summarized as follows:

$$\begin{cases} \sigma_0 = b\alpha^s + e_0(\alpha), \\ \sigma_1 = b\alpha^s + e_1(\alpha), \\ \sigma \;\, = b\alpha^{3s} + e_0(\alpha^3) + e_1(\alpha^3), \\ d \;\, = e + e_1(1). \end{cases} \tag{24}$$

The quadruple $\Sigma \equiv (\sigma_0, \sigma_1, \sigma, d)$ is conventionally termed the *syndrome* of r.

We now give a lemma which is based on rather well-known results of the theory of finite fields.[3]

LEMMA 8. *The set Θ of all $\theta \in \mathrm{GF}(2^{n-1})$ for which $y^2 + y + \theta = 0$ has solutions over $\mathrm{GF}(2^{n-1})$ is a vector space of dimension $(n - 2)$, given by the even linear combinations of a normal basis $\beta, \beta^2, \beta^4, \cdots, \beta^{2^{n-2}}$ of $\mathrm{GF}(2^{n-1})$.*

Proof. It is well-known (see, e.g., Albert (1956) p. 121) that there are bases of $\mathrm{GF}(2^{n-1})$ consisting of complete sets of conjugates (normal basis): let $\beta, \beta^2, \cdots, \beta^{2^{n-2}}$ be one such set of linearly independent conjugates. Then every $\gamma \in \mathrm{GF}(2^{n-1})$ is uniquely expressible as

$$\gamma = c_0\beta + c_1\beta^2 + \cdots + c_{n-2}\beta^{2^{n-2}} \quad (c_j \in \mathrm{GF}(2)).$$

Since

$$\beta^{2^{n-1}} = \beta, \quad \text{then} \quad \gamma^2 = c_{n-2}\beta + c_0\beta^2 + \cdots + c_{n-3}\beta^{2^{n-2}}$$

and

$$\gamma^2 + \gamma = d_0\beta + d_1\beta^2 + \cdots + d_{n-2}\beta^{2^{n-2}} \quad (d_j \in \mathrm{GF}(2)) \tag{25}$$

with $d_j = c_j + c_{j-1}$ (the subscripts are modulo $n - 1$). But the right side of (25) is the generic element of Θ: assume then that $d_0, d_1, \cdots, d_{n-2}$ are given. We then have $c_{n-2} = d_0 + c_0 = d_0 + d_1 + c_1 = \cdots =$

[3] The argument given here is substantially borrowed from Albert (1956). A very similar theorem was proved by Berlekamp et al. (1962, Thm. 1). A particularly illuminating reference is Berlekamp (1968), which also contains a generalization of the lemma (p. 166). Since the statement given here is particularly geared to subsequent considerations, lemma and proof are given in full.

$d_0 + \cdots + d_{n-2} + c_{n-2}$, i.e.,

$$d_0 + d_1 + \cdots d_{n-2} = 0$$

i.e., the number of nonzero d_j's is even. Q.E.D.

This lemma provides a rule for testing whether $\gamma \in GF(2^{n-1})$ is a member of Θ. In fact, we must first find a normal basis $\beta, \beta^2, \cdots, \beta^{2^{n-2}}$ of $GF(2^{n-1})$, (see, e.g., Berlekamp (1968), pp. 253–254). Let $\boldsymbol{\gamma}$ denote the column vector representation over GF(2) of $\gamma \in GF(2^{n-1})$ with respect to the basis $1, \alpha, \cdots, \alpha^{n-2}$, and let $M \triangleq [\boldsymbol{\beta}, \cdots, \boldsymbol{\beta}^{2^{n-2}}]$, a nonsingular $(n-1) \times (n-1)$ matrix. Then $\boldsymbol{\gamma}$ is related to the representation $[d_0, \cdots, d_{n-2}]$ of γ with respect to the basis $\beta, \beta^2, \cdots, \beta^{2^{n-2}}$ by

$$\boldsymbol{\gamma} = M\,[d_0, \cdots, d_{n-2}]^T$$

i.e., $M^{-1}\boldsymbol{\gamma} = [d_0, \cdots, d_{n-2}]^T$. Premultiplying both sides by the row vector $\mathbf{u} = [1, 1, \cdots, 1]$ we have the condition

$$\mathbf{u}\cdot[d_0, \cdots, d_{n-2}]^T = \begin{cases} 0 & \text{if} \quad \gamma \in \Theta \\ 1 & \text{if} \quad \gamma \notin \Theta \end{cases}$$

which, denoting by $\boldsymbol{\lambda}^T$ the row sum of M^{-1}, is translated into

$$\boldsymbol{\lambda}^T\boldsymbol{\gamma} = \begin{cases} 0 & \text{if} \quad \gamma \in \Theta \\ 1 & \text{if} \quad \gamma \notin \Theta. \end{cases} \tag{26}$$

The following lemma provides some insight into the distance relationship between the generic vector $\mathbf{r}$ and the members $\mathbf{w}$ of $\mathcal{K}_n$.

LEMMA 9. *Given any vector* $\mathbf{r} = [r_0(x), r, r_1(x)]$ *there exists a* $\mathbf{w} \in \mathcal{K}_n$ *such that* $\mathbf{r} + \mathbf{w} = [0, e, e(x)]$ with $W[e(x)] \leqq 3$.

Proof. Let $\{t(x)\}$ be the double-error-correcting BCH code generated by $g_1(x)g_3(x)$. We decompose $r_0(x)$ as $r_0(x) = m_0(x) + q_0(x)$ and form $r_1^*(x) = r_1(x) + m_0(x) + (m_0(1) + r)u(x) + q_0(x)f(x)$. Next $r_1^*(x)$ is decomposed as $r_1^*(x) = t(x) + e(x)$, where $t(x) \in \{t(x)\}$ and $e(x)$ is a minimum weight coset leader of $\{t(x)\}$: it is known (Gorenstein, et al. (1960)) that $W[e(x)] \leqq 3$. It is also of immediate verification that $(t(x) + t(1)u(x)) \in \{s(x)\}$. We then form the code word

$$\begin{aligned}\mathbf{w} = [&m_0(x) + q_0(x), r + t(1), m_0(x) + (m_0(1) \\ &+ r + t(1))u(x) + q_0(x)f(x) + t(x) + t(1)u(x)] \\ = [&r_0(x), r + t(1), r_1(x) + e(x)]\end{aligned}$$

Letting $t(1) = e$, $\mathbf{r} + \mathbf{w} = [0, e, e(x)]$. Q.E.D.

Hereafter the subscript j of σ_j or $e_j(x)$ is to be considered modulo 2. We define $\rho \triangleq \sigma + (\sigma_0 + \sigma_1)^3$ and prove the following basic Lemma.

LEMMA 10. *The conditions* $\rho + \sigma_j^3 = 0$ $(j = 0$ *and* $1)$, $d = 0$ *hold if and only if* $\mathbf{r} \in \mathcal{K}_n$, i.e., *they characterize the code* $\mathcal{K}_n$.

Proof. From Lemma 9, we can assume without loss of generality that the discrepancy between $\mathbf{r}$ and some $\mathbf{w} \in \mathcal{K}_n$ be of the form $[0, e, e(x)]$, $W[e(x)] \leqq 3$. Then relations (24) become

$$\begin{cases} \sigma_0 = b\alpha^s, \\ \sigma_1 = b\alpha^s + e(\alpha), \\ \sigma = b\alpha^{3s} + e(\alpha^3), \\ d = e + e(1). \end{cases} \tag{27}$$

The direct statement follows immediately by setting $e(x) = 0$, $e = 0$ in (27). To prove the converse, assume that $\rho + \sigma_j^3 = 0$ $(j = 0, 1)$. This implies $\sigma_0^3 = \sigma_1^3$, and, due to the uniqueness of the cubic root in $GF(2^{n-1})$, n even, $\sigma_0 = \sigma_1$. From (27) it follows that $e(\alpha) = 0$. We then have: $\rho + \sigma_j^3 = \sigma + \sigma_j^3 = e(\alpha^3) = 0$. Since $[H_1^T, H_3^T]$ is the parity check matrix of a double-error-correcting code, and $W[e(x)] \leqq 3$ (Lemma 9), from $e(\alpha) = e(\alpha^3) = 0$ we conclude that $e(x) = 0$. Finally $d = 0$ yields $e = e(1) = 0$. Q.E.D.

We readily recognize that $\rho + \sigma_j^3 = 0$ $(j = 0, 1)$, $d = 0$ are equivalent to

$$\sigma_0 = \sigma_1, \qquad \sigma = \sigma_0^3 = \sigma_1^3, \qquad d = 0 \tag{28}$$

which characterize the code.

Following is a sequence of three theorems (2.1, 2.2 and 2.3) which establish a correspondence between sets of syndromes and sets of correctable error configurations. The statements and the relative proofs follow an almost identical pattern. The necessary condition ("if") is demonstrated by showing through relations (24) that an error configuration of the prescribed type produces a syndrome of the prescribed type. The converse ("only if") is demonstrated as follows: we form a "correction" vector $\mathbf{c} = [c_0(x), c, c_1(x)]$ which is a function of the syndrome Σ alone and such that $c + W[c_0(x)] + W[c_1(x)] \leqq 2$; then we show that $\mathbf{r} + \mathbf{c} \in \mathcal{K}_n$, since the syndrome $\Sigma^* \equiv (\sigma_0^*, \sigma_1^*, \sigma^*, d^*)$ calculated for $(\mathbf{r} + \mathbf{c})$, that is

$$\begin{cases} \sigma_j^* = \sigma_j + c_j(\alpha) & (j = 0, 1), \\ \sigma^* = \sigma + c_0(\alpha^3) + c_1(\alpha^3), \\ d^* = d + c + c_1(1) \end{cases} \tag{29}$$

satisfies the conditions of Lemma 10 (or, equivalently, (28)); finally, due to the distance properties of $\mathcal{K}_n$, we conclude that $\mathbf{e} = \mathbf{c}$ is the only correctable error configuration which could have produced $\mathbf{r}$. Clearly each of these theorems also yields a decoding rule, embodied by the calculation of the vector $\mathbf{c}$ from Σ. After this introduction the proof of each theorem will be simply sketched.

THEOREM 2.1. *For correctable* $\mathbf{e}$ *the condition* $\rho + \sigma_j^3 = 0$ *is verified for exactly one value of* $j = 0, 1$ *if and only if* $W[e_0(x)] + W[e_1(x)] = 1$.

Proof. "*If*": $e_j(x) = x^k$, $e_{j+1}(x) = 0$ give $\sigma_j = b\alpha^s + \alpha^k$, $\sigma_{j+1} = b\alpha^s$, $\sigma = b\alpha^{3s} + \alpha^{3k}$, whence $\rho = b\alpha^{3s}$. Then $\rho + \sigma_{j+1}^3 = 0$ and

$$\rho + \sigma_j^3 = \alpha^{3k}[(b\alpha^{s-k})^2 + b\alpha^{s-k} + 1] \neq 0$$

since $z^2 + z + 1 \neq 0$ for any value of $z \in \mathrm{GF}(2^{n-1})$, n even.

"*Only if*": If $\rho + \sigma_j^3 \neq 0$, $\rho + \sigma_{j+1}^3 = 0$, we calculate $\sigma_0 + \sigma_1 = \alpha^h$; then we set $c_j(x) = x^h$, $c_{j+1}(x) = 0$, $c = d + c_1(1)$ and compute Σ^*, i e.,

$$\sigma_j^* = \sigma_j + \alpha^h = \sigma_{j+1} = \sigma_{j+1}^*, \qquad d^* = 0,$$

$$\sigma^* = \sigma + \alpha^{3h} = \sigma + (\sigma_0 + \sigma_1)^3 = \rho = \sigma_{j+1}^3 = \sigma_{j+1}^{*3}$$

i.e., (28) are satisfied with $c + W[c_0(x)] + W[c_1(x)] \leqq 2$. Q.E.D.

Theorem 2.1 yields the following decoding rule:

Rule 1. If $\sqrt[3]{\rho} = \sigma_j$, $\sqrt[3]{\rho} \neq \sigma_{j+1}$, then $c_{j+1}(x) = x^h$ and $c = d + c_1(1)$ where $\alpha^h = \sigma_0 + \sigma_1$.

THEOREM 2.2. *For correctable* $\mathbf{e}$ *the conditions* $\rho + \sigma_j^3 \neq 0$ $(j = 0, 1)$, $d = 1$ *hold if and only if* $e = 0$ *and* $W[e_j(x)] = 1$ $(j = 0, 1)$.

Proof. "*if*": $e_j(x) = x^{k_j}$, $e = 0$ give $\sigma_j = b\alpha^s + \alpha^{k_j}$, $\sigma = b\alpha^{3s} + \alpha^{3k_j} + \alpha^{3k_{j+1}}$, $d = 1$, whence $\rho = b\alpha^{3s} + \alpha^{k_j}\alpha^{k_{j+1}}(\alpha^{k_j} + \alpha^{k_{j+1}})$. We then have

$$\rho + \sigma_j^3 = \alpha^{3k_j}\left[\left(\frac{\sigma_{j+1}}{\alpha^{k_j}}\right)^2 + \frac{\sigma_{j+1}}{\alpha^{k_j}} + 1\right] \neq 0$$

in $GF(2^{n-1})$, n even.

"*Only if*". If $d = 1$, $\rho + \sigma_j^3 \neq 0$ $(j = 0, 1)$ we calculate $\rho' \triangleq \sigma + \sigma_0\sigma_1(\sigma_0 + \sigma_1)$ and obtain $\sigma_{j+1} + \sqrt[3]{\rho'} = \alpha^{k_j}$. Then we set $\mathbf{c} = [x^{k_0}, 0, x^{k_1}]$ and compute Σ^*, i.e.,

$$\sigma_j^* = \sigma_j + \alpha^{k_j} = \sigma_j + \sigma_{j+1} + \sqrt[3]{\rho'} = \sigma_{j+1}^*$$

$$\sigma^* = \sigma + \alpha^{3k_j} + \alpha^{3k_{j+1}} = \sigma + (\sigma_{j+1} + \sqrt[3]{\rho'})^3 + (\sigma_j + \sqrt[3]{\rho'})^3$$

$$= (\sigma + \sigma_j\sigma_{j+1}(\sigma_{j+1} + \sigma_j) + \rho') + (\sigma_j + \sigma_{j+1} + \sqrt[3]{\rho'})^3 = \sigma_j^{*3}$$

since $\sigma + \sigma_j\sigma_{j+1}(\sigma_{j+1} + \sigma_j) = \rho'$. Finally $d^* = d + c_1(1) = 0$. Relations (28) are satisfied with $c + W[c_0(x)] + W[c_1(x)] = 2$. Q.E.D.

We have therefore the following decoding rule:

Rule 2. If $\sqrt[3]{\rho} \neq \sigma_0$, $\sqrt[3]{\rho} \neq \sigma_1$, $d = 1$, then $c = 0$ and $c_j(x) = x^{k_j}$, where $\alpha^{k_j} = \sigma_{j+1} + \sqrt[3]{\sigma + \sigma_0\sigma_1(\sigma_0 + \sigma_1)}$.

Before giving Theorem 2.3, we notice that subject to $(\sigma_0 + \sigma_1) \neq 0$ the functions $\tau_j \triangleq (\rho + \sigma_j^3)/(\sigma_0 + \sigma_1)^3 (j = 0, 1)$ are related by

$$\tau_j + \tau_{j+1} + \frac{\sigma_0\sigma_1}{(\sigma_0 + \sigma_1)^2} + 1 = 0.$$

Expressing these elements of $GF(2^{n-1})$ as column vectors with respect to the basis $1, \alpha, \cdots, \alpha^{n-2}$ and premultiplying by $\boldsymbol{\lambda}^T$ (see (26)) we have

$$\boldsymbol{\lambda}^T\boldsymbol{\tau}_j + \boldsymbol{\lambda}^T\boldsymbol{\tau}_{j+1} = 1$$

since $\boldsymbol{\lambda}^T \cdot \mathbf{1} = 1$ and $(\sigma_0\sigma_1)/(\sigma_0 + \sigma_1)^2 \in \Theta$ (in fact $\sigma_0/(\sigma_0 + \sigma_1)$ solves the equation $y^2 + y + [\sigma_0\sigma_1/(\sigma_0 + \sigma_1)^2] = 0$). This proves the following lemma.

LEMMA 11. *If* $(\sigma_0 + \sigma_1) \neq 0$, *then exactly one of the two functions* τ_j, τ_{j+1} *belongs to* Θ.

THEOREM 2.3. *For correctable* $\mathbf{e}$ *the conditions* $\rho + \sigma_j^3 \neq 0$ $(j = 0, 1)$, $d = 0$, $(\sigma_0 + \sigma_1) \neq 0$ *hold if and only if* $e_j(x) = 0$, $W[e_{j+1}(x)] = 2$, $e = 0$.

Proof. "*If*": $e = 0$, $e_j(x) = 0$, $e_{j+1}(x) = x^{k_1} + x^{k_2}$ give $\sigma_j = b\alpha^s$, $\sigma_{j+1} = b\alpha^s + e_{j+1}(\alpha)$, $\sigma = b\alpha^{3s} + e_{j+1}(\alpha^3)$, whence $(e_{j+1}(\alpha) \neq 0)$

$$\rho + \sigma_j^3 = e_{j+1}^3(\alpha) + e_{j+1}(\alpha^3)$$

$$\rho + \sigma_{j+1}^3 = e_{j+1}^3(\alpha) + e_{j+1}(\alpha^3) + e_{j+1}^3(\alpha)\left[1 + \frac{\sigma_j}{e_{j+1}(\alpha)} + \left(\frac{\sigma_j}{e_{j+1}(\alpha)}\right)^2\right].$$

Recalling that $e_{j+1}^3(\alpha) + e_{j+1}(\alpha^3) = \alpha^{k_1}\alpha^{k_2}(\alpha^{k_1} + \alpha^{k_2})$ and letting $\gamma \triangleq \alpha^{k_1}/\alpha^{k_2} \neq 0$, $y \triangleq \sigma_j/e_{j+1}(\alpha)$ we have $\rho + \sigma_j^3 = \alpha^{3k_2}\gamma(1 + \gamma) \neq 0$ since $\gamma \neq 1 (k_1 \neq k_2)$, and

$$\rho + \sigma_{j+1}^3 = \alpha^{3k_2}(1 + \gamma)^3\left(y^2 + y + 1 + \frac{\gamma}{1 + \gamma^2}\right) \neq 0$$

since $1 \notin \Theta$ and $\gamma/(1 + \gamma^2) \in \Theta$ imply: $1 + \gamma/(1 + \gamma^2) \notin \Theta$. Moreover, $d = 0$ and $\sigma_0 + \sigma_1 = e_{j+1}(\alpha) \neq 0$.

"*Only if*": If $d = 0$, $\rho + \sigma_j^3 \neq 0$ $(j = 0, 1)$, $(\sigma_0 + \sigma_1) \neq 0$ we obtain

$\alpha^{k_1}/(\sigma_0 + \sigma_1)$ and $\alpha^{k_2}/(\sigma_0 + \sigma_1)$ as the solutions of

$$y^2 + y + \frac{\rho + \sigma_j^3}{(\sigma_0 + \sigma_1)^3} = 0 \tag{30}$$

(that (30) has solutions over $GF(2^{n-1})$ for exactly one value of j is guaranteed by Lemma 11). Set $c_{j+1}(x) = x^{k_1} + x^{k_2}$, $c_j(x) = 0$, $c = 0$ and compute Σ^*, i.e.,

$$d^* = d + e_1(1) = 0,$$

$$\sigma_{j+1}^* = \sigma_{j+1} + \frac{\alpha^{k_1} + \alpha^{k_2}}{\sigma_j + \sigma_{j+1}} (\sigma_j + \sigma_{j+1}) = \sigma_j = \sigma_j^*,$$

$$\begin{aligned}\sigma^* &= \sigma + \frac{\alpha^{3k_1} + \alpha^{3k_2}}{(\sigma_0 + \sigma_1)^3} (\sigma_0 + \sigma_1)^3 \\ &= \sigma + (\sigma_0 + \sigma_1)^3 \left(1 + \frac{\alpha^{k_1}\alpha^{k_2}}{(\sigma_0 + \sigma_1)^2} \frac{\alpha^{k_1} + \alpha^{k_2}}{\sigma_0 + \sigma_1}\right) \\ &= \sigma + (\sigma_0 + \sigma_1)^3 \left(1 + \frac{\rho + \sigma_j^3}{(\sigma_0 + \sigma_1)^3}\right) = \sigma_j^3 = \sigma_j^{*3}\end{aligned}$$

since $(\alpha^{k_1} + \alpha^{k_2})/(\sigma_0 + \sigma_1) = 1$ and $\alpha^{k_1}\alpha^{k_2}/(\sigma_0 + \sigma_1)^2 = (\rho + \sigma_j^3)/(\sigma_0 + \sigma_1)^3$, being the sum and the product of the solutions of (30), respectively. Relations (28) are satisfied with $c + W[c_0(x)] + W[c_1(x)] = 2$. Q.E.D.

This yields the following decoding rule:

Rule 3. If $\sqrt[3]{\rho} \neq \sigma_0$, $\sqrt[3]{\rho} \neq \sigma_1$, $d = 0$, $\sigma_0 + \sigma_1 \neq 0$, then set $c = 0$, $c_j(x) = 0$ and $c_{j+1}(x) = x^{k_1} + x^{k_2}$, where α^{k_1} and α^{k_2} are the solutions of $z^2 + (\sigma_0 + \sigma_1)z + (\rho + \sigma_j^3)/(\sigma_0 + \sigma_1) = 0$.

Rules 1, 2, 3 constitute an algorithm which encompasses the correction of all the correctable error patterns. What is the behavior of this algorithm when the received $\mathbf{r}$ is at distance $\geqq 3$ from any $\mathbf{w} \in \mathcal{K}_n$? The answer to this question is implicitly provided by the previous three theorems, which give necessary and sufficient conditions for the existence of a code word within distance 2 from the received word $\mathbf{r}$. Therefore $\mathbf{r}$ lies at distance $\geqq 3$ from any code word if and only if $\rho + \sigma_j^3 \neq 0$ $(j = 0, 1)$ (Theorem 2.1), $d = 0$ (Theorem 2.2) and $\sigma_0 + \sigma_1 = 0$ (Theorem 2.3). When Σ satisfies these conditions, clearly we can no longer perform the correction. In fact, while the distance properties of $\mathcal{K}_n$ guarantee that an existing correction vector $\mathbf{c}$ of weight $\leqq 2$ is also unique, more than one **c** of weight 3 can be constructed when Rules 1, 2 and 3 are inapplicable. This is shown by the following argument. Assume that the conditions $\sqrt[3]{\rho} \neq \sigma_j$ $(j = 0, 1)$, $d = 0$, $\sigma_0 = \sigma_1$ hold for **r**. We determine α^h such that $(1 + (\sigma + \sigma_0^3)/\alpha^{3h}) \in \Theta$: there are 2^{n-2} values of h which meet this requirement, since α^3 generates the multiplicative group of $\mathrm{GF}(2^{n-1})$ and, for fixed $(\sigma + \sigma_0^3)$, $(1 + (\sigma + \sigma_0^3)/\alpha^{3h})$ spans the set $\{0, \alpha, \alpha^2, \cdots, \alpha^{2^{n-2}}\}$ which contains Θ for even n (Lemma 8). We then form a correction vector **c** as follows: $c = 0$, $c_0(x) = x^h$, $c_1(x) = x^{k_1} + x^{k_2}$, where $\alpha^{k_i} = \sigma_0 + z_i\alpha^h$ $(i = 1, 2)$ and z_1, z_2 are the solutions of

$$z^2 + z + 1 + \frac{\sigma + \sigma_0^3}{\alpha^{3h}} = 0.$$

We notice that $\alpha^{k_1} + \alpha^{k_2} = (z_1 + z_2)\alpha^h = \alpha^h$ since $z_1 + z_2 = 1$. Recalling that $\sigma_0 = \sigma_1$, the syndrome Σ^* yields (see (29))

$$\sigma_0^* = \sigma_0 + \alpha^h = \sigma_1 + \alpha^{k_1} + \alpha^{k_2} = \sigma_1^*,$$

$$\sigma^* = \sigma + \alpha^{3h} + \alpha^{3k_1} + \alpha^{3k_2} = \sigma + \alpha^{k_1}\alpha^{k_2}(\alpha^{k_1} + \alpha^{k_2})$$

$$= \sigma + \alpha^h(\sigma_0 + z_1\alpha^h)(\sigma_0 + z_2\alpha^h) = (\sigma_0 + \alpha^h)^3 = \sigma_0^{*3},$$

$$d^* = c + c_1(1) = 0$$

i.e., $\mathbf{r} + \mathbf{c} \in \mathcal{K}_n$. This discussion proves that there are several[4] code words at distance 3 from **r** (but none at distance <3) and yields the following error detection rule:

Rule 4. If $\sqrt[3]{\rho} \neq \sigma_0$, $\sqrt[3]{\rho} \neq \sigma_1$, $d = 0$, $\sigma_0 + \sigma_1 = 0$, then the received **r** is at distance $\geqq 3$ from any code word.

An "extra bonus" of the same discussion is that given any **r** there are code words at distance $\leqq 3$ from **r**: this property is analogous to the one found by Gorenstein et al. (1960) for BCH double-error-correcting codes.

In Figure 2 we sketch a possible organization of a decoder for a $\mathcal{K}_n$ code. The serially received message is stored in three recirculating registers SR1, SR2, SR3, corresponding to the homologous registers of Figure 1. The received message is also fed to the SYNDROME COMPUTER, which, once reception is completed, stores the functions d, σ, σ_0, σ_1, i.e., the syndrome Σ. These functions constitute the inputs of combinational networks, which we now describe (in the illustration heavy

[4] Another weight 3 correction vector is obtained through Rule 2, i.e., $\mathbf{c} = [\alpha^k, 1, \alpha^k]$ where $\alpha^k = \sigma_0 + \sqrt[3]{\sigma} = \sigma_1 + \sqrt[3]{\sigma}$.

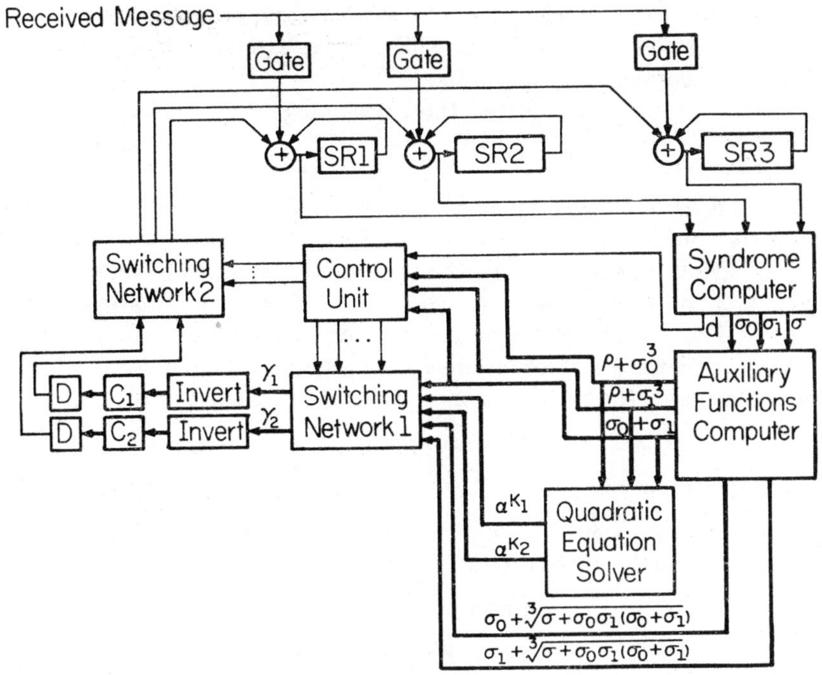

FIG. 2. Decoder for the $\mathcal{K}_n$ code.

lines denote bundles of $(n-1)$ binary lines). The AUXILIARY FUNCTIONS COMPUTER produces $\rho + \sigma_0^3$, $\rho + \sigma_1^3$, $\sigma_0 + \sigma_1$, $\sigma_0 + \sqrt[3]{\sigma + \sigma_0\sigma_1(\sigma_0 + \sigma_1)}$ and $\sigma_1 + \sqrt[3]{\sigma + \sigma_0\sigma_1(\sigma_0 + \sigma_1)}$. Of these, $\rho + \sigma_0^3$, $\rho + \sigma_1^3$, $\sigma_0 + \sigma_1$, together with d, are fed to the CONTROL UNIT, which determines which decoding rule must be applied. In parallel, $\rho + \sigma_0^3$, $\rho + \sigma_1^3$, $\sigma_0 + \sigma_1$ are fed to the QUADRATIC EQUATION SOLVER, where $\tau_0 = (\rho + \sigma_0^3)/(\sigma_0 + \sigma_1)^3$ and $\tau_1 = (\rho + \sigma_1^3)/(\sigma_0 + \sigma_1)^3$ are computed and tested for membership in Θ. If $\tau_j \in \Theta$, $T\tau_j$ and $(1 + T\tau_j)$ are the solutions of the equation $y^2 + y + \tau_j = 0$, where T is an appropriate square matrix (see, e.g., Berlekamp et al., (1967)); $T\tau_j$ and $(1 + T\tau_j)$ are then multiplied in GF(2^{n-1}) by $(\sigma_0 + \sigma_1)$ in order to obtain the solutions α^{k_1} and α^{k_2} of $z^2 + (\sigma_0 + \sigma_1)z + (\rho + \sigma_j^3)/(\sigma_0 + \sigma_1) = 0$ (see Rule 3). Since $W[c_0(x)] + W[c_1(x)] \leqq 2$, at most two correction bits must be produced. This is accomplished as follows: 1) $\sigma_0 + \sigma_1$, $\sigma_0 + \sqrt[3]{\sigma + \sigma_0\sigma_1(\sigma_0 + \sigma_1)}$, $\sigma_1 + \sqrt[3]{\sigma + \sigma_0\sigma_1(\sigma_0 + \sigma_1)}$, α^{k_1}, α^{k_2} are fed to the SWITCHING NETWORK 1: here signals from the control unit govern the selection of two correction functions γ_1, γ_2 in the form of the vector representations of two elements of $GF(2^{n-1})$; 2) the combinational circuit INVERT computes the inverse of $\gamma_j = \alpha^{h_j}$ (if $\gamma_j = 0$, the output of INVERT is conventionally 0) and α^{-h_j} is loaded in the Galois field counter C_j. It must be noticed that, assuming no delay in the combinational elements, loading of C_j with $\alpha^{-h_j}(j = 1, 2)$ occurs simultaneously with the production of d, σ, σ_0, σ_1. At this point the contents of SR2 and SR3 are recirculated synchronously with the stepping of C_1 and C_2: once the condition $10 \cdots 0$ is detected in C_j a time unit duration signal is generated by D and routed through the SWITCHING NETWORK 2 to perform the required correction of the contents of the registers. The decoding operation therefore terminates $(2^{n-1} - 1)$ time units after the serial reception of the message is completed.

This completes the presentation of the decoding procedure.

6. ACKNOWLEDGMENT

The author gratefully acknowledges the very valuable suggestions of E. R. Berlekamp and J..P. Robinson.

RECEIVED: June 28, 1968; revised September 11, 1968

REFERENCES

ALBERT, A. A. (1956), "Fundamental Concepts of Higher Algebra." University of Chicago Press, Chicago.

VASIL'EV, YU. L. (1962), O negruppovykh plotno upakovannykh kodakh (On nongroup close packed codes). *Problemy Kibernetiki*. **8,** 337–339.

BERLEKAMP, E. R., RUMSEY, H. AND SOLOMON, G. (1967). On the solution of algebraic equations over finite fields. *Inform. Control*. **10,** 553–564.

BERLEKAMP, E. R. (1968), "Algebraic Coding Theory." McGraw-Hill, New York.

GORENSTEIN, D., PETERSON, W. W. AND ZIERLER, N. (1960). Two-error correcting Bose–Chaudhuri codes are quasi-perfect. *Inform. Control*. **3,** 291–294.

GREEN, M. V. (1966). Two heuristic techniques for block-code construction. *IEEE Trans. on Inform. Theory*. **IT-12,** 273.

JOHNSON, S. M. (1962). A new upper-bound for error-correcting codes. *IRE Trans. on Inform. Theory*. **IT-8,** 203–207.

NADLER, M. (1962). A 32-point $n = 12$, $d = 5$ Code. *IRE Trans. on Inform. Theory*. **IT-8,** 58.

NORDSTROM, A. W. AND ROBINSON, J. P. (1967). An optimum nonlinear code. *Inform. Control*. **11,** 613–616.

PETERSON, W. W. (1961). "Error Correcting Codes." M.I.T. Press and John Wiley & Sons, New York.

PREPARATA, F. P. (1968a). An alternate description and a new decoding procedure of Nordstrom–Robinson optimum code. Proc. 2nd Princeton Conf. on Information Sciences and Systems, 131–134.

PREPARATA, F. P. (1968b). Weight and distance structure of Nordstrom–Robinson Quadratic Code. *Inform. Control*. **12,** 466–473. (see also Erratum, **13,** no. 7).

Shannon Codes

IX

Editor's Comments on Papers 34 and 35

34 Elias: *Error Free Coding*

35 Justesen: *A Class of Constructive Asymptotically Good Algebraic Codes*

Since coding theory was inspired by the work of Shannon, it is appropriate to conclude this collection with those papers containing the most successful attempts at constructing codes which meet asymptotically the performance specified in Shannon's theorems. I call any class of codes that was constructed specifically with this aim in mind a Shannon code. Although there is laxity in this definition, the intent is clear. There are precisely two such classes of codes and these are described in the following papers.

In the first paper Elias [34] introduces the idea of a product code in which he constructs a third code from two given codes, the important properties of the third code being derivable from the properties of the two given codes. Working with Hamming codes he uses this technique to construct a class of codes for the binary symmetric channel which, for a strictly positive rate, attain a vanishingly small probability of error as the block length tends to infinity. The rate, unfortunately, is well below the channel capacity and d/n, the ratio of code distance to block length, tends to zero.

From the Gilbert bound it is known that there exist codes such that d/n is bounded away from zero for a fixed ratio k/n, the code rate. This is a stronger condition than having the probability of error tend to zero for a fixed rate. Until recently there was no known coding system for which this property held, and for many years coding theorists assumed that any such code would have a complex structure and turned their attention to proving this. As an example, it is possible to show that, for any BCH code with rate k/n, the ratio d/n must tend to zero as $n \to \infty$ [Berlekamp (1968)] and, as far as is known, the same is true for any single class of codes except those of Elias discussed above.

It is remarkable that in this atmosphere Justesen [35] constructed a class of codes, using the concatenation scheme of Forney (1966), such that for any fixed rate $0 < R < 1$, there exists a sequency of binary (n,k) codes with $k/n > R$ and the ratio d/n bounded away from zero. Actually a lower bound on d/n was obtained which, for a given rate R, tended to a value considerably less than that indicated possible by the Gilbert bound. Thus, after twenty-five years of coding effort, the vision of Shannon finally seems attainable. Hopefully, the work of Justesen will inspire the search for practical codes that actually achieve this promise.

34

ERROR-FREE CODING*

Peter Elias

Department of Electrical Engineering and Research Laboratory of Electronics
Massachusetts Institute of Technology
Cambridge, Massachusetts

Introduction

This paper describes constructive procedures for encoding messages to be sent over noisy channels so that they may be decoded with an arbitrarily low error rate. The procedures are a kind of iteration of simple error-correcting codes such as those of Hamming[1] and Golay[2]; any additional systematic codes which may be discovered, such as those discussed by Reed[3] and Muller[4], may be iterated in the same way.

The procedures are not ideal; that is, the capacity of a noisy channel for the transmission of error-free information using such coding is smaller than information theory says it should be. However, the procedures do permit the transmission of error-free information at a positive rate. They also have these two properties.

(1) The codes are "systematic" in Hamming's sense: they are what Golay calls "digit codes" rather than "message codes." That is, the transmitted symbols are divided into so-called "information digits" and "check digits." The customer who has a message to send supplies the information digits which are transmitted unchanged. Periodically the coder at the transmitter computes some check digits, which are functions of past information digits, and transmits them. The customer with a short message does not have to wait for a long block of symbols to accumulate before coding can proceed, as in the case of codebook coding, nor does the coder need a codebook memory containing all possible symbol sequences. The coder needs only a memory of the past information digits it has transmitted and a quite simple computer.

(2) The error probability of the received messages is as low as the receiver cares to make it. If the coding process has been properly selected for a given noisy channel, the customer at the receiver can set the probability of error per decoded symbol (or the probability of error for the entire sequence of decoded symbols transmitted up to the present, or the equivocation of part or all of the decoded symbol sequence) at as low a value as he chooses. It will cost him more delay to get a more reliable message, but it will not be necessary to alter the coding and decoding procedure when he raises his standards, nor will it be necessary for less particular and more impatient customers using the same channel to put up with the additional delay. This is again unlike codebook processes, in which the codebook must be rewritten for all customers if any one of them raises his standards.

Perhaps the simplest way to indicate the basic behavior of such codes is to describe how one would work in a commercial telegraph system. A customer entering the telegraph office presents a sequence of symbols which are sent out immediately over a noisy channel to another office, which immediately reproduces the sequence, adds a note "the probability of error per symbol is 10^{-1}, but wait till tomorrow," and sends it off to the recipient. Next day the recipient receives a note saying "For 'sex' read 'six'. The probability of error per symbol is now 10^{-2}, but wait till next week." A week later the recipient gets another note: "For 'lather' read 'gather'. The probability of error per symbol is now 10^{-4}, but wait till next April." This flow of notes continues, the error probability dropping rapidly from note to note, until the recipient gets tired of the whole business and tells the telegraph company to stop bothering him.

Since these coding procedures are derived by an iteration of simple error-correcting and detecting codes, their performance depends on what kind of code is iterated. For a binary channel with a small and symmetric error probability, the best choice among the available procedures is the Hamming-Golay single-error-correction double-error-detection code developed by Hamming[1] for the binary case and extended by Golay[2] to the case of symbols selected from an alphabet of M different symbols, where M is any prime number. The analysis of the binary case will be presented in some detail and will be followed by some notes on diverse modifications and generalizations.

---

*This work was supported in part by the Signal Corps; the Office of Scientific Research, Air Research and Development Command; and the Office of Naval Research.

Iterated Hamming Codes

First-Order Check

Consider a noisy binary channel, which transmits each second either a zero or a one, with a probability $(1 - p_o)$ that the symbol will be received as transmitted, and a probability p_o that it will be received in error. Error probabilities for successive symbols are assumed to be statistically independent.

Let the receiver divide the received symbol sequence into consecutive blocks, each block consisting of N_1 consecutive symbols. Because of the assumed independence of successive transmission errors, the error distribution in the blocks will be binomial: there will be a probability

$$P(o) = (1 - p_o)^{N_1}$$

that no errors have occurred in a block, and a probability P(i)

$$P(i) = \frac{N_1 !}{i ! (N_1 - i) !} p_o^i (1 - p_o)^{N_1 - i} \tag{1}$$

that exactly i errors have occurred.

If the expected number of errors per received block, $N_1 p_o$, is small, then the use of a Hamming error-correction code will produce an average number of errors per block, $N_1 p_1$, after error correction, which is smaller still. Thus p_1, the average probability of error per position after error correction, will be less than p_o. An exact computation of the extent of this reduction is complicated, but some inequalities are easily obtained.

The single-error-correction check digits of the Hamming code give the location of any single error within the block of N_1 digits, permitting it to be corrected. If more errors have occurred, they give a location which is usually not that of an incorrect digit, so that altering the digit in that location will usually cause one new error, and cannot cause more than one. The double-error-detection check digit tells the receiver whether an even or an odd number of errors has occurred. If an even number has occurred and an error location is indicated, the receiver does not make the indicated correction, and thus avoids what is very probably the addition of a new error.

The single-correction double-detection code, therefore, will leave error-free blocks alone, will correct single errors, will not alter the number of errors when it is even, and may increase the number by at most one when it is odd and greater than one. This gives for the expected number of errors per block after checking

$$N_1 p_1 \leq \sum_{\text{even } i \geq 2}^{\leq N_1} i P(i) + \sum_{\text{odd } i \geq 3}^{\leq N_1} (i+1) P(i)$$

$$\leq P(2) + \sum_{i=3}^{N_1} (i+1) P(i)$$

$$\leq \sum_{i=0}^{N} (i+1) P(i) - P(0) - 2P(1) - P(2)$$

$$\leq 1 + N_1 p_o - P(0) - 2P(1) - P(2). \tag{2}$$

Substituting the binomial error probabilities from (1), expanding and collecting terms, gives, for $N_1 p_o \leq 3$,

$$N_1 p_1 \leq N_1(N_1 - 1) p_o^2,$$

$$p_1 \leq (N_1 - 1) p_o^2 < N_1 p_o^2. \tag{3}$$

The error probability per position can therefore be reduced by making N_1 sufficiently small. The shortest code of this type requires $N_1 = 4$, and the inequality (3) suggests that a reduction will therefore not be possible if $p_o \geq 1/3$. The fault is in the equation, however, and not the code: for $N_1 = 4$ it is a simple majority-rule code which will always produce an improvement for any $p_o < 1/2$.

A Hamming single-correction double-detection code uses C of the N positions in a block for checking purposes and the remaining N - C positions for the customer's symbols, where

$$C = \left[\log_2 (N-1) + 2\right]. \tag{4}$$

(Here and later, square brackets around a number denote the largest integer which is less than or equal to the number enclosed. Logarithms will be taken to the base 2 unless otherwise specified.)

Higher Order Checks

After completing the first-order check, the receiver discards the C_1 check digits, leaving only the $N_1 - C_1$ checked information digits, with the reduced error probability p_1 per position. (It can be shown that the error probability after checking is the same for all N_1 positions in the block, so that discarding the check digits does not alter the error probability per position for the information digits.) Now some of these checked digits are made use of for further checking, again with a Hamming code. The receiver divides the checked digits into blocks of N_2; the C_2 checked check digits in each block enable it, again, to correct any single error in the block, although multiple errors may be increased by one in number. In order for the checking to reduce the expected number of errors per second-order block, however, it is necessary to select the locations of the N_2 symbols in the block with some care.

The simplest choice would be to take several consecutive first-order blocks of $N_1 - C_1$ adjacent checked information digits as a second-order block, but this is guaranteed not to work. For if there are any errors at all left in this group of digits after the first-order checking, there are certainly two or more, and the second-order check cannot correct them. In order for the error probability per place after the second-order check to satisfy the analog of (3), namely,

$$p_j \leq (N_j - 1)\, p_{j-1}^2 < N_j\, p_{j-1}^2, \tag{5}$$

it is necessary for the N_2 positions included in the second-order check to have statistically independent errors after the first check has been completed. This will be true if, and only if, each position was in a <u>different</u> block of N_1 adjacent symbols for the first-order check.

The simplest way to guarantee this independence is to put each group of $N_1 \times N_2$ successive symbols in a rectangular array, checking each row of N_1 symbols by means of C_1 check digits, and then checking each column of already checked symbols by means of C_2 check digits. The procedure is illustrated in Fig. 1. The transmitter sends the $N_1 - C_1$ information digits in the first row, computes the C_1 check digits and sends them, and proceeds to the next row. This process continues down through row $N_2 - C_2$. Then the transmitter computes the C_2 check digits for each column and writes them down in the last C_2 rows. It transmits one row at a time, using the first $N_1 - C_1$ of the positions in that row for the second-order check, and the last C_1 digits in the row for a first-order check of the second-order check digits.

After the second-order check, then, the inequality (5) applies as before, and we have for p_2, the probability of error per position,

$$p_2 < N_2\, p_1^2 < N_2\, N_1^2\, p_o^4. \tag{6}$$

The N_3 digits to be checked by the third-order check may be taken from corresponding positions in each of N_3 different $N_1 \times N_2$ rectangles, the N_4 digits in a fourth-order block from corresponding positions in N_4 such collections of $N_1 \times N_2 \times N_3$ symbols each, and so on ad infinitum. At the kth stage this gives

$$p_k < N_k^{2^0} \cdot N_{k-1}^{2^1} \cdots N_{k-j}^{2^j} \cdots N_1^{2^{k-1}} \cdot p_o^{2^k}. \tag{7}$$

It is now necessary to show that not all of the channel is occupied, in the limit, with checking digits of one order or another so that some information can also get through. The fraction of symbols used for information at the first stage is $\left[1 - (C_1/N_1)\right]$. At the kth stage, it is

$$F_k = \prod_1^k \left(1 - \frac{C_j}{N_j}\right). \tag{8}$$

It is now necessary to find a sequence of N_j for which p_k approaches zero and F_k does not. as k increases without bound. A convenient sequence is

$$N_1 = 2^n$$

$$N_j = 2^{j-1} N_1 = 2^{j+n-1}. \tag{9}$$

This gives for p_k, from (7),

$$p_k < \left(N_1 \cdot 2^{k-1}\right)^{2^0} \dots \left(N_1 \cdot 2^{k-j}\right)^{2^{j-1}} \dots \left(N_1 \cdot 2^0\right)^{2^{k-1}} p_o^{2^k}$$

$$< \frac{1}{N_1} (2N_1 p_o)^{2^k} \cdot 2^{-(k+1)}. \tag{10}$$

The right side of (10) approaches zero as k increases, for any $N_1 p_o \leqslant 1/2$. Thus the error probability can be made to vanish in the limit. Note that the inequality gives a much weaker kind of approach to zero for the threshold value $N_1 p_o = 1/2$ than for any smaller value of errors per first-order block.

For the same sequence of N_j, a lower bound on F_∞ can be computed. From equations (8) and (4) we have

$$F_\infty = \prod_1^\infty \left(1 - \frac{C_j}{N_j}\right) = \prod_1^\infty \left(1 - \frac{\log_2 N_j + 1}{N_j}\right)$$

$$= \prod_1^\infty \left(1 - \frac{j+n}{2^{j+n-1}}\right). \tag{11}$$

Let

$$\sigma_j = \frac{C_j}{N_j}$$

$$\sigma = \sum_1^\infty \sigma_j. \tag{12}$$

Then σ_j is monotonic decreasing in j and is less than 1 for all constructable Hamming codes, that is, for $N_1 = 2^n \geqslant 4$. This makes it possible to write the following inequalities:

$$e^{-\sigma} > F_\infty > (1 - \sigma_1)^{\sigma/\sigma_1} > 1 - \sigma. \tag{13}$$

Here the last term on the right is one of the Weierstrasse inequalities for an infinite product; the other terms are useful when $\sigma > 1$, and show that for $\sigma_1 < 1$ and $\sigma < \infty$, F_∞ is strictly positive.

Evaluating σ in the present case gives

$$\sigma = \sum_{j=1}^\infty \frac{j+n}{2^{j+n-1}} = \frac{n+2}{2^{n-1}} = \frac{2 \log 4N_1}{N_1}. \tag{14}$$

At threshold, that is, at $N_1 p_o = 1/2$, this gives

$$\sigma = 4 p_o \log \frac{2}{p_o} < 4 \left\{ p_o \log \frac{1}{p_o} + (1 - p_o) \log \frac{1}{1 - p_o} \right\} = 4 E \tag{15}$$

where E is the equivocation of the noisy channel. Thus for p_o small, from (13) we have

$$F_\infty > 1 - 4E. \tag{16}$$

That is, under the specified conditions ($N_1 p_o = 1/2$, $N_1 = 2^n \geqslant 4$) the number of check digits required is never more than four times the number that would be required for an ideal code, provided that an ideal code of the check-digit type exists, which is not obvious. When E is > 1/4, the interior inequality shows that F_∞ is still positive.

Equivocation

Feinstein[3] has shown that it is possible to find ideal codes for which not only the probability of error, but the total equivocation, vanishes in the limit as longer and longer symbol sequences are used. This property is also true for the coding processes described here. This is a very important result in the case of codebook codes, where the message becomes infinite in the limit. For the codes under discussion here, it is a less important property, since any finite message can be received without an infinite lag, and its equivocation vanishes with the error probability per position.

The total number of binary digits checked by the kth checking stage is

$$M_k = \prod_1^k N_j . \tag{17}$$

Of these $F_k M_k$ are information digits and the remainder are checks. Using the values (9) for the N_j, we have

$$M_k = N_1^k \, 2^{\frac{k(k-1)}{2}} . \tag{18}$$

The bound (10) limits the probability of error per position. Multiplying this by M_k gives a bound on the mean number of errors per M_k digits, which is also a bound on the fraction of sequences of M_k digits which are in error after checking – a gross bound, since actually any such sequence which is in error must have many errors, and not just one. Thus for Q_k, the probability that a checked group of M_k digits is in error, we have

$$Q_k < p_k M_k \leqslant \frac{1}{4}\left(\frac{N_1}{2}\right)^{k-1} (2N_1 p_o)^{2^k} \, 2^{\frac{k(k-1)}{2}} . \tag{19}$$

At threshold ($N_1 p_o = 1/2$) this inequality does not guarantee convergence, but for $N_1 p_o < 1/2$, Q_k certainly approaches zero as k increases.

The equivocation E_k per sequence of M_k terms is bounded by the value it would have if any error in a block made all possible symbol sequences equally likely at the receiver, that is,

$$E_k < Q_k \log \frac{1}{Q_k} + (1 - Q_k) \log \frac{1}{1 - Q_k} + Q_k M_k . \tag{20}$$

Again at threshold convergence is not guaranteed, but for $N_1 p_o < 1/2$, E_k, the absolute equivocation of the block, will also vanish as k increases.

Distance Properties

At the kth stage of this coding process, a sequence of M_k binary digits has been selected as a message. Because the check digit values are determined by the information digit values, there are only $2^{F_k M_k}$ possible message sequences, rather than 2^{M_k}. Any two of these possible messages will have a "distance" from one another, defined as the number of positions in which they have different binary symbols, and the smallest such distance will be 4^k for the iterated single-correction, double-detection code. This means that by using this set of codes with a codebook, any set of errors less than one-half of the minimum distance in number can be corrected by choosing as the transmitted message the message point nearest to the received sequence.

It is easy to see that for the coding procedure just described this error-correction capability will not be realized. Any set of 2^k errors which are at the corners of a k-dimensional cube in the k-dimensional rectangle of symbol positions will not be corrected by this process, since each check will merely indicate a double error which it cannot correct. By inspecting any two of the sets of check digits at once, these errors could be located, but they will not have been corrected by the process as described above. The effective minimum of the maximum number of errors which will be corrected is therefore $2^k - 1$, rather than $2^{2k-1} - 1$.

This shows a loss of error-correction capability because of the strictly sequential use of the checking information. Without going to the extreme memory requirement of codebook techniques, a portion of this loss may be recouped by not throwing the low-order check digits away but using them to recheck after higher order checking has been done. This does not increase the maximum number of errors for which correction is always guaranteed, but it does reduce the average error probability at each stage; the exact amount of this reduction is, unfortunately, difficult to compute. This behavior, however, points up a significant feature of the coding process. If the maximum number of errors for which correction is always guaranteed were the maximum number of errors for which correction was ever guaranteed, the procedure could not transmit information at a nonzero rate; that is, the minimum distance properties of the code are inadequate for the job. It is average error-correction capability that makes transmission at a nonzero rate possible.

The Poisson Limit

Much of the above analysis has assumed that $N_j = 2^{j-1} N_1$, and part of it has further assumed that $N_1 = 2^n$. However, any series of N_j which increases rapidly enough so that σ is finite will lead to a coding process that is error-free for sufficiently small values of $N_1 p_o$. In particular, any other approximately geometric series may be used, for which

$$N_j \approx b^{j-1} N_1, \quad b > 1. \tag{21}$$

The approximation is necessary if b is not an integer. The expression for p_k analogous to (10) is then

$$p_k < \frac{1}{N_1} (bN_1 p_o)^{2^k} b^{-(k+1)}, \tag{22}$$

with a threshold at $N_1 p_o = 1/b$. The value of σ can also be bounded for this series. At threshold, the bound corresponding to (15) is

$$\sigma \leqslant \frac{b^2 p_o}{b-1} \log\left(\frac{4}{p_o\, b^{\frac{b-2}{b-1}}}\right). \tag{23}$$

Again, for $N_1 p_o$ below threshold, Q_k and E_k approach zero as k increases.

For very small p_o, the value of b that minimizes σ is b = 2. This leads to the maximum value of F_∞ given by (16). However, for very small p_o, N_1 may be made very large. The distribution of errors in the blocks then approaches the Poisson distribution, for which the probability that just i errors have occurred in a block is

$$P(i) = e^{-N_1 p_o} \cdot \frac{(N_i p_o)^i}{i!}. \tag{24}$$

This equation may be used to derive an iterative inequality on the mean number of errors per block after single-detection, double-correction coding.

$$N_j p_j \leqslant 1 + N_j p_{j-1} - e^{-N_j p_{j-1}} \left\{1 + 2N_j p_{j-1} + \frac{(N_j p_{j-1})^2}{2!} + \frac{(N_j p_{j-1})^4}{4!} + \ldots\right\}$$

$$\leqslant 1 - N_j p_{j-1}\left(2e^{-N_j p_{j-1}} - 1\right) - \frac{1}{2}\left(1 - e^{-2N_j p_{j-1}}\right). \tag{25}$$

Keeping $N_j p_j$ constant gives the geometric series (21) for N_j. A joint selection of $N_1 p_o$ and b for the minimization of the bound on σ gives $N_1 p_o \approx 0.75$, $b \approx 1.75$, and an effective channel capacity

$$F_\infty \sim 1 - 3.11\,E, \tag{26}$$

where E is the equivocation of the binary channel. This is an improvement over (16).

Iteration of Other Codes

The analysis in the preceding sections has dealt only with iteration of the Hamming single-error-correction, double-error-detection code. Other kinds of codes may also be iterated; nor is it necessary to use the same type of code at each stage in the iterative process. The only requirement is that each code be of the check-digit, or systematic, type, so that its check digits may be computed on the basis of the preceding information digits and added on to the message.

First, the final parity check digit of a Hamming code may be omitted, destroying the double-detection feature of the code. This leads to the inequality

$$p_j \leqslant \frac{3}{2}(N_j - 1)\, p_{j-1}^2 < \frac{3}{2} N_j\, p_{j-1}^2, \tag{27}$$

in place of (5). Iterating this code alone gives a bound on σ that is only slightly smaller than (15), but the threshold becomes $N_1 p_o = 1/3$ rather than $N_1 p_o = 1/2$, and the effective channel capacity for small p_o is bounded by

$$F_\infty > 1 - 6\,E, \tag{28}$$

where E is the equivocation of the binary channel.

Second, the Golay[2] analogs to both kinds of Hamming code may be constructed, for M-ary channels, where M is a prime number. If there is a probability $(1 - p_o)$ that any symbol will be received correctly, and if the consecutive errors are statistically independent, the results of the binary case carry over quite directly. The inequalities (5) and (27) still hold for the two kinds of codes, since the errors as a whole are still binomially distributed in blocks. At threshold, inequalities (16) and (28) still hold for the effective channel capacity, where E is now the equivocation of a symmetrical M-ary channel; that is, of a channel in which the probability of an error taking any given symbol into any other different symbol is $p_o/(M-1)$. The result (26) for the Poisson limit also applies, with the same interpretation of E.

Third, the Reed[4]-Muller[5] codes may be treated as check digit codes, and may be iterated to give an error-proof system. For these codes, the average error-reduction capability is not known; only the minimum distance is known. Certain of the codes, such as the triple-correction quadrupole-detection code for blocks of 32 binary symbols, might provide a good starting point for an iteration which proceeds by iteration of Hamming codes. The Golay triple-correction quadruple-detection code for blocks of 24 symbols might be used in the same way. It will take considerable computation to evaluate such mixed iteration schemes.

It is not, at present, profitable to use the Reed-Muller codes for later stages in the iteration. The reason is that an efficient triple-correction quadruple-detection code should require about $C = 2 \log N$ check digits for a block of length N. The Reed-Muller codes require about $C = 1 + \log N + 1/2 \log N (\log N - 1)$ check digits for this purpose. For large N, therefore, the effective channel capacity is reduced by the large number of check digits required. There is a similar inefficiency in the Reed-Muller codes with greater error-correction capabilities, which might be removed if the average error-correction capabilities of these codes were known.

Nonrectangular Iteration

The problem of assuring statistical independence among the N_k digits checked by a kth order check, so that the inequality (5) derived on the basis of statistical independence can be used as an iterative inequality, was solved above by what might be called rectangular iteration. Each of the N_k digit positions in a check group are selected from a different sequence of M_{k-1} consecutive symbols. Thus until the kth order checking has been carried out, no two of them have been associated by lower order checking procedures in any way. This iteration solves the problem, but it makes M_k a function that grows very rapidly with k. When N_j is the geometric series (21), then

$$M_k \approx N_1^k\, b^{\frac{1}{2}k(k-1)} \tag{29}$$

This means that p_k, Q_k, and E_k decrease quite rapidly as functions of k, but much more slowly as functions of the length of the message M_k, or its information content $F_k M_k$.

Roughly speaking, if $H_k = F_k M_k$ is the total number of information digits transmitted at the kth stage,

$$p_k \sim A e^{-2^{a(\log H_k)^{1/2}}}, \quad A > 0, \quad a > 0. \tag{30}$$

This is a much slower decrease of error probability than Feinstein's result[3] which is

$$p_k \sim B e^{-b \cdot 2^{\log H_k}} = B e^{-bH_k}, \quad B > 0, \ b > 0. \tag{31}$$

A less stringent requirement on the choice of digits checked in a single group is that no two of them have been together in any lower order check group. This requires that there be at least N_k different groups of order k - 1 from which to select digit positions. Thus

$$M_k \geqslant N_k N_{k-1}. \tag{32}$$

If it is possible to approximate equality in (32), and if the statistical dependence so introduced does not seriously weaken inequality (5), then it might be possible to get the result

$$p_k \sim D e^{-2^{d \log H_k}}, \quad D > 0, \ d > 0, \tag{33}$$

which is closer to Feinstein's result.

Conclusion

From a practical point of view, this coding procedure has much to recommend it. A question of both theoretical and practical interest is the extent to which the convenience associated with a computable and error-free code is compatible with ideal coding, or the smallest price that must be paid for the convenience if the two are incompatible. No answer to this question is in sight at present. However, the existence of the error-free process, despite its lack of ideality, puts the burden of efficient coding on the first stage of the coding process. For if a coding process succeeds in reducing the equivocation in a received message to some small but positive value E, the remaining errors may always be eliminated at a cost of 4 E (or 3.11 E) in channel capacity: an error-proof termination is available, at a price, to take care of the residual errors left by any other error-correcting scheme.

Acknowledgment

The iterative approach used in this paper was suggested by a comment of Dr. Victor H. Yngve, of the Research Laboratory of Electronics, M.I.T., on the fact that redundancy in language was added at many different levels, a point that he discusses in reference 6.

References

(1) R. W. Hamming, Error Detecting and Error Correcting Codes, Bell System Tech. J. 29, pp. 147-160 (1950).

(2) M. J. E. Golay, Notes on Digital Coding, Proc. I.R.E. 37, p. 657 (1949).

(3) A. Feinstein, Some New Basic Results in Information Theory, these transactions.

(4) I. S. Reed, A Class of Multiple-Error-Correcting Codes and the Decoding Scheme, Technical Report No. 44, Lincoln Laboratory, M.I.T., (1953). See also the paper under this title in these transactions.

(5) D. E. Muller, "Metric Properties of Boolean Algebra and their Applications to Switching Circuits," Report No. 46, Digital Computer Laboratory, University of Illinois (1953).

(6) V. H. Yngve, "Language as an Error-Correcting Code," pp. 73-74, Quarterly Progress Report, Research Laboratory of Electronics, M.I.T., April 15, 1954.

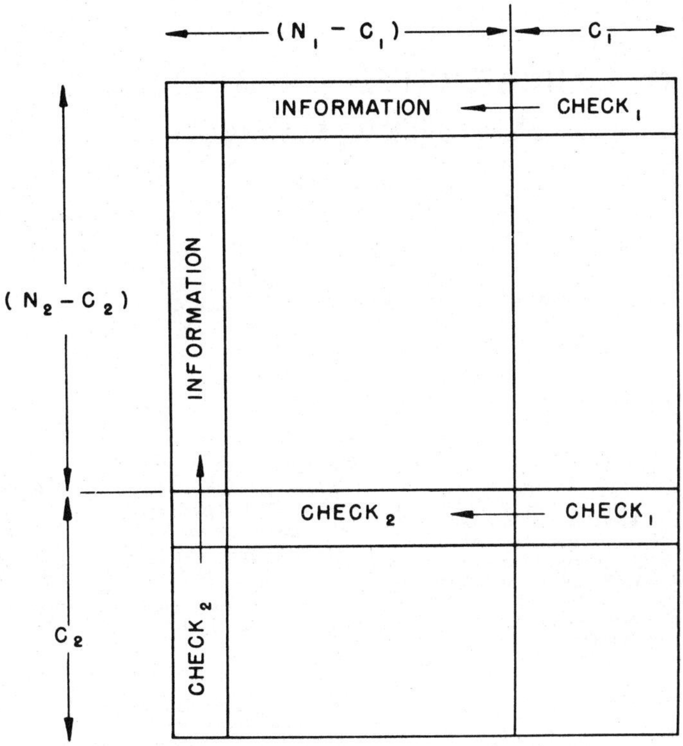

Fig. 1 - Organization of First- and Second-Order Check Digits.

35

A Class of Constructive Asymptotically Good Algebraic Codes

JØRN JUSTESEN

***Abstract*—For any rate R, $0 < R < 1$, a sequence of specific (n,k) binary codes with rate $R_n > R$ and minimum distance d is constructed such that**

$$\liminf_{n\to\infty} \frac{d}{n} \geq (1 - r^{-1}R)H^{-1}(1 - r) > 0$$

(and hence the codes are asymptotically good), where r is the maximum of $\frac{1}{2}$ and the solution of

$$R = \frac{r^2}{1 + \log_2 [1 - H^{-1}(1 - r)]}.$$

The codes are extensions of the Reed–Solomon codes over $GF(2^m)$ with a simple algebraic description of the added digits. Alternatively, the codes are the concatenation of a Reed–Solomon outer code of length $N = 2^m - 1$ with N distinct inner codes, namely all the codes in Wozencraft's ensemble of randomly shifted codes. A decoding procedure is given that corrects all errors guaranteed correctable by the asymptotic lower bound on d. This procedure can be carried out by a simple decoder which performs approximately $n^2 \log n$ computations.

I. Introduction

The Gilbert–Varsharmov [1], [2] bound ensures the existence of binary (n,k) codes of arbitrary dimension k and arbitrary block length $n(0 < k < n)$ such that their minimum distance d_n satisfies

$$H(d_n/n) \geq 1 - R_n,$$

where $R_n = k/n$ is the rate and $H(x) = -x \log_2 x - (1 - x) \log_2 (1 - x)$ is the binary entropy function. Thus, at least in principle, it is possible for any rate R, $0 < R < 1$, to select a sequence of linear codes with increasing block length n, minimum distance d_n, and rate $R_n \geq R$ such that

$$\liminf_{n\to\infty} (d_n/n) \geq H^{-1}(1 - R), \qquad (1)$$

where we require $0 < H^{-1}(1 - R) < \frac{1}{2}$ so that the inverse function is uniquely defined.

Hitherto, however, no constructive sequence of binary (n,k) codes with $R_n \geq R$ and

$$\liminf_{n\to\infty} (d_n/n) > 0 \qquad (2)$$

had been given for any R, $0 < R < 1$. We shall call a countably infinite sequence of (n,k) codes constructive if it can be specified in terms of entities generally accepted as known in a manner that requires no searching. Thus, in this sense of the word, the cyclic binary Hamming codes are a constructive sequence of codes. We shall call a sequence of codes that satisfy (1) *asymptotically good.*

In Section II of this paper we specify for any rate R, $0 < R < 1$, a constructive countably infinite sequence of (n,k) codes with rates $R_n \geq R$ such that (2) is satisfied. The construction is based on Forney's concept [3] of concatenated codes in which the m information digits of an inner binary code are treated as single digits of an outer Reed–Solomon [4] code over $GF(2^m)$, but we have generalized the concept to allow variation of the inner code. The inner codes are given by a simple algebraic description and shown to be equivalent to the $2^m - 1$ distinct codes in the ensemble of randomly shifted codes described by Massey [5] and attributed to Wozencraft. Alternatively, the codewords of the binary codes we shall construct can be considered as a pair of codewords in different Reed–Solomon codes over $GF(2^m)$.

In Section III we present a decoding method for the codes constructed in Section II that is closely related to Forney's [3] generalized minimum distance decoding as applied by Reddy and Robinson [6] and Weldon [7] to the decoding of iterated codes. It is proved that all error patterns with weight guaranteed correctable by the lower bound on minimum distance in Section II will be corrected by this procedure, and that the number of bit operations required for decoding is proportional to $n^2 \log n$.

Manuscript received February 7, 1972; revised May 1, 1972. This paper was presented at the IEEE International Symposium on Information Theory, Asilomar, Calif., February 2, 1972.
The author is with the Laboratory for Communication Theory, Technical University of Denmark, Lyngby, Denmark.

II. The Codes

Let the information sequence of an (N/K) Reed–Solomon (RS) code of block length $N = 2^m - 1$ and rate $r_n = K/N$ over $GF(2^m)$ be

$$\boldsymbol{i} = [i_0, i_1, \cdots, i_{K-1}], \qquad i_j \in GF(2^m)$$

and define the associated polynomial as

$$i(x) = i_0 + i_1 x + \cdots + i_{K-1} x^{K-1}.$$

Let α be a primitive element of $GF(2^m)$. The encoded sequence may be written as the N-tuple [8]

$$\boldsymbol{a} = [a_0, a_1, \cdots, a_{N-1}], \qquad a_j \in GF(2^m)$$

or equivalently as the polynomial

$$a(x) = a_0 + a_1 x + \cdots + a_{N-1} x^{N-1},$$

where

$$a_j = i(\alpha^j), \qquad 0 \leq j \leq 2^m - 2.$$

Since $GF(2^m)$ is a vector space over $GF(2)$, the elements of $GF(2^m)$ can, and will hereafter, be taken as binary m-tuples.

Now consider the binary (n,k) code $\mathscr{C}_m$ with codewords

$$\boldsymbol{c} = [c_0, c_1, \cdots, c_{N-1}], \tag{3}$$

where

$$c_j = [a_j, \alpha^j a_j], \qquad 0 \leq j \leq N - 1. \tag{4}$$

This binary code has length $n = 2mN$, dimension $k = mK$, and rate $R_n = \frac{1}{2} r_N$. $\mathscr{C}_m$ is a linear code over $GF(2^m)$ when the c_j are considered as pairs of elements from $GF(2^m)$, and is consequently also linear over $GF(2)$ when the c_j are expressed as 2^m-tuples over $GF(2)$, as we have done.

The code $\mathscr{C}_m$ may be interpreted as a concatenated code with a Reed–Solomon outer code and a varying inner code. For each j, (4) specifies a $(2m,m)$ binary code; these are all distinct and in fact constitute the $R = \frac{1}{2}$ codes in Wozencraft's ensemble of randomly shifted codes. These codes have the interesting property that any $2m$-tuple a,b with $a \neq 0$ and $b \neq 0$ appears in exactly one code, namely the code for which $\alpha^j = b/a$. This property was used by Massey [5] to show that the ensemble contains codes that meet the Gilbert–Varsharmov bound.

Alternatively the codewords of $\mathscr{C}_m$ may be interpreted as pairs $[\boldsymbol{a},\boldsymbol{b}]$ where $\boldsymbol{a}$ and $\boldsymbol{b} = [b_0, b_1, \cdots, b_{N-1}]$ are both codewords of RS codes. Since the RS codes form a subclass of the Bose–Chaudhuri–Hocquenghem (BCH) codes [8], these codes are cyclic, and the RS code described above is easily seen to have the generator polynomial

$$g(x) = (x + \alpha)(x + \alpha^2) \cdots (x + \alpha^{N-K}).$$

The vector $\boldsymbol{b}$ has the associated polynomial

$$b(x) = a_0 + \alpha a_1 x + \cdots + \alpha^{N-1} a_{N-1} x^{N-1} = a(\alpha x)$$

and it is consequently a codeword of the RS code with generator polynomial

$$g_b(x) = (x + 1)(x + \alpha) \cdots (x + \alpha^{N-K-1}).$$

It is customary to accept a sequence of codes specified in terms of primitive field elements of successively larger fields $GF(2^m)$ as constructive. Thus the sequence of codes $\mathscr{C}_m$ as given above is constructive in the same sense as the cyclic Hamming codes or the RS codes. However, following a suggestion by McEliece [9], the specification of a primitive element of $GF(2^m)$, or equivalently a primitive polynomial of degree m over $GF(2)$, can be avoided in the specification of $\mathscr{C}_m$ as follows. Let the $m = 2 \cdot 3^l$-tuple

$$[f_0, f_1, \cdots, f_{m-1}]$$

correspond to

$$f_0 + f_1 \sigma + \cdots + f_{m-1} \sigma^{m-1},$$

where σ is a root of the irreducible, but not primitive polynomial [10]

$$F_l(x) = x^{2 \cdot 3^l} + x^{3^l} + 1,$$

i.e.,

$$\sigma^{2 \cdot 3^l} + \sigma^{3^l} + 1 = 0.$$

If we replace α^j everywhere by the $m = 2 \cdot 3^l$-tuple whose digits are the radix-two form of j, $0 \leq j < 2^{2 \cdot 3^l}$, the codes thus defined for the particular values of $m = 2 \cdot 3^l$ are just permutations of the corresponding subsequence of the codes as defined by (3) and (4). This subsequence of codes is, we believe, incontestably constructive.

The codes (without McEliece's modification) are most easily encoded by a RS encoder that forms $a_{N-1}, a_{N-2}, \cdots, a_0$, followed by an encoder for the inner codes. The inner encoder stores a multiplier that is initially α^{N-1} and is scaled by a factor α^{-1} after each encoding, and so computes the product $b_j = \alpha^j a_j$. These $2N$ $GF(2^m)$ multiplications in the inner encoder are a small addition to the approximately $N(N - K)$ multiplications required by the RS encoder.

Theorem 1: For any given rate R, $0 < R < \frac{1}{2}$, the sequence of binary $(2mN,mK)$, $m = 1,2,\cdots$, codes $\mathscr{C}_m$, with $R_n \doteq \frac{1}{2} r_N = \frac{1}{2} K/N$ chosen as the smallest rate not less than R, satisfy

$$\liminf_{n \to \infty} (d_n/n) \geq (1 - 2R) H^{-1}(\tfrac{1}{2}) \simeq 0.11(1 - 2R), \tag{5}$$

where d_n is the minimum distance of $\mathscr{C}_m$.

The lower bound (5) is plotted in Fig. 1 together with the Gilbert–Varsharmov bound for comparison. For example, with $R = \frac{1}{4}$, the codes of Theorem 1 have minimum distance at least 5.5 percent of the block length, compared to 21.5 percent, which the Gilbert–Varsharmov bound indicates as possible.

In the proof of Theorem 1 we shall make use of the following.

Lemma: Let $o_2(L) \to 0$ as $L \to \infty$. Then for any γ, $0 < \gamma < 1$, and any δ, $0 < \delta < 1$, the total Hamming weight W of $M_L = [\gamma - o_2(L)](2^{L\delta} - 1)$ distinct nonzero binary L-tuples satisfies

$$W \geq \gamma L[H^{-1}(\delta) - o_1(L)](2^{L\delta} - 1). \tag{6}$$

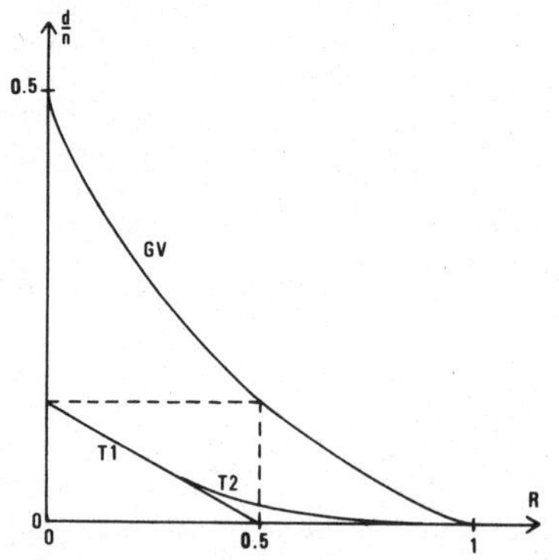

Fig. 1. Comparison of the bounds of Theorem 1 (T1) and Theorem 2 (T2) with the Gilbert–Varsharmov bound (GV).

Proof: We make use of the well-known inequality [8]

$$\sum_{i=1}^{\lambda L} \binom{L}{i} \le 2^{LH(\lambda)}, \qquad 0 \le \lambda \le \tfrac{1}{2}. \tag{7}$$

From this, it follows that the fraction f_L of the M_L specified nonzero L-tuples that have Hamming weight λL or less satisfies

$$f_L \le M_L^{-1} 2^{LH(\lambda)}, \qquad 0 \le \lambda \le \tfrac{1}{2}$$

or

$$f_L \le [\gamma - o_2(L)]^{-1}(2^{L\delta} - 1)2^{LH(\lambda)} \tag{8}$$

If we choose

$$\lambda = H^{-1}(\delta - 1/\log L),$$

where L is large enough so that $\delta > 1/\log L$, (8) becomes

$$f_L \le [\gamma - o_2(L)]^{-1} 2^{-L/\log L}$$

and thus, $f_L = o_3(L)$ where $o_3(L) \to 0$ as $L \to \infty$. Hence the total weight of the M_L L-tuples satisfies

$$W \ge (1 - f_L)M_L H^{-1}(\delta - 1/\log L)L$$

or

$$W \ge [1 - o_3(L)][\gamma - o_2(L)](2^{L\delta} - 1)H^{-1}(\delta)[1 - o_4(L)]L, \tag{9}$$

which by defining

$$\gamma - o_1(L) = [1 - o_3(L)][1 - o_4(1)][\gamma - o_2(L)]$$

gives (6).

We note that since no two randomly shifted codes have any nonzero codewords in common, this lemma may be interpreted as stating that the fraction f_L of M_L randomly shifted codes that have minimum distance $LH^{-1}(\delta - 1/\log L)$ or less vanishes as $L \to \infty$.

Proof of Theorem 1: Consider any nonzero codeword $\boldsymbol{c}$ of $\mathscr{C}_m$. Then $\boldsymbol{a}$ is also nonzero, and, since it is a codeword of an RS code, it contains at least $N - K + 1 > N - K = N(1 - r_N) = N(1 - 2R_n) = (2^m - 1)[1 - 2R - o_2(m)]$ nonzero digits, where $o_2(m) \to 0$ as $m \to \infty$. Since no $2m$-tuple appears in more than one inner code, we may apply the lemma with $L = 2m$, $\delta = \frac{1}{2}$, and $\gamma = 1 - 2R$. The Hamming weight of $\boldsymbol{c}$ is lower bounded by

$$W \ge [H^{-1}(\tfrac{1}{2}) - o_1(2m)](1 - 2R)(2^m - 1)2^m.$$

Consequently the minimum distance d_n of $\mathscr{C}_m$ satisfies

$$d_n/n = d_n/[2m(2^m - 1)] \ge (1 - 2R)[H^{-1}(\tfrac{1}{2}) - o_1(2m)]. \tag{10}$$

Since $m \to \infty$ as $n = 2m(2^m - 1) \to \infty$, (5) follows immediately from (10) and the theorem is proved.

We note that this result could not have been obtained by simply considering the RS code as a binary code. Let d_0 be the minimum distance of the RS code over $GF(2^m)$ and let i be the smallest integer such that $d_0' = 2^i - 1 \ge d_0$. It was proved by Kasami *et al.* [11] that the minimum distance of the BCH code of length $2^m - 1$ and design distance d_0' has minimum distance exactly d_0'. Now as shown in [3] the BCH codes with design distance at least d_0 are subcodes of the RS code with minimum distance d_0, and hence their minimum distance is an upper bound on the minimum distance d_{BRS} of the RS code considered as a binary code of length $n = mN$. Since $d_0' \le 2d_0'$, we have

$$d_{\mathrm{BRS}}/n = d_{\mathrm{BRS}}/(mN) < 2(N - K)/(mN)$$

and consequently

$$d_{\mathrm{BRS}}/n \to 0, \qquad \text{as } m \to \infty.$$

In order to extend the construction to higher rates, we puncture the codes previously constructed by deleting the last s digits from each inner code. We shall write

$$r_n = m/(2m - s), \qquad 0 \le s < m$$

for the resulting rate of the inner code.

For a given value of R, $0 < R < 1$, and any choice of r_n, we choose the rate of the RS outer code as the smallest rate such that

$$R_n = r_n r_M \ge R.$$

We again note that any nonzero codeword $\boldsymbol{c}$ in the punctured code $\mathscr{C}_m$ will contain

$$M > (2^m - 1)(1 - r_N) = (2^m - 1)[1 - (R/r_n) - o_2(m)]$$

nonzero a_j. The nonzero (a_j, b_j) are distinct, but after puncturing there may be as many as 2^s identical nonzero c_j, since any c_j is a codeword in exactly 2^s inner codes. Thus at least

$$2^{-s}M > (2^{m-s} - 1)[1 - (R/r_n) - o_3(m)]$$

nonzero $(2m - s)$-tuples are distinct and we can apply the lemma with $L = 2m - s$, $\delta = (m - s)/L = 1 - r_n$, and $\gamma = 1 - (R/r_n)$. We obtain the following lower bound for the Hamming weight of $\boldsymbol{c}$:

$$W \ge [1 - (R/r_n)](2m - s)[H^{-1}(1 - r_n) - o_1(m)](2^{m-s} - 1)2^s. \tag{11}$$

But if r_n approaches r, $\frac{1}{2} \le r < 1$, the ratio s/m is bounded away from 1 and (11) implies

$$\liminf_{n\to\infty} (d_n/n) \ge [1 - (R/r)]H^{-1}(1 - r). \quad (12)$$

The minimum distance may be maximized by setting the derivative of the right-hand side of (12) with respect to r equal to zero. The condition satisfied by the maximizing choice of r is

$$R = \frac{r^2}{1 + \log_2 [1 - H^{-1}(1 - r)]} \quad (13)$$

except that $r = \frac{1}{2}$ must be chosen when the solution of (13) would yield $r < \frac{1}{2}$, since our construction permits only inner code rates at least $\frac{1}{2}$.

We summarize the foregoing as Theorem 2.

Theorem 2: For any rate R, $0 < R < 1$, the binary (n,k) codes $\mathscr{C}_m$ punctured so that r_n approaches r, where r is the maximum of $\frac{1}{2}$ and the solution to (13), satisfy

$$\liminf_{n\to\infty} (d_n/n) \ge [1 - (R/r)]H^{-1}(1 - r). \quad (14)$$

The bound (14) can be viewed as the envelope of a family of straight lines constructed in the following way. The point $[r,H^{-1}(1 - r)]$ on the Gilbert–Varsharmov bound is projected on the axes to give the points $[r,0]$ and $[0,H^{-1} \cdot (1 - r)]$. These two points are then joined by a straight line. Because of the restriction $r \ge \frac{1}{2}$, the bound of Theorem 2 is identical to that of Theorem 1 for $0 < R < 0.30$. For higher rates the bound (14) coincides with the lower bound on the minimum distance for the ensemble of concatenated codes obtained by Zyablov [12]. For rates $r < 0.30$ the bound is inferior to Zyablov's bound because the construction of Section II requires a good ensemble of inner codes with at most $2^m - 1$ codes and we cannot constructively specify such an ensemble for rates less than $\frac{1}{2}$.

We note that, for rates near unity, the maximizing value of r is

$$r \simeq R_n \simeq R^{1/2}.$$

III. A Decoding Procedure

All error patterns of weight less than half the value of our asymptotic lower bound on minimum distance (14) can be corrected by the version of generalized minimum distance decoding which was applied to product codes by Reddy and Robinson [6] and Weldon [7].

It was pointed out earlier that all but a vanishing fraction of the inner codes have minimum distance d_i satisfying

$$\frac{d_i}{2m - s} \ge \frac{D}{2m - s} = H^{-1}(1 - r_n),$$

where D is the lower bound on minimum distance given by (1).

For our decoding procedure, we select the scheme shown in Fig. 2 in which the received block

$$r = [r_0,r_1,\cdots,r_{N-1}],$$

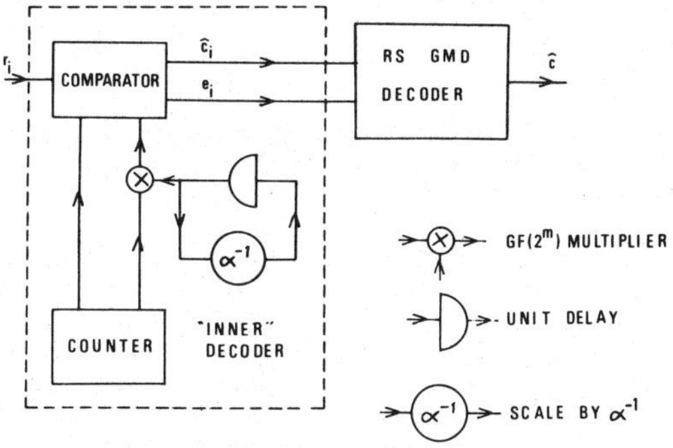

Fig. 2. Block diagram of the decoder for the (n,k) codes of Theorem 2.

where r_i is the received $(2m - s)$-tuple corresponding to the inner codeword c_i, is first decoded by an inner decoder whose decisions are then passed to a generalized minimum distance (GMD) decoder [3] for the RS outer code. The inner decoder may be constructed as shown in Fig. 2 such that, for each received inner word r_i, the decoder (by means of an m-stage binary ring counter) generates all 2^m codewords in the inner code and feeds these to a comparator that also receives r_i. The inner decoder decodes r_i into $\hat{c}_i$ as soon as a codeword $\hat{c}_i$ is found at distance less than $D/2$ from r_i. Since this decoding process must be done $N = 2^m - 1$ times, the inner decoder performs $2^m(2^m - 1)$, or approximately $(n/\log n)^2$ multiplications in all.

Let the output of the comparator be the number e_i together with the estimate $\hat{c}_i$ of c_i, where e_i is defined by

$$e_i = \begin{cases} \text{weight } (\hat{c}_i + r_i), & r_i \text{ decoded} \\ D/2, & \text{otherwise (in this case } \hat{c}_i \text{ may be arbitrary).} \end{cases}$$

Define the normalized weight of the error at position i as

$$\beta_i = \begin{cases} \dfrac{e_i}{D}, & \hat{c}_i = c_i \\ \dfrac{D - e_i}{D}, & \hat{c}_i \ne c_i. \end{cases} \quad (15)$$

According to the theory of GMD [3], the RS "outer" decoder will decode correctly whenever

$$\sum_{i=0}^{N-1} \beta_i < \frac{N - K + 1}{2}. \quad (16)$$

Noting that, because of the lemma, the fraction of the r_i yielding errors because $d_i < D$ vanishes as $N \to \infty$ and that the number of errors required to cause a given value of β_i for an inner code with minimum distance at least D is $t_i \ge \beta_i D$, it then follows from (16) that the minimum number of errors that can cause a decoding error asymptotically satisfies

$$t = \sum_{i=0}^{N-1} t_i \ge \sum_{i=0}^{N-1} \beta_i D > (D/2)(N - K + 1) > \tfrac{1}{2}DN(1 - r_N). \quad (17)$$

But, asymptotically, our lower bound d_B on the minimum distance d is

$$d_B = ND(1 - r_N)$$

so that (17) shows that our decoding procedure corrects all the errors guaranteed correctable by the asymptotic lower bound d_B.

We also note that, to do GMD decoding for the outer RS code the number of trials required equals the number of distinct values of e_i. Since there are $D/2$ allowed values of e_i and since errors-and-erasures decoding of an RS code with Berlekamp's iterative algorithm [13] requires a number of multiplications proportional to 2^{2m}, it follows that the total number of multiplications in our decoding procedure is proportional to $m2^{2m}$. This corresponds to a number of bit operations proportional to $m^3\ 2^{2m}$ or approximately $n^2 \log n$.

IV. Remarks

The code construction given in Theorem 2 for binary codes is easily extended to the construction of asymptotically good codes over any finite field $GF(q)$. In this case, one uses the RS code over the extension field $GF(q^m)$, and chooses the α in (4) as a primitive element of $GF(q^m)$. McEliece's modification cannot be used to remove the last possible objection to constructivity for $q > 2$, however, since we know no expression for irreducible polynomials of arbitrarily large degree over an arbitrary field $GF(q)$.

Finally, we remark that, for any particular block length $n = (2m - s)2^m$, a search for the best fixed inner code requires less than n^2 steps. Even though a good concatenated code with the best fixed inner code can thus be found with a very moderate amount of searching, we have rejected this approach as nonconstructive, since we cannot specify the inner code *a priori*. It is an open question whether the actual distance of our constructive codes with varying inner codes is better or worse than that obtained using the best fixed inner code.

Acknowledgment

The author wishes to thank Prof. J. L. Massey for his help in preparing the final draft of this paper.

References

[1] E. N. Gilbert, "A comparison of signalling alphabets," *Bell Syst. Tech. J.*, vol. 31, pp. 504–522, 1952.
[2] R. R. Varsharmov, "Estimate of the number of signals in error correcting codes," *Dokl. Akad. Nauk. SSSR*, vol. 117, pp. 739–741, 1957.
[3] G. D. Forney, Jr., *Concatenated Codes*. Cambridge, Mass.: M.I.T. Press, 1966.
[4] I. S. Reed and G. Solomon, "Polynomial codes over certain finite fields," *J. Soc. Ind. Appl. Math.*, vol. 8, pp. 300–304, 1960.
[5] J. L. Massey, *Threshold Decoding*. Cambridge, Mass.: M.I.T. Press, p. 21, 1963.
[6] S. M. Reddy and J. P. Robinson, "Random error and burst correction by iterated codes," *IEEE Trans. Inform. Theory*, vol. IT-18, pp. 182–185, Jan. 1972.
[7] E. J. Weldon, Jr., "Decoding binary block codes on Q-ary output channels," *IEEE Trans. Inform. Theory*, vol. IT-17, pp. 713–718 (Appendix), Nov. 1971.
[8] W. W. Peterson, *Error Correcting Codes*. Cambridge, Mass.: M.I.T. Press and New York: Wiley, 1961.
[9] R. J. McEliece, private communication, 1972.
[10] S. W. Golomb, *Shift Register Sequences*. San Francisco: Holden-Day, 1967, p. 96.
[11] T. Kasami, S. Lin, and W. W. Peterson, "Some results on weight distributions of BCH codes" (Abstract), *IEEE Trans. Inform. Theory*, vol. IT-12, p. 274, Apr. 1966.
[12] V. V. Zyablov, "On estimation of complexity of construction of binary linear concatenated codes," *Probl. Peredach. Inform.*, vol. 7, pp. 5–13, 1971.
[13] E. R. Berlekamp, *Algebraic Coding Theory*. New York: McGraw-Hill, 1968, p. 184.

References

Berlekamp, E. R. (1968). *Algebraic Coding Theory,* McGraw-Hill Book Company, New York.

——— (1966). "Nonbinary BCH Decoding," Institute of Statistics Mimeo Series 502, Dept. Statistics, University of North Carolina, Chapel Hill, N.C.

Berman, S. D. (1967). "On the Theory of Group Codes," *Kibernetika,* **3**, 31–39.

Bose, R. C. (1947). "Mathematical Theory of the Symmetrical Factorial Design," *Sankhya,* **8**, 107–166.

Chien, R. T. (1964). "Cyclic Decoding Procedures for Bose-Chaudhuri-Hocquenghem Codes," *IEEE Trans. Inform. Theory,* **IT-10**, 357–363.

Delsarte, P. (1970). "On Cyclic Codes That Are Invariant Under the General Linear Group," *IEEE Trans. Inform. Theory,* **IT-16**, 760–769.

Dobrushin, R. L. (1972). "Survey of Soviet Research in Information Theory," *IEEE Trans. Inform. Theory,* **IT-18**, 703–724.

Elias, P. (1955). "Coding for Noisy Channels," *I.R.E. Nat. Conv. Record,* Part 4, 37–45.

Fisher, R. A. (1942). "The Theory of Confounding in Factorial Experiments in Relation to the Theory of Groups," *Ann. Augenics,* **11**, 341–353.

Forney, G. D. (1965). "On Decoding BCH Codes," *IEEE Trans. Inform. Theory,* **IT-11**, 549–557.

——— (1966). *Concatenated Codes,* M.I.T. Press Research Monograph 37, The MIT Press, Cambridge, Mass.

Goethals, J. M., and P. Delsarte (1968). "On a Class of Majority Logic Decodable Cyclic Codes," *IEEE Trans. Inform. Theory,* **IT-14**, 182–189.

Goppa, V. D. (1971). "Rational Presentation of Codes and (*L,g*)-Codes," *Probl. Peredachi Informatsii,* **7**, 41–49.

Gore, W. C., and A. B. Cooper (1970). "Comments on Polynomial Codes," *IEEE Trans. Inform. Theory,* **IT-16**, 635–638.

Green, M. W. (1966). "Two Heuristic Techniques for Block Code Construction," *IEEE Trans. Inform. Theory,* **IT-12**, 273. (Abstract only.)

Kautz, W. H., and K. N. Levitt (1969). "A Survey of Progress in Coding Theory in the Soviet Union," *IEEE Trans. Inform. Theory,* **IT-15**, No. 1, Part II.

Massey, J. L. (1963). *Threshold Decoding,* MIT Press Research Monograph 20, The MIT Press, Cambridge, Mass.

Mitani, N. (1951). "On the Transmission of Numbers in a Sequential Computer," National Convention of the Institute of Electrical Communication Engineers of Japan, November.
Nadler, M. (1962). "A 32-point n = 12, d = 5 Code," *IRE Trans. Inform. Theory,* **IT-8**, 58.
Peterson, W. W., and E. J. Weldon (1972). *Error-Correcting Codes,* MIT Press, 2nd Edition, Cambridge, Mass.
Prange, E., (1957). "Cyclic Error-Correcting Codes in Two Symbols," Rept. AFCRC-TN-57-103, USAF Cambridge Research Center, Bedford, Mass.
——— (1958). "Some Cyclic Error-Correcting Codes with Simple Decoding Algorithms," Rept. AFCRC-TN-58-156, USAF Cambridge Research Center, Bedford, Mass.
Rao, C. R., (1947). "Factorial Experiments Derivable from Combinatorial Arrangements of Arrays," *J. Roy. Statist. Soc. Suppl. 1.* **9**, 128–139.
Rudolph, L. D. (1964). "Geometric Configuration and Majority Logic Decodable Codes," M.E.E. Thesis, University of Oklahoma, Oklahoma.
Semakov, N. V., and V. A. Zinoviev (1968). "Equidistant *q*-ary Codes with Maximum Distance and Resolvable Balanced Incomplete Block Designs," *Probl. Inform. Transmission,* **4**, 3–10.
——— and V. A. Zinoviev (1969). "Fixed-Composition Codes and Tactical Configurations," *Probl. Inform. Transmission,* **5**, 28–36.
———, V. A. Zinoviev, and G. V. Zaitsev (1969). "A Class of Maximal Equidistant Codes," *Probl. Inform. Transmission,* **5**, 84–87.
Shannon, C. E. (1948). "A Mathematical Theory of Communication," *Bell System Tech. J.,* **27**, Part I, 379–423, Part II, 623–656.
Tietäväinen, A., and A. Perko (1971). "There Are No Unknown Perfect Binary Codes," *Ann. Univ. Turku., Ser. A,* 148, 3–10.
——— (1973). "On the Nonexistence of Perfect Codes over Finite Fields," *SIAM J. Appl. Math.,* **24**, 88–96.
Weldon, E. J. (1967). "Euclidean Geometry Cyclic Codes," Chap. 23 in *Proceedings of the Conference on Combinatorial Mathematics and Its Applications,* R. C. Bose and T. Dowling, eds., University of North Carolina Press, Chapel Hill, N.C.
Zaremba, S. K. (1951). "A Covering Theorem for Abelian Groups," *J. London Math. Soc.,* **26**, 71–72.
——— (1952). "Covering Problems Concerning Abelian Groups, " *J. London Math. Soc.,* **27**, 242–246.

Author Citation Index

Subject Index